Thomas Kirk

The students' flora of New Zealand and the outlying islands

Thomas Kirk

The students' flora of New Zealand and the outlying islands

ISBN/EAN: 9783337271572

Printed in Europe, USA, Canada, Australia, Japan

Cover: Foto ©berggeist007 / pixelio.de

More available books at **www.hansebooks.com**

THE STUDENTS' FLORA OF NEW ZEALAND AND THE OUTLYING ISLANDS.

BY THOMAS KIRK, F.L.S.

WELLINGTON, N.Z.:
BY AUTHORITY: JOHN MACKAY, GOVERNMENT PRINTER.
LONDON:
EYRE AND SPOTTISWOODE, FLEET STREET, E.C.

THE STUDENTS' FLORA OF NEW ZEALAND

AND

THE OUTLYING ISLANDS.

WELLINGTON, NZ.
By Authority: John Mackay, Government Printer.
London: Eyre and Spottiswoode, Fleet Street, E.C.

TABLE OF CONTENTS.

INTRODUCTORY NOTICE.

THE death of Professor Kirk before the "Flora" was completed involves, among other serious losses to scientific progress, the publication of the completed portion of the work without such an introductory chapter on geographical distribution, on the wider questions of affinity, and on the historical aspect of Botany and botanical research in New Zealand as he intended to write, and was pre-eminently qualified to write. The same sad cause accounts for the fact that the glossary and other addenda to the work have been compiled without his supervision.

The Government have in view the necessity for making arrangements for the completion of the work. The plates selected to illustrate the "Flora" will form a separate volume.

Through the kindness of the Trustees of the British Museum, who sent out complete sets of the plants collected by Sir Joseph Banks and Dr. Solander, the great advantage was gained of an examination of the specimens actually collected by these botanists. The valuable help of the Trustees did not end here, as they also gave permission to use the beautiful copper-plates engraved for Sir Joseph Banks, and done with marvellous care and accuracy. With regard to these plates, the following extract from Sir Joseph D. Hooker's biographical sketch of Sir Joseph Banks, appended to the "Journal of the Right Hon. Sir Joseph Banks," 1896, will be of interest: "About seven hundred plates were engraved on copper, in folio, at Banks's expense, and a few prints or proofs were taken, but they were never published. Five folio books of neat manuscript, and the coppers, rest in the hands of the Trustees of the British Museum. The question arises, Why were they never utilised? . . . This has always been regarded as an insoluble problem." The writer goes on to show that, in all probability, it was Solander's death that arrested publication.

A large number of the plates that will illustrate this work are thus of great historic interest, as well as of great scientific value. Others of the plates are taken, by permission, given upon payment of a small royalty, of Messrs. L. Reeves and Co., with the ready concurrence of Sir Joseph D. Hooker, from "Flora Antarctica," from "Flora Novæ Zealandiæ," and from "Flora Tasmanica."

Throughout the work will be found acknowledgment of the assistance readily and heartily given to the author by botanical collectors and observers in all parts of the colony and beyond it. This acknowledgment refers often to assistance given many years ago, before the inception of the present work, but found useful now. A list of those whose help is acknowledged would be too long to be here reproduced, and the numerous and enthusiastic helpers will understand that no discourtesy or ingratitude is shown by the fact that a long list that would cover several pages is not published to supplement the acknowledgments penned by the author in the body of the work. Special acknowledgment is, however, made of the ready help given by the authorities at Kew, especially by the Director, Dr. Thiselton Dyer, and by Mr. N. E. Brown, and of that given by Mr. J. Britten, of the British Museum. First of all, probably, would the author have acknowledged gratefully the help given at all times by Sir Joseph D. Hooker, and special acknowledgment would certainly have been made of the assistance of Bishop Williams, of Waiapu, of that of Sir James Hector, of the late Rev. W. Colenso, of Messrs. D. Petrie, T. F. Cheeseman, and G. H. Thomson. The courtesy and readiness to help shown at all times by the officers of the Government Printing Office is gratefully acknowledged.

Note.—The names of authorities for the occurrence of plants in any given locality are printed in italics. A note of exclamation after a name so printed indicates that the author had the opportunity of examining specimens sent from that locality by the collector. Native and popular names are printed in italics.

Education Department,

Wellington, 10*th April,* 1899.

THE STUDENTS'

FLORA OF NEW ZEALAND

AND

THE OUTLYING ISLANDS.

Class I.—DICOTYLEDONS.

Division I.—POLYPETALAE.

Order I.—RANUNCULACEAE.

Herbs with radical or alternate leaves, rarely shrubs with opposite leaves; stipules adnate to the petiole or 0. Flowers usually regular, perfect or rarely unisexual. Sepals 5 or more, rarely 2–4, deciduous, imbricate or rarely valvate in bud. Petals 5 or more, rarely 3 or 0. Stamens numerous or rarely few, hypogynous; anthers adnate, dehiscing laterally. Carpels numerous or rarely 3–4 or solitary, free on a torus which is sometimes elongated, ovules 1 or many, anatropous. Fruit of 1-seeded achenes, or many-seeded follicles. Seeds with copious endosperm and a minute embryo.

Distribution.—Plentiful in temperate and cold regions, rare in the tropics. Genera, 30. Species, 520. In several exotic genera the flowers are irregular, sometimes with a single carpel, or with united carpels, or with baccate fruit.

Key to the Genera.

Tribe I. CLEMATIDEAE.—Sepals valvate. Carpels indehiscent, with a pendulous ovule in each. Climbing shrubs, with opposite leaves.

1. Clematis. Petals 0.

II. ANEMONEAE.—Sepals imbricate. Carpels indehiscent, with a pendulous seed in each. Leaves radical or alternate, or forming an involucre below the flower.

2. Myosurus. Leaves all radical, entire. Achenes numerous, forming a slender, elongated spike.

III. RANUNCULEAE.—Sepals imbricate. Carpels with an erect seed in each.

3. Ranunculus. Sepals deciduous. Petals 3, 4, 5, or ∞.

Tribe IV. HELLEBOREAE.—Sepals imbricate, petaloid. Carpels usually dehiscing along the inner face. Ovules or seeds, 2 or more, erect.

4. CALTHA. Sepals 5–7, petaloid. Petals 0.

* AQUILEGIA. Sepals 5–6, petaloid. Petals spurred.

* NIGELLA. Sepals 5, petaloid. Petals small.

1. CLEMATIS, Linn.

Climbing shrubs, with slender flexuous branches and opposite compound leaves. Flowers in few- or many-flowered axillary panicles, dioecious in the New Zealand species. Sepals 4–8, petaloid, valvate. Petals 0. Stamens 4–70. Carpels numerous, each with 1 pendulous ovule. Achenes in dense heads, the elongated and persistent style forming a long plumose awn.

ETYM. From the Greek, signifying *a vine-shoot*, in reference to the habit of growth.

The genus comprises upwards of 100 species, generally diffused throughout the temperate regions, but rare in the tropics. Many exotic species have showy flowers, a few have minute petals, others have erect or herbaceous stems, a few have entire leaves. All the New Zealand species are endemic, and have unisexual flowers, although the female flowers usually contain a few abortive stamens, and more rarely one or two abortive carpels are produced in the male flowers. The species are remarkably variable, and in some cases seem to pass one into the other in the mature state.

KEY TO THE SPECIES.

Sepals white.

Leaflets usually entire. Flowers 2in.–4in. diam.	1. *C. indivisa.*
Leaflets toothed or lobed. Flowers 1in.–1½in. diam.	2. *C. hexasepala.*
Leaflets pinnate or pinnatifid	3. *C. australis.*

Sepals yellow or greenish-yellow; purplish in 7.

Stems leafless. Sepals 4	4. *C. afoliata.*
Leaflets toothed or lobed. Sepals 5–7, silky	5. *C. Colensoi.*
Leaflets 2-ternate or pinnatifid	5. var. *rutaefolia.*
Leaflets small, entire or lobed. Sepals 4	6. *C. marata.*
Leaflets small, linear. Sepals 4, purplish	7. *C. quadribracteolata.*
Leaflets large, 1in.–1½in. long. Sepals 6–8, tomentose	8. *C. foetida.*
Leaflets ½in.–1in. long. Sepals 6–8, linear. Anthers ovate	9. *C. parviflora.*

1. **C. indivisa,** *Willd., Sp. Plant.* ii. 1291. A stout woody climber. Leaves 3-foliolate, coriaceous, glabrous, 1in.–4in. long, petiolulate, oblong-ovate or ovate-cordate, usually entire. Flowers 2in.–4in. in diameter, in large axillary panicles, fragrant, white. Sepals 6–8, oblong. Anthers oblong. Achenes downy.—Hook., Bot. Mag. t. 4398; DC., Prod. i. 5; A. Rich., Fl. N.Z. 288; A. Cunn., Precurs. n. 635; Raoul, Enum. Pl. N.Z. 47; Hook. f., Fl. N.Z. i. 6, and Handbk. N.Z. Fl. 2. *C. integrifolia,* Forst., Prod. n. 231.

Var. *b.* **lobulata.** Leaflets toothed or lobulate. Flowers smaller.

Var. *c.* **decomposita.** Leaflets 2-ternate.

Var. *d.* **linearis.** Leaflets narrow-linear, 3in.–7in. long, ¼in.–½in. broad, entire or with two lateral lobes at the base.

NORTH and SOUTH Islands: Common in lowland districts. THREE KINGS Islands. STEWART Island: Rare. Ascends to 2,500ft. Native name, *Puawananga.* Aug. to Nov.

The varieties are very local, *c* and *d* being apparently confined to the North Island, and should perhaps be considered as temporary states rather than permanent forms: var. *linearis* occasionally passes into *decomposita,* and that again into the typical form.

2. **C. hexasepala,** *DC. in Syst.* i. 146. Smaller in all its parts than *C. indivisa.* Leaves 3-foliolate, coriaceous, glabrous; leaflets petiolulate, ovate or ovate-cordate, acute or acuminate, irregularly toothed or lobed, rarely entire, 1in.–2in. long. Flowers 1in.–1½in. in diameter; peduncles pubescent. Sepals 6, broadly linear, obtuse, downy. Anthers linear, obtuse. Achenes ovate, pilose.—DC., Prod. i. 5; Raoul, Enum. 47; Hook. f., Handbk. N.Z. Fl. 2; *C. odorata,* Banks and Sol. MSS.; *C. hexapetala,* Forst., Prod. n. 230; *C. Forsteri,* Gmel., Syst. 873; *C. Colensoi,* Hook. f., Fl. N.Z., i. 6, t. 1 (not of Handbk.).

NORTH and SOUTH Islands: From the Kaipara River to the Bluff, but rare and local in the South. Ascends to nearly 3,000ft. *Pikiarero.* Sept. to Nov.

Readily distinguished from *C. indivisa* by the smaller flowers and toothed pale-green leaves. In young plants the leaves are sometimes ternately divided, but this state is not permanent.

3. **C. australis,** *n. s.* Stems and branches slender, pubescent at the tips. Leaves slightly coriaceous, glabrous, 3-foliolate; leaflets ¾in.–1in. long, pinnate or pinnatifid; secondary leaflets and segments deeply lobed or toothed. Peduncles pubescent, 1in.–4in. long, very slender. Flowers solitary on the peduncles or in 2–4-flowered panicles, rarely exceeding 1in. in diameter. Sepals 5–8, downy. Achenes narrowed at the apex, pilose.

SOUTH Island: Mountain districts in Nelson and Canterbury, 1,500ft. to 3,000ft. November. Allied to *C. hexasepala,* from which it differs in the slender habit, decompound leaves, solitary flowers, and small heads of fusiform achenes. I have not seen the female flower.

4. **C. afoliata,** *Buch. in Trans. N.Z.I.* iii. (1870) 211. Stems and branches wiry, finely grooved; leafless, or with a few 3-foliolate leaves in the young state, when the leaflets are entire, ovate, acute, on long petioles; nodes distant. Leaves represented by petioles in the mature state. Flowers in fascicles of 2–5 in the axils of the petioles; peduncles slender, pilose, each with a pair of ovate pilose bracteoles on the lower half. Sepals 4, ovate-lanceolate, acute or obtuse, silky. Achenes ovate, silky.

NORTH Island: Mr. Colenso informed me that he collected this species in the North Island many years ago, but had forgotten the locality. SOUTH Island: Nelson, Marlborough, Canterbury, and Otago, but local. October.

A singular plant, often forming thick masses of leafless stems or branches, 9ft. or 10ft. in length, bound together by the twining petioles. The sepals are greenish-white.

Var. *b.* **aphylla,** Colenso, sp., Trans. N.Z.I. xix. (1886) 259. Flowers hermaphrodite, stamens 10. NORTH Island: Puketapu, Hawke's Bay, *H. Hill.* I have not seen flowering specimens.

5. **C. Colensoi,** *Hook. f., Handbk. N.Z. Fl., p.* 3. Stems slender, much branched. Leaves 3-foliolate, membranous or slightly coriaceous; leaflets petioled, ½in.–1½in. long, ovate or ovate-cordate, crenate or irregularly toothed or deeply 3-partite. Panicle short, much branched or consisting chiefly of 1-flowered peduncles; peduncles slender, pubescent. Flowers ¾in.–1½in. in diameter. Sepals 5–7, silky, oblong. Anthers linear. Achenes silky.—*C. hexasepala,* Hook. f., Fl. N.Z. i. p. 7 (not of DC.).

NORTH Island: Hawke's Bay, local; Wellington, frequent. SOUTH Island: Nelson, rare and local.

Var. b. *rutaefolia,* Hook. f., Fl. N.Z. i. 7. Usually smaller, leaves 2-ternate or 2-pinnate; secondary leaflets often petioled. NORTH Island: Wellington, frequent. SOUTH Island: Nelson.

6. **C. marata,** *J. B. Armst. in Trans. N.Z.I.* xiii. (1880) 335. Stems and branches weak, often trailing or forming bundles bound together by the twining petioles, pubescent. Leaves 3-foliolate, usually pubescent on both surfaces; leaflets ¼in.–⅓in. long, petioled, entire or lobulate or 3-partite. Panicle short, often reduced to two or more 1–3-flowered peduncles; bracteoles large, often foliaceous. Male flowers ½in. in diameter. Sepals 4, oblong, silky. Stamens linear, oblong. Female peduncles rarely exceeding 1in. in length. Achenes with a thickened margin, glabrous or puberulous.

NORTH and SOUTH Islands: East Cape to Otago, but often local; chiefly in mountain districts. Ascends from sea-level to 3,000ft. Oct., Nov. The brown colour and depauperated aspect of this species distinguish it from all others, except *C. quadribracteolata*, which has linear leaflets and sepals.

7. **C. quadribracteolata,** *Colenso in Trans. N.Z.I.* xiv. (1881) 329. A very slender species, with much-branched stems 1ft.–3ft. long, glabrous except at the tips. Leaves few, ½in.–2in. long, 3-foliolate; leaflets ¼in.–½in. long, petioled, linear, linear-oblong, or ovate-acuminate, bluntly apiculate, 3-lobate. Peduncles solitary or in pairs, usually shorter than the petioles, pubescent, 1-flowered; bracteoles in two pairs, connate. Sepals 4, linear, obtuse or acute, silky, ⅓in. long. Anthers linear. Sepals of female flower short, oblong, obtuse. Achenes sparingly silky, awns short.—*C. foetida, β depauperata*, Hook. f., Fl. N.Z. i. 7.

NORTH Island: Hawke's Bay, on the margins of marshes, &c. Lake Rotoatara, *Colenso*. Low ground between the Ngaruroro and Tukituki Rivers, *Sturm*. Petane, *A. Hamilton!*

Leafless specimens with male flowers collected by Mr. Petrie at Tuapeka Mouth appear to belong to this species, although the stems are somewhat stouter; but the material is insufficient for positive identification.

8. **C. foetida,** *Raoul, Choix de Plantes N.Z.* 23, *t.* 22. Young shoots, petioles, and leaves pubescent on both surfaces. Leaves slightly coriaceous, 3-foliolate, leaflets ovate or ovate-cordate, acute or acuminate, entire or with 2–4 irregular teeth or lobes on each side, 1in.–1½in. long. Panicle much branched, rhachis and pedicels densely tomentose; bracteoles connate. Flowers ½in.–¾in. in diameter. Sepals 6–8, narrow-linear, acute or obtuse, usually tomentose on the outer surface. Anthers linear-oblong. Achenes ovate, silky, awns short.—Hook. f., Fl. N.Z. i. 7; Handbk. N.Z. Fl. 2. *C. Parkinsoniana*, Colenso in Trans. N.Z.I. xii. (1878), is a very slender form with smaller panicles than in the type.

NORTH and SOUTH Islands; THREE KINGS Islands. Not uncommon in lowland districts. Sept. to Nov.

A robust species, completely hiding the shrubs or small trees over which it grows to the height of from 15ft. to 25ft., and varying greatly in texture and the degree of hairiness at different seasons. In autumn the upper surface of the leaf is often glabrous. Easily distinguished from all other species by the white or fulvous pubescence of the branches and leaves, the dense panicles of yellowish flowers which are produced in vast profusion, and the tomentose sepals. The flowers emit a strong odour, but are not foetid.

9. **C. parviflora,** *A. Cunn., Precurs. n.* 636. A slender pubescent species. Leaves 3-foliolate, usually pubescent on both surfaces, slightly coriaceous; leaflets petioled, ovate or ovate-cordate, ½in.–1¼in. long, usually entire, subacute. Panicle 1in.–3in. long, rhachis and peduncles pubescent;

bracteoles linear, minute. Flowers ¾in. in diameter. Sepals 6–8, narrow-linear, silky, subacute, spreading. Anthers short, broadly elliptic, with a minute rounded appendage at the apex. Achenes ovate, slightly compressed, villous or silky.—Raoul, Enum. 47; Hook. f., Fl. N.Z. i. 7; Handbk. N.Z. Fl. 2.

NORTH and SOUTH Islands: Local—Whangaroa, Bay of Islands, Whangarei, Great Omaha, East Cape, northern parts of Hawke's Bay. Oct., Nov.

A graceful species, easily distinguished by the slender wiry stems, entire leaflets, and especially the minute rounded apical appendage to the short broad anthers. The typical form appears to be confined to the northern part of the North Island. *C. marata* has been distributed under this name from Canterbury, and Mr. Buchanan informs me that *C. foetida* has been mistaken for it in Otago.

Var. *b.* **depauperata.** Handbk. N.Z. Fl. 2. Leaflets very small. Sepals narrowed into long slender points. SOUTH Island: Nelson, *W. T. L. Travers.*

Var. *c.* **trilobata.** Leaves submembranous; leaflets deeply 3-lobate, segments entire or lobed. Flowers smaller. Sepals more pubescent. NORTH Island: Bay of Islands, *T. Kirk.* SOUTH Island: Okarito, *A. Hamilton!*

2. MYOSURUS, Linn.

Annual herbs, with linear radical leaves and naked 1-flowered scapes. Sepals 5–7, each with a minute spur at the base. Petals 0 in the N.Z. species. Stamens 5–7. Carpels numerous; ovule 1, pendulous. Achenes arranged in a dense spike formed by the elongated floral receptacle, each with a raised nerve at the back; style persistent.

ETYM. From the Greek, signifying *mouse* and *tail*, in reference to the elongated fruit-spike.

SPECIES, 2—the present, and another found in northern temperate regions, Chili, and Australia.

1. **M. aristatus,** *Benth. in Hook. Lond. Journ. Bot.* vi. 459. Leaves erect, linear or linear-spathulate, ½in.–2in. long, $\frac{1}{25}$in. wide or filiform. Scapes slender, ½in.–3in. long, naked, 1-flowered. Flower minute. Sepals with a minute spur. Petals 0. Fruiting torus ¼in.–¾in. long. Achenes numerous, dense.—Hook. f., Fl. N.Z. i. 8; Handbk. 3.

NORTH Island: Palliser Bay, *Colenso.* Ocean Beach, Cook Strait, *Buchanan.* SOUTH Island: Lake Tekapo, *Cheeseman!* Otago: Hyde, Ida Valley, Speargrass Flat, Alexandra, *Petrie!* Gimmerburn, *T. Kirk.* Sea-level to 2,000ft. Usually in places where water has stagnated during the winter. It has not been collected in the North Island for many years. It extends to California and South Chili, where it ascends to 11,500ft. Nov., Dec.

3. RANUNCULUS, Linn.

Sepals usually 5, rarely 4 or 3. Petals 4–20, usually with 1–3 small nectariferous pits at the base. Carpels usually numerous, each with a short persistent straight or hooked style and one ascending ovule. Herbs, mostly with radical entire or divided leaves, and yellow or white flowers.

SPECIES, 175, chiefly distributed through the temperate and cold regions of both hemispheres. The New Zealand species comprise several of the most beautiful known. The majority are endemic; four species extend to Australia. Several European species are naturalised.

1. **Hecatonia.** Carpels glabrous or rarely silky. Perennial or rarely annual herbs, tufted, or creeping or floating. Flowers white or yellow.

A. Stem erect, without creeping stolons.

* *Achenes villous or silky.*

Leaves large, peltate, entire, crenate. Flowers white	1. *R. Lyallii.*
Leaves large, peltate, lobed, 2-crenate. Flowers white	2. *R. Traversii.*
Leaves rounded cordate. Flowers yellow	3. *R. insignis.*
Leaves broadly oblong. Achenes with few hairs	4. *R. Godleyanus.*
Leaves deeply 3–5-lobed or dissected. Flowers white	5. *R. Buchanani.*
Leaves rounded reniform, deeply crenate	8. *R. Monroi* var. *sericeus.*

** *Achenes glabrous, tumid or angled, not compressed.*

a. *Stems branched*:—

Leaves rounded reniform, 3–7-lobed	6. *R. nivicola.*
Leaves broadly reniform, 3–5-lobed	7. *R. geraniifolius.*
Leaves rounded reniform. Petals exceeding the sepals	8. *R. Monroi.*
Leaves much divided. Stem leafy. Achenes minute	* *R. sceleratus.*

b. *Stems simple; rarely branched in R. Enysii*:—

Leaves rounded reniform. Petals shorter than the sepals	9. *R. pinguis.*
Leaves few, multifid, coriaceous; cauline involucrate	10. *R. Haastii.*
Leaves numerous, 3-partite. Scape naked, shorter than the leaves ..	11. *R. chordorhizos.*
Leaves few, 3-partite. Scape naked, stout, equalling the leaves ..	12. *R. pauciflorus.*
Leaves multifid. Scape naked, shorter than the leaves	13. *R. crithmifolius.*
Leaves ovate-oblong, multifid. Scape exceeding the leaves	14. *R Sinclairii.*
Leaves pinnate, with lobed segments, or 2-ternate, hairy	15. *R. gracilipes.*
Leaves 3 pinnatisect, silky or glabrate. Scape usually naked ..	16. *R. sericophyllus.*
Leaves orbicular-reniform, 3-partite. Scape naked	17. *R. Berggrenii.*
Leaves ternately divided. Scape naked, exceeding the leaves ..	18. *R. Novae-Zelandiae.*
Leaves 3-foliolate, exceeding the scape	19. *R. recens.*
Leaves glabrous, 3-foliolate; leaflets entire or 3-ternate	20. *R. Enysii.*
Leaves glabrous, broadly reniform, 3-lobate. Scape with verticilate linear bracts	21. *R. verticillatus.*
Leaves broadly reniform, 3–5-foliolate. Style spirally curved ..	22. *R. tenuicaulis.*

*** *Achenes compressed, with a thickened margin; style short, hooked; leaves hirsute.*

Stem branched. Leaves 3-foliolate or pinnately 2-ternate. Sepals deflexed	23. *R. hirtus.*
Leaves entire, 3-lobed or toothed. Achenes numerous	24. *R. foliosus.*
Leaves entire, lobed or 3-foliolate. Achenes, 3–6	25. *R. Kirkii.*
Stem simple. Leaves usually entire, lobed, or partite	26. *R. lappaceus.*
Leaves triangular-ovate, 3-foliolate or partite. Peduncles 1–4 ..	27. *R. subscaposus.*
Leaves orbicular-ovate, 3-foliolate or partite. Peduncles 1–3 ..	28. *R. Hectori.*
Erect, much branched. Leaves 3–7-partite	* *R. acris.*
Decumbent, giving off stout runners. Leaves 3-foliolate, segments acute	* *R. repens.*
Erect, stem swollen at the base. Leaves 3-foliolate, lobes obtuse ..	* *R. bulbosus.*

**** *Achenes glabrous, compressed, but not margined.*

Leaves silky-strigose, orbicular, 3-fid. Scape 2-leaved, 1-flowered ..	29. *R. Aucklandicus.*

B. Stems creeping or with creeping stolons.

Achenes glabrous. (See also 29. *repens.*)

Stems robust, much branched, and rooting at the nodes. Leaves 3-toothed	30. *R. Cheesemanii.*
Leaves multifid, hirsute. Stolons short. Scapes naked, 1-flowered ..	31. *R. depressus.*
Stem creeping, and rooting at the nodes. Leaves reniform, 3-partite	32. *R. crassipes.*

Stems fistulose, creeping, and rooting at the nodes. Leaves 3–5-partite	33. *R. macropus.*
Stems creeping, and rooting at the node. Flowering-stems slender, erect	34. *R. rivularis.*
Flowering-stems floating..	var. *subfluitans.*
Leaves 3-foliolate. Scapes 1-flowered, naked, shorter than the leaves	35. *R. acaulis.*
Leaves 3-foliolate, or leaflets ternately divided. Stem filiform, creeping, and rooting	36. *R. ternatifolius.*
Stem prostrate, stout. Leaves cuneate, exceeding the 1-flowered scapes	37. *R. pachyrrhizus.*
Stems filiform. Leaves linear-spathulate. Flowers tetramerous ..	38. *R. Limosella.*

2. **Echinella.** Achenes tuberculate, or muricate, or hispid on the sides. Annual.

Erect, hairy. Carpels with a row of intramarginal tubercles	* *R. philonotis.*
Stems slender, decumbent. Leaves orbicular, 3-partite ..	* *R. parviflorus.*
Stems filiform, prostrate. Leaves 3-lobed or partite ..	var. *australis.*
Erect, glabrous. Achenes with hooked spines	* *R. arvensis.*
Suberect or spreading. Achenes with straight spines ..	* *R. muricatus.*

3. **Ceratocephalus.** Achenes with a swollen empty cavity on each side at the base; the beak falcate, and greatly elongated.

Small, depressed. Leaves multifid	* *R. falcatus.*

1. **R. Lyallii,** *Hook. f., Handbk. N.Z. Fl.* 4. Rootstock very stout. Stem 1ft.–3ft. high or more, paniculately branched. Radical leaves on long stout petioles, peltate, orbicular, slightly funnel-shaped, coriaceous, crenate, glabrous or with few scattered hairs, 6in.–15in. in diameter. Cauline leaves few, sessile, lobed, connate. Peduncles stout, with linear bracts, villous. Flowers numerous, white or cream-coloured, 2in.–3in. in diameter. Sepals 5, pilose. Petals 5–10 or more, broadly cuneate, with a narrow gland at the base. Anthers oblong, on short filaments. Receptacle cylindrical, hairy, elongating after flowering. Achenes villous; style subulate, flexuous.

SOUTH Island: Spencer Mountains to Otago. STEWART Island: Mount Anglem. 2,000ft. to 5,000ft. *Shepherd's lily.* Nov. to Jan.

The leaves of seedling plants vary in shape from cuneate-rhomboid to reniform, but are not peltate. The noblest species of the genus.

2. **R. Traversii,** *Hook. f., Handbk.* 4. Similar to *R. Lyallii*, but smaller. Leaves 4in.–7in. in diameter, margins crenate or doubly crenate or lobed, and with one or two deep incisions at the base forming a closed sinus. Carpels not seen.

SOUTH Island: Canterbury: Hurunui Mountains. Otago: Mount Earnslaw.

This appears to be a variety of the preceding; but I do not care to unite the two until the carpels have been examined.

3. **R. insignis,** *Hook. f., Fl. N.Z.* i. 8, *t.* 2. An erect robust paniculate species, 1ft.–2ft. high, villous in all its parts; brown or reddish-brown when dry. Radical leaves on stout sheathing petioles, rounded-ovate or cordate, rarely reniform, very coriaceous, crenate or lobed, 4in.–6in. broad. Panicle much branched; peduncles numerous, bracts linear, oblong. Flowers

1½in. in diameter, yellow. Sepals broadly oblong, woolly. Petals 5 or more, emarginate or obcordate, with one or two glands near the base. Achenes numerous, viscid, tumid; style long, slender.—Hook. f., Handbk. N.Z. Fl. 4. *R. Ruahinicus*, Colenso in Trans. N.Z.I. xviii. (1885) 256, has a single glandular depression at the base of the petal. *R. sychnopetala*, Col., *l.c.* xxv. (1892) 324, and xxvi. 313, is a monstrous state, with from 40 to 45 petals.

NORTH Island: Hikurangi, East Cape; Ruahine and Tararua Ranges; Tongariro, Ruapehu, &c. SOUTH Island: Mountains of Nelson, not unfrequent. Marlborough: Kaikoura Range, &c. 2,000ft. to 6,000ft.

Var. lobulatus. Leaves membranous, suborbicular, deeply lobed or sinuate, with few weak hairs; rarely peltate or subpeltate. Flowers not seen. Marlborough: Kowhai River, 500ft. Mount Fyffe, 3,000ft. *T. K.* Oct., Nov.

A magnificent species, originally discovered by Colenso. The long sheathing-base of the petiole is ciliated at the margins.

4. **R. Godleyanus,** *Hook. f., Handbk.* 723. A stout, erect, glabrous species. Leaves all radical, petioles ¾in.–1in. broad, 3in.–6in. long, shortly sheathing at the base. Blade broadly oblong or almost reniform, rounded at the apex, coarsely crenate, coriaceous or fleshy, with radiating nerves. Scape stout, exceeding the leaves, naked below, with two or more sessile or stalked crenate bracts. Peduncles numerous, springing from the axils of the bracts, 2in.–4in. long, naked or with one or two secondary bracts above the middle. Flowers 1½in. in diameter or more. Sepals 5, broadly oblong. Petals 5, broadly obcordate, with 1–3 naked pits at the base. Receptacle pilose. Achenes very numerous, with few hairs, forming a dense head, narrowed into the slender curved style.

SOUTH Island: Canterbury: Whitcomb's Pass. 4,300ft. *Haast! Enys! J. B. Armstrong!*

A noble species. The flowers are less numerous than in *R. insignis*, but more highly coloured.

5. **R. Buchanani,** *Hook. f., Handbk.* 5. Root-fibres stout, fleshy. Scapes stout, erect, 6in.–12in. high or more, leafy above, 1-flowered. Whole plant more or less villous or rarely glabrate. Radical leaves on rather stout petioles, 2in.–5in. long, with short, broad, scarious sheaths; blade broadly reniform, 2in.–6in. broad, ternatisect, the main divisions petiolate, broad, cuneate-lobed or incised, or coarsely toothed, or often cut into narrow linear segments; rarely entire. Cauline leaves on very short petioles or sessile, excessively divided often to the base. Flowers 1½in.–2½in. in diameter. Sepals 5, villous. Petals numerous, linear-oblong, rounded at the apex, shortly clawed with a gland near the base. Receptacle oblong, papillose. Achenes turgid, pilose, with the dorsal margin acute; style long, flexuous.

SOUTH Island: Otago: Lake district, *Buchanan!* Mounts Bonpland, Tyndall, and Aspiring, *Petrie!* Dec., Jan.

A remarkable species. The cauline leaves often present the appearance of whorls of narrow linear bracts. Flowers white, rarely yellow.

6. **R. nivicola,** *Hook., Ic. Pl. t.* 571, 572. Stem erect, branched, robust or slender, 1ft.–2½ft. high, more or less clothed with soft white spreading hairs or nearly glabrous. Radical leaves on stout petioles 4in.–12in. long, with

short membranous sheaths; blade 3in.–5in. in diameter, with an open sinus, broadly reniform, deeply 3–7-lobed; primary lobes broadly cuneate. Cauline leaves deeply lobed or incised, sessile or shortly petiolate. Flowers 1in.–1½in. in diameter, in a lax open panicle. Sepals 5, oblong or linear-oblong, hirsute. Petals 10–15, with a gland at the base, emarginate or rounded at the apex, yellow. Achenes forming a rounded head, glabrous; style long, straight, hooked at the apex.—Hook. f., Fl. N.Z. i. 8, and Handbk. 5; Raoul, Enum. 47. *R. reticulatus*, Colenso in Trans. N.Z.I. xx. (1887) 188.

NORTH Island: Mount Egmont ranges, &c.; Ngauruhoe, Tongariro, Ruapehu, Tauhara, and other mountains in the Taupo district. 2,500ft. to 5,000ft. Nov. to Jan.

A striking and beautiful plant, with golden-yellow flowers. I am indebted to Mr. Colenso for the loan of his type-specimen of *R. reticulatus*, which, in the absence of flowers or fruit, offers no character enabling me to distinguish it from *R. nivicola*, although its author suggests that its nearest ally is *R. pinguis*, Hook. f.

7. **R. geraniifolius,** *Hook. f., Fl. N.Z.* i. 9, *t.* 3. Slender, erect, sparingly branched, 9in.–30in. high, glabrous or more or less villous. Radical leaves on slender petioles 3in.–9in. long; blade 2in.–3in. in diameter, broadly reniform, or truncate or cuneate at the base, deeply 3–5-lobed or ternatisect; segments crenate or crenate-lobed. Cauline leaves usually sessile, lobed or ternatisect. Flowers ½in.–1½in. in diameter, yellow. Sepals oblong, glabrous or sparingly hairy. Petals 10–15, twice as long as the sepals, with a basal gland. Achenes forming a small spherical head, turgid, glabrous; style short, flexuous. —Handbk. N.Z. Fl. 5.

NORTH Island: Hikurangi, East Cape; Ruahine Range, *Colenso*. Tararua Range, *Arnold! Buchanan!* SOUTH Island: Marlborough: Mount Stokes, &c., *J. Macmahon, T. K.* Nelson mountains, not unfrequent, *Cheeseman!* &c. 3,000ft. to 5,000ft. Dec., Jan.

Less robust than *R. nivicola*, and varying greatly in the division of the leaves, the degree of hairiness, &c. Fragmentary specimens from Mount Arthur (*Rev. F. H. Spencer*) and from Mount Owen (*Dr. Gibbs*) may be different.

8. **R. Monroi,** *Hook. f., Fl. N.Z.* ii. 323. Rootstock short, stout, clothed with the ragged bases of old petioles. Leaves all radical, subcoriaceous or almost fleshy; petioles 2in.–6in. long or more; blade rounded-reniform or nearly orbicular or ovate, glabrate or sometimes silky or villous, coarsely crenate or crenate-dentate. Scapes branched, rarely simple; bracts deeply lobed. Flowers yellow, ¾in.–1½in. in diameter. Sepals linear-oblong, obtuse, glabrous or silky. Petals 5 or more, twice as long as the sepals, with a naked pit at the base. Achenes forming a small rounded head, usually glabrous, turgid, faintly keeled at the back; style long, filiform, straight or recurved.—T. Kirk in Trans. N.Z.I. xxvii. (1894) 349. *R. pinguis, a.* Hook f., Handbk. N.Z. Fl. 5. *R. Muelleri*, Buch. in Trans. N.Z.I. xix. (1886) 215, t. 16.

NORTH Island: Wellington: Tararua Range, *Buchanan!* SOUTH Island: Nelson: Wairau Mountains, Tarndale, Spencer Mountains, &c. Marlborough: Kaikoura Range. Canterbury: Mount Torlesse, Porter River, mountains above the Broken River Basin, &c. 2,000ft. to 6,300ft.

Var. **sericeus.** Achenes clothed with silky hairs. Kaikoura Range, *Buchanan!* Dec., Jan.

Var. **dentatus.** Leaves broadly ovate or ovate-lanceolate, crenate, lobed or dentate; clothed on both surfaces with strigose ferruginous pubescence. Kaikoura Range; Mount Torlesse; Broken River, &c.

2

This species is nearly related to *R. pinguis*, but differs in the branched scapes, in the petals being twice the length of the sepals, and especially in the filiform style. It varies greatly in the shape of the leaf and the amount of pubescence.

* **R. sceleratus**, *L.*, *Syst. Pl.* 551. Annual; stems hollow; erect, 6in.–18in. high, much branched, nearly glabrous. Lower leaves petiolate, deeply 3-partite or 3-foliolate; segments rounded, toothed or crenate. Cauline leaves nearly sessile, with oblong toothed or entire segments. Flowers small. Sepals 5, reflexed, nearly equalling the petals. Receptacle oblong. Achenes very small, rounded, slightly wrinkled; style minute.

NORTH and SOUTH Islands: Naturalised in damp places. *Celery-leaved crowfoot.* Nov. to March. · Europe.

9. **R. pinguis**, *Hook. f.*, *Fl. Antarc.* i. 3, *t.* 1. Usually stout and rather fleshy, 2in.–10in. high, glabrous or with few weak hairs on petiole and scape. Rootstock stout. Leaves all radical on long broad petioles, reniform with an open sinus, deeply crenately lobed. Scape exceeding the leaves, naked or rarely with one or two bracts above the middle, thickened upwards, 1-flowered. Sepals 5–6, linear-oblong. Petals with 1–3 glands at the base, narrow-oblong, shorter than the sepals. Receptacles broadly ovoid. Achenes very numerous, small; style subulate, straight, with 3 narrow wings.—*R. pinguis β*, Hook f., Handbk. 5. Dec.

Var. **rhomboideus.** 1in.–3in. high, rhomboid-cuneate, 3-lobed or toothed. Receptacle conical, with a ring of hairs at the base.

AUCKLAND and CAMPBELL Islands: Ascends to 2,000ft. Dec.

Easily distinguished from all states of *R. Monroi* by the simple scapes, short petals, and stout subulate style.

10. **R. Haastii**, *Hook. f.*, *Handbk.* 6. Usually glabrous, stout, fleshy or coriaceous, 3in.–6in. high. Petioles and scapes tapering downwards, grooved when dry. Rootstock 2in.–6in. long, ½in.–1in. in diameter, horizontal, viscid and milky when bruised. Radical leaves 1 or 2; petioles 2in.–6in. long; blade 2in.–4in. in diameter, reniform or orbicular reniform, often with a closed sinus, palmately 5–7-lobed, or partite to the base, the divisions lobed or irregularly cut into narrow blunt segments. Petioles shortly sheathing, often villous at the base. Scape naked below, with 1–3 sessile or petioled deeply lobed or incised leaves forming a kind of involucre to the flowers. Peduncles 1–3, naked. Flowers 1½in. in diameter. Sepals 5, oblong or ovate-oblong, usually glabrous. Petals 10–15. Achenes ¼in. long, forming a rounded head ¾in. in diameter, glabrous, turgid; style long, flattened upwards, subulate. Receptacle globose, papillose.

SOUTH Island: Nelson: On shingle-slips, Wairau Gorge and Mount Captain Range, *T. K.* Canterbury: Mount Torlesse and Ribbon-wood Range, *Haast!* Broken River and Leith Hill, *Enys!* Ashburton Mountains, *T. H. Potts!* Otago: Mount Kyehuru, Mount St. Bathan's, &c., *Petrie!* 2,500ft. to 6,000ft. Dec., Jan.

This appears to be the only New Zealand species with a stout fleshy rootstock. The Otago specimens have broader petals and less divided leaves, while the scapes are more villous at the base, and the apex of the rootstock is usually clothed with the remains of old petioles. I have not seen achenes.

11. **R. chordorhizos**, *Hook. f.*, *Handbk.* 723. Rootstock short, stout, with numerous vertical rootlets. Leaves glabrous, thick, coriaceous, 1in.–2in.

long, suborbicular, 3–5-lobed or 3-partite to the base; segments often petiolulate, obovate or obovate-spathulate, crenate or crenate-dentate; pitted on the upper surface when dry. Sheath of petiole short, broad, scarious. Scape usually solitary, shorter than the petiole, naked, 1-flowered. Sepals 5, narrow-elliptic. Petals 5, twice as long as the sepals. Receptacle small, globose. Achenes turgid, with a flexuous slender style equalling or exceeding the achene.

SOUTH Island: Canterbury: Macaulay River and Mount Somers Range, *Haast* in Handbk. Otago: *Buchanan!* Mount Kyeburn and Mount St. Bathan's, *Petrie!* 3,000ft. to 5,000ft. Dec.

The description in the Handbook apparently includes both this species and the next, while the thick rhizomes belong only to *R. Haastii*. It varies greatly in the division of the leaves, which completely hide both flowers and fruit. I have not seen Canterbury specimens.

12. **R. paucifolius,** *n. s.* Rootstock short, stout, with thick vertical fibres 6in.–8in. long. Whole plant glabrous. Leaves 1 or 2, radical, 1in.–2in. long, spreading, petiolate, suborbicular, cuneate or almost reniform at base, nearly entire or 3–5-lobed or partite nearly to the middle; segments overlapping, minutely crenate or subserrate, not pitted above. Petiole with a broad sheath for half its length. Scape solitary, stout, naked, 1-flowered, equalling the petioles. Sepals 5, ovate-oblong, subacute, deflexed. Petals 5. Achenes few, turgid, with a straight subulate beak.

SOUTH Island: Amongst limestone gravel, Broken River, Waimakariri, *J. D. Enys!* Dec.

Nearly related to *R. chordorhizos*, but easily distinguished by the leaves never exceeding 2 or 3, and being carried on longer sheathing spreading petioles, while the scape equals or exceeds the petioles, so that the flower is fully exposed. Better material is wanted to allow of a good diagnosis being drawn, as I have only seen a single flower. Haast's plant from the Macaulay River and Mount Somers Range, doubtfully referred to *R. chordorhizos*, may be identical with this.

13. **R. crithmifolius,** *Hook. f., Handbk.* 6. A small glabrous, fleshy, glaucous species. Rootstock short, stout, horizontal, with thick fleshy fibres. Leaves all radical, on recurved petioles 1in.–2in. long; blade ½in.–1in. broad, reniform in outline, biternately multifid; segments short, linear, $\frac{1}{10}$in. long, obtuse. Scape stout, fleshy, erect, shorter than the leaves, 1-flowered. Flowers small. Sepals linear-oblong. Petals not seen. Achenes in a globose head ⅓in. in diameter, turgid, keeled; style sharp, straight, subulate.

SOUTH Island: On shingle-slips, Wairau Gorge. 6,000ft. *Travers.*

A very singular plant, easily recognised by its glaucous fleshy habit, finely-divided leaves, and short 1-flowered scapes. Only a solitary specimen observed.

14. **R. Sinclairii,** *Hook. f., Handbk.* 6. Glabrous, or with few weak hairs on the petioles and scapes; 2in.–6in. high. Rootstock prostrate or erect. Leaves all radical, 1in.–4in. long, tufted, oblong or ovate-oblong, 2-pinnatisect or multifid; segments short, narrow-linear or slightly ovate, acute or subacute, primary in 3 or 4 pairs, opposite; petioles slightly sheathing. Scapes slender, exceeding the leaves, 1-flowered. Sepals 5. Petals 5, twice as long as the sepals, with a deep gland. Receptacle small, conical. Achenes few, small, slightly turgid; style very short, straight, subulate.

SOUTH Island: Nelson: Wooded peak, &c., Wairau Gorge, *Travers!* Raglan Mountains, *Cheeseman!* Tarndale, *Sinclair!* Canterbury: Mountains above Broken River, *Enys!* Otago: Lake district, *Buchanan!* Dec., Jan.

Var. **angustatus.** Leaves narrow-oblong, 1in.–1½in. long, ¼in. broad, spreading, pinnae deeply incised, rhachis more hairy than in the type. Scapes very slender, hairy, spreading. Flowers small. Otago: Maungatua, *Petrie!*

The typical form is easily recognised by its soft, finely-cut leaves, which sometimes have capillary segments, and are always shorter than the scapes. Var. *angustatus* should probably be placed under the next species, with which it agrees in the narrow leaves and slender hairy scapes; but the carpels are unknown. Hooker, *l.c.*, mentions a plant with less divided leaves and long hairs on scape and petiole sent in a flowerless state from the Ruahine by Colenso.

15. **R. gracilipes,** *Hook. f., Handbk.* 8. Rootstock short, with rather stout fibrous roots. Leaves all radical, glabrate or villous, 1in.–6in. long, ¼in.–½in. broad; petioles slender; blade linear, oblong pinnate or twice pinnate. Leaflets in from 2–7 pairs, sessile or petioled, nearly entire or lobed, or 3-fid or 3-partite or 2-ternate, the segments cuneate at the base, usually acute or subacute at the tip. Scapes 1–5, 2in.–7in. high, slender, naked, villous, 1-flowered. Sepals 5, narrow-ovate, hairy, spreading. Petals 5–10, narrow, obovate, narrowed into a slender claw with a gland above the base. Stigma short, slender, oblique. Receptacle conical. Achenes not seen.

SOUTH Island: Canterbury: Shores of Lake Ohau, *Haast!* Otago: *Buchanan!* Dunstan Mountains, Mounts Ida, Bonpland, and Kyeburn; Mount Pisa Range; Kurow Flat; Old-Man Range; *Petrie!* 3,000ft. to 4,500ft.

In the Handbook this species is described as perfectly glabrous, but the fine series of specimens in Mr. Petrie's herbarium, and those for which I am indebted to him, are pubescent or villous, rarely glabrate, never glabrous, and the petals are never retuse. Occasionally the leaves are rather fleshy and the rhachis is flattened.

16. **R. sericophyllus,** *Hook. f., Handbk.* 6. The entire plant excessively silky, rarely glabrate. Rootstock short, stout. Leaves all radical, 1in.–2in. long; petiole broadly sheathing; blade ovate or broadly ovate, 3-pinnatisect; segments very short, subacute or acute, tipped with a pencil of silky hairs. Scape exceeding the leaves, stout, naked or with an entire or divided bract, 1-flowered. Flowers 1½in. in diameter. Sepals 5, broadly oblong or linear. Petals 5–10, obovate-cuneate, with naked glands. Receptacle ovoid. Achenes forming a globose head, slightly turgid, faintly keeled, with a filiform flexuous style as long as the achene.

SOUTH Island: Mountains of Canterbury, Westland, and Otago. Browning's Pass, Mount Brewster, Hopkins River, and source of the Rakaia, *Haast!* Mount Cook, *S. H. Dixon!* Otago, *Buchanan!* Matukituki Valley and hill opposite Mount Aspiring, *Petrie!* 3,500ft. to 7,000ft. Dec., Jan.

A singular plant, with golden-yellow flowers. Mr. Petrie's specimens are glabrate or almost glabrous.

17. **R. Berggrenii,** *Petrie in Trans. N.Z.I.* xix. (1886) 325. A stemless species, glabrous in all its parts. Rootstock rather stout, rarely horizontal, but not creeping, with thick vertical rootlets. Leaves all radical, on flattened petioles ½in.–1in. long; blade ⅓in.–¾in. in diameter, orbicular or orbicular-reniform, with an open sinus, unequally 3-partite to the middle; segments lobed, crenate. Scapes 1, rarely 2, naked, 1-flowered. Sepals 5, broadly ovate, margins scarious. Petals 5, narrow-obovate, exceeding the sepals, gland near the base; style subulate, straight. Achenes not seen.

SOUTH Island: Otago: Carrick Range, *Petrie!* 4,000ft. Dec.

18. **R. Novae-Zelandiae,** *Petrie in Trans. N.Z.I.* xxvi. (1893) 267. Stemless, glabrous. Rootstock rather stout, clothed with the bases of old petioles; rootlets thick. Leaves on flattened petioles, ½in.–1in. long, sheathing at the base; blade ¾in.–1in. long or more, 3-foliolate, slightly coriaceous; the lower leaflets usually sessile, the upper distant, petiolulate, more or less deeply divided into three 3-lobed crenate segments. Scapes 1–3, naked, 1-flowered, 1in.–2in. long. Sepals 5, oblong, shorter than the petals, often deflexed. Petals 5, with a broad gland at the base; style short, subulate, straight. Achenes not seen.

SOUTH Island: Otago: Rough and shingly places at the summit of the Rock and Pillar Range, opposite Middlemarch, and on the Old-Man Range. *Petrie!* 4,000ft. and upwards.

Closely related to *R. Berggrenii*, of which it may possibly prove to be a variety; but this can only be determined by the discovery of the ripe achenes of both species.

19. **R. recens,** *n. s.* Stemless, depressed, less than 1in. high. Leaves rosulate, ¾in.–1in. long including the stout sheathing petiole, 3-foliolate, coriaceous; leaflets deeply lobed or pinnatifid; segments subacute or obtuse, glabrous or with a few scattered hairs on the upper surface; petiole strigose or glabrate. Scapes about ½in. high, strigose, 1-flowered. Flowers not seen. Achenes orbicular-ovate, slightly compressed, faintly keeled; style minute, scarcely recurved.

SOUTH Island: Otago: Alpine. *Buchanan! Petrie!*

The leaves approach *R. Novae-Zelandiae*, but the segments are smaller and more fleshy; the achenes are unlike those of any other species. I have only three small specimens.

20. **R. Enysii,** *T. Kirk in Trans. N.Z.I.* xii. (1879) 394. Glabrous, scapes erect, 10in.–15in. high. Rootstock rather stout, with strong spreading rootlets. Radical leaves on grooved petioles, 4in.–8in. long; blade digitately 3–5-foliolate or 2-ternate; leaflets petiolulate, obliquely-rounded, cuneate, shortly lobed, with coarsely crenate or crenate-serrate margins, or 2-ternate or quinate, the terminal segments irregularly lobed or cut. Scapes 2–5, exceeding the leaves, simple or rarely with 1 or 2 branches, naked or with a single bract or petioled leaf. Flowers ¾in. in diameter. Sepals broadly ovate, acute. Petals 5–10, broadly obovate, with a deep gland near the base. Achenes small, in dense globose heads, glabrous, turgid, minutely reticulate; style short, slender, curved.—*R. tenuis*, Buch. in Trans. N.Z.I. xx. (1887) 255, t. 12.

SOUTH Island: Canterbury: Mount Torlesse, Broken River, Coleridge Pass, &c., *J. D. Enys!* Otago: Mountains above Lake Harris, *T. K.* East Taieri hills, *Buchanan, l.c.* 2,000ft. to 3,500ft. Dec., Jan.

A very distinct species; the margins of the leaves are often thickened and the teeth acute.

21. **R. verticillatus,** *n. s.* Leaves all radical on slender petioles 3in. long, broadly reniform, 3-lobate, margins lobed or crenate, glabrous. Scape simple, with one or two whorls of linear bracts on the upper portion. Flowers solitary. Sepals broadly ovate. Petals linear-oblong, ¾in. long. Achenes not seen. Scape and bracts pubescent above.

SOUTH Island: Canterbury: Near Lake Ohau, *Buchanan!*

I have only seen a very imperfect specimen of this plant, which appears distinct from all other New Zealand species.

22. **R. tenuicaulis,** *Cheeseman in Trans. N.Z.I.* xvii. (1884) 235. A slender, erect species, 3in.–18in. high. Rootstock short. Leaves all radical on slender petioles 2in.–6in. long; blade broadly reniform or subreniform, divided to the base into 3, rarely 5, broadly cuneate divisions, each of which is deeply 2–3-lobed; the lobes narrow, toothed, glabrate or with a few scattered hairs. Scapes very slender, with two or three simple or deeply-divided bracts below the solitary flower, grooved, glabrate or almost strigose. Flowers small. Sepals not seen. Petals 5, linear, acute. Achenes 5–20, spreading, flask-shaped, glabrous, stipitate, gradually narrowed into a long spirally-curved style.

SOUTH Island: Arthur's Pass: *T. K.* (1876), *Cheeseman!* Craigieburn Mountains, *L. Cockayne!* Otago: Swampy Hill, Dunedin; Mount Kyeburn, &c., *Petrie!* 2,000ft. to 4,000ft. Nov., Dec.

Easily distinguished by the remarkable achenes and spirally-curved styles. I have only seen a single flower, and that in a very imperfect condition.

23. **R. hirtus,** *Banks and Sol.* ex *Forst. Prod. n.* 525. Usually erect, 6in.–18in. high, slender, sparingly branched, hirsute, hairs spreading or rarely appressed. Radical leaves on long petioles, pinnately 3–5-foliolate; leaflets petiolulate, broadly ovate, entire or deeply lobed or toothed, rounded or rarely cuneate at the base. Cauline leaves smaller. Peduncles slender, glabrous or hairy. Flowers about ½in. in diameter. Sepals 5, shorter than the petals, reflexed, fugacious. Petals 5, narrow, with a gland near the base. Receptacle pilose. Achenes glabrous, more or less compressed; style short, recurved.—*R. hirtus*, DC., Syst. Veg. i. 289; A. Cunn., Precurs. n. 634; Raoul, Enum. 47; Hook. f., Fl. N.Z. i. 9. *R. plebeius*, Hook. f., Handbk. 7 (not of R. Br.). *R. acris*, A. Rich., Fl. N.Z. 289 (not of L.).

NORTH CAPE to STEWART Island; THREE KINGS Island; CHATHAM Islands. Ascends to 4,000ft. Nov. to Jan.

Var. **robustus.** Erect, stout, much branched. Cauline leaves usually 3-foliolate; leaflets narrowed below. Heads of achenes larger than in the type. In subalpine places.

Var. **stoloniferus.** Stems slender, procumbent and rooting at the nodes. Leaves small, 3-fid or 3-partite. Flowers small. Achenes small. In subalpine situations.

Var. **membranifolius.** Stems capillary, 3in.–5in. long, suberect. Radical leaves on long slender petioles, subreniform, 3-lobed. Flowers and achenes very small. SOUTH Island: Westland: Teremakau, *Petrie!*

Sub sp. **plebeius.** Suberect or erect, sparingly villous or silky, slender. Radical leaves on long petioles, 3-foliolate; leaflets shortly stalked, ovate-cuneate, 3-lobed or toothed. Peduncles slender. Sepals appressed or spreading, rarely reflexed. Petals narrow-obovate, close. Achenes glabrous, with a slender hooked beak.—*R. plebeius*, R. Br., in DC. Syst. Veg. 288. SOUTH Island: Hilly and subalpine localities, ascending to 4,000ft. Nov. to Jan. This differs from the Australian plant in the leaves never being digitately divided.

24. **R. foliosus,** *n. s.* Tufted, 3in.–6in. high, strigose, hirsute or pilose; stems simple or sparingly branched, stout. Radical and cauline leaves on long petioles, blades ¾in.–½in. long, obovate-cuneate, 2–4-lobed or toothed, strigose on both surfaces. Scapes 1–2-flowered. Achenes usually hidden by the leaves, small, slightly turgid; style short, subulate, recurved. Flowers not seen.—*R. subscaposus*, Hook. f., Handbk. 7 (not of Fl. Antarc.).

SOUTH Island: Fowler's Pass, Amuri, *T. K.* Hopkins River, Canterbury, *Haast.* 3,000ft. to 5,000ft. Dec., Jan.

Most nearly allied to *R. lappaceus*. Mr. N. E. Brown, who has compared my plant with the Hopkins River specimen at Kew, considers them identical. I refer a plant collected at the Broken River by Cheeseman to this species, although it is more robust, and has broader leaves.

25. **R. Kirkii,** *Petrie in Trans. N.Z.I.* xix. (1886) 323. Glabrate or almost strigose, 1in.–5in. high. Root-fibres fleshy. Radical leaves on slender petioles, 1–3-foliolate; leaflets petiolulate, rounded, ovate, or rarely cuneate, 3-lobed or 3-fid. Scapes simple or branched, peduncles naked or with a solitary bract. Sepals oblong-lanceolate. Petals obtuse. Achenes 3–6, glabrous, slightly compressed, faintly keeled, orbicular-ovate, narrowed at the base; style short, subulate, shortly hooked at the apex.

SOUTH Island: In mountain districts; Arthur's Pass, &c. STEWART Island: Not uncommon. Sea-level to 3,500ft. Dec., Jan.

Only distinguished from states of *R. lappaceus* by the larger achenes, which are very shortly recurved.

26. **R. lappaceus,** *Smith in Rees Cyclopedia* xxix. (1815) *n.* 61. Rootstock short; whole plant more or less hairy. Leaves usually all radical on long petioles; blade ovate or rounded-ovate or cuneate, entire or 3–5-lobed or partite, rarely palmate or pinnate, the lobes sometimes cut into narrow segments, coarsely crenate. Scapes usually leafless, 1-flowered, rarely 2 or more flowered, and sparingly leafy; 1in.–8in. high. Sepals 5, usually hairy, spreading. Petals 5, obovate, with a small basilar gland. Achenes forming a globose head, compressed, glabrous, margined; style short, hooked or recurved.—Hook. f., Fl. Tasm. i. 6; A. Gray, Bot. U.S. Expl. Exped. 9; Benth., Fl. Austr. i. 12.

NORTH Island, SOUTH Island, STEWART Island: From Hawke's Bay southwards. Ascends to 4,500ft.

Var. **macrophyllus**. Leaves all radical, petioles 2in.–4in. long, blade ¾in.–1½in. in diameter, hairy, rounded-ovate, truncate or cordate at the base, obscurely 3-lobed, margins crenate or coarsely toothed. Scapes 3in.–8in. high, naked, erect. Flowers large. SOUTH Island: Nelson, &c.

Var. **multiscapus**. Scapes numerous, spreading, petioles shorter; blade ¼in.–½in. long, ovate or ovate-orbicular, cuneate below, 3-fid, 3-lobed or toothed. Achenes few, much compressed.—Hook. f., Handbk. 7. *R. multiscapus*, Hook. f., Fl. N.Z. i. 9, t. 5, and *R. muricatulus*, Colenso in Trans. N.Z.I. xxiii. (1890) 381, differ in the smaller leaves, which are almost entire, and in the achenes being rather turgid. Hawke's Bay to Stewart Island.

Var. **villosus**. 1in.–2in. high, villous or silky in all its parts. Leaves all radical, spreading, 3-lobed or partite, or nearly entire. Scape silky, shorter than the petioles, 1-flowered. Achenes slightly turgid. *R. subscaposus*, var. *Canterburiensis*, J. B. Armst. in Trans. N.Z.I. xiii. 333, appears to be a form of this. SOUTH Island: In alpine situations.

A plant sent by Petrie from Cardrona, Waipahi, &c., is referred here for the present. It has rounded cordate 3-partite or 3-foliolate leaves on long petioles, the segments cuneate at base, and the silky scapes are 2–3-flowered; but the material is insufficient for the exact determination of its position.

27. **R. subscaposus,** *Hook. f., Fl. Antarc.* i. 5. Rootstock stout, short, fibrous, erect or suberect, hairy or almost hispid, fulvous when dry, 6in.–18in. high. Radical leaves on slender petioles, 3in.–8in. long; blade broadly triangular-ovate, slightly cordate, 3-foliolate or 3-partite to the base; leaflets cuneate at base, and more or less deeply incised or toothed, or rarely entire with the margins deeply cut. Cauline similar. Scape usually much longer than the leaves, rarely shorter, 1–3 flowered; peduncles usually much

longer than the leaves, rarely shorter. Sepals 5, spreading. Petals 5, narrow. Receptacle ovoid, pilose. Achenes margined, glabrous: beak short, stout, subulate, scarcely recurved.—Handbk. 7.

CAMPBELL Island: Rare. *Lyall*, 1840; *Rathouis!* 1874; *T. K.*, 1890. Closely related to *R. hirtus*; but the leaves are altogether different, the calyx is spreading, while the achenes are more turgid and have a stouter beak. My specimens are in poor condition.

28. **R. Hectori,** *n. s.* Erect, 6in.–15in. high; whole plant more or less clothed with strigose or appressed hairs. Rootstock short. Leaves chiefly radical, reticulate above when fresh, fleshy, hairy on both surfaces; petioles 4in.–7in. long, slightly sheathing at base; blade 1in.–1½in. long and broad, ovate-orbicular, 3-lobed to below the middle, truncate or slightly cordate at base, lobes acute or subacute. Scapes 1 or 2; peduncles 2 or 3. Cauline leaves petiolate, 3-partite, the segments sparingly lobed or toothed. Receptacle ovate or conical, papillose, sparingly hairy. Flowers not seen. Achenes glabrous, narrowed below, oblique, slightly turgid, faintly keeled or margined; style shortly subulate, slightly recurved.

AUCKLAND Islands: *Sir J. Hector!* (1895). More and better specimens are required to furnish a full description. Its nearest ally is *R. subscaposus.*

*__R. acris,__ *L., Sp. Pl.* 554. Stem erect, slender, branched, hairy, 1ft.–2ft. high. Radical leaves on long petioles, 3–5-partite; segments cuneate at the base, deeply cut into oblong or linear lobes. Cauline leaves small, divided into linear lobes. Sepals 5, spreading, pubescent. Petals 5, with a gland covered by a small scale. Achenes compressed, margined; style recurved.

NORTH and SOUTH Islands: Naturalised in pastures, &c. Nov., Dec. Europe.

*__R. repens,__ *L., Sp. Pl.* 554. Stem creeping and rooting at the nodes. Radical leaves petioled, hairy, 3-partite or 3-foliolate, or ternately pinnatisect; segments cuneate, lobed or toothed. Sepals 5, spreading. Petals 5, gland covered by a scale. Achenes glabrous, compressed, minutely pitted; style recurved.

NORTH and SOUTH Islands; STEWART Island. Naturalised in pastures, waste places, &c. *Creeping buttercup.* Nov. to Jan. Europe.

*__R. bulbosus,__ *L., Sp. Pl.* 554. Stems erect, hairy, 9in.–12in. high, from a swollen base. Leaves on long petioles, 3-foliolate; leaflets ternatisect; segments cuneate, lobed or toothed. Flowers ¾in.–1in. broad. Sepals 5, reflexed. Petals 5–8, gland with a small scale. Achenes compressed; style short, scarcely recurved.

NORTH and SOUTH Islands: Naturalised in pastures, but not common. Oct. to Dec. Europe.

29. **R. Aucklandicus,** *A. Gray, Bot. U.S. Expl. Exped.* i. 8. Radical leaves strigose-hirsute, on slender petioles, 3in.–5in. long or more, slightly sheathing at the base, rounded reniform or the upper rounded truncate, or almost subcordate, 1in.–1¼in. in diameter, 3-cleft to or beyond the middle, mostly with a closed sinus; the broad lobes again 2–3-lobed, or coarsely toothed. Scapes 1–3, strict, 1-flowered, 6in.–9in. high, strigose-hirsute, with 2, or rarely 3, leaves near the base. Flowers not seen. Fruiting receptacle cylindric or subclavate, ⅓in. long, minutely hairy. Achenes narrowed below, almost obovate, much compressed, glabrous, with a short subulate straight style. —Hook. f., Handbk. 723.

AUCKLAND Islands. 1,800ft. Dec.

This species resembles *R. lappaceus*, var. *macrophyllus*, but differs widely in the obovate achene and short straight style.

30. **R. Cheesemanii,** *n. s.* Much branched. Stem and branches stout, grooved, prostrate, often rooting at the nodes, sparingly strigose, especially at the base of the petioles. Radical and cauline leaves similar, about 1½in. long including the petiole, which is broadly sheathing at the base; blade obovate, cuneate, glabrate or glabrous, 3-lobed or toothed at the apex. Peduncles axillary, ½in.–1in. long. Flowers not seen. Receptacle minute, globose, papillose. Achenes few, scarcely stipitate, turgid, glabrous; style much recurved when young, obtuse when fully mature.

SOUTH Island: Near Fowler's Pass, &c., in places where water has stagnated. 3,000ft. *T. K.*

The cauline leaves are often opposite. The horizontal stems and branches, the small glabrous axillary receptacle, which resembles that of the submerged aquatic section, and the turgid scarcely-beaked achenes, render it easy to distinguish this species from all others.

31. **R. depressus,** *T. Kirk in Trans. N.Z.I.* xii. (1879) 393. Tufted, and forming matted patches. Rootstock short, often giving off short stolons. Leaves and scapes clothed with long straight hairs. Petioles 1in.–1½in. long, depressed, spreading; blade broadly ovate in outline, 3-foliolate; leaflets 3-lobed, or toothed, or pinnatisect; segments narrow-linear, obtuse. Scapes solitary, ¾in.–1in. long. Sepals 5, membranous. Petals 5, scarcely exceeding the sepals, with a gland near the base. Carpels few, hidden amongst the leaves, slightly turgid, with a minute beak.

SOUTH Island: Canterbury: In swamps, Broken River Basin, *J. D. Enys* and *T. K.* Otago: Mount Cardrona, *Petrie!* 2,000ft. to 4,000ft.

A singular plant, easily overlooked. The short stout scapes hidden amongst the leaves, the small flowers and turgid carpels, distinguish it from *R. Sinclairii*, with which it has been confused.

Var. **glabratus.** Smaller, almost glabrous. Sheath of petiole longer. Leaves 3-lobate, segments flat. Scape shorter. Achenes not seen. Otago: Mount Cardrona, 4,000ft. *Petrie!*

32. **R. crassipes,** *Hook. f., Fl. Antarc.* ii. 224, *t.* 81. Glabrous, stems creeping and rooting at the nodes. Leaves on petioles 1in.–4in. long, almost fleshy when fresh, reniform-cordate, unequally 3-partite or -fid, the lower segments lobed and deeply toothed or crenate. Peduncles 1-flowered, axillary, shorter than the leaves. Flowers small. Sepals 4 or 5, ovate, membranous. Petals 4 or 5, scarcely exceeding the sepals, 3-nerved, with a gland just below the middle. Achenes forming a small globose head, broadly ovate, turgid, faintly margined; style short, strict.

MACQUARIE Island: *A. Hamilton!* Also on KERGUELEN'S Land and MARION Island.

Mr. Hamilton states that the carpels become scarlet when fully ripe, and render the plant very conspicuous.

33. **R. macropus,** *Hook. f. in Hook. Ic. Pl. t.* 634. Glabrous stems, creeping and rooting at the nodes, fistulose, 3in.–12in. high. Radical leaves on petioles 3in.–15in. long; blade 1in.–2in. in diameter, flabellate, 3–5-partite to the base; leaflets cuneate at base, margins irregularly cut into obtuse lobes or

teeth. Peduncles naked, 3in.–10in. long, axillary or leaf-opposed. Sepals 5, broadly ovate. Petals 5–10, usually shorter than the sepals, narrow, with a basal gland. Receptacle shortly oblong, tumid. Achenes turgid, glabrous, smooth or faintly muriculate; beak long, subulate, nearly straight.—Fl. N.Z. i. 10; Handbk. 7.

NORTH and SOUTH Islands: From the Auckland Isthmus to Otago: In pools and swamps in lowland districts.

A succulent plant, varying according to the depth of water in which it grows. Petioles may be seen over 20in. in length, with the blade 3in. in diameter and scarcely toothed.

34. **R. rivularis,** *Banks and Sol.* ex *Forst. Prod. n.* 524. Glabrous, creeping and stoloniferous, producing tufts of radical leaves and erect stems at each node, or floating and branching irregularly. Leaves on slender petioles 1in.–6in. long, suborbicular or ovate, ½in.–1in. in diameter, divided into 3, 5, or 7 leaflets, linear or narrow-cuneate, ternatisect or 3-lobed, rarely entire. Peduncles exceeding the leaves. Sepals 5, spreading. Petals 5–10, usually exceeding the sepals, narrow, with a gland below the middle. Achenes glabrous, sometimes muricatulate or wrinkled, slightly turgid; style short, slender, straight or recurved.—DC., Syst. Veg. i. 270; A. Cunn., Precurs. n. 630; Hook. f., Fl. N.Z. i. 11, and Handbk. 8; A. Gray, Bot. U.S. Expl. Exped. 7; Benth., Fl. Austr. i. 13.

NORTH and SOUTH Islands; STEWART Island; CHATHAM Islands: In swamps and streams, &c. Ascends to 2,000ft. Also in Australia. Oct. to Jan.

Var. **major**. Suberect, 2in.–12in. high. Leaves tufted; blade dissected. Stems simple or sparingly branched. Beak of achenes longer.—Benth., Fl. Austr. i. 14; Hook. f., Handbk. 8. *R. incisus*, Hook. f., Fl. N.Z. i. 10, t. 4. *R. amphitrica*, Col. in Trans. N.Z.I. xvii. (1884) 237.

Var. **subfluitans**. Leaves small, less divided. Peduncles short.—Benth., *l.c.*; Hook. f., *l.c.* *R. inundatus*, R. Br. in DC. Syst. Veg. i. 269; A. Gray, Bot. U.S. Expl. Exped. 9.—Floating or creeping in streams or swamps.

Var. **inconspicuus**. Very small and slender, suberect. Leaves dissected. Flowers minute. Achenes few.—Benth., Fl. Austr. i. 13. *R. inconspicuus*, Hook. f., Fl. Tasm. i. 9, t. 2B. NORTH Island: Pencarrow Lagoon, *T.K.*

35. **R. acaulis,** *Banks and Sol.* ex *DC. Syst. Veg.* i. 270. A small glabrous stoloniferous species, slightly fleshy. Leaves with sheathing petioles 1in.–3in. or more, 3-foliolate or 3-lobate; leaflets obovate or oblong, entire or lobed or toothed, sessile. Scapes shorter than the leaves, naked, 1-flowered. Sepals 5, broadly ovate, membranous. Petals 5–8, spathulate, 3-nerved, with a gland near the middle of the petal. Achenes forming a globose head, turgid, glabrous; style short, straight.—DC., Prod. i. 34; A. Cunn., Prod. n. 631; Hook. f., Fl. Antarc. i. 4, t. 2; Fl. N.Z. i. 11, and Handbk. 8; A. Gray, Bot. U.S. Expl. Exped. 7. *R. stenopetalus*, Hook., Ic. Pl. 677.

NORTH and SOUTH Islands; STEWART Island; CHATHAM Islands; AUCKLAND Islands: On sea-beaches, but often local. Only known inland at Rotorua and Tarawera Lakes, 1,100ft., before the eruption of 1886. Also in Chili. The creeping scions are almost filiform and often subterranean. A very distinct species, easily recognised by its fleshy 3-foliolate leaves exceeding the scapes.

36. **R. ternatifolius,** *T. Kirk in Trans. N.Z.I.* x. (1877), *App.* xxix. Stems filiform, tufted, procumbent, sometimes matted and rooting at the nodes.

Lower leaves on slender petioles 1in.–3in. long, 3-foliolate or 2-ternate; segments entire or 3-lobed, acute, glabrate or with long scattered hairs on both surfaces. Peduncles ¼in.–1in. long, axillary or opposite the petioles. Flowers minute. Sepals 5, ovate, membranous, fugacious. Petals 5, narrow-oblong, scarcely exceeding the sepals, with a minute basal gland. Achenes 5–10, slightly turgid, faintly keeled; beak short, straight or recurved.—*R. trilobatus*, T. Kirk in Trans. N.Z.I. ix. (1876) 547 (not of Kit.).

SOUTH Island: Canterbury: Source of the Broken River, *Cheeseman!* Otago: Swampy Hill, Dunedin; Catlin's River; Kelso; Heriot; *Petrie!* Makarewa, Winton, Centre Hill, &c., *T. K.* Ascends to 3,500ft. In damp places, &c. Dec., Jan.

A very distinct little species, which forms matted patches in places where water has stagnated during the winter. The stems are sparingly rooted at the nodes. Originally discovered by Mr. Petrie.

37. **R. pachyrrhizus,** *Hook. f., Handbk.* 8. Forming dense patches 1in.–1½in. high; petioles, undersurface of leaves, and scapes clothed with long weak hairs. Rhizomes robust, branched, creeping. Leaves all radical, succulent; petiole stout, ¼in.–½in. long; blade narrow-cuneate or obovate-cuneate, with 3–4 obtuse or acute lobes or teeth. Scape stout, naked, 1-flowered, ½in.–1in. high. Sepals 5, linear-oblong, spreading, obtuse. Petals 10–15, obovate-spathulate, with a gland near the base. Receptacle shortly ovoid, slightly hairy. Achenes glabrous, turgid, rounded; style subulate, incurved.

SOUTH Island: Otago: Lake district, *Hector* and *Buchanan!* Old-Man Range, Mount Cardrona, Mount Tyndall, Mount Pisa, &c., *Petrie!* 4,000ft. to 6,000ft.

A singular little plant, differing in habit from all other New Zealand species. The achenes approach those of *R. pinguis*.

38. **R. Limosella,** *T. Kirk* ex *F. Muell. in Trans. N.Z.I.* iii. (1871) 177. Small, glabrous, stems filiform, creeping and rooting at the nodes; often matted. Leaves solitary or rarely in pairs, ½in.–3in. long, narrow linear-spathulate, nerveless. Flowers ¼in. in diameter, solitary, axillary, on filiform peduncles much shorter than the leaves, tetramerous. Sepals broadly ovate, with membranous margins. Petals narrow-linear, thrice as long as the sepals, revolute at the lips, and with a gland near the base. Stamens 8–12. Carpels 8. Achenes few, rounded; style slender, straight or shortly recurved.—*R. limoselloides*, F. Muell. in Hook. Ic. Pl. t. 1081 (not of Turc.).

NORTH Island: Lower Waikato lakes, Auckland, *T. K.* Taranaki: Between Opunake and Normanby, *T. K.* SOUTH Island: Lake Lyndon, Broken River Basin, Lake Pearson, &c., *T. K.* Otago: Roxburgh, *E. W. Bastings!* Sea-level to 3,000ft. In muddy or watery places; often mixed with *Crantzia*, *Limosella*, &c., so that it is easily overlooked. Dec., Jan.

Easily distinguished from all other species by its narrow linear-spathulate leaves and quaternary flowers, which are never developed on submerged plants. Sir Joseph Hooker remarks that it approaches *R. hydrophilus*, Gaud., of the Falkland Islands in its habit and the form of its leaves.

* **R. sardous,** *Crantz, Stirp. Austr. ed.* i. 84. Erect, 1ft.–2ft. high, stem and petioles clothed with spreading hairs. Lower leaves 3-foliolate or 3-partite; leaflets stalked, 3-fid; segments lobed, obtuse. Sepals 5, reflexed. Gland of the petal covered. Achenes glabrous, rounded with a series of intramarginal tubercles; style short, recurved.—*R. hirsutus*, Curt., Fl. Lond. Fasc. 2, t. 40.

NORTH and SOUTH Islands: Naturalised in cool pastures. When growing in swamps it sometimes attains 3ft. in height, with stout stems, large flowers, and less divided almost glossy leaves. Nov., Dec.

* **R. parviflorus,** *L.*, *Sp. Pl.* 780, *ed.* 2. A tufted annual with weak decumbent or suberect stems 6in.–18in. long. Radical leaves on long petioles, orbicular or reniform, 3–5-lobed, margins toothed, sparingly hairy on both surfaces. Cauline leaves smaller. Peduncles short, opposite the leaves or axillary. Flowers small. Sepals 5, equalling the petals, which have a gland near the base. Achenes somewhat compressed, rough, with short hooked spines on the sides; style short, stout, hooked.

NORTH and SOUTH Islands: Naturalised in cornfields, pastures, and waste places. Oct. to Dec. Europe.

Var. **australis,** Benth., Fl. Austr. i. 14. A fugacious annual with filiform or capillary stems 1in.–5in. long, suberect or decumbent and intricate. Radical and cauline leaves on petioles about 1in. long; blade ¼in.–½in. broad, very membranous, with few weak hairs, 3–5-lobed or toothed, lobes acute. Flowers minute, sessile. Achenes slightly compressed; style very short, straight.—Hook. f., Handbk. 8. *R. sessiliflorus*, R. Br. in DC. Syst. Veg. i. 302; Hook. f., Fl. N.Z. i. 11. *R. collinus* and *R. Pumilio*, R. Br. in DC. Syst. Veg. i 271.

NORTH Island: Amongst rocks, &c., Auckland Isthmus. Originally discovered by Mr. Colenso. Doubtfully indigenous. Aug., Sept.

* **R. arvensis,** *L.*, *Sp. Pl.* 555. An erect almost glabrous annual, sparingly branched above. Radical leaves obovate, 3-fid, toothed. Cauline leaves deeply 3-partite, the segments cut into narrow linear lobes. Receptacle hairy. Sepals 5, spreading. Achenes large, compressed, with a strong spinous margin, sides with hooked tubercles; style stout, hooked.

NORTH and SOUTH Islands: Naturalised in cornfields, but not abundant. Jan. Europe.

* **R. muricatus,** *L.*, *Sp. Pl.* 555. Annual, tufted, 3in.–12in. high, glabrous. Radical leaves on long petioles, glossy when fresh, 3–5-lobed or partite, segments incised or toothed. Peduncles opposite the leaves. Sepals 5, ovate, shorter than the petals. Achenes large, with spinous tubercles on the sides, margin stout but not spinous; style broadly subulate, stout.

NORTH and SOUTH Islands: Naturalised in pastures and waste places. Nov., Dec. Europe.

* **R. falcatus,** *L.*, *Sp. Pl.* 556. Annual, ½in. high. Leaves all radical, spreading in flower, later curled over the fruit, ½in.–¾in. long, narrowed into a flat petiole below, deeply cut into 3–5 linear lobed toothed or entire segments above, hairy. Scapes 1–3, very short, 1-flowered. Sepals 5. Petals narrow-obovate, abruptly narrowed into a claw at the base, with a gland just above the claw. Achenes in a dense elongated spike, with a gibbosity on each side at the base forming an empty cavity; beak 4–5 times as long as the achene, stout, canaliculate, straight or incurved.

SOUTH Island: Otago: Kurow, Bald Hill Flat, &c., *Petrie!* Naturalised in dry sheep-country. Europe.

[*R. areolatus*, Petrie in Trans. N.Z.I. xxii. (1889) 439, said to have been found at Lake Wakatipu by A. C. Purdie, consists of poor specimens of the Scandinavian *R. pygmaeus*, Wahlb., mixed with scraps of *R. subfluitans*.]

4. CALTHA, Linn.

Sepals 4 or 5, imbricate, petaloid. Petals 0. Stamens numerous. Carpels sessile, capitate; ovules numerous in two series on the ventral suture. Follicles many-seeded, opening along the inner face. Seeds with a thickened funicle and prominent rhaphe. Glabrous tufted perennials, with radical leaves.

SPECIES, about 10, restricted to the temperate and cold regions of both hemispheres.

ETYM. From the Greek, signifying *a cup*.

1. **C. Novae-Zelandiae,** *Hook. f.*, *Fl. N.Z.* i. 12, *t.* 6. A small tufted perennial herb, 1in.–5in. high. Rootstock stout, with fleshy rootlets. Leaves on spreading sheathing petioles ½in.–4in. long; blade ovate-oblong, retuse, cordate and auricled at the base, the auricles usually folded and appressed to the upper surface of the leaf, crenulate. Scapes ½in.–5in. long, naked, 1-flowered. Stamens with unequal filaments. Carpels 5–8 or more, gibbous; style short, hooked.—Handbk. N.Z. Fl. 9. *C. marginata*, Col. in Trans. N.Z.I. xxiii. (1890) 382.

NORTH Island: Ruahine and Tararua Ranges, &c. SOUTH Island: In alpine situations, Nelson to Southland. STEWART Island: Summit of Rakiahua, &c. 2,000ft. to 5,000ft. Oct. to Jan.

When growing under the shelter of shrubs this plant presents a delicate appearance, with long petioles and short sheaths; but when growing in exposed situations the petioles are short, stout, with ample sheaths, and the blade is much thicker. Stewart Island specimens are less than ½in. high Flowers yellow or whitish.

*NIGELLA, Linn.

Sepals 5, imbricate, petaloid, fugacious. Petals 5, small, 2-fid at the apex. Carpels 3, 5, or more, more or less coherent, dehiscing internally at the apex when ripe. Seeds numerous. Annual herbs with pinnatisect cauline leaves.

* **N. damascena,** *L.*, *Sp. Pl.* 584. Annual, slender, 6in.–10in. high. Leaves pinnate; leaflets dissected; segments filiform or capillary. Flowers terminal, with a dissected leafy involucre. Carpels 5, coherent for nearly their entire length: each carpel is spuriously 2-celled, owing to the separation of the outer wall into two layers. Styles persistent, free.

NORTH Island: Auckland Isthmus and other places. Sparingly naturalised. *Fennel-flower.* Dec., Jan. Europe.

*AQUILEGIA, Linn.

Sepals 5, petaloid, imbricate. Petals 5, spurred behind; lower stamens abortive. Carpels 5, many-ovuled. Follicles 5, erect, many-seeded. Embryo minute. Herbs with paniculate flowers and ternately-divided leaves.

* **A. vulgaris,** *L.*, *Sp. Pl.* 583. Rootstock stout. Stem 1ft.–2ft. high, slender, paniculately branched. Radical leaves on slender petioles, 2-ternate; leaflets large, lobed or crenate, usually glabrous. Flowers in lax corymbs, pendulous. Petals obvolute, with a curved spur, convolute at the tip. Follicles erect, hairy.

NORTH Island: Sparingly naturalised near Auckland and Wellington, &c. *Columbine.* Oct. to Dec. Europe.

ORDER II.—MAGNOLIACEAE.

TRIBE—WINTEREAE.

Flowers regular, perfect. Sepals and petals in 2 or 3 or several series, imbricated, fugacious. Stamens ∞, hypogynous, with thick filaments and adnate anthers. Carpels few; ovules 2 or more, attached to the ventral suture. Stigma sessile. Fruit a small drupe, a follicle, or a berry. Seeds few, glossy, with copious endosperm; embryo small. Aromatic exstipulate trees or shrubs, with alternate leaves.

The order contains numerous species with showy flowers and handsome foliage; some species attain a large size and afford valuable timber. The preceding description refers to the tribe *Winterae* alone, which contains the only genus represented in the colony.

1. DRIMYS, Forst.

Sepals 2–4, membranous, united at the base and forming a calyx with 4 or 5 unequal lobes. Petals 5 or 6 or more, in two series, spreading. Filaments clavate; anther-cells diverging. Carpels few. Fruit a berry.

1. **D. axillaris,** *Forst., Gen. t.* 42. An evergreen shrub or small tree with black bark, 10ft.–30ft. high. Leaves entire, shortly petioled, elliptic, ovate, or oblong-lanceolate, green on both surfaces or glaucous below, 1in.–4in. long. Flowers perfect or rarely unisexual, in 3–6-flowered fascicles, springing from the axils of the leaves or leaf-scars, pedicellate. Calyx 2–4-lobed. Petals 5–6, linear, spreading. Stamens about 10 in 3 series; filaments short, clavate. Carpels 4 or 5. Fruit a berry, 4–6-seeded; seeds angular.—DC., Syst. Veg. i. 443; A. Rich., Fl. N.Z. 290; A. Cunn., Precurs. n. 629; Hook., Ic. Pl. t. 576; Hook. f., Fl. N.Z. i. 12; Handbk. 10; T. Kirk, Forest Fl. N.Z. t. 1. *Wintera axillaris,* Forst., Prod. n. 229.

NORTH and SOUTH Islands: Bay of Islands to Banks Peninsula. Sea-level to 2,500ft. *Horopito.* Oct. to Dec.

Var. **colorata.** Smaller, pungent. Leaves yellowish, blotched with red, glaucous beneath. Flowers in 2–3-flowered fascicles. Calyx saucer-shaped, not lobed. Berries 2-seeded.—T. Kirk, Forest Fl. N.Z. t. 2. *D. colorata,* Raoul in Ann. Sc. Nat. ii. 121; Choix, t. 23. NORTH and SOUTH Islands: Ohinemutu to Southland. STEWART Island. *Pepper-tree.*

The leaves vary greatly in size. The typical form is rarely or never pungent.

Order—* PAPAVERACEAE.

Sepals 2, rarely 3. Petals 4, crumpled. Stamens ∞, hypogynous. Ovary 1-celled, with parietal placentation, many-ovuled. Fruit a capsule or pod, dehiscing by pores. Seeds minute, with oily or fleshy endosperm; embryo minute, basilar. Herbs with alternate exstipulate leaves and milky juice.

* Papaver. Stigmas forming a radiating disk.

* Glaucium. Stigmas deflexed.

* Eschscholtzia. Stigmas linear, erect.

* PAPAVER, Linn.

Erect. Sepals 2, fugacious. Petals 4. Stamens ∞. Ovary shortly stipitate, usually 1-celled, septate by the intruded placentas, which bear ovules over their entire surface. Style short and thick, or 0. Stigma with radiating lobes. Capsule globular or clavate. Seeds minute, pitted; endosperm copious, oily. Erect annual herbs with alternated lobed or dissected leaves. Buds drooping.

Leaf-segments narrow. Capsule roundly obovate .. .. * *P. Rhoeas.*
Leaf-segments narrow. Capsule clavate .. .. * *P. dubium.*
Leaves amplexicaul. Capsule globular .. .. * *P. somniferum.*

* **P. Rhoeas,** *L., Sp. Pl.* 507. Erect, hispid. Leaves 2-pinnatifid, deeply cut into linear segments, each with a bristle at the apex. Peduncles with spreading or rarely appressed hairs. Flowers large. Petals unequal, crimson. Capsule glabrous, rounded-obovate, stipitate. Stigma convex. Rays 8–12.

NORTH and SOUTH Islands: Naturalised in cornfields and cultivated land. Not common. *Corn-poppy.* Dec. to Feb. Europe.

***P. dubium,** *L., Sp. Pl.* 1196. Erect, branched. Leaves pinnatifid or twice pinnatifid; segments obtuse. Peduncles with appressed hairs. Capsule sessile, glabrous, obovoid. Lobes of stigmatic disc projecting.—*P. Lamottei,* Boreau.

SOUTH Island: Cornfields and cultivated land, Otago; local. *Petrie!* Europe.

***P. somniferum,** *L., Sp. Pl.* 508. Erect, glaucous, 1ft.–3ft. high, usually glabrous except a few bristles on the capsule. Leaves amplexicaul, margins waved or toothed. Capsule ovoid or globose, stipitate. Filaments slightly dilated upwards.

NORTH and SOUTH Islands: Sparingly naturalised. Chiefly an escape from cultivation. *Poppy.* Jan., Feb.

*GLAUCIUM, Tourn.

Sepals 2. Petals 4. Ovary imperfectly 2-celled. Style short or 0. Stigmatic lobes 2, deflexed. Ovules ∞. Placentas connected by a spongy dissepiment. Capsule elongated, 2-valved. Seeds ∞. Erect or spreading glaucous herbs, with large flowers.

***G. flavum,** *Crantz, Stirp. Austr.* ii. 131. 1ft.–2ft. high, branched. Radical leaves petioled, lyrate or 2-pinnatifid, glaucous and more or less hispid. Cauline leaves amplexicaul or nearly so, irregularly lobed. Buds erect. Flowers large, on short grooved peduncles, golden-yellow. Capsules linear, 8in.–12in. or more, curved or flexuous. Seeds sunk in the spongy septum.

NORTH Island: Castlepoint and Whanganui to Cook Strait. Naturalised on sea-beaches. *Horned poppy.* Nov. to Jan. Europe.

*ESCHSCHOLTZIA, Cham.

Sepals 3, coherent, forming a calyptrate calyx. Petals 4, perigynous. Stamens ∞, perigynous. Ovary 1-celled; stigmatic lobes 4–8, linear. Capsule elongated, linear, 10-furrowed, dehiscing to base. Seeds ∞.

***E. californica,** *Cham. in Nees. Hor. Phys. Berol.* 73. A suberect, almost glaucous, much-branched herb. Leaves 2- or 3-pinnately divided into short linear obtuse segments or lobes. Flowers on long axillary peduncles, orange-yellow. Stigma unequally 4-lobed.

NORTH and SOUTH Islands: Naturalised in cultivated ground, river-beds, waste places, &c. Auckland, Wellington, Canterbury, Otago. Dec. to Feb. California.

ORDER—* FUMARIACEAE.

Corolla irregular. Sepals 2, scale-like. Petals 4, in two dissimilar pairs; the two outer longer, one or both of them gibbous and spurred at the base; the two inner erect, smaller, often united at the tips. Stamens 6, in two sets opposite the outer petals, the middle anther of each set being 2-celled, the others 1-celled. Stigmas obtuse or lobed, placentas parietal. Seeds with endosperm. Fruit a 2-valved many-seeded capsule, or an indehiscent nut. Weak herbs with much-divided exstipulate leaves and watery juice.

*FUMARIA, Linn.

Sepals 2. Petals 4, erect, connivent, the uppermost spurred, the lower flat, the 2 lateral narrow, cohering at the tips. Stamens 6, diadelphous. Ovary 1-celled. Fruit globose, small, 1-seeded. Weak herbs with much-dissected leaves, sometimes climbing. Flowers in terminal racemes or spikes or opposite the leaves.

Climbing by the twining petioles .. .. * *F. muralis.*
Decumbent or suberect .. .. * *F. officinalis.*

***F. muralis,** *Sond. in Koch Syn. Fl. Germ.* 1017, *ed.* 2. Stem lax, diffuse, climbing by means of the twisting petioles or prostrate. Leaves alternate, pinnate, ternate or 2-ternate; segments lobed or cut. Flowers in short racemes opposite the leaves. Sepals 2, as broad as the corolla-tube and one-third as long, ovate, toothed. Corolla black at its apex, lower petal narrowed from its middle upwards. Fruit small, globose, minutely rugose, with two faint apical pits; neck narrower than the top of the pedicel.

NORTH and SOUTH Islands: Naturalised in cultivated and waste land. Oct. to Feb. Europe.

* **F. officinalis,** *L., Sp. Pl.* 790. Stems weak, diffuse, not climbing. Flowers in elongated racemes. Sepals ovate-lanceolate, narrower than the corolla-tube. Lower petal spathulate, rose-coloured. Pedicel exceeding the bract. Fruit spherical, depressed, retuse, with a large spherical pit at its apex, rugose.

NORTH and SOUTH Islands: Naturalised in cultivated land, but not common. *Smoke-weed.* Nov. to Jan. Europe.

ORDER III.—CRUCIFERAE.

Sepals 4, free, deciduous. Petals 4, free, forming two opposite pairs, the limb usually spreading. Stamens 6, tetradynamous, rarely 1, 2, or 4. Ovary with two parietal placentas. Pod usually 2-celled by a membranous plate which forms a false septum, or sometimes divided into superimposed cellules by transverse septa, rarely 1-celled, 2-valved. Seeds campylotropous, without endosperm, the radicle turned up towards the edges of the cotyledons (accumbent), or towards the back of a cotyledon (incumbent), or the cotyledons folded upon themselves and round the radicle (conduplicate). Herbs with alternate leaves and racemose or corymbose flowers, rarely unisexual.

A large order, most abundant in Europe and the temperate parts of Asia; less frequent in the Southern Hemisphere. *Pachycladon* and *Notothlaspi* are endemic; the other indigenous genera are widely distributed, and fifteen exotic genera have become naturalised. The order contains numerous plants of economic value on account of their stimulant and antiscorbutic properties.

The following arrangement of the genera is adopted from Bentham and Hooker's "Genera Plantarum":—

A. POD DEHISCING THROUGH ITS ENTIRE LENGTH, ELONGATE OR SHORT, FLAT OR TURGID, NOT COMPRESSED AT RIGHT ANGLES TO THE SEPTUM.

Tribe I. ARABIDEAE.—Pod much longer than broad. Cotyledons accumbent. Seeds compressed, 1-seriate.

* MATTHIOLA. Pod round or compressed. Stigmas lobed, gibbous at the back.

* *Stigma small, terminal, simple.*

* CHEIRANTHUS. Pod compressed. Lateral sepals saccate.
1. NASTURTIUM. Pod terete. Seeds irregularly 2-seriate.
* BARBAREA. Pod 4-angled, with a prominent rib.
2. CARDAMINE. Pod compressed, valves flat.

II. ALYSSINEAE.—Pod shorter than broad. Cotyledons accumbent. Seeds 2-seriate.

* ALYSSUM. Pod compressed, orbicular, 2-4 seeded.
* EROPHILA. Pod oblong, flat. Petals 2-fid.
* COCHLEARIA. Pod inflated, very convex. Petals entire.

III. SISYMBRIEAE. — Pod elongate. Cotyledons incumbent. Radicle dorsal. Seeds 1-seriate, compressed.

* HESPERIS. Pod subcompressed. Stigma lobed or decurrent.
* MALCOLMIA. Pod terete or subterete. Stigma conical.
3. SISYMBRIUM. Pod terete or tetragonous. Stigma discoid.

IV. CAMELINEAE.—Pod oblong, ovoid, or globose. Seeds in two series. Cotyledons incumbent.

* CAMELINA. Pod subovate, ventricose. Style persistent.

V. BRASSICEAE.—Pod elongate. Cotyledons conduplicate.

* BRASSICA. Pod beaked, terete or angled. Seeds 1-seriate, spherical.
* DIPLOTAXIS. Pod compressed. Seeds 2-seriate, oblong or oval.

B. POD SHORT, DEHISCING THROUGH ITS ENTIRE LENGTH, BROAD, FLAT, OR TURGID, COMPRESSED AT RIGHT ANGLES TO THE SEPTUM (INDEHISCENT IN *Senebiera*).

Tribe VI. LEPIDINEAE.—Cotyledons incumbent, straight, incurved or longitudinally folded.

4. PACHYCLADON. Pod compressed laterally, not winged.

5. CAPSELLA. Pod dehiscent, many-seeded.

* SENEBIERA. Pod indehiscent, 2-seeded.

6. LEPIDIUM. Pod dehiscent, winged, 2–4-seeded.

7. NOTOTHLASPI. Pod much flattened, broadly winged, many-seeded.

VII. THLASPIDEAE.—Cotyledons decumbent, straight. Pedicels horizontal.

* IBERIS. Outer petals larger than the inner. Pod ovate.

C. POD ELONGATE, INDEHISCENT, 1-CELLED, MANY-SEEDED, OR TRANSVERSELY ARTICULATE, CELLS 1-SEEDED.

Tribe VIII. RAPHANEAE.

* RAPHANUS.

*MATTHIOLA, R. Br.

Herbaceous or suffruticose, pubescent or hoary, hairs stellate. Leaves entire. Sepals erect, saccate at the base. Stigma lobed, gibbous. Pod elongate, rounded or compressed. Septum thick. Seeds 1-seriate, winged.

* **M. incana,** *R. Br. in Ait. Hort. Kew., ed.* 2, iv. 119. Erect, suffruticose. Leaves oblong-lanceolate, entire or obscurely toothed, hoary. Flowers purple. Pod compressed, eglandular. Seeds orbicular, winged.

NORTH Island: Naturalised on almost inaccessible rocks at Castlepoint, East Coast. *Queen-stock.* Nov., Dec. Levant, North Africa, &c.

*CHEIRANTHUS, Linn.

Suffruticose herbs. Leaves entire or toothed. Flowers large, yellow or reddish. Sepals erect, the lateral saccate at the base. Petals with long claws. Stigmatic lobes diverging. Pod compressed or slightly tetragonous. Seeds 1-seriate.

* **C. cheiri,** *L., Sp. Pl.* 661. Stems shrubby, branched. Leaves lanceolate, acute, entire, clothed with 2-partite adpressed hairs. Pods tetragonous.

NORTH and SOUTH Islands: Sparingly naturalised. *Wallflower.* Oct. to Dec. North and Central Europe.

1. NASTURTIUM, R. Br.

Branched herbs, glabrous or clothed with simple hairs. Leaves entire or lobed or pinnate. Flowers small, on spreading pedicels. Sepals short, equal, patent. Petals slightly clawed. Stamens usually tetradynamous. Stigma simple or 2-lobed. Pods terete, curved; valves convex. Seeds 2-seriate, turgid.

A genus comprising about 20 species, chiefly natives of temperate and warm regions.

1. **N. palustre,** *DC., Syst.* ii. 191. Erect or decumbent, 6in.–24in. high, glabrous, pubescent, or rarely pilose. Leaves auriculate, entire, lobed, pinnatifid or lyrate; the segments sinuate-toothed. Racemes short, ebracteate. Flowers small, petals scarcely equalling the calyx. Pods oblong, turgid, obliquely curved.—Hook. f., Handbk. 10; Benth., Fl. Austr. i. 65. *N. terrestre,* R. Br. in Ait. Hort. Kew. ix. 110; Sm. E. B. t. 1747; Hook. f., Fl. N.Z. i. 14. *N. semipinnatifidum,* Hook., Journ. Bot. i. 246. *N. sylvestre,* A. Rich., Fl. Nouv.-Zel. 309; A. Cunn., Precurs. n. 635.

NORTH and SOUTH Islands: Damp places in lowland districts, but often local. Ascends to 2,000ft. Widely distributed in temperate and subtropical regions. Nov. to Jan.

4

* **N. officinale,** *R. Br. in Ait. Hort. Kew., ed.* 2, iv. 111. Glabrous. Stems 6in.–6ft. long or high. Leaves pinnate; leaflets ovate or oblong, subcordate, sinuate-dentate. Flowers in elongated racemes, white. Sepals half the length of petals. Pods patent.

NORTH and SOUTH Islands: Abundantly naturalised in watery places, rivers, &c., often impeding drainage. *Watercress.* Oct. to Feb. Europe, &c.

*BARBAREA, R. Br.

Erect, leafy, usually glabrous. Leaves entire, lobed or pinnate. Sepals sub-erect. Petals clawed. Stigma capitate or 2-lobed. Pod elongate, erect, obscurely 4-angled. Seeds 1-seriate.

* **B. praecox,** *R. Br. in Ait. Hort. Kew., ed.* 2, iv. 109. A strict or sparingly-branched biennial. Stems angled and grooved. Lower leaves lyrate, the terminal division rounded or ovate, lateral in 4–8 pairs. Flowers yellow. Pods ascending, slightly spreading. Pedicels as thick as the pods.—*B. australis*, Hook. f., Fl. N.Z. i. 14. *B. vulgaris*, Hook. f., Handbk. 11.

NORTH and SOUTH Islands: Naturalised in pastures, waste places, &c. *American cress.* Oct. to Dec. Europe.

The description of *B. australis* in Fl. N.Z. i. 14 was drawn from Tasmanian specimens, those collected by Mr. Colenso being too young to admit of exact identification. The true *B. vulgaris*, R. Br., has not been observed in New Zealand.

2. CARDAMINE, Linn.

Glabrous or rarely pubescent annuals or perennials. Leaves entire, lobed or pinnate. Sepals erect or spreading. Pod linear, flattened; valves opening from the base elastically. Seeds numerous, flattened; funiculus slender. Cotyledons accumbent.

SPECIES, about 60. Common in temperate and cold regions; often alpine or subalpine.

Leaves pinnate. Flowers small	1. *C. hirsuta.*
Leaves reduced to a single pinnule	var. *uniflora.*
Leaves spathulate, entire or lobed. Flowers small	2. *C. depressa.*
Leaves deeply lobed near the base. Flowers large	3. *C. bilobata.*
Tall. Leaves entire or pinnatifid. Flowers in elongated racemes	4. *C. stylosa.*
Stem short, stout, branched. Pods curved, narrow	5. *C. fastigiata.*
Stem short, stout, branched. Pods broad	6. *C. latesiliqua.*
Stem short, branched from the base. Flowers forming a dense corymb	7. *C. Enysii.*

1. **C. hirsuta,** *L., Sp. Pl.* 655. A slender much-branched annual or perennial, 1in.–18in. high, glabrous or pubescent. Leaves pinnate; leaflets rounded, shortly stalked, rarely toothed or angled; cauline pinnatifid, with linear lobes. Flowers usually small. Petals narrow. Stamens sometimes tetrandrous. Pedicels slightly spreading. Pods slender, erect, narrow-linear. Seeds smooth.—DC., Syst. Veg. ii. 259; Sm. E. B. t. 492; Hook. f., Fl. N.Z. i. 13, Handbk. 12; Benth., Fl. Austr. i. 70. *C. parviflora*, L., Sp. Pl. 919; Banks and Sol. *ex* DC. Syst. Veg. ii. 265.

NORTH and SOUTH Islands; THREE KINGS Islands; STEWART Island; CHATHAM Islands; AUCKLAND, CAMPBELL, and MACQUARIE Islands. Sea-level to 6,000ft. *Land-cress.* Sept. to April.

Var. **hirsuta.** Leaves rosulate. Stamens usually 4.—*C. flexuosa*, With.

Var. **debilis.** Suberect or decumbent. Branches slender. Leaflets in 2 or 3 pairs, rounded or subcordate, sinuate. Pods with long slender pedicels. Styles long and slender.—*C. debilis*, Banks and Sol. MSS.; DC., Syst. Veg. ii. 265; A. Cunn., Precurs. n. 626. *Sisymbrium heterophyllum*, Forst., Prod. n. 250. On the margins of lowland woods.

Var. **corymbosa.** Smaller, suberect or prostrate. Leaflets in 1 or 2 pairs, or reduced to a terminal leaflet. Flowers in few-flowered corymbs or solitary, pedicels short.—Hook. f., Handbk. 12; sp. Hook. f., Fl. Antarc. i. 6; Hook., Ic. Pl. t. 686. SOUTH Island, STEWART Island, ANTARCTIC Islands, &c. Chiefly in the mountains.

Var. **subcarnosa,** Hook. f., Fl. Antarc. i. 5. Leaflets in 3–6 pairs, narrow, rather fleshy. Flowers larger. Sepals broader. Pods slender. Styles long or short. SOUTH Island: In mountain districts. AUCKLAND, CAMPBELL, and MACQUARIE Islands.

Var. **uniflora,** Hook. f., Handbk. Small. Leaves reduced to a single pinnule. Flowers on slender 1-flowered scapes; sometimes ½in. in diameter. NORTH and SOUTH Islands; STEWART Island. Often littoral, but most frequent in mountain districts.

One of the most frequent plants, the varieties being largely dependent upon situation.

2. **C. depressa,** *Hook. f., Fl. Antarc.* i. 6, *t.* 3 *and* 4. Perennial, glabrous or pilose, stemless. Leaves rosulate, 1in.–2in. long, spathulate or ovate-spathulate, rounded at the apex or retuse. Flowers on slender 1-flowered scapes, ½in. long or more, crowded; rarely on short few-flowered leafy scapes. Pods rather stout, erect; style short, stout.—Handbk. 12.

Var. **depressa,** Handbk. Glabrous or pilose, larger. Leaves usually lobulate.—*C. depressa*, Hook. f., Fl. Antarc. t. 3 and 4B.

Var. **stellata,** Handbk. Glabrous or pilose. Leaves quite entire or obscurely crenate. —*C. stellata*, Hook. f., Fl. Antarc. t. 4A.

SOUTH Island: Var. *depressa*: Marlborough: Mount Mouatt, 4,000ft., *T. K.* Lake Tennyson and Wairau Mountains, Nelson, *Travers.* Hopkins River and Lake Ohau, Canterbury, *Haast.* Lake district, Otago, *Hector* and *Buchanan.* Both forms in the AUCKLAND and CAMPBELL Islands, ascending to nearly 2,000ft. *Hooker f.; T. K.*

The short stout erect pods with their very short styles best distinguish this from all forms of *C. hirsuta*.

3. **C. bilobata,** *n. s.* Rootstock rather stout. Leaves all radical, on slender petioles 1in.–4in. long, glabrous; blade ½in.–1in. long, oblong or obovate, entire or with 1 or 2 pairs of lateral lobes at the base. Scapes very slender, 5in.–10in. high, naked, simple or with a single short branch. Flowers few, on slender pedicels ½in.–1in. long or more, large. Stamens tetradynamous. Pedicels capillary, spreading in fruit. Pods ¾in.–1in. long, narrow, longer or shorter than the pedicels; style very slender.

SOUTH Island: Kurow and Naseby, Mount Ida Range, Hector Mountains, Otago. 2,000ft. to 3,000ft. *Petrie!*

A much larger plant than *C. depressa*, with more slender pods and spreading pedicels. The flowers almost equal those of *C. pratensis*, L.

4. **C. stylosa,** *DC., Syst. Veg.* ii. 24. A rather stout glabrous herb, suberect or decumbent, 1ft.–3ft. high; branches leafy at the base. Leaves 3in.–5in. long, oblong-lanceolate or spathulate, sinuate-toothed, entire or lobed or pinnatifid at the base; the lower on long petioles, the upper sessile, auriculate or sagittate. Racemes elongated, 1ft.–2ft. long or more. Flowers small. Pedicels short, stout. Pods distant, spreading, 1in.–1½in. long; style stout, valves convex.—Hook. f., Handbk. 12, Fl. Tasm. i. 18; Benth., Fl. Austr. i. 68. *C. divaricata*, Hook. f., Fl. N.Z. i. 13. *Arabis gigantea*, Hook., Ic. Pl. t. 259. *Sisymbrium divaricatum*, Banks and Sol. MSS.

NORTH Island: Not unfrequent in the Auckland District and outlying islands. KERMADEC Islands, *Cheeseman.* SOUTH Island: Marlborough: Picton, *J. Rutland!* Mount Stokes, *J. McMahon!* Queen Charlotte Sound, *Banks and Solander!*

Easily distinguished from all other species by the coarse foliage and elongated racemes.

5. **C. fastigiata,** *Hook. f., Handbk.* 13. An erect glabrous biennial. Rootstock stout, fusiform, ½in. in diameter. Radical leaves very coriaceous, densely rosulate, 1in.–3in. long, narrow lanceolate-spathulate, coarsely incised or dentate; cauline similar but smaller. Stem branching from the base, 6in.–18in. high. Flowers large, numerous. Pods suberect on short slender pedicels, $\frac{1}{20}$in.–$\frac{1}{10}$in. wide, narrow-linear, curved, spreading, 1in.–2in. long; style short. Seeds compressed, oblong.—*Arabis fastigiata*, Hook. f., Fl. N.Z. ii. 324. *Pachycladon elongata*, Buch. in Trans. N.Z.I. xix. (1886) 216. *Notothlaspi Hookeri*, Buch. in Trans. N.Z.I. xx. (1887) 255, t. 13!

SOUTH Island: Nelson: Wairau Gorge, *Sinclair*, &c. Marlborough: Upper Awatere, *T. K. McRae's run, Monro.* Canterbury: River-bed of the Macaulay, *Haast.* Otago, *Buchanan!* 2,500ft. to 3,500ft.

Rare; a remarkably local alpine species, easily distinguished from any other by the narrow curved pods. The seeds have a reticulate testa.

6. **C. latesiliqua,** *Cheeseman in Trans. N.Z.I.* xv. (1882) 298. Rootstock stout, spongy, often branched at the top. Stems usually numerous, spreading or erect, 6in.–24in. high; branches sparingly leafy at the base. Radical leaves 3in.–6in. long, ⅓in.–⅔in. broad, narrow linear-spathulate to obovate-spathulate, gradually narrowed below, coarsely serrate above, more or less villous or the margins ciliate. Upper cauline leaves smaller, less toothed. Flowers numerous, forming a large corymbose mass. Sepals broadly ovate. Fruiting pedicel slender. Pods erect or suberect, 1½in.–2½in. long, curved or straight, slightly turgid, ⅛in.–¼in. broad. Seeds compressed.

SOUTH Island: Nelson: Mount Arthur, Mount Owen, Raglan Mountains. *Cheeseman! Gibbs! Bryant!* 3,000ft. to 5,000ft.

A more robust plant, approaching *C. fastigiata*, but distinguished by the villous leaves, larger flowers, and broader pods, which are less falcate.

7. **C. Enysii,** *Cheeseman, MS.* 2in.–3in. high. Rootstock stout. Radical leaves rosulate, hairy on both surfaces, ¾in.–1½in. long, oblong-lanceolate or ovate, acute or rounded at the apex, coarsely serrate, narrowed into a winged petiole. Scape much branched from the base, forming a dense corymb. Flowers large, white. Pedicels ¼in.–⅓in. long, slender. Pods immature.

SOUTH Island: Canterbury: Broken River Basin, *Cheeseman.* Otago: Mount Ida Range, *Petrie!*

A remarkable plant, of which much better specimens are wanted before a good diagnosis can be drawn.

*ALYSSUM, Tourn.

Erect or spreading herbs or rarely small shrubs, with simple leaves. Sepals short, equal. Petals short, entire or 2-fid. Filaments sometimes toothed. Pod nearly orbicular or oblong; valves flat or turgid. Stigma entire. Seeds 2–10 in each cell. Cotyledons accumbent.

Erect, hoary. Leaves linear-spathulate .. .. * *A. calycinum.*
Decumbent, pubescent or glabrous. Leaves linear-lanceolate .. .. * *A. maritimum.*

* **A. calycinum,** *L., Sp. Pl. ed.* ii. 908. A small hoary annual, 3in.–6in. high, branched from the base, hairs stellate, appressed. Leaves linear-spathulate. Flowers in terminal racemes. Petals exceeding the calyx. Pods shortly pedicelled, orbicular, strongly margined, 2–4-seeded.

NORTH and SOUTH Islands: Naturalised in many localities. Ascends to 3,000ft. Nov. to Jan. Central and Southern Europe.

* **A. maritimum,** *Lam.*, *Encyc.* i. 98. Suberect or decumbent, often woody below, spreading, pubescent or nearly glabrous. Leaves linear-lanceolate, acute. Flowers on spreading pedicels, white, fragrant. Pods orbicular, flattened, 2-seeded.

NORTH and SOUTH Islands: Naturalised in places near the sea. Bay of Islands, *A. Cunningham!* Especially abundant on cliffs at New Plymouth. *Sweet alyssum.* Nov. to Feb. South Europe.

* EROPHILA, DC.

Small annuals with entire spreading radical leaves. Flowers few. Sepals equal. Petals obovate, 2-fid. Pod oblong, compressed; valves membranous, flat or convex. Seeds 2-seriate.

* **E. vulgaris,** *DC.*, *Syst. Veg.* ii. 356. Leaves all radical, lanceolate, spreading. Scapes filiform, 1in.–3in. high, naked, 2–5-flowered. Pods on spreading capillary pedicels, obovate-oblong, more than twice as long as broad.

SOUTH Island: Naturalised in several parts of Otago; Balclutha, Sowburn, Lake Wakatipu, *Petrie!* *Whitlow-grass.* Oct., Nov. Europe.

* COCHLEARIA, Tourn.

Sepals short, equal or nearly so, spreading. Petals shortly clawed. Stamens tetradynamous. Pods oblong or globose, sessile or shortly stalked; valves turgid, with or without a dorsal nerve. Seeds 2-seriate. Perennial herbs, often littoral.

* **C. Armoracia,** *L.*, *Sp. Pl.* 648. Rootstock long, thick, cylindrical, fleshy, often branched at the top. Radical leaves oblong or linear-oblong, 1ft. long or more on long petioles, unequal at the base, crenate-serrate; cauline leaves narrower, smaller, usually sessile. Flowers small. Pod ovoid.

NORTH and SOUTH Islands: Naturalised on the sites of deserted homesteads, railway-banks, waste places, &c. I have not seen ripe pods produced in the colony, and the flowers are comparatively rare. *Horse-radish.* Europe.

* HESPERIS, Linn.

Sepals erect, the lateral gibbous at the base. Pod elongate, terete or obscurely 4-angled; valves 3-nerved. Stigma lobed, erect. Seeds in one row in each cell, numerous. Cotyledon incumbent. Erect biennial or perennial herbs.

* **H. matronalis,** *L.*, *Sp. Pl.* 663. Erect, branched, 1ft.–3ft. high. Leaves usually petiolate, lanceolate, acuminate, serrate. Flowers in terminal racemes, large, purple; pedicels ½in.–1in. long, ascending. Pods 2in.–4in. long, slender, terete, irregularly contracted.

NORTH and SOUTH Islands: Naturalised in several localities, chiefly on the sites of deserted gardens. Ohiro, *T. K.* Oamaru, *Petrie!* *Dame's rocket.* Nov., Dec. Europe.

* MALCOLMIA, R. Br.

Sepals erect, linear, the lateral slightly saccate. Petals unguiculate. Stamens tetradynamous. Stigma conical. Pod subterete. Seeds 1-seriate. Fruiting pedicels shorter than the pod. Weak annual herbs.

* **M. maritima,** *R. Br. in Hort. Kew.*, *ed.* 2, iv. 121. A much-branched pubescent annual. Stems weak, suberect or decumbent. Leaves on slender petioles, elliptical or rarely obovate, narrowed below. Flowers in short terminal racemes. Pod 1in. long, subterete.

NORTH Island: Sparingly naturalised in several localities, but not common. Plentiful on railway-banks near Plimmerton. *Virginian stock.* Sept. to June. South Europe, &c.

3. SISYMBRIUM, Tourn.

Sepals suberect. Petals clawed. Pod terete, compressed or angled. Stigma simple, 2-lobed or cup-shaped. Seeds numerous, 1-seriate. Cotyledons incumbent. Annual or biennial herbs with racemose flowers.

SPECIES, about 80, distributed through temperate and cold regions; chiefly in the Northern Hemisphere.

ETYM. An ancient Greek name for a plant supposed to have belonged to this family.

Leaves nearly all radical, short. Pods spreading .. .. .. 1. *S. Novae-Zelandiae.*
Leaves chiefly cauline, finely dissected .. .. * *S. Sophia.*
Leaves strongly toothed or pinnatifid. Pods appressed to the stem .. * *S. officinale.*

1. **S. Novae-Zelandiae,** *Hook. f., Handbk.* 11. Erect, slender, 6in.–18in. high, simple or sparingly branched, glabrous or clothed with minute stellate pubescence. Leaves mostly radical, ½in.–2in. long, narrow-obovate, oblong, or lanceolate, quite entire or sinuate-toothed or lobed or pinnatifid; petiole usually short. Cauline leaves few, smaller. Racemes terminal. Flowers on slender pedicels. Stigma shortly lobed. Pods 1in.–2in. long or more, $\frac{1}{20}$in.–$\frac{1}{15}$in. broad; style very short, obtuse, glabrous. Cotyledons obliquely incumbent.

SOUTH Island: Nelson: Wairau Gorge, *Travers, Rough!* Canterbury: Porter's Pass, Coleridge Pass, and Broken River Basin, *Enys* and *Kirk.* Otago: Cape Whanbrow, Kurow, Waitaki, Mount Ida and other ranges, *Petrie!* Sea-level to 3,000ft. Dec., Jan.

Except in Otago this is a remarkably local species. It bears considerable resemblance to *S. Thaliana*, Hook., but is more robust and certainly perennial.

***S. Sophia,** *L., Sp. Pl.* 659. An erect sparingly-branched annual, more or less pubescent. Leaves 2–3-pinnatifid; segments narrow-linear, obtuse. Flowers in terminal racemes, yellow. Pods on slender spreading pedicels, ascending, terete, narrow; style short.

SOUTH Island: Naturalised in fields near Naseby and Cromwell, Otago. *Petrie! Flixweed.* Dec., Jan. Europe.

***S. officinale,** *Scop., Fl. Carn. ed.* 2, 824. An erect herb, 1ft.–2ft. high, with more or less rigid, leafy, flexuous stems and branches; pubescent. Leaves runcinate-toothed, or unequally pinnatifid; terminal lobe large, hastate, often sinuate. Flowers yellow, small. Pods on very short pedicels, subulate, adpressed to the stem; style shortly lobed.

NORTH and SOUTH Islands; STEWART Island: Naturalised in fields, waste places, roadsides, &c. *Hedge-mustard.* Nov. to Jan. Europe.

*CAMELINA, Crantz.

Erect annuals with simple leaves. Sepals equal at the base. Petals obovate. Pods obovate, inflated; valves with a linear prolongation, which is produced upwards until it becomes confluent with the persistent base of the style. Stigma entire. Seeds 2-seriate. Cotyledons incumbent.

***C. sativa,** *Crantz, Stirp. Austr. ed.* i. *fasc.* i. 14. Erect, annual, branched above. Lower leaves petioled, fugacious; cauline oblong-lanceolate, with lobed auricles, obtuse. Flowers in terminal racemes, white or yellow. Pod rounded at the apex or truncate. Seeds few.

NORTH and SOUTH Islands: In cultivated fields, Auckland and Otago. Not infrequent, but scarcely naturalised. *Gold of pleasure.* Dec., Jan. Europe.

*BRASSICA, Tourn.

Sepals erect or spreading, equal or the lateral saccate at the base. Pods linear, nearly terete or 4-angled, with a stout indehiscent 1-seeded beak. Seeds globose, 1-seriate; valves 1–5-nerved. Cotyledons incumbent. Leaves often large, entire, lobed, lyrate or pinnatifid. Flowers in long terminal racemes.

*1. Sepals erect. BRASSICA.

Leaves thick, glaucous, somewhat fleshy	* *B. oleracea.*
Radical leaves glaucous, hispid. Stem leaves auricled	* *B. campestris.*

2. Sepals spreading. SINAPIS.

* *Pods adpressed to the stem. Valves 1-nerved.*

Pod subulate, beak short, seedless	* *B. nigra.*
Pod subcylindric, beak clavate, 1-seeded	* *B. adpressa.*

** *Pods spreading. Valves 3-nerved.*

Pods longer than the compressed beak	* *B. Sinapistrum.*
Pods beaded, equalling the beak	* *B. alba.*

* **B. oleracea**, *L.*, *Sp. Pl.* 667. Rootstock stout. Leaves glabrous, glaucous below, lower lyrate or sinuate, upper oblong, sessile. Racemes elongated. Flowers large. Pods 2in.–3in. long, spreading, slightly compressed; valves keeled; beak short.

NORTH and SOUTH Islands: Naturalised on sea-cliffs in many places. *Maori cabbage.* Nov., Dec. Europe.

* **B. campestris**, *L.*, *Sp. Pl.* 666. Erect. Lower leaves glaucous-hispid, lyrate, pinnate, lobed or dentate; cauline leaves glabrous, oblong or ovate-lanceolate, auriculate. Flowers corymbose. Beak of pod seedless. Valves 1-nerved.

Var. **campestris.** Root tuberous. Radical leaves hispid. Petals persistent until the corymb lengthens. *Swede.*

Var. **Napus.** Root fusiform. Leaves glabrous. Petals falling before the corymb lengthens. *Rape.*

Var. **Rapa.** Root tuberous, fleshy. Lower leaves hispid (not glaucous), upper glabrous and glaucous. *Turnip.*

NORTH and SOUTH Islands: Not uncommon in cultivated districts. Var. *c* the most frequent. Dec., Jan. Europe.

* **B. nigra**, *Koch in Roehl. Deutschl. Fl., ed.* 3, iv. 713. Glabrous, erect, 2ft.–3ft. high, much branched. Lower leaves petioled, rough, lyrate, or irregularly lobed, toothed or entire; upper lanceolate, glabrous, entire. Pedicels short, stout. Pods tetragonous, adpressed, ½in.–¾in. long; beak short, seedless; valves 1-nerved.

NORTH and SOUTH Islands: Naturalised in cultivated fields and waste places, but often local. *Black mustard.* Nov., Dec. Europe.

* **B. adpressa**, *Boiss*, *Voy. Espagne* ii. 38. An erect, slender, branched herb, 6in.–24in. high, hispid, hairy or glabrate. Leaves lyrate or lyrate-pinnatifid; cauline few, linear-lanceolate. Pods closely adpressed to the stem, ⅓in.–½in. long, on short pedicels; beak 1-seeded, sometimes half as long as the pod.

NORTH and SOUTH Islands: Naturalised in waste and cultivated ground. Auckland, Taranaki, Wellington, and Nelson. *T. K.* South Europe.

* **B. Sinapistrum**, *Boiss*, *Voy. Espagne* ii. 39. Stem and leaves hispid. Lower leaves petiolate, lyrate-pinnatifid; upper toothed or entire. Flowers subcorymbose, yellow, large. Pods setose, rarely glabrous, subcylindrical, longer than the 2-edged sterile or 1-seeded beak.

NORTH and SOUTH Islands: Naturalised in cultivated ground, but somewhat local. *Charlock.* Nov., Dec. Europe.

* **B. alba,** *Boiss, Voy. Espagne* ii. 39. Erect; 1ft.–2ft. high. Leaves hispid, with reflexed hairs, pinnate or lyrate-pinnatifid; segments cut or lobed. Flowers pale-yellow. Fruiting pedicels slender, spreading. Pods ½in.–2in. long, hispid, knotted; beak as long as the pod, 1-seeded; valves strongly ribbed, often convex, 1–3-seeded.

NORTH and SOUTH Islands: Naturalised in cultivated land, but local. Auckland, Canterbury, &c. *White mustard.* Nov. to Jan. Europe, North Africa, North and West Asia, &c.

* DIPLOTAXIS, DC.

Sepals spreading, equal. Pod elongated, compressed; stigma entire. Seeds 2-seriate, compressed. Cotyledons incumbent. Herbs with sinuate or pinnatifid leaves and yellow flowers.

* **D. muralis,** *DC., Syst. Veg.* ii. 634. Annual, hispid or glabrate. Stems simple, slender, leafy below. Leaves usually petioled, ovate-lanceolate or oblong, entire, sinuate-dentate or pinnatifid. Flowers in terminal racemes. Pods 1in.–2in. long, compressed, narrowed above; style long, straight; valves flat, nerves obscure.

Var. **muralis**, proper. Leaves deeply pinnatifid; stem-leaves few, slender. Petals twice or thrice as long as the sepals. Styles scarcely narrowed below.

Var. **viminea.** Leaves nearly entire. Scapes slender, leafless or nearly so. Petals twice the length of the sepals. Style narrowed below.

NORTH Island: Var. *muralis*: naturalised on waste land, Napier. Var. *viminea*: on sand-hills near New Plymouth and Porirua Harbour. *Wall-mustard.* Dec. to Feb. West and South Europe.

* ERUCA, Tourn.

Sepals erect, lateral saccate at base. Petals clawed. Pod oblong, nearly terete, turgid, with a long seedless beak; valves 3-nerved. Stigma obscurely 2-lobed. Seeds 2-seriate. Cotyledon incumbent. Annual or biennial branching herbs with pinnatifid leaves.

* **E. sativa,** *Lam., Fl. Fr.* ii. 496. Stem branching, glabrate or hairy. Leaves lyrate-pinnatipartite; segments toothed or lobed. Sepals fugacious. Petals strongly veined.

Naturalised on KAIKOURAS Island, Port Fitzroy. Nov., Dec. South Europe.

4. PACHYCLADON, Hook. f.

Sepals 5, equal. Stamens 6. Pods laterally compressed, elliptic; valves boat-shaped; septum incomplete; style short, straight. Stigma 2-lobed. Seeds 3–5 in each cell; funicle short. Cotyledons incumbent. A densely-tufted herb, with stout taproots and radical rosulate leaves.

ETYM. From the Greek, in reference to the stout fleshy branches of the rootstock.

A monotypic genus restricted to the extreme southern parts of the colony.

1. **P. Novae-Zelandiae,** *Hook. f., Handbk.* 724. Root long, stout, fleshy, branching into several stems above, each ½in.–¾in. in diameter, clothed with remains of old leaves below, and crowned with a dense head of small imbricating horizontal leaves clothed with 2-fid or stellate hairs. Leaves ¼in.–1in. long, blade oblong, narrowed into a short flat petiole, pinnatifidly lobed; cauline with longer petioles digitately lobed or entire, linear. Scapes numerous, springing from below the leaves and exceeding them, erect or spreading, 2–5-flowered. Petals linear-obovate or spathulate, twice as long as the sepals.

Pods shortly pedicelled, $\frac{1}{8}$in.–$\frac{1}{4}$in. long, laterally compressed; valves boat-shaped; style minutely 2-lobed. Seeds 3–5 in each valve. Septum incomplete. —Hook. f. in Hook. Ic. Pl. t. 1009; Buch. in Trans. N.Z.I. xiv. (1881) t. 24, f. 1. *Braya Novae-Zelandiae*, Hook. f., Handbk. 13.

SOUTH Island: Otago: Mount Alta, *Hector* and *Buchanan!* Hector Mountains, Mount Pisa and other high mountains further west, Dunstan Range, Mount St. Bathan's, Old-Man Range, &c. 4,500ft. to 6,000ft. *Petrie!*

Var. **glabra.** Leaves larger, glabrate or pubescent, lobes ascending.—Sp. Buchanan, *l.c.*, t. 24, f. 2. SOUTH Island: Canterbury: mountain range, head of Lake Ohau. 5,000ft. Mr. Buchanan remarks: "The present plant may probably be considered as only a form of *P. Novae-Zelandiae* produced by climatic causes."

5. CAPSELLA, Medicus.

Sepals spreading, equal at the base. Petals short. Pod ovoid, oblong, obcuneate or obcordate, laterally compressed; valves turgid, keeled below; septum thin; style short or 0. Seeds several, 2-seriate. Cotyledons incumbent. Annual herbs with entire lobed or pinnatifid leaves.

ETYM. Diminutive of *Capsula*.

A small genus, comprising about 10 species, found in the north and south temperate regions.

1. **C. elliptica**, *C. A. Meyer, Verz der Pflan. v. Caucas.* 191. Glabrous, much branched, slender, 1in.–5in. high. Leaves lanceolate or oblong, the lower petiolate, lobed, pinnatifid or lyrate-pinnatifid, $\frac{1}{2}$in.–$\frac{3}{4}$in. long; cauline smaller, sessile, often entire. Racemes suberect, elongating in fruit. Pedicels spreading, filiform. Flowers white. Petals scarcely exceeding the sepals. Pod ovoid; valves turgid, lanceolate. Seeds 10–15 in each cell.—F. Muell., 2nd Cens. Austr. Pl. 10. *C. procumbens*, Fries., Novit. Fl. Lucc. Mant. i. 14; Benth., Fl. Austr. i. 81. *Stenopetalum incisaefolium*, Hook. f. in Hook. Ic. Pl. t. 276.

SOUTH Island: Otago: on cliffs moistened by sea-spray. Oamaru, Forbury Heads, *Petrie!* Sept., Oct. Also in Australia, Europe, West and Central Asia, North-west America, temperate South America.

* **C. bursa-pastoris**, *Medic., Pflanzeng.* 85. Erect, 6in.–18in. high, simple or sparingly branched. Leaves sinuate-pinnatifid, rarely entire; cauline leaves small, auricled. Flowers in erect racemes, minute. Pods on slender ascending pedicels, triangular-obcordate, much compressed; style short. Seeds numerous.

Naturalised in cultivated and waste ground everywhere, except the Antarctic Islands. *Shepherd's purse.* Sept. to April. Europe.

* SENEBIERA, Poiret.

Sepals 4, short, spreading. Stamens 2, 4, or 6. Flowers white, in short racemes opposite the leaves. Pod small, didymous, indehiscent, compressed laterally; stigma sessile, slightly reniform or almost 2-lobed; valves rugose or crested. Cotyledons incumbent. Leafy annuals or biennials.

Flowers in small fascicles .. .. * *S. Coronopus.*
Flowers in slender racemes .. .. * *S. didyma.*

* **S. Coronopus**, *Poir., Dict.* 7, 76. A glabrous annual with short branches closely appressed to the ground. Leaves oblong or obovate, 2-pinnatifid. Flowers in small sessile or shortly-pedicellate fascicles. Lobes of fruit not separating, deeply wrinkled. Seeds 1 in each cell. Style reduced to a subulate point.

NORTH and SOUTH Islands: Naturalised in waste places, roadsides, &c. *Hog's cress.* Nov. to April. Europe.

*S. **didyma**, *Persoon*, *Syn.* ii. 185. Annual. Stems weak, suberect, diffuse, sparingly hairy. Leaves more deeply divided than in the preceding species, 2-pinnatifid, lobes narrow. Flowers small, in slender racemes. Pods didymous, lobes separating, valves wrinkled.

NORTH and SOUTH Islands; STEWART Island: Abundantly naturalised in waste places, especially near the sea. Dec. to Feb. Southern parts of South America.

6. LEPIDIUM, Linn.

Sepals short, equal at the base. Petals short or rarely 0. Stamens 2, 4, or 6. Pod ovate, obovate, obcordate, or rarely orbicular, laterally compressed; septum narrow; valves keeled or winged; style short or 0; stigma notched. Seeds 1 in each cell. Cotyledons incumbent in the New Zealand species. Erect or spreading rarely suffruticose herbs, annual or perennial. Flowers rarely unisexual.

ETYM. From the Greek, in reference to the scale-like form of the pods.

SPECIES, about 85. Distributed through temperate and warm regions. All the indigenous species are endemic, but several exotic species have become naturalised.

* *Stems leafy. Pods entire.*

Erect. Leaves entire, serrate or incised .. .. 1. *L. oleraceum.*

** *Stems leafy. Pods emarginate.*

Prostrate. Leaves entire, crenate or incised .. .. 2. *L. obtusatum.*
Suberect, flexuous. Leaves incised. Pods cordate .. .. .. 3. *L. Forsteri.*
Suberect. Leaves pinnatifid. Racemes lateral .. .. 4. *L. flexicaule.*
Erect, strict, pubescent. Cauline leaves clasping . .. . * *L. campestre.*
Erect or suberect. Styles exceeding the notch .. .. * *L. Smithii.*
Erect, strict. Leaves pinnate or pinnatifid, pilose, dioecious .. .. 5. *L. Matau.*
Erect, spreading. Leaves pinnate; leaflets pinnatifid .. .. .. 6. *L. Kawarau.*
Erect. Lower leaves 2-pinnatifid; segments linear .. .. .. * *L. ruderale.*
Erect. Lower leaves lobed, pinnate or 2-pinnate .. .. * *L. sativum.*

*** *Leaves chiefly radical. Stems short. Cauline leaves few or 0.*

Small, glabrous. Leaves entire. Stems filiform .. .. 7. *L. Kirkii.*
Leaves rosulate, hairy, pubescent. Scapes suberect, dioecious .. .. 8. *L. sisymbrioides.*
Leaves rosulate, pubescent. Stems weak. Pods orbicular .. .. .. 9. *L. tenuicaule.*

1. **L. oleraceum**, *G. Forst., Prod. n.* 248. Glabrous, erect or suberect, stout or slender. Stems sometimes as thick as the finger, often suffruticose. Leaves elliptic-oblong or broadly cuneate-oblong, narrowed into a winged petiole, serrate or incised. Flowers in terminal simple or branched racemes, 2in.–4in. long, usually numerous and subcorymbose. Pedicels slender, spreading. Stamens 4. Pods ovate, subacute, wingless, entire; style slender.—Pl. Esc. 38; DC., Syst. Veg. ii. 547; A. Rich., Fl. 310; Willd., Sp. Pl. iii. 437; A. Cunn., Precurs. n. 628; Hook. f., Fl. N.Z. i. 15, and Handbk. 14; T. Kirk in Trans. N.Z.I. xiv. (1881) 379.

Cook's scurvy-grass. Nov. to Feb.

Var. **frondosum**, (*sp.*), Banks and Sol., MSS. and Ic. Robust, leaves large, fleshy, broadly cuneate-oblong or oblong, sometimes 3in.–5in. long and 1in. wide, sessile or narrowed into a broad petiole, serrate.

Var. **acuti-dentatum.** Stems with slender leafy branches. Leaves 1in.–1½in. long, narrow, cuneate or oblong-spathulate, the upper portion acutely serrate or almost dentate. NORTH and SOUTH Islands; STEWART Island; the SNARES; AUCKLAND Islands; CHATHAM Islands. In places near the sea. Var. *a*, restricted to small islands in the north, has become very rare. The plant is everywhere destroyed by cattle and sheep. It has a strong disagreeable odour.

2. **L. obtusatum,** *T. Kirk in Trans. N.Z.I.* xxiv. (1891) 423. Glabrous, much branched, leafy. Stems prostrate, 6in.–12in. long. Lower leaves elliptic or elliptic-oblong, gradually narrowed into a naked or winged petiole, crenate or coarsely serrate. Cauline leaves oblong or obovate, shortly petioled or sessile. Racemes numerous, terminating short leafy branches. Pedicels slender, ascending. Flowers small. Stamens 4. Pods ovate-cordate, slightly winged above, with a narrow notch; style never exceeding the notch; stigma capitate. Seeds obliquely ovate.

NORTH Island: Sea-cliffs, Titirangi, *Cheeseman.* Maritime rocks at the entrance to Port Nicholson, *Miss Kirk!* Nov. to March.

This species differs from the preceding in the prostrate habit and emarginate shortly-winged pods with the included style. The hypogynous glands are very short and obtuse.

3. **L. Banksii,** *n. s.* Glabrous. Stems much branched, terete, flexuous, suberect, 1ft.–1½ft. long. Leaves distant, narrow, oblong-cuneate or oblong-spathulate, acutely toothed or incised above, sessile or very shortly petioled. Racemes terminal. Stamens 4. Pedicels strict, slender, more than twice as long as the pods. Pods ovate, cordate at the base, truncate at the apex, with a broad notch, slightly keeled and winged; style slightly exceeding the notch.—*L. oleraceum,* A. Rich. Fl. N.Z. 310, t. 35 (not of Forst.).

SOUTH Island: Queen Charlotte Sound and Astrolabe Harbour, *A. Richard, l.c.*

Var. **ovatum.** Pods broadly ovate, almost truncated at the base but not cordate; style equalling the broad notch. SOUTH Island: Pelorus Sound, *J. Rutland!* Kenepuru, *J. McMahon!*

This species is distinguished from *L. oleraceum* by the emarginate winged pod, and from *L. obtusatum* by its narrow leaves and the cordate or broadly-ovate pod, which is never narrowed below. The typical form has not been observed of late years: it appears to combine the habit of *L. flexicaule* with the leaves of *L. oleraceum,* var. *acutidentatum.*

4. **L. flexicaule,** *T. Kirk in Trans. N.Z.I.* xiv. (1881) 380. Glabrous. Stems numerous, flexuous, decumbent. Lower leaves 2in.–3in. long, on long naked petioles, linear-oblong, pinnatifid or rarely pinnate, pinnules in 4 to 6 pairs, lobed and toothed, near the apex obtuse; cauline smaller, gradually narrowed into a winged petiole or sessile, entire, narrow linear-spathulate or cuneate, coarsely toothed. Racemes 1in.–2in. long, lateral and terminal, leaf-opposed, each with a solitary flower below its base. Flowers perfect. Sepals ovate. Petals narrow-linear, obtuse. Stamens 2. Pods on slender erect pedicels, ovate-cordate, narrowed below, shortly winged and notched at the apex; style wholly included in the notch.—*L. incisum,* Hook. f., Fl. N.Z. i. 15; Handbk. 14 (not of Roth. Neue Beitr. i. 224); Banks and Sol. MSS!

NORTH Island: Manukau and Waitemata Harbours, *T. K.* Rangitoto Island, *Cheeseman.* Mercury Bay, *Banks and Solander.*

This is distinguished from all other New Zealand species by the lateral racemes and diandrous flowers. The racemes are at first terminal, but are quickly reduced to a lateral position by a strong usurping shoot which overtops them. The plant resembles *Senebiera didyma* in its inflorescence.

*__L. campestre__, *R. Br. in Ait. Hort. Kew. ed.* 2, iv. 88. Stems erect, pubescent, simple or branched from the base. Radical leaves oblong, narrowed into the petiole, entire or pinnatifid; cauline lanceolate or oblong-lanceolate, auricled and clasping, toothed or rarely entire. Racemes terminal. Flowers small. Pedicels spreading. Pods broadly ovate, papillose, winged, notched at the apex; style scarcely exceeding the notch. Anthers usually yellow.

NORTH and SOUTH Islands: Sparingly naturalised in cornfields and waste places, &c. Europe.

*__L. hirtum__, *Sm. Comp. Br. Fl. ed.* 3, 98. Similar to *L. campestre*, but more robust, frequently branched from the base, diffuse, pubescent or rarely villous. Leaves broader. Pod ovate, glabrous, notched or entire, winged above; style long. Seeds small. Anthers violet.

NORTH and SOUTH Islands: Sparingly naturalised in cornfields, &c., roadsides, but rare and local. Plentiful at Bendigo, Otago. Dec., Jan. West Europe.

5. **L. Matau**, *Petrie in Trans. N.Z.I.* xix. (1886) 323. A small erect dioecious apetalous herb, grey with hispid hairs. Stems 1in.–3in. high. Radical leaves coriaceous, scabrid, 1in.–1½in. long, narrow-oblong, pinnate or pinnatifid; segments rarely incised, rounded or oblong; rhachis broad and flat. Cauline leaves few, oblong or ovate, entire, sessile; branches spreading. Racemes short, pedicel spreading. Male flowers: Sepals pilose, whitish; stamens 4. Female: Pedicels patent or decurved. Pod ovate, narrowed below, retuse; style exceeding the notch, very short. Seeds ovate, yellow, excessively mucilaginous.

SOUTH Island: Otago: Alexandra South, *Petrie!*

A remarkable species, easily recognised by its greyish aspect, strict habit, scabrid coriaceous leaves, and apetalous dioecious flowers.

6. **L. Kawarau**, *Petrie in Trans. N.Z.I.* xvii. (1884) 270. Dioecious, erect, 6in.–12in. high, somewhat spreading, glabrous or with few white simple or 2-fid hairs. Radical leaves oblong, 3in.–5in. long, with sheathing petioles, pinnate; rhachis broad, flattened; leaflets linear, entire or once or twice pinnatifid. Cauline smaller, uppermost narrow-linear, entire. Racemes forming a much-branched subcorymbose panicle. Flowers on slender ascending pedicels. Male: Sepals with membranous margins, lateral sepals narrow; petals 0; stamens 4. Female: Anthers 0 or abortive; petals 0. Pods oblong-ovate, truncate and notched at the apex, stigma slightly exceeding the notch. Seeds oblong, brown.

SOUTH Island: Otago: Kawarau River, Nevis Bluff, Cromwell, *Petrie!*

Mr. Petrie describes the male flower as having 4 small petals and the female as usually apetalous. I am unable to find any trace of petals in either.

? Var. **dubium.** Erect, very leafy, scabrid, 1ft. high or more, sparingly branched above; the branches short, or long and spreading. Radical leaves similar to those of the type or more divided; cauline shorter, sessile on broad bases, pinnatifidly toothed. Racemes lax; pedicels patent. Stamens 4-6. Petals 2-4, narrow. Pods oblong-ovate. Seeds broadly oblong, brown. Otago: Earthquakes near Duntroon, *Petrie!* My specimens are in very poor condition; better material might render it necessary to raise this puzzling plant to specific rank.

*__L. ruderale__, *Sp. Pl.* 645. Annual. Erect or suberect, slender, much branched, 1ft.–1½ft. high, glabrous. Lower leaves 2-pinnatifid, segments linear; upper entire or lobed. Racemes terminal. Flowers minute. Petals 0. Stamens 2.

Pods very small, much compressed, narrowed at both ends, broadly notched at the apex and narrowly winged; style included.

NORTH and SOUTH Islands: Abundantly naturalised from sea-level to 2,500ft. *Sheep's cress.*

*L. **sativum**, *L.*, *Sp. Pl.* 644. Erect, annual, glabrous. Stems pale or whitish, much branched above, branches ascending. Lower leaves lobed, pinnate or 2-pinnate; upper entire, linear, sessile. Flowers small. Pod rounded-ovate, compressed, winged, shortly notched at the apex. Cotyledons 3-partite.

NORTH Island: Sparingly naturalised, but rarely seen in large quantity. *Garden-cress.*

7. **L. Kirkii,** *D. Petrie in Trans. N.Z.I.* xxii. (1889) 439. Small, glabrous or rarely glabrate, with prostrate filiform sparingly-branched scapes 2in.–4in. long. Radical leaves entire, narrow-linear or linear-spathulate, obtuse, ½in.–1in. long, sheathing at the base; cauline smaller. Stems few or many, filiform, flexuous. Racemes short, terminal; pedicels slender. Flowers minute. Sepals broadly ovate, inflated. Petals shorter than sepals, rarely retuse. Stamens 4, rarely 6. Pods distant, equalling the slender spreading pedicels, ovate-orbicular, very strongly notched; style minute.

SOUTH Island: Otago: Gimmerburn, Maniototo Plains, *Petrie!* Dec.

A very distinct species. The hypogynous glands, 6 in number, are unusually prominent, forming a ring at the base of the ovary.

8. **L. sisymbrioides,** *Hook. f., Handbk. N.Z. Fl.* 14. Dioecious. Rootstock stout and fleshy, much divided above. Leaves all radical or nearly so, rosulate, glabrate or hairy, narrow-lanceolate or oblong, 1in.–1½in. long; petioles dilated at the base, pinnatifid; the pinnules narrow, frequently lobed above. Scapes slender, suberect, branched, naked or with a few entire leaves at the base. Flowers minute. Male. Petals narrow; stamens 4. Female: Petals 0. Pods about as long as the slender pedicels, sparsely hairy, broadly subquadrate-ovate, narrowed at both ends, minutely notched at the apex, narrowly winged; style exceeding the notch. Seed brown, obovate.— *L. Solandri*, T. Kirk in Trans. N.Z.I. xiv. (1881) 380.

SOUTH Island: Canterbury: Broken River Basin, *Enys* and *Kirk*. Lake Ohau, *Haast!* Mackenzie Country, *J. F. Armstrong!* Otago: Waitaki Valley, Lake Wanaka, *Buchanan!* Kurow, *Petrie!* 600ft. to 2,500ft. Dec., Jan.

The stout tap-root is often from 2ft.–4ft. long, and, owing to its branching just below the surface, forms a crown of leaves a foot or more in diameter. It is easily distinguished by the remarkable pod, which is narrowed at both ends.

9. **L. tenuicaule,** *T. Kirk in Trans. N.Z.I.* xiv (1881) 381. Leaves chiefly radical, more or less clothed with soft whitish hairs, rarely glabrous, narrow-oblong, pinnate or pinnatifid, 1in.–4in. long; leaflets sometimes petiolulate, incised, or sharply serrate on the upper margin, often piliferous. Scapes numerous, slender, prostrate or suberect, 6in.–12in. long, simple or branched, naked or with few small entire leaves below. Flowers very numerous, small, usually apetalous. Stamens 4. Ovary winged. Pod minute, orbicular, much shorter than the slender erect pedicel, narrowly winged above; style scarcely

exceeding the minute notch. Seed large, broadly oblong; testa finely muriculate.

SOUTH Island: Otago: littoral, Cape Whanbrow, Awamoa (18 miles inland), *Petrie!* STEWART Island, Dog Island, Ruapuke Island, *T. K.* Dec., Jan.

Var. **australe**, (*sp.*), T. Kirk, *l.c.* Erect, branched; branches spreading. Radical leaves mostly pinnate; leaflets petiolulate. Cauline leaves numerous, pinnatifid, toothed, lobed or entire. Pods orbicular or orbicular-ovate. Cape Whanbrow, *T.K.*

This species is readily distinguished by its prostrate stems and minute orbicular pods. The Dog Island specimens have almost fleshy glabrous leaves and suberect stems, but present no other difference. In the original description the pods were erroneously described as wingless.

7. NOTOTHLASPI, Hook. f.

Sepals erect, equal. Stamens 6. Pods excessively compressed and broadly winged; cells many-seeded. Funicle slender. Radicle incumbent, sometimes very long. Annual or biennial herbs, with rosulate fleshy radical leaves.

The genus is restricted to the SOUTH Island of New Zealand.

Scape stout, simple. Style short .. .. .. 1. *N. rosulatum.*
Stem much branched. Style long .. .. .. 2. *N. australe.*

1. **N. rosulatum,** *Hook. f., Handbk.* 15. Erect, 3in.–10in. high, stemless or nearly so. Leaves all radical, fleshy, imbricated, forming a densely-crowded rosette, petioled, spathulate or spathulate-oblong, crenate or rarely dentate, subacute, clothed with caducous stellate cellular hairs when young, glabrate or glabrous when old. Scape ½in.–¾in. in diameter. Flowers densely crowded, forming a conical or pyramidal raceme, fragrant. Sepals elliptical. Petals obovate-spathulate. Ovary obovate, sessile. Pod ½in.–1in. long, obovate, broadly winged, deeply emarginate; style very much shorter than the notch; stigma broadly 2-lobed. Seeds ∞, reniform; testa pitted. Radicle long and slender, folded first upwards, then downwards and backwards over the back of the cotyledons.—*N. notabilis*, Buch. in Trans. N.Z.I. xiv. (1881) 344, t. 25.

SOUTH Island: Mountains of Nelson and Canterbury, but rare and local. St. Arnaud Range, Wairau Gorge, Lake Tennyson &c., Upper Waimakariri, Porter's Pass, Mount Torlesse, Broken River Basin, Ribbon Range, Lake Ohau, &c. Otago: Mount Ida, *P. Goyen.* 2,000ft. to 6,000ft. *Pen-wiper plant.* Dec., Jan.

One of the most remarkable plants known; now becoming very rare owing to the ravages of sheep. The leaves are often ciliated, with flat ribbon-like hairs, and before flowering form a dense imbricating cushion 2in.–3in. high. The flowers have the fragrance of orange-blossoms.

2. **N. australe,** *Hook. f., Handbk.* 15. Usually much branched from the base; branches 1in.–3in. long. Radical leaves forming a rosette, glabrous or with few cellular hairs, petiolate, linear-lanceolate or spathulate or oblong-spathulate, fleshy, crenate or rarely entire. Flowers numerous, corymbose, smaller than in the preceding species. Ovary oblong, usually with a few short hairs; style stout, long. Fruiting pedicels ½in.–1in. long. Pod much shorter, broadly elliptic, winged, scarcely retuse; style one-third as long as the pod. Seeds numerous; testa pitted. Radicle slender.—*Thlaspi australe*, Hook. f., Fl. N.Z. ii. 325.

SOUTH Island: Nelson Mountains, 3,000ft. to 5,000ft. Frequent. Originally discovered by Captain Rough and Sir David Monro.

Var. **stellatum.** Not branched. Leaves narrow linear-spathulate; petioles pubescent. Flowers numerous, on long 1-flowered peduncles. Mount Rintoul, *F. G. Gibbs! W. H. Bryant!*

In the autumn leafy stems 3in.-5in. long are occasionally developed.

*IBERIS, Linn.

Sepals equal at the base. Petals 4, the two outermost longer than the others. Pod orbicular or ovate, much compressed; valves keeled or winged. Stigma notched. Seeds 1 in each cell. Cotyledons accumbent; radicle horizontal or ascending. Annual or perennial herbs, sometimes suffruticose. Flowers corymbose.

***I. amara,** *L., Sp. Pl.* 649. Annual, erect, 6in.-12in. high, glabrous. Leaves lanceolate or oblong-lanceolate, toothed, sessile. Racemes terminal, elongating. Pods orbicular, on spreading pedicels, flat, emarginate, narrowly winged; style exceeding the notch.

NORTH and SOUTH Islands: Sparingly naturalised in waste places; rarely in cornfields. *Candytuft.* Nov. to Jan. Western Europe.

*RAPHANUS, Linn.

Sepals erect, lateral saccate at base. Petals narrowed into long claws. Stamens tetradynamous. Pod elongated, terete, indehiscent, or moniliform, smooth or ribbed; beak slender. Seeds ∞. Cotyledons conduplicate. Annual or biennial branched herbs; root usually fleshy or succulent.

***R. sativus,** *L., Sp. Pl.* 669. Root fleshy. Stem much branched above. Leaves lyrate-pinnatifid; segments few, distant, nearly glabrous or with scattered hairs. Pods on long erect pedicels, cylindrical, inflated, faintly constricted; beak long and narrow.

NORTH and SOUTH Islands: Naturalised in many places; often abundant. *Radish.* Dec. to Feb. Europe.

Order—* RESEDACEAE.

Flowers often irregular. Calyx 4-7-partite, imbricate in bud. Petals 4-7, hypogynous, entire or lobed, equal or the posterior larger. Disc 1-sided, hypogynous, between the petals and stamens. Stamens 3-∞, inserted on the disc. Ovary of 2-6 coherent carpels; ovules few or many, amphitropous or campylotropous, on 2-6 parietal placentas. Capsule 3-6-lobed, opening at the top before the seeds are matured. Seed without endosperm. Herbs, rarely shrubs, with alternate leaves. Stipules 0 or glandular.

*RESEDA, Linn.

Flowers in terminal racemes. Calyx irregular. Petals 4-7, cleft, unequal. Disc broad, dilated behind. Stamens 10-40, on one side of the flower. Leaves entire, lobed or pinnatifid.

***R. Luteola,** *L., Sp. Pl.* 448. An erect herb, 1ft.-2ft. high. Leaves entire, linear-lanceolate. Flowers in erect terminal racemes, yellow. Pedicels longer than the calyx. Calyx 4-partite. Petals 4, the upper 3-5-fid, the lateral 3-fid, the lower entire, linear. Stigmas 3.

NORTH and SOUTH Islands: Naturalised in waste places, &c., but not general. *Dyer's weed. Weld.* Dec., Jan. Europe.

***R. alba,** *L., Sp. Pl.* 449. Erect, 1½ft.-2ft. high. Leaves pinnatifid, glaucous. Flowers in terminal racemes, white. Pedicels shorter than the calyx. Sepals 5-6. Petals 5-6, 3-fid, exceeding the calyx.

SOUTH Island: Naturalised, but local. *Tree-mignonette.* South Europe.

Order IV.—VIOLARIEAE.

Flowers regular or irregular, usually perfect. Sepals 5, imbricate. Petals 5, imbricate, the lower one often spurred or larger. Stamens 5, hypogynous; the anthers erect and connivent round the pistil, sessile or on very short filaments, the connective often very broad. Anthers introrse. Ovary superior, 1-celled, with 2–5 parietal placentas. Fruit a capsule or berry. Seeds with fleshy endosperm; embryo axile. Herbs, shrubs, or trees. Flowers axillary, solitary or in cymes or panicles.

A large order, generally distributed in the tropical and temperate regions of both hemispheres. Genera, 21. Species, 250.

Herbs with irregular flowers. Fruit a capsule.

1. Viola. Lower petal spurred at the base. Anthers united.

* Ionidium. Lower petal large, saccate or gibbous at the base. Anthers free.

Shrubs or trees. Flowers regular. Fruit a berry.

2. Melicytus. Shrubs or trees. Anthers free.

3. Hymenanthera. Shrubs. Anthers connate. Placentas 2.

1. VIOLA, Linn.

Sepals 5, produced into a short expansion at the base. Petals unequal, spreading, lowest spurred at the base. Anthers 5, nearly sessile, the connective produced into a thin flat membrane, the two lower often with a short dorsal appendage or spur. Style thickened or dilated at the top or incurved with the stigma in front. Capsule with 3 elastic valves, each with 1 parietal placenta. Herbs with short woody rootstocks and erect or trailing stems. Leaves alternate, stipulate.

The genus comprises over 100 species, distributed through the temperate regions. All the New Zealand species are endemic except *V. Cunninghamii*, which extends to Tasmania.

Etym. The old Latin name.

In many species the flowers are of two kinds—a large pedunculate form with conspicuous petals, and later in the season inconspicuous cleistogamic flowers, which are apetalous, and produce seeds freely, while the ordinary form is often sterile.

Leaves cordate.

Stems slender. Stipules and bracts lacerate .. .. .. 1. *V. filicaulis.*

Stems slender. Leaves cordate. Stipules and bracts entire .. 2. *V. Lyallii.*

Stems short. Leaves ovate. Stipules and bracts entire .. .. 3. *V. Cunninghamii.*

Leaves lyrate.

Stems branched. Stipules pinnatifid .. .. * *V. tricolor.*

1. **V. filicaulis,** *Hook. f., Fl. N.Z.* i. 16. Glabrous. Stems slender, filiform, prostrate. Leaves alternate, on slender petioles, orbicular-cordate or ovate-cordate or almost reniform, usually rounded or obtuse, crenate. Stipules laciniate, teeth often tipped with a gland. Peduncles solitary or geminate, exceeding the leaves, slender, with two linear-toothed or lacerate bracts below the middle. Sepals linear-lanceolate. Spur short. Lower anthers spurred.—Handbk. 16.

NORTH and SOUTH Islands; STEWART Island: from Cape Colville southwards. Ascends to 3,000ft. Nov., Dec. Flowers yellowish. This species produces cleistogamous flowers in Jan. and Feb.

Var. **hydrocotyloides** (*sp.*), J. B. Armst. in Trans. N.Z.I. xiv. (1881) 360. Small leaves, ⅛in.–½in. in diameter. Peduncles very short. In sandy soil. Otago *Petrie*. Stewart Island, *T. K.* Merely a stunted condition.

2. **V. Lyallii,** *Hook. f., Handbk.* 16. Stems very slender. Leaves broadly ovate or orbicular, ovate-cordate or subreniform, obscurely crenate. Stipules entire, linear. Bracts linear-acute, above the middle of the peduncle. Sepals broad, acute. Lower stamens shortly spurred.—*V. Cunninghamii*, var. *gracilis*, Hook. f., Fl. N.Z. i. 16. *Erpetion spathulatum*, A. Cunn., Precurs. n. 622.

NORTH and SOUTH Islands: in lowland districts from the Bay of Islands to Canterbury, but often local.

The position of the bracts distinguishes this species from the preceding. Oct., Nov.

3. **V. Cunninghamii,** *Hook. f., Fl. N.Z.* i. 16. Glabrous except the petioles. Rootstock usually woody, creeping below. Stem short or 0. Leaves tufted, triangular-ovate or ovate-oblong, truncate at the base or narrowed into the petiole, crenate; petiole often more or less pubescent. Stipules broadly adnate at base of the petiole; apex free, acute. Peduncles exceeding the leaves, with a pair of acute bracts above the middle. Sepals linear-oblong, obtuse or acute. Lateral petals bearded; spur short, obtuse. The lower anthers shortly spurred.—Handbk. 16.

NORTH and SOUTH Islands; STEWART Island: from the East Cape district southwards. Ascends to 4,000ft. Oct. to Jan.

This species varies greatly in size; the peduncles are ½in.–4in. in length or more. It produces cleistogamous flowers in the lowlands.

Var. **per-exigua,** Col. (*sp.*) in Trans. N.Z.I. xvi. (1883) 326. Very small. Leaves broadly cordate or orbicular. Sepals rather large, oblong-ovate, acute with scarious margins. NORTH Island: between Matamau and Dannevirke, *Colenso*.

* **V. tricolor,** *L., Sp. Pl.* 935. Stems branched from the base, angled. Leaves on long petioles, lanceolate, oval or ovate-oblong, lyrate, crenate or crenate-serrate. Stipules leafy, pinnatifid. Flowers large, flat, showy. Sepals auricled, lanceolate, acute. Capsule ovoid.

NORTH and SOUTH Islands: naturalised in waste and cultivated ground, but local. *Pansy. Heart's-ease.* Nov. to Feb. North Europe, North Africa, North and West Asia, North-west India, &c.

Var. **arvensis.** Small, slender. Flowers very small. Petals not exceeding the sepals, white or pale-yellow. Capsule globose. Naturalised in cornfields and cultivated land.

* IONIDIUM, Vent.

Flowers irregular. Sepals not prolonged at the base. Petals spreading, the lowest much larger than the others, contracted in the middle, produced into a membranous expansion. Style thickened, incurved, with the stigma in front. Fruit a 3-celled capsule. Perennial branching herbs or rarely small shrubs. Leaves alternate or opposite.

* **I. filiforme,** *F. Muell., Pl. Vict.* i. 66. Stems slender, wiry, 6in.–12in. high, glabrous. Leaves entire, narrow-linear, 1in.–1½in. long, mostly alternate or the upper opposite. Flowers in slender axillary naked 4–5-flowered racemes. Sepals narrow lanceolate-acute, shorter than the lateral petals. Lower petals about ½in.

long, narrowed into a concave claw, saccate at the base; lateral petals very short. Anthers with 2 minute subulate processes in the cells.

NORTH Island: naturalised in grassy places. Lake Takapuna, Auckland, *Miss Rolleston!*

2. MELICYTUS, Forst.

Flowers regular, dioecious. Sepals 5, coherent at the base. Petals 5, spreading. Anthers 5, free, sessile; connective produced into a short membranous expansion with a minute dorsal scale. Stigma discoid, almost sessile or style 3–5-fid. Berry with 3–6 placentas. Seeds few or several, angled or minutely tuberculate. Shrubs or trees with alternate petioled minutely-stipulate leaves.

A small genus, of which all the species are endemic in New Zealand except one, which extends to Norfolk Island and the Tonga Islands.

ETYM. From the Greek, signifying *honey* and *a cavity*, in reference to the scales behind the anthers, termed "nectaries" by Forster.

Leaves oblong or oblong-lanceolate, serrate .. 1. *M. ramiflorus.*
Leaves obovate, large, obscurely serrate .. .. 2. *M. macrophyllus.*
Leaves linear-lanceolate, sharply serrate .. .. 3. *M. lanceolatus.*
Leaves small, orbicular, ovate, sinuate .. .. 4. *M. micranthus.*

1. **M. ramiflorus**, *Forst., Char. Gen. t.* 62. A large shrub or tree, 25ft.–30ft. high; trunk 1½ft.–2ft. in diameter; bark white. Leaves 3in.–5in. long, oblong-lanceolate or lanceolate, obtuse or shortly acuminate, serrate or obscurely serrate. Flowers in axillary fascicles or on the naked branches. Pedicels very slender, about ¼in. long, with minute bracts. Calyx-teeth minute, obtuse. Anthers obtuse. Stigma 5–6-lobed, sessile. Berry violet-coloured, ⅛in. in diameter. Seeds few.—Prod. n. 371; Lamk. Ill. t. 812, f. 1; DC., Prod. i. 257; A. Rich., Fl. N.Z. 313; A. Cunn., Precurs. n. 623; Raoul, Enum. 48; Hook. f., Fl. N.Z. i. 18, and Handbk. 17; Gray, Bot. U.S. Expl. Exped. 97; T. Kirk, Forest Fl. N.Z. t. 3. *Tachytes umbellifera*, Banks and Sol. MSS.

NORTH and SOUTH Islands; STEWART Island; THREE KINGS Islands: common. Ascends to 3,000ft. *Mahoe; hinahina.* Dec., Jan. Also in Norfolk Island and the Tonga Islands. The twigs are very brittle.

2. **M. macrophyllus**, *A. Cunn., Precurs. n.* 624. A sparingly-branched shrub, 6ft.–15ft. high. Leaves obovate or oblong, 3in.–5in. long, acute, coriaceous, obscurely sinuate, serrate. Male: Flowers in 4–8-flowered fascicles; pedicels decurved, with 1 or 2 bracts just beneath the flower; calyx-lobes broadly rounded; petals strap-shaped, spreading, more than twice as long as the calyx; anthers apiculate. Female: Petals scarcely exceeding the calyx; style short, stout, discoid, 2–3-lobed. Berry ⅓in. in diameter. Style persistent. Seeds about 4, angled, smooth.—Raoul, Enum. 48; Hook. f., Fl. N.Z. i. 19, and Handbk. 17; Gray, Bot. U.S. Expl. Exped. 97.

NORTH and SOUTH Islands: not uncommon in lowland situations in the Auckland District. Only known elsewhere from Waikari Creek, near Dunedin. *G. M. Thomson!* Ascends to 1,500ft. Attains its greatest luxuriance in the Kaipara. Sept., Oct.

Larger in all its parts and more robust than *M. ramiflorus*; pedicels longer and stouter.

3. **M. lanceolatus,** *Hook. f., Fl. N.Z.* i. 18, t. 8. A slender shrub, 5ft.–12ft. high. Branches usually few, brittle. Leaves 3in.–6in. long, narrow linear-lanceolate or oblong-lanceolate, acute or acuminate, membranous, sharply serrate. Flowers in 2–4-flowered fascicles; pedicels slender, decurved, with bracts above the middle. Calyx-lobes oblong. Petals erect, recurved at the apex. Connective elongated into a flat subulate point. Style long; stigmas 3. Berry spherical, violet-coloured. Seeds about 10, angled, minutely tuberculate.—Handbk. 17.

NORTH and SOUTH Islands; STEWART Island: from Whangarei southwards. Chiefly in lowland districts, but often local. Ascends to upwards of 2,000ft. Plentiful in Westland. Oct., Nov.

Easily distinguished by the narrow leaves, 3-fid style, and minutely tuberculate seeds. The leaves sometimes attain the extreme length of 10in.

4. **M. micranthus,** *Hook. f., Fl. N.Z.* i. 18. A shrub or small tree, 6ft.–15ft. high; trunk 4in.–5in. in diameter. Branchlets numerous, crowded, forming a compact head, tortuous or straight, often pubescent at the tips or almost setose. Leaves scattered, small, coriaceous, oblong or orbicular-ovate, sometimes contracted in the middle, sinuate, often with minute teeth, narrowed into the short puberulous petiole, rarely lobed. Flowers minute, inconspicuous, axillary. Peduncles solitary, pubescent, longer or shorter than the petioles, bracteolate. Calyx-lobes usually 5, ciliated, shorter than the spreading orbicular petals. Anthers broad, rounded; connective short, flat. Female: Anthers abortive; style short, stout; stigma 4-lobed. Berry minute, red. Seeds 2 or 3, scarcely angled.—Handbk. 17. *M. microphyllus*, Col., N.Z.I. xix. (1886) 260 and xx. 189. *Elaeodendron micranthus*, Hook. f. in Lond. Journ. Bot. iii. 228, t. 8.

NORTH and SOUTH Islands: from Whangarei to Otago. In lowland forests. Nov., Dec.

Easily distinguished by its small leaves, rigid habit, and minute 2–3-seeded berries. In general appearance it resembles states of *Melicope simplex*, *Myrsine divaricata*, and *Panax anomalum*. The female flowers have imperfect anthers. In the toughness of the branches and the form of the leaves it approaches *Hymenanthera*.

3. HYMENANTHERA, R. Br.

Flowers regular, perfect or unisexual. Sepals 5, coherent at the base. Petals 5, short, rounded. Anthers sessile, connate into a tube; connective produced into a crested or fimbriate membrane with a scale at the back. Ovary 1-celled, with a short style and 2-fid stigma. Fruit a 2- rarely 3- or 1-seeded berry. Placentas 2. Seeds with copious endosperm, angled on the inner face. Shrubs. Leaves small, alternate, minutely stipulate.

A small genus, extending to Tasmania, New South Wales, and Norfolk Island.

Branchlets numerous. Leaves smooth. Flowers fascicled.

Bark white. Leaves linear-obovate, often fascicled .. .. .. 1. *H. crassifolia.*
Bark dark. Leaves linear-oblong .. .. 2. *H. dentata*, var. *angustifolia.*

Branchlets few. Leaves smooth.

Bark green. Leaves obovate or oblong, solitary .. .. 3. *H. obovata.*
Bark red. Leaves oblong-obovate. Flowers solitary .. .. .. 4. *H. Traversii.*

Branches few. Leaves reticulate on both surfaces.

Leaves ovate-lanceolate, sinuate. Flowers perfect .. .. .. 5. *H. latifolia*.
Leaves lanceolate, sharply toothed. Flowers dioecious .. .. .. 6. *H. Chathamica*.

1. **H. crassifolia,** *Hook. f., Fl. N.Z.* i. 17, *t.* 8. A low rigid spreading shrub, with short stout tortuous branches. Bark white, furrowed. Branchlets pubescent. Leaves alternate or fascicled, very coriaceous, linear-spathulate or linear-obovate, ⅓in.–½in. long, entire or sinuate, toothed or lobed, rounded or retuse; petioles very short. Flowers small, axillary, solitary or in small fascicles. Peduncles shorter than the flowers, curved, with 2 ovate bracts below the middle. Sepals much shorter than the petals, minutely erose. Petals linear-oblong, recurved at the apex. Anthers forming a tube round the ovary; connective with a dorsal scale, fimbriate. Ovary 1-celled; placentas 2. Berry purple and white, 2-seeded. Seeds 2, convex on the outer face.—Handbk. 18. *Scaevola? Novae-Zelandiae*, A. Cunn., Precurs. n. 429.

NORTH Island: maritime rocks opposite the Cavallos Islands, *R. Cunningham*. Northern shore of Cook Strait, from Cape Terawhiti to Cape Palliser, &c., Port Nicholson, *T. K.* SOUTH Island: Nelson: coast between the Boulder Bank and Croixelles Harbour, *T. K.* Marlborough: Pelorus Sound, &c., *J. Rutland!* Canterbury: coast of Banks Peninsula, *J. B. Armstrong!* Otago, *Petrie*. STEWART Island, *T. K.* Oct., Nov.

Erect specimens with broader leaves are sometimes developed in sheltered situations.

2. **H. dentata,** *R. Br.*, var. **angustifolia,** *Benth., Fl. Austr.* i. 104. A much-branched leafless shrub, 2ft.–8ft. high. Branchlets terete, intricate, about as thick as whipcord, occasionally spinescent, closely dotted with minute lenticels. Leaves on young plants shortly petioled, narrow-linear-oblong, entire, lobed or pinnatifid, often fascicled, cuneate at the base, membranous or subcoriaceous, entire or sinuate or shortly lobed, rounded at the apex, ½in.–¾in. long. Flowers perfect or dioecious, almost sessile, solitary or germinate. Male flowers not seen. Female almost sessile; petals narrow; anthers abortive. Style short; stigmas 2, spreading. Berry 2-seeded. Seeds oblong, flat on the inner face, convex on the outer, with a small discoid strophiole.—*H. angustifolia*, R. Br. in DC. Prod. i. 315; Hook., Comp. to Bot. Mag. i. 274; Hook. f., Fl. Tasm. i. 27. *H. Banksii*, F. Muell., Pl. Vict. i. 69.

NORTH Island: Wellington: Turangarere, *A. Hamilton!* Upper Rangitikei, *Petrie!* SOUTH Island: Nelson: Wairoa Valley, *Bryant!* Canterbury: Alps, *J. B. Armstrong!* Otago: Paradise, Mount Earnslaw, *T. K.* Nov.

Easily distinguished by the lenticillate bark, slender leafless twigs, which are flexuous when grown in sheltered places, and the dioecious flowers. The Tasmanian plant is said to produce hermaphrodite flowers. At present I have only had the opportunity of examining the female flowers of the New Zealand plant.

Var. **alpina.** Depressed, 1ft.–2ft. high. Branches very short, rigid, stout, usually terminating in a stout spine. Bark whitish, lenticellate. Leaves less than ⅓in. long, usually fascicled, very coriaceous, oblong-obovate; petioles very short. Flowers on very short straight peduncles. Sepals erosulate-ciliate. Petals broad, recurved. Anthers very broad, connective much produced, nearly entire; dorsal scale broadly cuneate. Style slender; stigmas 2, spreading. Fruit not seen. SOUTH Island: Alps of Canterbury and Otago. 2,000ft. to 4,000ft. A remarkable plant, forming a mass of very short stout spinous branches. Possibly a distinct species. Dec., Jan.

3. **H. obovata,** *T. Kirk in Trans. N.Z.I.* xxvii. (1894) 350. An erect glabrous shrub, 4ft.–8ft. high. Branches few, slender, ascending. Bark pale.

Leaves in the young state obovate-cuneate, 3-lobed or toothed, membranous; mature very coriaceous, 1in.–2in. long, obovate or oblong, narrowed into a slender petiole below, rounded or retuse above, rarely apiculate; margins slightly recurved, rarely entire. Flowers: Male not seen. Female: Sepals broadly ovate, rounded at apex. Stigmas 2. Fruits ovoid, solitary or twin, on very short curved peduncles, purple, 2-seeded. Seeds nearly ovate, slightly concave on the inner face, convex on the outer. Strophiole cupular, thin.

SOUTH Island: *Buchanan!* Marlborough: Queen Charlotte Sound, *Banks and Solander!* Nelson: Graham River, Mount Owen, *Cheeseman!* Between Takaka and Riwaka, *T. K.* Canterbury: Broken River Basin, *J. D. Enys* and *T. K.* (1876). Ashburton mountains, *T. H. Potts!* Chiefly on limestone rocks. 2,000ft. to 4,000ft.

The rather slender branches and the strict habit distinguish this species from all others.

4. **H. Traversii,** *Buch. in Trans. N.Z.I.* xv. (1882) 339, *t.* 28. A spreading shrub, 1ft.–2ft. high. Twigs with reddish longitudinally-rugose bark, viscid when fresh. Leaves rather crowded, coriaceous, oblong-obovate or oblong-spathulate, about 1in. long, narrowed into a rather stout appressed petiole, obtuse or subacute; margins recurved; nerves obscure. Flowers few, solitary on short decurved pedicels in the axils of the upper leaves. Sepals coherent at the base, subacute. Petals linear-oblong, narrowed below, spreading. Fruit not seen.

SOUTH Island: Nelson: in the forest near Gouland Downs, near Collingwood, *H. H. Travers.*

My knowledge of this plant is confined to Mr. Buchanan's description, and to a small flowerless specimen for which I am indebted to him. It is distinguished from all other species by the red bark and the rugose leaves with appressed peduncles. Good specimens in flower and fruit are much to be desired, as the anthers and ovary are not mentioned in the original description.

5. **H. latifolia,** *Endlich., Fl. Insul. Prod. Norf. n.* 127. A sparingly branched shrub, 2ft.–10ft. high, erect or straggling. Leaves ovate-lanceolate or oblong-lanceolate, narrowed into the petiole, 2in.–4in. long, 1in. broad, obtuse, coriaceous, entire, sinuate or sinuate-serrate; marginal nerve stout. Flowers not seen. Sepals ovate in the fruiting state, scarcely coherent at the base. Fruiting peduncles very short, erect or curved. Berry 2-seeded. Seeds ovoid, flat on the inner face, with irregular longitudinal striae on the outer convex surface. Strophiole large, cupular.—*H. latifolia*, var. *Tasmanica*, T. K. in Trans. N.Z.I. iii. 163.

NORTH Island: Auckland: Tapotopoto Bay, *T. K.* Mount Camel, *Buchanan.* Whangapoua and Flat Island, Taranga Islands, Great Barrier Island, Arid Island, Little Barrier Island, Waiheki Island, *T. K.* Three Kings Islands, *Cheeseman.* Also on Norfolk Island.

The New Zealand plant has not been seen in flower. Endlicher describes the Norfolk Island plant as having hermaphrodite and female flowers, the latter with abortive stamens. He also states that the ovary is 2-celled, and the stigma capitate: both statements appear to be erroneous. The leaves are strongly reticulate on both surfaces.

6. **H. Chathamica,** *T. Kirk in Trans. N.Z.I.* xxviii. (1895) 514. An erect shrub with furrowed lenticellate bark. Leaves lanceolate or oblong-lanceolate, narrowed at the base, acute, 3in.–5in. long, ¾in.–1½in. broad, very coriaceous, strongly reticulate on both surfaces, sharply toothed. Flowers in crowded fascicles, dioecious. Pedicels slender, longer than the flowers, decurved.

Sepals coherent at the base, narrow-ovate. Petals very long, obovate with a broad base, revolute at the apex. Anthers with a narrow lanceolate jagged connective, more than half as long as the anther-cells; dorsal gland cuneate-spathulate, rounded above. Female flowers not seen. Berry ovoid or almost globose, white, 4- rarely 3-seeded. Stigmas 4-lobed. Seeds angled, curved towards the point, convex on the outer surface; strophiole very small.—*H. latifolia*, var. *Chathamica*, F. Muell., Veg. Chat. Isds. 9.

NORTH Island: Wellington: Patea, *Hector!* (flowers and fruit not seen). CHATHAM Islands: originally discovered by Captain Gilbert Mair. *Mahoe.* Sept., Oct.

Distinguished from all other species by the long lanceolate sharply-toothed leaves, strictly dioecious flowers, tetramerous stigma-lobes, and 4-seeded berries. Occasionally the leaves are linear-lanceolate and less than ½in. in breadth.

ORDER V.—PITTOSPOREAE.

Flowers regular, perfect, rarely unisexual. Sepals 5, imbricate. Petals 5, imbricate, the base narrowed into a claw, forming a tube, the limb recurved or spreading. Stamens 5, hypogynous, free. Ovary 1-celled or imperfectly 2–5-celled, with 2–5 parietal placentas. Styles simple; stigmas 2–5-lobed. Ovules in 2 rows on each placenta. Fruit a capsule or berry. Seeds usually several in each cell. Endosperm hard. Embryo next the hilum, minute. Herbs, shrubs, or trees.

A small order, all the genera of which, except *Pittosporum*, are restricted to Australia. GENERA, 9. SPECIES, 110.

1. PITTOSPORUM, Banks.

Sepals 5, free. Petals 5, free, connivent at their base, usually recurved. Stamens pentandrous. Ovary sessile or rarely stipitate, 1-celled or imperfectly 2–5-celled, silky. Capsule globose obovate or oblong, with 2–5 woody valves. Seeds globular or angled, immersed in a viscous fluid. Shrubs or trees with entire exstipulate often whorled leaves and branches.

SPECIES, about 70, distributed through Africa and the warmer parts of Asia, the Pacific Islands, Australia, and New Zealand. The genus is more largely developed in the North Island than elsewhere. All the New Zealand species are endemic.

ETYM. From the Greek, signifying *pitch*, and *a seed*, in allusion to the viscid secretion in which the seeds are immersed.

A. *Flowers axillary, solitary, rarely fascicled and terminal. Peduncles very short, except in 4.*

Leaves oblong, 1in.–1½in. long. Capsule 3-valved .. .. .. 1. *P. tenuifolium.*
Leaves oblong, coriaceous. Flowers solitary or fascicled .. .. var. *fasciculatum.*
Leaves elliptic-oblong, subcoriaceous, 2in.–5in. long .. .. .. 2. *P. Buchanani.*
Leaves broadly oblong, 2in.–5in. long. Capsule 3-valved .. .. 3. *P. Huttonianum.*
Leaves broadly obovate, rounded. Flowers solitary or fascicled. Capsules terminal .. .. 4. *P. intermedium.*
Leaves small, rounded or obcordate .. .. 5. *P. obcordatum.*
Leaves small, narrow-obovate, lobed or entire .. .. 6. *P. rigidum.*

B. *Flowers in terminal umbels or fascicles.*

Leaves narrow linear-acuminate or linear-oblong .. .. .. 7. *P. pimeleoides.*
Leaves narrow-obovate, whorled. Epiphytic .. .. 8. *P. cornifolium.*
Leaves narrow linear-obovate. Capsules elliptic, large .. .. 9. *P. Kirkii.*
Leaves broadly lanceolate-oblong. Capsules 4-lobed .. .. .. 10. *P. umbellatum.*

Leaves linear-obovate or oblong. Capsule rounded	var. *cordatum*.
Leaves narrow linear-oblong, pinnatifid, lobed. Capsule globose, 2-valved	11. *P. patulum*.
Leaves linear-lanceolate or elliptic, ferruginous	12. *P. virgatum*.
Leaves broadly oblong, obovate, white beneath. Capsule 3-valved ..	13. *P. Ralphii*.
Leaves narrow-obovate, thick. Capsule large, decurved	14. *P. crassifolium*.
Leaves elliptic-obovate. Capsule corrugated and pitted	15. *P. Fairchildii*.
Leaves elliptic or oblong-lanceolate, ferruginous	16. *P. ellipticum*.
Leaves obovate-cuneate, ferruginous	var. *ovatum*.

C. *Flowers in compound terminal umbels or corymbs.*

A tree with white bark. Petals linear	17. *P. eugenioides*.

1. **P. tenuifolium,** *Banks and Sol. in Gaert. Fruct.* i. 286, *t.* 59, *f.* 7. A shrub or small tree, 20ft.–40ft. high, with black bark. Leaves and twigs mostly pubescent when young. Leaves 1in.–3in. long, pale beneath, shortly petioled, lanceolate oblong or elliptic-obovate, obtuse, acute or acuminate, often undulate, submembranous or coriaceous. Flowers pedunculate, axillary, solitary or rarely fascicled. Ovary silky. Capsule downy when young, 3-valved, glabrous and minutely rugose when mature.

An extremely variable plant, of which the chief forms pass so gradually into each other that it is impossible to draw sharp lines of distinction.

Var. **tenuifolium,** *verum*. Branchlets very slender. Leaves 1in.–1½in. long, lanceolate, oblong or elliptic, obovate, submembranous, often undulate; petioles slender. Flowers solitary. Peduncles straight, about as long as the calyx, usually pubescent. Sepals ovate, oblong, or almost subulate, obtuse or rarely acute, silky. Capsule broadly obovoid, valves rather thin.—DC., Prod. i. 347; A. Cunn., Precurs. n. 615; Putterlich, Syn. Pitt. 14; Hook. f., Fl. N.Z. i. 21, and Handbk. 19; A. Gray, Bot. U.S. Expl. Exped. 222; T. Kirk, For. Fl. N.Z. t. 46. *Trichilia monophylla*, A. Rich., Fl. Nov.-Zel. 306, t. 34, *bis*. NORTH and SOUTH Islands: common. Ascends to 3,000ft. Oct., Nov. Very variable in the form of leaves and sepals. The leaves are remarkable for their pale-green colour, especially in the young state.

Var. **colensoi,** T. Kirk in Trans. N.Z.I. iv. (1871) 262. Larger and more robust than *tenuifolium*. Branchlets stouter. Leaves 1½in.–3in. long, coriaceous, often waved, lanceolate, broadly oblong-lanceolate, more acute, rarely obtuse. Peduncles very short, usually decurved, glabrous or pubescent, the scarious bracts often persistent. Sepals usually oblong, glabrous or pubescent. Capsule more globose, and valves much thicker; inner reticulations stronger.—*P. Colensoi*, Hook. f., Fl. N.Z. i. 22, and Handbk. 19. NORTH and SOUTH Islands; STEWART Island: from the East Cape southward. Ascends to 3,500ft. The larger-leaved forms are most frequent in the South.

Var. **fasciculatum,** T. Kirk in Trans. N.Z.I. iv. (1871) 262. Habit of *Colensoi*. Leaves usually elliptic-oblong or elliptic-obovate, acute or rarely obtuse, coriaceous. Flowers in terminal or lateral fascicles or solitary. Peduncles usually short, glabrate or silky. Sepals ovate or oblong, silky or tomentose, obtuse or acute. Valves of capsules very stout.—*Sp.* Hook. f., Handbk. 20. NORTH and SOUTH Islands: from Rotorua to Preservation Inlet, but often local.

Sub-var. **cymosum.** Leaves oblong-obovate or elliptic-obovate, subacute, more membranous. Flowers in terminal cymes. Pedicels longer. Sepals longer, lanceolate-subulate, tomentose. Capsule not seen. NORTH Island: Anaura, East Cape (with large leaves); Takeke, Waikare (leaves small). *Bishop Williams!*

2. **P. Buchanani,** *Hook. f., Handbk.* 725. A shrub, 8ft.–12ft. high, with slender spreading or ascending branches; glabrous except the young shoots and leaves, which are silky-pubescent. Leaves scattered, spreading, submembranous, elliptic-oblong or lanceolate or broadly oblong, 2in.–5in. long, ½in.–1½in. broad, acuminate or acute, flat; petioles slender. Flowers axillary, solitary, perfect or unisexual, often polygamous. Peduncles very slender, ascending, usually solitary, glabrous or silky-pubescent, 1- rarely 2-flowered; bracts small, acute. Sepals ovate-oblong, acute or obtuse. Petals linear, with

a long claw. Filaments and ovary silky-pubescent. Capsules 3-valved on spreading peduncles.

NORTH Island: Auckland: Kaitaia and Mongonui, *Buchanan!* Taranaki: near Mount Egmont, *Hector*. Near Wellington, *T. K.* Oct.

Closely related to *P. tenuifolium*, from which it is best distinguished by the longer corolla-tube and stamens, the small acute bracts, long peduncles, small capsules, and spreading leaves.

3. **P. Huttonianum,** *T. Kirk in Trans. N.Z.I.* ii. (1869) 92. A laxly-branched shrub or small tree, 12ft.–25ft. high. Bark black. Branchlets slender, and with the young leaves clothed with white floccose tomentum. Leaves broadly oblong or elliptical-oblong, rarely broadly obovate, obtuse or acute, rarely acuminate, 3in.–5in. long, coriaceous; petioles slender. Flowers axillary, solitary or rarely in terminal cymes. Peduncles erect, longer than the sepals. Sepals oblong or lanceolate, acute, downy, often spreading. Fruiting peduncles decurved. Capsules large, 3- or rarely 2-valved, globose or almost pyriform, fully ½in. in diameter, downy or rarely glabrous.

NORTH Island: Auckland: Great and Little Barrier Islands, Cape Colville Peninsula, &c., *T. K.*, *Cheeseman!* Oct.

Mr. Cheeseman's specimens approach the type more closely than the inland and southern form.

Var. **fasciatum.** Flowers in terminal cymes. Peduncles white with floccose tomentum. NORTH Island: Kauaeranga Creek, Thames, *T. K.*

Var. **viridifolium.** Branchlets very slender. Leaves oblong-obovate, cuneate below, acute, scarcely acuminate, glabrous. Flowers axillary, solitary. NORTH Island: Auckland: Rotorua, &c., *T. K.* Taranaki: Urenui, *Cheeseman!* SOUTH Island: Milford Sound, &c., *T. K.*

4. **P. intermedium,** *T. Kirk in Trans. N.Z.I.* iv. (1871) 266. A small tree with black bark, in habit and foliage resembling large specimens of *P. tenuifolium*. Young shoots, leaves, and sepals pubescent. Leaves 1½in.–2in. long, broadly obovate, rounded at the apex or shortly acuminate, narrowed into rather long slender petioles, membranous. Flowers terminal or confined to the axils of the uppermost leaves, solitary or in 2–3-flowered clusters. Pedicel very short. Sepals oblong, acute or obtuse. Petals oblong, spathulate, recurved. Fruiting peduncles always terminal, decurved, usually solitary. Capsule ovate-acuminate, downy, 2–3-valved; valves 2-lobed.

NORTH Island: Kawau Island, *T. K.* Oct.

Best distinguished from all forms of *P. tenuifolium*, which it resembles in general appearance, by the decurved terminal fruiting peduncles and larger capsules.

5. **P. obcordatum,** *Raoul in Ann. Sc. Nat. ser.* 3, ii. (1844) 121. A glabrous shrub or small tree, sometimes 15ft. high, with divaricating branches. Bark pale. Leaves small, distant, rarely 2 or 3 together, rounded or obcordate, entire or crenate, ½in. long, abruptly contracted into a very short petiole. Flowers axillary, solitary or geminate, campanulate. Peduncles very short, puberulous. Sepals very short, ovate-lanceolate, acute, ciliate. Petals linear-lanceolate, with spreading tips. Ovary glabrous. Capsule not seen.—Raoul, Choix de Plantes 25, t. 26; Hook. f., Fl. N.Z. i. 23, and Handbk. 20.

SOUTH Island: woods near Akaroa, *Raoul.* Dec.

This singular plant has not been collected by any living botanist.

6. **P. rigidum,** *Hook. f., Fl. N.Z.* i. 22, *t.* 10. A rigid depressed much-branched shrub, with short stout tortuous intricate spreading branches, or rarely with slender erect branches, 5ft.–6ft. high. Leaves $\frac{1}{8}$in.–$\frac{3}{4}$in. long, often fascicled, submembranous or coriaceous, with recurved margins, shortly petioled, oblong, narrow-obovate or cuneate, entire or sinuate, dentate or deeply lobed both in the young and mature states, usually glabrous. Flowers solitary, axillary or terminating arrested branchlets, sessile or on very short peduncles, campanulate. Sepals very short, narrow-ovate, acute. Ovary hirsute. Capsule small, broadly ovoid, acute, 2-valved, pilose when young.—Handbk. 20.

NORTH Island: mountains above Waikare Moana, and Ruahine Range, *Colenso.* Tararua Range, *Arnold! H. Travers!* SOUTH Island: Mount Stokes, *Macmahon!* Matai Creek, *F. H. Spencer!* Dun Mountain Range, *Cheeseman.* Lake Guyon, *Travers!* Lake Grasmere, *T. K.* Acheron Island, Dusky Bay, *Hector* and *Buchanan.* Ascends to fully 3,000ft., but is rare at low levels. Nov., Dec.

In the Tararua specimens the young shoots and leaves are clothed with ferruginous pubescence.

7. **P. pimeleoides,** *R. Cunn. in A. Cunn. Precurs. n.* 618. A slender much-branched shrub, 1ft.–5ft. high. Branchlets almost filiform, sparingly pilose in the young state. Leaves usually crowded or verticillate, ascending patent or recurved, $\frac{3}{4}$in.–1$\frac{1}{2}$in. long, $\frac{1}{12}$in.–$\frac{1}{4}$in. broad, membranous, linear-oblong, obtuse or acute, rarely denticulate. Flowers in few-flowered terminal umbels or solitary. Peduncles capillary, loosely pilose. Sepals subulate-acuminate. Petals linear-acuminate, tips spreading. Ovary silky. Capsule erect, ovoid-acuminate, 2-valved; style recurved.—*P. pimeleoides,* R. Cunn., Hook. f., Fl. N.Z. i. 24, and Handbk. 21.

Var. **reflexum.** Leaves usually crowded, linear-lanceolate, acuminate, $\frac{1}{12}$in.–$\frac{1}{8}$in. broad.—*P. pimelioides,* R. Cunn. in A. Cunn. Precurs. n. 617. *P. radicans,* R. Cunn. *l.c.* n. 617. *P. pimelioides* var. *reflexum,* Hook. f., Fl. N.Z. i. 25. *P. reflexum,* Hook. f., Handbk. 19. *P. Gilliesianum,* T. Kirk in Trans. N.Z.I. i. (1868) 143. *P. crenulatum,* Patt. Synop. Pittosp. 15.

NORTH Island: rare and local. Auckland: Mongonui, &c., *T. K.* Whangaroa, *R. Cunningham!* Kawakawa River, Bay of Islands, *Hooker f.* Both forms usually intermixed. March.

A plant which must probably be referred here grows with the ordinary forms at the Bay of Islands, but is characterized by elliptic leaves, rounded at the apex, and narrowed below into a slender petiole. Flowers and fruit not seen.

8. **P. cornifolium,** *A. Cunn. in Bot. Mag. t.* 3161, *Precurs. n.* 616. Shrub, 2ft.–4ft. high, glabrous except the young leaves and peduncles, which are silky-pubescent. Branches forked or verticillate. Leaves verticillate, very shortly petioled, coriaceous, narrow-obovate or elliptic-lanceolate, acute, 1$\frac{1}{2}$in.–2$\frac{1}{2}$in. long. Flowers often unisexual, in 3–4-flowered terminal umbels; umbels solitary or fascicled. Peduncles filiform, drooping. Sepals linear, subulate, exceeding the corolla-tube, acuminate. Petals saccate at the base, spreading at the tips, linear-acuminate. Capsule erect, broadly obovoid, 3-lobed and valved; valves orange inside. Seeds large.—Hook. f., Fl. N.Z. i. 23, and Handbk. 21; A. Gray, Bot. U.S. Expl. Exped. 223. *Pittosporoides verticillata,* Banks and Sol. MSS.

NORTH Island: most plentiful on the eastern side. Epiphytal. Ascends to 2,500ft. SOUTH Island: Pelorus Sound and Titi Island, *J. Rutland! Karo.* Aug., Sept.

An elegant species, growing amongst the masses of *Astelia,* &c., common on rocks or trees in the northern forests. Never truly terrestrial.

9. **P. Kirkii,** *Hook. f. ex T. Kirk in Trans. N.Z.I.* ii. (1869) 92. A glabrous loosely-branched shrub, 3ft.–15ft. high. Branches stout, ascending. Bark reddish-purple. Leaves erect, crowded or whorled, narrow-linear, obovate, acute or obtuse, 2in.–5in. long, narrowed into short broad petioles, excessively coriaceous. Flowers in terminal 3–7-flowered umbels. Sepals lanceolate, acuminate, slightly exceeding the corolla-tube. Petals narrow-ligulate, acuminate, the sharply-recurved limb longer than its tube. Styles very short; stigma capitate, 2-lobed. Fruiting peduncles stout, erect. Capsules elliptic, 1½in. long, 2- rarely 3-valved, glabrous, compressed, cuspidate.—Trans. N.Z.I. iv. (1872) 264.

NORTH Island: Auckland: Whangape North. Kauaeoruruwahine forests, Whangape North, *T. K.* Bay of Islands, *A. Cunningham!* Whangarei, &c., *Buchanan.* Hirakimata Range, Great Barrier Island; Great Omaha and Cape Colville Peninsula, *T. K.* Titirangi, *Cheeseman.* Thames Goldfield, *J. W. Hall!* Taranaki: Mount Egmont ranges, *Cheeseman!* Ascends to nearly 3,000ft. Dec., Jan.

The most striking and beautiful of the New Zealand species. North of Hokianga epiphytal specimens may be seen several yards in diameter.

10. **P. umbellatum,** *Banks and Sol. in Gaertn. Fruct.* i. 286, t. 59. A shrub or small tree, 10ft. to 20ft. high, glabrous except the young shoots, leaves, peduncles, and sepals, which are more or less silky. Branches sometimes whorled. Leaves coriaceous, broadly lanceolate, oblong, elliptic-oblong or obovate-oblong, 2in.–4in. long, narrowed into slender petioles, obtuse or acute. Flowers in many-flowered terminal umbels. Peduncles filiform. Sepals ovate-lanceolate. Petals slightly recurved. Fruiting peduncles slender, decurved. Capsule tetragonous, 2-valved; valves 2-lobed, woody.—DC., Prod. i. 347; A. Cunn., Precurs. n. 613; Putterl. Syn. Pittosp. 12; Hook. f., Fl. N.Z. i. 24, and Handbk. 21.

NORTH Island: North Cape to Poverty Bay. Littoral. Oct., Nov.

Var. **cordatum,** T. Kirk, Trans. N.Z.I. iv. 264. Leaves linear-obovate or obovate-spathulate, acute, narrowed into the petiole. Capsules rounded, cordate-acuminate; valves not lobed. Flowers not seen. Haratoanga, Great Barrier Island. Easily distinguished from the type by the narrow leaves and rounded capsules.

11. **P. patulum,** *Hook. f., Handbk. N.Z. Fl.* 19. A glabrous bush or erect shrub, 3ft.–15ft. high. Branchlets stout. Leaves in the young state 1in.–2in. long, $\frac{1}{12}$in.–⅛in. broad, closely pinnati-lobed or -fid, gradually passing into the narrow linear-oblong serrate crenate or quite entire coriaceous mature state, narrowed into very short petioles, shining. Flowers in 4–6-flowered terminal umbels. Sepals and petals not seen. Peduncles slender, ¾in. long, spreading or decurved in fruit. Capsules about ¼in. in diameter, globose, compressed, 2-valved, finely granulated.

SOUTH Island: Nelson: Wairau Mountains, *Sinclair*, 5,000ft. Lake Guyon, *Travers!* Glacier Gully, Spencer Mountains, *T. K.!* 2,800ft. A very distinct species, and, so far as known, remarkably local.

12. **P. virgatum,** *T. Kirk in Trans. N.Z.I.* iv. (1871) 264. A slender tree, 20ft.–25ft. high. Bark black. Young shoots, leaves, peduncles, and sepals clothed with ferruginous pubescence. Leaves linear-lanceolate or elliptic-

lanceolate or narrow-obovate, entire or deeply lobed or pinnatifid, especially in the young state. Flowers terminal in 2–3-flowered umbels or solitary. Sepals linear-lanceolate, acute, fully equalling the corolla-tube. Petals shortly recurved. Capsules erect, globose, slightly depressed, 2-valved, hirsute when young, glabrous in the mature state.—Trans. N.Z.I. ed. 2 (1875) 90.

Var. **sinuatum**. Leaves broader, elliptic-oblong or nearly obovate, acute, sinuate.

NORTH Island: Whangaroa North and Great Barrier Island, *T. K.* Kennedy Bay, &c., *Cheeseman!* Oct.

Most readily distinguished by the slender rather spreading twigs, ferruginous pubescence, linear acute sepals, and globose depressed capsule.

13. **P. Ralphii,** *T. Kirk in Trans. N.Z.I.* iii. (1870) 161. A laxly-branched shrub, 8ft.–14ft. high, with slender spreading branches. Leaves spreading on long slender petioles, broadly oblong-obovate, 3in.–5in. long, 1in.–2in. broad, obtuse or acute, coriaceous, white with appressed tomentum beneath. Flowers in terminal 3–8-flowered umbels. Sepals ovate-acuminate, tomentose. Petals spreading. Fruiting peduncles slender, not decurved. Capsules ovoid, depressed, 3-valved; valves lobed, downy.—*P. crassifolium*, Banks and Sol. MSS.

NORTH Island: Tolaga Bay, *Banks and Solander!* Patea, *Dr. Ralph!* Upper part of the Whanganui River, *H. C. Field!* Petane River, *A. Hamilton!* Between Cape Kidnappers and Pourerere, *T. K.* Nov.

Best distinguished from *P. crassifolium* by the broad slender leaves, shorter peduncles, and smaller capsules with less woody valves.

14. **P. crassifolium,** *A. Cunn., Precurs. n.* 612. A shrub or tree, sometimes exceeding 30ft. in height. Bark black. Branches strict, erect. Young shoots, leaves below, peduncles, and sepals white with appressed tomentum. Leaves 2in.–3in. long, narrow-obovate or oblong, obtuse, very coriaceous; margins recurved; petioles short, stout. Inflorescence in terminal 2–5-flowered umbels, often solitary. Sepals linear-oblong, tomentose. Petals recurved. Fruiting peduncle strongly decurved. Capsule subglobose, 3- rarely 2- or 4-valved; valves lobed, downy.—Putterl. Syn. Pittosp. 12; Hook. f., Fl. N.Z. i. 23, and Handbk. 20; T. Kirk, Forest Fl. N.Z. t. 14.

NORTH Island: littoral. KERMADEC Islands, *Cheeseman.* North Cape to Poverty Bay. Most plentiful on the East Coast and outlying islands. *Karo.* Sept.

Var. **strictum**. Leaves narrow-oblong or oblong-obovate. Peduncles strict, 1½in. long. Capsule smaller. Little Barrier Island, *T. K.* East Cape, *Bishop Williams!*

15. **P. Fairchildii,** *Cheeseman in Trans. N.Z.I.* xx. (1887) 147. A shrub, 8ft.–15ft. high, with slender branches. Young shoots, leaves, and peduncles clothed with loose white tomentum. Leaves often crowded, obovate elliptic-obovate or elliptic-oblong, obtuse or acute, narrowed into a very short petiole, coriaceous but not thick. Flowers terminal in 2–4-flowered fascicles. Pedicels slender, ½in.–¾in. long, pendulous. Sepals linear-oblong, acute, tomentose, much shorter than the corolla-tube. Petals shortly recurved. Fruiting peduncles decurved, 3-valved, ¾in.–1in. in diameter, shortly obovoid, depressed, broader than long, glabrous. Valves woody, finely corrugated and granulated.

THREE KINGS Islands: *Cheeseman!* Aug.

Distinguished from *P. crassifolium* by the broad flat leaves and the smaller depressed glabrous brown capsule; from *P. umbellatum* by the larger 3-valved capsule. It is nearly allied to *P. bracteolatum*. Endlich. of Norfolk Island.

16. **P. ellipticum,** *T. Kirk in Trans. N.Z.I.* iv. (1871) 266. A small tree with black bark, 20ft.–25ft. high. Young shoots, leaves, peduncles, sepals, and capsules clothed with ferruginous pubescence. Leaves 2in.–3½in. long, ¾in.–1in. broad, elliptic or oblong-lanceolate, coriaceous, subacute or acute, rarely obtuse. Flowers in terminal 2–5-flowered umbels. Peduncles short, spreading. Sepals narrow-ovate or oblong-lanceolate, acute, pilose. Petals shortly recurved. Capsule shortly decurved, 2-valved, ovate, globose, slightly compressed, fully ½in. in diameter; valves faintly 2-lobed.

NORTH Island: Whangaroa North, Mount Manaia &c., Whangarei, *T. K.* Waitakerei Ranges, *Cheeseman!* Oct.

Var. **ovatum.** Leaves more spreading, broadly elliptical or obovate-cuneate, broadly rounded at the apex. Flowers not seen. Whangaroa North, Manaia Hills, western part of the Titirangi district, *T. K.*

The midrib in the typical form is often curiously flattened beneath, but in var. *ovatum* it is very slender.

17. **P. eugenioides,** *A. Cunn., Precurs. n.* 614. A glabrous much-branched twiggy tree, 20ft.–40ft. high. Trunk 1ft.–2ft. in diameter. Leaves alternate or sometimes verticillate, 2in.–4in. long, ¾in.–1½in. broad, elliptical or broadly oblong, acute or subacute, often undulate, subcoriaceous. Flowers in terminal branched many-flowered compound umbels or corymbs, often unisexual. Pedicels and peduncles spreading, silky-pubescent. Sepals ovate, acute or acuminate. Petals linear, much longer than the sepals, recurved at the tips. Capsules ¼in. in diameter, erect, ovoid, acute, 2- rarely 3-valved.—Hook. f., Fl. N.Z. i. 22, and Handbk. 21; T. Kirk, Forest Fl. N.Z. t. 49. *P. elegans,* Raoul in Ann. Sc. Nat. ser. 3, ii. (1844) 121. *P. microcarpum,* Putterl., Syn. Pittosp. 15.

NORTH and SOUTH Islands: North Cape to Otago. *Tarata.* Sept., Oct.

The only New Zealand species with compound inflorescence. Flowers fragrant, varying greatly in size. Bark white.

ORDER—* POLYGALEAE.

Flowers hermaphrodite, irregular. Sepals 5, free, imbricate in bud, the two inner petaloid. Petals 2 or 4, rarely 3 or 5, hypogynous; usually one or two pairs coherent at the base, with the lower hooded petal or keel into a tube split behind, or the two outer petals free. Stamens 8, rarely 5 or 4; filaments forming a cleft sheath adnate to the petals. Anthers usually 1-celled, opening by pores. Disk small. Ovary superior, 2-celled; style simple, usually curved at the top; stigma entire or 2-lobed; ovules 1 in each cell, pendulous, anatropous. Seeds with fleshy endosperm, pendulous; hilum strophiolate; embryo straight. Herbs or shrubs with alternate exstipulate leaves.

* POLYGALA, Linn.

The 2 inner sepals large and petaloid, 2 outer green, smaller. Petals 3, united below with the staminal sheath, the keel bearing a crest at the back near the apex. Stamens 8; filaments coherent to above the middle; anthers opening by transverse pores. Capsule thin, flattened, usually notched at the apex, loculicidal. Seeds hairy downy or pubescent; rhaphe crested. Herbs or shrubs with terminal or lateral spicate or racemose inflorescence.

* **P. myrtifolia**, *L.*, *Sp. Pl.* 703. A shrub, 4ft.–8ft. high, with slender pubescent branchlets. Leaves obovate, slightly mucronate. Racemes 3-4-flowered. Pedicels short. Flowers rosy-purple. Keel falcate.

Naturalised near Auckland. Cape of Good Hope.

ORDER VI.—CARYOPHYLLEAE.

Sepals 4 or 5, free or coherent, imbricate in bud. Petals 4–5 or rarely 0, hypogynous or perigynous, imbricate or contorted in bud. Stamens not exceeding 10, inserted with the petals. Styles 2–5, rarely united. Ovary usually 1- rarely 3–5-celled at the base. Ovules 2 or many, on slender funicles attached to the base or to the central column of the ovary. Capsule many-seeded; valves as many or twice as many as the styles. Seeds with mealy endosperm; embryo usually curved, cylindrical. Herbs, rarely woody at the base, with tumid nodes and opposite entire leaves.

A large order, common in temperate and cold regions. GENERA, 35. SPECIES, 1,250. Many naturalised species are found in the colony.

Tribe I. SILENEAE.—Calyx 4–5-lobed. Petals and stamens inserted on the elongated disk. Styles free from the base. Stipules 0.

- * DIANTHUS. Calyx cylindrical, many-nerved, with 2 or more bracts at the base. Seeds peltate.
- 1. GYPSOPHILA. Calyx broadly or obscurely 5-nerved. Styles 2. Capsule deeply 4-valved.
- * SAPONARIA. Capsule shortly 4-valved. Styles 3.
- * LYCHNIS. Calyx 10-nerved. Styles 4 or 5.

II. ALSINEAE.—Sepals free or united at their base by the small disk. Styles free. Stipules 0.

- * CERASTIUM. Petals notched. Capsule cylindric.
- 2. STELLARIA. Petals notched. Capsule globose.
- * ARENARIA. Petals entire. Styles 3 or 4.
- * SAGINA. Petals entire or 0. Styles 4 or 5.
- 3. COLOBANTHUS. Petals 0. Styles as many as the sepals.

* *Stipules scarious.*

- * SPERGULA. Styles 5. Capsule 5-valved. Leaves clustered.
- 4. SPERGULARIA. Styles 3. Capsule 3-valved. Leaves opposite.

III. POLYCARPEAE.—Sepals free. Petals small or 0. Stamens not exceeding 5. Styles coherent at the base, short. Stipules scarious.

- * POLYCARPON.

* DIANTHUS, Linn.

Calyx cylindrical, 5-toothed, with 2 or more imbricate bracteoles at the base. Petals 5. Stamens 10. Disk elongated. Ovary 1-celled; styles 2. Capsule 4-valved at the apex. Seeds imbricate on the columnar placenta; embryo straight. More or less tufted herbs with fascicled flowers.

Stems few or solitary. Leaves narrow.. .. .. * *D. Armeria.*
Stems numerous. Leaves broad .. .. .. : .. .. * *D. barbatus.*

* **D. Armeria**, *L.*, *Sp. Pl.* 410. Erect, 1ft. high, strict. Leaves linear, the lowest obtuse, upper acute. Fascicles dichotomous. Bracts lanceolate, subulate, as long as the calyx. Petals small, toothed, red.

NORTH and SOUTH Islands: naturalised in dry places, but local. Annual. *Deptford pink.* Dec., Jan. Europe, Western Asia.

* **D. barbatus,** *L., Sp. Pl.* 409. Stems numerous, thickened at the nodes. Leaves large, lanceolate, glabrous. Fascicles broad, much branched. Bracts exceeding the calyx-tube, filiform. Petals toothed.

NORTH and SOUTH Islands: Naturalised on the sites of deserted homesteads and waste places. *Sweet-william.* Nov. to Jan. Central Europe.

1. GYPSOPHILA, Linn.

Calyx campanulate or nearly turbinate, 5-toothed and nerved. Petals 5, with a narrow claw. Disk small. Stamens 10. Styles 2. Capsule globose or ovoid, 4–5-valved to the middle or below. Seeds numerous. Mostly annual, with the flowers solitary in the forks of the stem, or paniculate.

SPECIES, about 50. With one exception, restricted to the shores of the Mediterranean and temperate parts of Asia.

ETYM. From the Greek, signifying *gypsum* and *to love.*

1. **G. tubulosa,** *Boiss., Diagn. Pl. Or.* i. 11. An erect or spreading annual, dichotomously branched, 2in.–5in. high or more, excessively glandular, pubescent, often viscid. Branches slender. Leaves linear-subulate, $\frac{1}{8}$in.–$\frac{1}{2}$in. long. Peduncles in the forks of the branchlets, slender, 1-flowered. Calyx narrow, campanulate, 5-toothed and -ribbed. Petals red or whitish-red, small, retuse. Capsule slightly exceeding the calyx, 5-valved. Seeds transversely rugose, pitted.—Hook., Fl. N.Z. ii. 325, and Handbk. 22; F. Muell., Pl. Vict. i. 155. *Dichoglottis tubulosa,* Joubert and Spach., Ill. Fl. Or. i. 14, t. 6. *D. australis,* Schlecht., Linnaea. xx. 631. *Saponaria tubulosa,* F. Muell., Nat. Pl. Vict. i. 136.

NORTH Island: East Coast, from Ahuriri to Cape Palliser, *Colenso.* SOUTH Island: Nelson: Tarndale, 4,000ft., *Travers.* Canterbury: Rangitata Valley, *Sinclair* and *Haast.* Lake Ohau, *Haast.* Lake Forsyth, sea-level, and Lake Lyndon, &c., 2,800ft., *T. K.* Otago: common in the interior, *Hector* and *Buchanan.* Nov. to Jan. The Nelson habitat requires confirmation. Also in South Europe and Asia Minor, South Australia, Victoria, New South Wales, and Queensland.

* SAPONARIA, Linn.

Calyx tubular, ovoid, 5-toothed, ebracteate. Petals 5, entire or notched. Stamens 10. Styles 2. Capsule oblong or ovoid, 2–4-celled at the base, 4-toothed at the apex. Seeds reniform; embryo annular. Annual or perennial herbs, with paniculate or fascicled flowers.

* **S. Vaccaria,** *L., Sp. Pl.* 409. A glabrous erect annual. Radical leaves spathulate; cauline leaves ovate, lanceolate or acuminate. Flowers in spreading corymbose cymes. Calyx inflated at base, 5-angled, expanded and winged in fruit.

NORTH Island: Occasionally found in cultivated land, Auckland and Wellington. Dec., Jan. Europe.

* SILENE, Linn.

Calyx inflated, 10- or many-nerved, 5-toothed. Petals 5, with a narrow claw. Stamens 10. Disk usually elongated. Ovary 1–3-celled below the middle. Styles 3, rarely 2–5. Capsule 6- or rarely 3-valved at the apex. Seeds reniform; embryo curved round the endosperm. Annual or perennial herbs.

Calyx inflated. Petals deeply cleft	.. * *S. Cucubalus.*
Calyx with 30 furrows, conical. Petals small, cleft	.. * *S. conica.*
Calyx tubular in flower. Petals small, entire ..	.. * *S. gallica.*
Calyx with 10 furrows, teeth short. Petals 2-fid	.. * *S. nocturna.*

* **S. Cucubalus,** *Wibel, Prim. Fl. Werth.* 241. An erect or spreading perennial, 1ft.–3ft. high, much branched, glabrous or downy. Leaves ovate, oblong, or lanceolate. Flowers drooping, in few- or many-flowered panicles. Calyx inflated. Petals deeply cleft, white; segments narrow. Capsule globose, conical.

NORTH and SOUTH Islands: naturalised in cultivated and grassy lands in Auckland and Otago, but local. *Bladder-campion.* Nov., Dec. Europe.

* **S. conica,** *L., Sp. Pl.* 418. An erect annual, stem simple or dichotomously branched, 6in.–12in. high, hirsute. Leaves linear, upper acute, often appressed to the stem. Flowers in the forks of the branches or terminal. Calyx with 30 furrows, conical; teeth subulate, acute, enlarged in fruit. Petals small, cleft. Capsule oblong-ovoid.

SOUTH Island: Otago: naturalised. *Petrie!* Europe.

* **S. gallica,** *L., Sp. Pl.* 417. An erect or spreading annual, 6in.–18in. high, hirsute, often viscid. Lower leaves obovate or spathulate; upper leaves oblong or linear, obtuse or acute. Flowers in terminal leafy racemose cymes. Pedicels very short. Calyx hirsute, at first tubular, afterwards ovoid and contracted at the mouth; teeth setaceous. Petals entire, small, white or pink.

Var. **quinquevulnera.** Petals entire, white with a broad red blotch.

NORTH Cape to STEWART Island: naturalised in cultivated and grassy land, &c. *Spotted catchfly.* Nov., Dec. Europe.

* **S. nocturna,** *L., Sp. Pl.* 416. An erect pubescent perennial herb. Radical and lower cauline leaves on long petioles, broadly elliptic-lanceolate, narrowed at both ends, acute, uppermost sessile. Flowers few in 1–2-flowered lateral cymes. Calyx 10-furrowed; teeth short, glandular, hirsute. Petals 2-fid. Capsule ovoid.

NORTH Island: naturalised, Karori, Wellington. Dec. Europe.

*LYCHNIS, Linn.

Calyx 5-toothed. Petals with an appendage. Styles 5, rarely 4. Teeth of capsule twice the number of styles. Erect annual or perennial herbs.

Calyx tubular. Petals 4-lobed .. .. * *L. Flos-cuculi.*
Calyx inflated. Petals 2-fid .. .. * *L. vespartina.*
Calyx coriaceous. Segments short .. .. * *L. coronaria.*
Calyx coriaceous. Segments foliaceous .. .. * *L. Githago.*

* **L. Flos-cuculi,** *L., Sp. Pl.* 436. An erect perennial herb, glabrous or pubescent above. Leaves petioled, oblong-lanceolate. Flowers in loose open panicles or cymes, drooping. Calyx tubular, 5-toothed; teeth acuminate. Petals cleft into 4 linear lobes, crowned.

NORTH Island: naturalised at Whangarei, Auckland. *Cheeseman. Ragged robin.* Oct., Nov. Europe, &c.

* **L. vespartina,** *Sibth., Fl. Oxon* 146. A viscid pubescent biennial, erect or spreading, usually diœcious, 1ft.–2ft. high. Lower leaves petiolate, upper lanceolate, sessile. Calyx with 5 linear teeth. Petals 5, white, expanding at sunset. Capsule conical, with 10 teeth.

SOUTH Island: sparingly naturalised in Otago. *Petrie! White campion.* Nov., Dec. Europe.

* **L. coronaria,** *Desr. in Lam. Encyc.* iii. 643. An erect dichotomously-branched perennial, clothed with loose white tomentum in all its parts. Radical leaves lanceolate or lanceolate-oblong; petioles as long as the blade. Calyx coriaceous, strongly ribbed; teeth short. Petals entire, large, crowned, rosy-purple.

NORTH Island: naturalised at Karori, Porirua, &c.; plentiful. SOUTH Island: near Blenheim, &c. Nov., Dec. South Europe.

* **L. Githago,** *Scop., Fl. Carn. ed.* 2, i. 310. An erect annual, 1½ft.–2ft. high, clothed with soft, appressed hairs. Leaves linear-lanceolate. Calyx coriaceous, silky, strongly ribbed; segments leafy, exceeding the petals. Petals crownless, entire, pale-purple.

NORTH Island: in cornfields and cultivated land, Auckland. *Corn-cockle.* Nov., Dec. Europe.

* CERASTIUM, Linn.

Sepals 5, rarely 4. Petals 5 or 4, usually 2-lobed or cleft, rarely entire or 0. Stamens 10 or 5. Ovary 1-celled. Styles 5 or 4. Capsule 1-celled, usually elongated, opening at the apex by twice as many teeth as there are sepals. Seeds compressed. Pubescent or glandulous or rarely glabrous herbs, with small leaves. Flowers white, in terminal dichotomous cymes.

Annual. Leaves ovate. Sepals acute .. .. * *C. glomeratum.*
Annual. Leaves oblong-lanceolate. Sepals subacute or obtuse .. .. * *C. triviale.*
Perennial. Petals greatly exceeding the calyx .. .. * *C. arvense.*

* **C. glomeratum,** *Thuill., Fl. Par. ed.* 2, 226. Annual. Glandular-hairy, suberect or decumbent. Leaves ovate. Cymes subcapitate at first. Sepals acute; margins and tips membranous, glabrous. Petals not exceeding the calyx, 2-fid. Capsule curved, twice as long as the calyx. Fruiting pedicels equalling or shorter than the calyx.

NORTH Cape to STEWART Island; AUCKLAND Islands: naturalised in fields and waste places, &c. *Mouse-ear chickweed.* Sept. to April. South Europe.

* **C. triviale,** *Link., Enum. Hort. Berol.* i. 433. Annual. Stems decumbent at the base. Leaves oblong-lanceolate, hairy, but with few glands. Cymes lax. Sepals obtuse or subacute, hairy throughout; margins membranous. Petals equalling the calyx, 2-fid. Capsule twice as long as the calyx. Fruiting pedicel equalling or exceeding the calyx.

NORTH Cape to STEWART Island; AUCKLAND Islands; MACQUARIE Island: naturalised. *Larger mouse-ear.* Oct. to April. Europe.

The teeth of the capsule are often truncate, when it is *C. amblyodontum*, Colenso in Trans. N.Z.I. xxvii (1894) 384. *C. truncatulum*, Col. *l.c.* xxv. (1892) 327 appears to be the same, but I have not seen specimens of either.

* **C. arvense,** *L., Sp. Pl.* 438. Perennial, tufted, erect or ascending, pubescent, downy or glabrate, 3in.–6in. high. Leaves linear-lanceolate. Cymes many-flowered; bracts with membranous margins. Sepals subacute or obtuse, margins and tips membranous. Petals obcordate, greatly exceeding the sepals. Capsule slightly longer than the calyx. Seeds tubercled.

SOUTH Island: naturalised on Swampy Hill, Dunedin. *Petrie!* Dec., Jan. Europe.

2. STELLARIA, Linn.

Sepals 4 or 5, usually spreading. Petals 4–5, 2-fid, rarely 0. Stamens usually 8 or 10, hypogynous. Styles 3, rarely 4 or 5. Capsule globose or ovoid, 1-celled, dehiscing to below the middle by twice as many valves as there are styles. Seeds few or many, often muricate. Annual or perennial herbs, usually of diffuse habit and solitary or cymose inflorescence.

SPECIES, 75. Chiefly in the temperate and cold regions of both hemispheres. The indigenous species are endemic, but the naturalised species are of wide distribution.

Stems capillary, creeping. Leaves suborbicular, acute .. .. 1. *S. parviflora.*
Creeping. Leaves ovate, obovate, or linear-lanceolate, acute .. .. 2. *S. decipiens.*
Minute. Leaves ovate, acute, soft. Flowers geminate .. .. 3. *S. minuta.*
Stem and branches with an interrupted hairy line .. .. .. * *S. media.*

Minute. Leaves linear-oblong. Flowers solitary .. .. 4. *S. elatinoides.*
Leaves linear-lanceolate, spreading. Cymes terminal. Flowers paniculate .. * *S. graminea.*
Leaves elliptic-lanceolate, ascending. Cymes few-flowered, lateral .. .. * *S. uliginosa.*
Erect, dichotomously branched. Flowers terminal .. .. 5. *S. Roughii.*
Leaves acerose, with recurved margins. Peduncles long, 1-flowered, erect .. 6. *S. gracilenta.*

1. **S. parviflora,** *Banks and Sol.* ex. *Hook. f., Fl. N.Z.* i. 25. Annual. Stems capillary, 6in.–12in. long, creeping, matted, weak, glabrous. Leaves flaccid, nearly orbicular, acute, 3 to 6 lines long, exceeding the petioles. Peduncles axillary, solitary or 2-flowered, with a pair of bracteoles at the base of the pedicels and on one or both pedicels. Sepals 5, lanceolate or oblong-acuminate or subacute, exceeding the petals. Petals 0 or 5, 2-lobed nearly to the base. Stamens 5–10. Styles 3. Capsule slightly exceeding the sepals or twice as long, deeply 6-valved. Seeds few, deeply pitted.—Handbk. 23.

Var. **oligosperma** (*sp.*), Colenso in Trans. N.Z.I. xviii. (1885) 257! Peduncles usually 2-flowered. Capsule twice as long as the sepals. Seeds finely marked, not pitted.

Var. **pellucida** (*sp.*), Colenso *l.c.* xxvii. (1894) 383. Leaves minutely punctiform. Seeds minutely muriculate.

NORTH Island; SOUTH Island; STEWART Island; CHATHAM Islands: common in lowland districts. Ascends to 5,000ft. on the Wairau Mountains. *Travers.*

2. **S. decipiens,** *Hook. f., Fl. Antarct.* i. 7. Annual. Stems weak, decumbent, much branched. Leaves orbicular or orbicular-ovate, acuminate or apiculate, narrowed into a broad short petiole which is sometimes ciliate; blade with a callous point and slightly thickened margin. Peduncles axillary, 2-flowered, rarely 3-flowered or solitary, longer or shorter than the leaves, with a pair of bracts at the base and another on one of the pedicels. Sepals 5, oblong-ovate, obtuse or subacute, glabrous. Petals 5 or 0, shorter than the sepals, deeply 2-fid. Stamens 5, 8, or 10. Styles 3. Capsule fully one-third longer than the sepals, oblong-ovoid, 5-valved nearly to the base, scarcely 2-fid at the tips. Seeds black, tuberculate.—Hook. f. in Hook. Ic. Pl. t. 680, and Handbk. 23.

AUCKLAND and CAMPBELL Islands, *Hooker f.* MACQUARIE Island, *A. Hamilton!*

Var. **angustata.** Leaves narrow, linear-lanceolate, acute or acuminate. ANTIPODES Island, *T. K.*

3. **S. minuta,** *n. s.* Annual. Stems ½in.–1in. high, narrowly winged, branched, glabrous, ciliate. Leaves ovate-acuminate or acute, narrowed into a short broad petiole; apex callous. Peduncles axillary, 1- or 2-flowered, with a pair of bracts at the base of the naked pedicels, not diverging. Sepals broadly oblong, obtuse. Petals 5, shorter than the sepals, 2-fid nearly to the base. Stamens 8, rarely 10. Capsule not seen.

SOUTH Island: Mount Stokes, 3,000ft., *J. Macmahon!* Westport, on the sea-beach, *Dr. Gase* (a scrap only).

Distinguished from all forms of *S. parviflora*, *S. decipiens*, and *S. elatinoides* by the broadly obtuse sepals, and from *S. media* by its solitary or geminate flowers and the absence of the hairy line on the stems and branches.

* **S. media,** *Cyrill. Char. Comm.* 36. A weak green decumbent annual. Stems and branches with a pubescent line which changes its position between the nodes.

Leaves ovate, acute or acuminate, the lower only shortly petioled. Flowers in axillary cymes. Pedicels very slender, spreading or reflexed in fruit, glandular, hairy. Petals shorter than the sepals, deeply 2-fid, rarely 0. Stamens 3–10. Capsule scarcely exceeding the sepals. Seeds punctate or tubercled.

NORTH and SOUTH Islands; STEWART Island; AUCKLAND and MACQUARIE Islands: naturalised in cool soils.

4. **S. elatinoides,** *Hook. f., Fl. N.Z.* i. 25. Annual. Glabrous, tufted. Stems ½in.–1½in. high, much branched. Leaves narrow-oblong or linear, narrowed into a short petiole, subacute, $\frac{1}{10}$in.–⅕in. long. Flowers equalling or slightly exceeding the peduncles, short, erect. Sepals lanceolate or narrow-ovate-lanceolate, acuminate, with prominent white membranous margins. Petals 0. Stamens 10. Capsule ovoid or globose, equalling the sepals, 6-valved to the middle; valves scarcely 2-fid. Seeds 6–12, tubercled.—Handbk. 23.

NORTH Island: Cape Kidnappers; Lake Rotoatara, Hawke's Bay; *Colenso*. SOUTH Island: Sowburn, 1,200ft.; Kurow; Tuapeka Mouth; *Petrie!* Very local.

This plant has not been collected in the North Island since its original discovery by Colenso. In the Handbook the ovary is said to be globose: it is clearly ovoid in all the Otago specimens.

* **S. graminea,** *L., Sp. Pl.* 422. A slender straggling glabrous perennial. Stems angular, 1ft.–2ft. long. Leaves linear-lanceolate, sessile, acute, ciliate. Flowers few or many in paniculate cymes. Pedicels slender; bracts ciliate. Sepals equalling or shorter than the petals, lanceolate, acute or acuminate, ciliate, 3-nerved. Petals 2-partite, rarely 0. Capsule ovoid, slightly exceeding the sepals.

NORTH and SOUTH Islands: naturalised in many localities, but rarely abundant. *Grassy-leaved chickweed.* Specimens from Otahuhu have slightly scabrid stems. Nov. to Jan. Europe.

* **S. uliginosa,** *Murray, Prod Stirp. Gott.* 55. Stems weak, smooth, angled, 2in.–10in. long. Leaves lanceolate or elliptical, ascending, acute; tip callous. Cymes terminal or axillary. Flowers 1–6. Sepals lanceolate, acute, 3-nerved, exceeding the 2-partite petals. Capsule ovoid, shorter than the calyx.

RUAPUKE Island, Foveaux Strait: naturalised in bogs. *Mrs. A. W. Traill! Swamp chickweed.* Nov., Dec. Europe.

5. **S. Roughii,** *Hook. f., Handbk. N.Z. Fl.* 23. Stems erect, 2in.–4in. high, dichotomously branched, glaucous, succulent. Leaves ½in.–1in. long, linear, acuminate, 1-nerved, fleshy. Flowers ½in. diameter or more, on stout terminal peduncles. Sepals very long, almost foliaceous, lanceolate, acuminate, strongly 3-nerved. Petals much shorter than the sepals, deeply 2-partite. Stamens 10. Capsule broadly oblong, much shorter than the petals, 6-valved to below half-way down; valves not 2-fid. Seeds 10–12 or more, large, strongly papillose.

SOUTH Island: Nelson: Dun Mountain Range, *Rough!* Wairau Gorge, *Travers*. Mount Captain, *T. K.* Canterbury: Mount Torlesse, *Haast*. Leith Hill, &c., *J. D. Enys!* 4,000ft. to 6,500ft. Nov. to Jan.

A remarkably local alpine plant, which cannot be mistaken for any other. Chiefly found on shingle slips.

6. **S. gracilenta,** *Hook. f., Fl. N.Z.* ii. 326. A wiry slender tufted plant. Stems 1ft.–4ft. high, nodose, scabrid, often matted, forming a more or less compact mass. Leaves ¼in. long, more or less recurved, acute or subacute or apiculate, usually with revolute margins; each axil with a small fascicle of

leaves. Peduncles axillary, 1-flowered, 1in.–4in. long, slender, strict, with a pair of bracts near the middle. Sepals linear-oblong, obtuse or acute, with scarious margins. Petals 5, 2-partite. Stamens 5–10. Capsule linear-oblong, one-third longer than the sepals, 6-valved nearly to the base. Seeds papillose, brown.—Handbk. 24.

SOUTH Island: mountains of Nelson, Canterbury, and Otago. Descends to sea-level at the mouth of the Waitaki. *Petrie.* Ascends to 4,500ft.

The usual number of stamens is 5; if more than 5, one or more are abortive. In some cases all the stamens are reduced to staminodia, with very short flat filaments.

*ARENARIA, Linn.

Sepals 5. Petals 5, scarcely retuse, rarely 0. Stamens 10, rarely fewer, inserted on the disk. Styles 3, rarely 4 or 2. Ovary 1-celled. Capsule short, splitting into as many valves as there are styles, or twice as many. Seeds many, rarely few, smooth or tubercled, compresed. Embryo annular. Small tufted annual or perennial herbs, with sessile leaves and white flowers.

* **A. serpyllifolia,** *L., Sp. Pl.* 423. Annual; spreading or erect. Stems often much branched, 2in.–6in. high. Leaves ovate, acuminate, sessile or subsessile, small, stiff, ciliate. Cymes leafy, many-flowered. Petals shorter than the sepals.

Var. **sphaerocarpa** (*sp.*), Tenore. Much branched, diffuse. Sepals ovate-lanceolate, 3–5-nerved. Fruiting pedicel erect or patent. Capsule exceeding the sepals, ovoid-oblong.

Var. **leptoclados** (*sp.*), Gussone. Stems very slender, erect, sparingly branched. Leaves more acute, smaller, 3-nerved. Fruiting pedicel curved at the top.

NORTH and SOUTH Islands: abundantly naturalised. Var. *leptoclados:* extending for miles in the Kaikoura district; Canterbury and Otago. Nov. to Jan. Europe.

*SAGINA, Linn.

Sepals 4 or 5. Petals 4 or 5, entire, retuse, longer or shorter than sepals or 0. Stamens as many or twice as many as the sepals, or fewer. Styles as many as the sepals. Capsule 4- or 5-valved to the base, the valves alternating with the sepals. Tufted or matted herbs, with minute subulate leaves and small flowers on slender peduncles.

All the branches floriferous .. .. * *S. apetala.*
Primary shoots and branches flowerless .. .. * *S. procumbens.*

* **S. apetala,** *L., Mant.* ii. 559. A slender filiform wiry annual, erect or spreading, all the branches flowering. Leaves connate, rosulate at the base, ciliate, glandular, pubescent or glabrous, mucronate. Pedicels ascending. Sepals ovate or ovate-lanceolate. Petals minute or frequently 0.

NORTH and SOUTH Islands: naturalised in dry situations. Oct. to Jan. Europe.

* **S. procumbens,** *L., Sp. Pl.* 128. A minute spreading perennial, with matted stems and flowerless primary shoots. Branches fascicled and procumbent. Leaves glabrous or ciliate, subacute or obtuse. Peduncles capillary, exceeding the leaves. Sepals 4 rarely 5, reflexed in fruit. Petals minute. Capsule exceeding the sepals. Fruiting peduncles usually hooked near the top.

NORTH, SOUTH, and STEWART Islands: naturalised in damp places. Oct. to March. Europe.

3. COLOBANTHUS, Bartling.

Sepals 4 or 5, usually cartilaginous, erect. Petals 0. Stamens 4 or 5, alternating with the sepals, perigynous. Styles as many as the sepals and opposite them. Capsule ovoid or oblong, opening by as many valves as there are sepals. Small glabrous densely-tufted herbs, with opposite imbricating cartilaginous or fleshy leaves and erect solitary terminal or axillary flowers.

SPECIES, 15; restricted to mountain regions in South America, Australia, New Zealand, and the Antarctic islands. The genus is more largely developed in the colony than elsewhere, and all the species except *C. quitensis*, *C. Billardieri*, and *C. Benthamianus* are endemic.

Flaccid. Stems leafy. Sepals 4	1. *C. quitensis.*
Flaccid. Leaves all radical. Sepals 5, ovate	2. *C. Billardieri.*
Rigid. Leaves rigid, recurved, concave. Sepals with acicular tips ..	3. *C. Muelleri.*
Rigid. Leaves short, recurved, canaliculate. Sepals ovate	4. *C. canaliculatus.*
Soft. Leaves subulate, spreading. Sepals 4, shorter than the capsule ..	5. *C. repens.*
Rigid. Leaves linear-subulate. Sepals 5, broadly ovate	6. *C. brevisepalus.*
Leaves rigid, with short acicular points. Sepals 5, ovate-subulate	7. *C. Benthamianus.*
Leaves linear-subulate, with long acicular points	8. *C. acicularis.*
Flowers axillary on slender peduncles	9. *C. Buchanani.*
Flaccid. Leaves obtuse. Flowers sunk among the leaves. Sepals 4 ..	10. *C. muscoides.*

1. **C. quitensis,** *Bart. in Presl. Reliq. Haenkianae* ii. 13. Tufted, 1in.–2in. high, soft, green, glabrous, excessively branched. Lower leaves about ½in. long, with broad membranous bases, concave above, acute; upper ⅛in.–¼in., connate. Peduncles very short, terminal. Sepals 4, ovate-oblong, obtuse, the two lateral smaller than the others, one-third longer than the ripe capsule. Disk narrow. —Hook. f., Handbk. N.Z. Fl. 24; Phil., Cat. Pl. Vasc. Chil. 27.

SOUTH Island: Nelson mountains; Kowhai River, Canterbury. Otago, *Buchanan!*

The leaves and sepals of the New Zealand plant are rarely mucronate, and never acicular. This species is found throughout the Andes, also on Amsterdam Island, &c.

2. **C. Billardieri,** *Fenzl. in Ann. des Wien Mus.* i. 49. Tufted, rarely exceeding 1in. in height, flaccid, grassy. Leaves ¼in.–¾in. long, linear-subulate, channelled above, tips acute or acicular. Peduncles ¾in.–1in. long, white. Sepals 5, ovate, acute or acuminate, scarcely exceeding the ovary. Disk narrow. —Hook. f., Fl. Antarc. i. 26, Fl. Tasm. i. 45, and Handbk. N.Z. Fl. 25; Benth., Fl. Austr. i. 161; Phil., Cat. Pl. Vasc. Chil. 87. *C. affinis,* Hook. in Ic. Pl. t. 266? *Stellaria apetala,* Labill., P. Nov. Holl. i. 112, t. 142; DC., Prod. i. 395. *S. uniflora,* Banks and Sol. MSS.

SOUTH Island: Nelson to Southland. STEWART Island; AUCKLAND and CAMPBELL Islands; ANTIPODES Island; MACQUARIE Island. Chiefly at sea-level in the South. Ascends to 3,000ft. in Nelson and Canterbury.

Var. **alpinus.** Much larger than the type, but equally flaccid, forming small tufts 1in.–4in. across, with spreading peduncles 1in.–4in. long. Sepals ovate-acuminate, with membranous margins, slightly exceeding the capsule. Leaves 1in.–2in. long or more, with long acicular tips. NORTH Island: Ruahine and Tararua Mountains. SOUTH Island: Southern Alps. 1,500ft. to 4,500ft. Also in Victoria, Tasmania, Chili, &c.

Colobanthus affinis, Hook., figured in Ic. Pl. t. 266, differs from any form of *C. Billardieri* found in New Zealand in the broadly ovate sepals, which are only half the length of the capsule, and in the broad disk.

3. **C. Muelleri,** *T. Kirk in Trans. N.Z.I.* xxvii. (1894) 356. A rigid glabrous plant, forming small tufts ¾in.–1½in. high. Leaves cartilaginous, usually recurved, broadly channelled, with short acicular tips and evident midrib, ½in.–¾in. long. Peduncles ¼in.–¾in. long, often hidden amongst the leaves. Sepals 5, ovate, abruptly narrowed into cartilaginous points with acicular tips,

channelled, about one-third longer than the capsule. Disk very narrow.—*C. Billardieri*, var. *platypoda*, F. Muell. in Veg. Chath. Isds. 11.

NORTH and SOUTH Islands: chiefly in lowland districts. Mount Camel to Stewart Island. CHATHAM Islands. Often rare and local. Most plentiful on shingly beaches.

This is readily distinguished from *C. Billardieri* by its rigid habit, and especially by the apiculate cartilaginous sepals, which are longer than the capsule. I venture to refer the plant collected by Buchanan at Mount Camel to this species, although I have not seen specimens.

Var. **multicaulis.** Rigid, much branched from the base. Branches naked below. Leaves somewhat lax, spreading, linear-subulate, apiculate, about ¼in. long. Peduncles scarcely exceeding the leaves, slender. Sepals narrow-ovate, acute or mucronate, equalling the capsule. SOUTH Island: interior of Otago, *Buchanan!* I have only a few scraps of this interesting form, picked from amongst other plants in Mr. Buchanan's herbarium.

4. **C. canaliculatus,** *T. Kirk in Trans. N.Z.I.* xxvii. 357. A small tufted much-branched plant, ½in. high. Branches spreading. Leaves patent or slightly recurved, 3/16in. long, canaliculate with stout marginal nerves, sharply apiculate. Flowers ⅛in. long, on very short stout peduncles, axillary or terminating short lateral branches. Sepals 5, ovate, acute or subacute, equalling the capsule. Disk thickened.

SOUTH Island: Central Otago, *Buchanan!* I have but few specimens of this curious plant.

5. **C. repens,** *Colenso in Trans. N.Z.I.* xix. (1886) 260. A small plant. Branches creeping, matted, 3in.–4in. long. Leaves about ½in. long, subulate, with acicular tips, not rigid, spreading, green. Scapes slightly exceeding the leaves. Sepals 4, broadly ovate, shorter than the capsule, slightly margined.

NORTH Island: near Norsewood, Hawke's Bay, *Colenso.*

I have not seen specimens of this, which appears to be intermediate between *C. Muelleri* and *C. quitensis.*

6. **C. brevisepalus,** *T. Kirk l.c.* 357, *t.* 27E. Tufted, rigid, ½in.–1in. high, much branched. Leaves densely imbricated, 1/16in.–1/12in. long, linear-subulate, with a broad base, concave above, rounded beneath, obtuse, mucronate. Flowers terminal, sunk amongst the leaves. Sepals 5, narrow-ovate, convex or almost keeled, mucronate, slightly exceeding the capsule.

SOUTH Island: Mount Mouatt, Awatere, 4,000ft., *T. K.* Otago: Gorge Creek and Kurow, *Petrie!*

7. **C. Benthamianus,** *Fenzl. in Ann. des Wien Mus.* i. 49. Densely tufted, about 1in. high. Leaves densely imbricated, rigid, ⅛in.–¼in. long, channelled above, with short acicular points, rounded below. Peduncles very short. Flowers exceeding the terminal leaves. Sepals 5, ovate-subulate, mucronate, rigid, equalling or slightly exceeding the capsule. Disk reduced to a mere line. —F. Muell., 2nd Cens. Austr. Pl. 46; C. Moore, Handbk. Fl. N.S.W. 100; Phil., Cat. Pl. Vasc. Chil. 87. *C. subulatus*, Hook. f., Fl. Antarc. i. 13, and ii. 247, t. 293; Handbk. N.Z. Fl. 25; Benth., Fl. Austr. i. 160. *C. pulvinatus*, F. Muell. in Trans. Phil. Soc. Vict. i. 201, and Pl. Vict. 213, t. 11. *Sagina subulata*, D'Urville, Fl. Ms. Mal. in Mem. Soc. Linn. par. iv. 618 (not of Wimm). *Sagina muscosa, β squarrosa*, and *γ laricifolia*, Banks and Sol. MSS. in Herb. Mus. Brit.

SOUTH Island: Marlborough: "Awatere Valley and Sinclair Range; alt. 4,000ft. Otago: Lake district."—Handbk. N.Z. Fl. 25. CAMPBELL Island. Jan. Also in Australia and Arctic America.

The above description is drawn exclusively from Campbell Island specimens, as all the South Island forms that have come under my notice appear to belong to *C. acicularis*, Hook. f., the leaves and sepals having longer acicular tips than those of the Campbell Island plant, the perianth of which is exactly represented by fig. 4, t. 93, in Fl. Antarc. The Falkland Island plant, from its more slender habit and tetramerous flowers, appears to be distinct.

8. **C. acicularis,** *Hook. f., Handbk. N.Z. Fl.* 25. Densely tufted, rigid. Stems 1in.–3in. long, shining, green or brown. Leaves densely imbricated, ¼in.–¾in. long or more, linear-subulate, with very long acicular points. Flowers hidden amongst the leaves, almost sessile. Sepals 5, linear-lanceolate, with long acicular tips, one-third longer than the capsule.

SOUTH Island: in rocky places, Nelson to Southland. 2,500ft. to 5,500ft.

A more robust species than the preceding, from which it differs in the longer stems and leaves and the narrower sepals.

9. **C. Buchanani,** *T. Kirk l.c.* 358, *t.* 27D. Laxly tufted. Stems slender, 1in.–3in. high. Leaves not imbricating, lax, ¼in.–$\frac{7}{16}$in. long, linear-subulate, with acicular points, membranous, concave above, convex beneath, patent or spreading. Flowers axillary, on slender peduncles exceeding the leaves, or rarely shorter. Sepals 5, narrow linear-subulate, acute, one-half longer than the capsule. Disk narrow.

SOUTH Island: Manuherikia Valley, Otago, *Buchanan!* A remarkably distinct species, of which I have only two specimens.

10. **C. muscoides,** *Hook. f., Fl. Antarc.* i. 14. A soft densely-tufted bright-green plant, forming amorphous or rounded masses 1in.–18in. in diameter or more, excessively branched and matted. Leaves glabrous, densely imbricated, patent or ascending, $\frac{1}{12}$in.–⅛in. long, linear with dilated bases, obtuse. Flowers on short peduncles sunk amongst the leaves. Sepals 4, ovate-lanceolate, obtuse, concave, equalling the capsule, the two lateral sepals larger than the others. Disk very large.—Handbk. 25; Decaisne, Voy. au Pôle Sud. Bot. Dicot. t. 17.

The SNARES: AUCKLAND, CAMPBELL, MACQUARIE, and ANTIPODES Islands.

As the stems become matted they give off rootlets in abundance, while the seeds sometimes germinate in the capsules, and at others are found buried amongst the decaying stems.

*SPERGULA, Linn.

Sepals 5. Petals 5. Stamens 5 or 10. Styles 5. Capsules 5-valved; valves entire, opposite the sepals. Seeds compressed, usually margined or winged. Embryo spirally annular. Annual herbs with fascicled branches and crowded apparently verticillate leaves. Flowers in terminal paniculate cymes, white.

* **S. arvensis,** *L., Sp. Pl.* 440. Annual, with more or less pubescent geniculate stems. 6in.–12in. high, branched from the root. Leaves linear, convex above, furrowed beneath. Flowers in terminal racemose cymes, often paniculate. Sepals ovate, rounded, longer or shorter than the petals. Seeds black, subglobose, girt with a raised ring round the middle, minutely punctulate or smooth.

NORTH and SOUTH Islands: naturalised in cultivated and waste land. *Corn-spurrey.* Oct. to Jan. Europe.

* **S. pentandra**, *L.*, *Sp. Pl.* 440. Of similar habit and appearance to *S. arvensis*. Seeds black, lenticular, surrounded by a white membranous wing, slightly scabrid or papillose.

Naturalised near Wellington, *T. K.* Europe.

4. SPERGULARIA, Persoon.

Sepals 5. Petals 5, entire, rarely 0. Stamens 5–10, rarely fewer, hypogynous. Styles 3. Capsule 3- rarely 5-valved. Prostrate or suberect spreading herbs, with linear leaves often clustered in the axils, presenting a verticillate appearance. Flowers pedicellate. Seeds compressed, with or without a membranous border.

* **S. rubra**, *J. and C. Presl.*, *Fl. Cech.* 94. Annual or perennial. Stems prostrate, much branched, glandular above, more or less pubescent, 3in.–10in. long. Leaves linear, flat, obtuse or apiculate. Stipules lanceolate, cleft or lacerate. Flowers on slender glandular pedicels, equalling or exceeding the leaves, axillary, solitary or forming lax elongated cymes. Sepals 5, lanceolate, obtuse, glandular. Petals 5, about as long as the sepals, purple. Stamens 5–10; if the larger number, 2 or more imperfect. Fruiting peduncles spreading or reflexed. Styles 3. Capsule 3-valved, as long as the sepals. Seeds plano-convex, gibbous, muricatulate or tubercled. Owing to the development of leaf-buds in the axils the leaves often appear verticillate.

NORTH and SOUTH Islands: naturalised by roadsides, waste places, &c. *Sand-spurrey.* Nov. to Feb. Europe.

1. **S. media**, *Presl.*, *Fl. Sic.* p. xvii. Usually a larger plant than *S. rubra*, with longer half-terete leaves. Stems suberect or prostrate, compressed. Stipules broadly ovate, acuminate. Flowers larger, whiter. Pedicels longer. Capsule much longer than the calyx. Seeds compressed, rounded, smooth or papillose, with or without a scarious margin.—*Spergularia rubra* var. *marina*, Hook. f., Handbk. 25; A. Cunn., Precurs. n. 619. *Arenaria media*, L., Sp. Pl. 606; Hook. f., Fl. N.Z. i. 26. *A. marina*, Sm. E. B. t. 958. *Buda marina*, Dum., Fl. Belg. 110.

NORTH Island; SOUTH Island; STEWART Island: a variable plant, common all round the coasts.

Var. **salina**. Pedicels equalling the foliaceous bracts. Seeds smooth. Dog Island, Foveaux Strait, *T. K.*

A common littoral plant in many temperate and subtropical countries.

* POLYCARPON, Linn.

Sepals 5, keeled, hooded at the apex. Petals 5, small, emarginate. Stamens 3–5. Style short, 3-fid. Capsule 3-valved, many-seeded. Annual herbs, with opposite or apparently whorled leaves, scarious stipules, and small flowers crowded in bracteate cymes.

* **P. tetraphyllum**, *L.*, *Syst. ed.* x. 881. Stems much branched, 1in.–4in. high. Leaves opposite or in fours, broadly obovate; petioles short, broad. Flowers triandrous with a 3-fid style, or rarely pentandrous with the style entire. Capsule shorter than the calyx.

NORTH and SOUTH Island; STEWART Island: abundantly naturalised in dry places. *Allseed.* Sept. to June. Europe.

Order VII.—PORTULACEAE.

Sepals 2, free or barely adnate to the ovary, imbricate. Petals 4 or 5, hypogynous or rarely perigynous, free or connate at the base, imbricate. Stamens 1–∞, often adherent with the base of the petals. Ovary 1-celled, free or half inferior. Style divided into 2 or 3 or more stigmatose branches. Ovules few or many. Capsule with as many valves as styles. Seeds few or many, with mealy endosperm. Embryo usually curved or annular. Herbs, usually glabrous and fleshy, rarely hirsute, with mostly exstipulate entire opposite or alternate leaves and hermaphrodite flowers.

A small order, most plentiful in South America, but extending to South Africa, Australia, New Zealand, and other countries. Genera, 15. Species, 130.

*PORTULACA, Linn.

Sepals 2, united, adherent to the base of ovary. Petals usually 5, perigynous. Stamens 6–20, perigynous. Style 2–8-parted. Ovary half inferior; ovules several. Capsule globular, membranous, dehiscing transversely, the free portion forming a lid. Seeds reniform, shining. Leaves alternate or opposite, sometimes forming an involucre round the flowers. Stipules scarious or reduced to hairs.

***P. oleracea,** *L.*, *Sp. Pl.* 445. A prostrate rather fleshy glabrous annual. Stems branched, 3in.–10in. long. Leaves cuneate-oblong or almost obovate, shortly petioled. Flowers terminal and sessile, solitary or forming small dichotomous cymes; floral leaves orbicular-truncate. Sepals 2–3 lines long, keeled. Petals 5, slightly coherent at the base, fugacious, bright-yellow. Stamens 7–12. Style short; stigmatic branches 5 or 6. Seeds minutely punctulate.

North Island: naturalised in cultivated land and by roadsides, especially in the Auckland District. *Purslane.* Nov. to Jan. Tropical and warm temperate countries.

*CALANDRINA, H.B. and K.

Sepals 2, usually persistent. Petals 5 or more, hypogynous. Stamens many or few, free or the base united with the petals. Styles 3 or 4, free or forming a single 3- or 4-lobed or cleft style. Capsule globose or oblong, almost indehiscent or 3–4-valved. Seeds reniform, shining, granulate. Embryo annular. Annual or perennial herbs. Leaves alternate, exstipulate. Flowers solitary, racemose or capitate.

***C. caulescens,** *H. B. and K.*, *Nov. Gen. et Sp.* vi. 78, *t.* 526. A small annual with petiolate linear-oblong radical and cauline leaves. Bracts foliaceous. Sepals ovate-acuminate. Petals 5, soon withering, but forming a persistent covering to the young fruit. Stamens 5, free. Stigmatic branches 3. Capsule exceeding the calyx, 3-valved.

North Island: naturalised at Otahuhu, *Cheeseman!* South Island: near Christchurch, *T. K.*

1. CLAYTONIA, Gronor.

Sepals 2, free, persistent. Petals 5, coherent at the base, hypogynous. Stamens 5, adhering to the base of the petals. Ovary superior, ovules few; style 3-cleft at the apex. Capsule globular or ovoid, 3-valved. Seeds reniform or orbicular. Embryo annular. Annual or perennial herbs with exstipulate leaves.

1. **C. australasica,** *Hook. f. in Hook. Ic. Pl.* t. 293. Stems very slender, creeping, ½in.–6in. long. Leaves alternate or in distant pairs, narrow-linear, spathulate, obtuse, tender, ¼in.–½in. long, with a wide sheathing base. Flowers axillary, solitary or in a few-flowered lax raceme or cyme, longer or shorter than the leaves, white or purplish, ¼in.–⅔in. in diameter. Sepals small. Capsule equalling or exceeding the sepals. Seeds 3, black, shining.—Fl. N.Z. i. 72; Fl. Tasm. i. 144; Handbk. 27; Benth., Fl. Austr. i. 177. Var. *biflora* and var. *racemosa*, Buch. in Trans. N.Z.I. iii. (1870) 210.

NORTH Island: Ruahine Range and Ruapehu, *H. Hill! W. E. Andrew!* SOUTH Island: common in mountain districts from Nelson to Southland. Ascending to 5,000ft.; descends to sea-level on sandhills near Dunedin.

* **C. perfoliata,** *Donn., Ind. Host. Cantab.* ex *Willd. Sp. Pl.* i. 1186. A tufted glabrous rather succulent annual, 4in.–9in. high. Radical leaves spathulate-rhomboid; cauline 2, connate immediately beneath the short terminal raceme, forming a flat almost orbicular involucre. Raceme subverticillate, the two basal flowers on long pedicels, the upper very short. Flowers few, small, white. Petals exceeding the calyx.

SOUTH Island: naturalised, Cheviot, *Haast!* North America.

2. MONTIA, Linn.

Sepals 2 or 3, persistent. Petals 5, connate at the base and split on one side, perigynous. Stamens 3 or 5, adnate to the petals. Ovary superior; ovules 3. Capsule 3-valved. Seeds nearly orbicular. A small glabrous herb with opposite leaves and axillary flowers. A monotypic genus, found in Europe, western North America, South America, Labrador, Greenland, Tasmania, Kerguelen Land, and South Georgia.

1. **M. fontana,** *L., Sp. Pl.* 87. Stems usually tufted or matted, simple or branched, 1in.–5in. high, slender, weak. Leaves opposite, linear-lanceolate, elliptic or obovate, subacute. Flowers solitary or in 2–3-flowered racemes, drooping. Petals scarcely exceeding the sepals.—Fl. N.Z. i. 74, Handbk. 27.

NORTH and SOUTH Islands: from Rotorua to Stewart Island. AUCKLAND, CAMPBELL, ANTIPODES, and MACQUARIE Islands. Sea-level to 4,000ft.

3. HECTORELLA, Hook. f.

Sepals 2, short, continuous with the short broad flat pedicel. Petals 5, coherent at the base, thickened near the tip. Stamens 5, adnate with the corolla-tube, alternating with the petals; filaments equalling the petals. Anthers linear-oblong, 2-celled. Ovary ovoid, membranous, veined; style erect, divided into 1–3 linear stigmatose lobes; ovules 4–5, erect from the base of the cell, amphitropous, slender. Capsule membranous, equalling the petals or nearly so. Seeds 2–4. A small densely-tufted glabrous plant, much branched from the base. Leaves densely imbricated, entire, coriaceous. Flowers nearly sessile or on very short peduncles amongst the terminal leaves.

1. **H. caespitosa,** *Hook. f., Handbk. N.Z. Fl.* 27. Stems with leaves ½in.–¾in. in diameter, 1in.–1½in. high. Leaves varying from broadly triangular-

ovate to oblong-triangular-ovate with a broad base, most densely imbricating, membranous at the base, coriaceous and keeled above, with thickened margins and tips; veins reticulated. Flowers often unisexual, the staminate being smaller than the others, very shortly pedunculate, forming a ring amongst the outer apical leaves. Peduncles with two small bracts. Sepals concave-truncate. Petals much longer than the sepals. Capsule globular, membranous, equalling the sepals. Seeds 2–4, rounded-ovate, smooth, shining, scarcely compressed.—Hook. f. in Hook. Ic. Pl. t. 1046.

SOUTH Island: Canterbury: Mount Cook, *F. G. Gibbs!* Mount Alta, Mount Aspiring, &c., *Hector* and *Buchanan!* Hector Mountains, Dunstan Mountains, Mount Cardrona and high peaks to the west, *Petrie!* 5,000ft. to 6,000ft.

Mr. Buchanan distinguishes the plant with longer narrower leaves as *H. elongata* in Trans. N.Z.I. xvi. (1883) t. 35, restricting *H. caespitosa* to the plant with ovate-deltoid leaves. Hooker's drawing in Ic. Pl. shows an intermediate form; but every degree of variation may be found on examining a series of specimens. The capsule was first described by F. v. Mueller in the Vict. Nat. for Aug. 1890.

ORDER VIII.—ELATINEAE.

Sepals 2–5, free, imbricate. Petals 2–5, free, hypogynous, imbricate, rarely 0. Stamens 2–10, hypogynous, free. Ovary superior, 2–5-celled. Styles 2–5, free from the base; stigmas capitate. Ovules several, attached to the inner angle of each cell, anatropous. Capsule septicidal, the valves falling and leaving most of the dissepiments with the seeds attached to the axis. Seeds straight or curved, with a crustaceous testa and terete embryo. Radicle superior. Endosperm 0. Herbs, usually aquatic, rarely woody at the base, with opposite stipulate leaves and small flowers.

A small order, comprising 2 genera; both widely distributed. SPECIES, 30.

1. ELATINE, Linn.

Flowers 2- 3- or 4-merous. Sepals membranous. Ovary globose. Capsule membranous. Small glabrous aquatic herbs with minute axillary solitary flowers. Seeds cylindrical or oblong, longitudinally striated and transversely barred or wrinkled.

SPECIES, about 10. Found in ponds, streams, and lakes throughout the temperate and subtropical regions of the earth.

1. **E. americana,** *Arnott in Edin. Journ. Nat. and Geogr. Sc.* i. (1830) 431; var. *australiensis,* Benth., Fl. Aust. i. 178. Stems 1in.–6in. long, often matted, creeping and rooting at the joints when growing on mud, elongated in water, bright-green or reddish, slightly succulent. Leaves shortly petioled, ovate obovate or broadly oblong, obtuse, usually with a few distant intramarginal glands. Stipules minute, usually fugacious. Flowers minute, sessile, 3-merous. Sepals obtuse. Petals longer than the sepals. Stamens 6. Styles 3, divergent. Seeds 8–10 in each cell, slightly curved, with 6–8 longitudinal ridges, closely and transversely wrinkled or barred.—*E. americana,* Arnott in Edin. Journ. Nat. Sc. i. (1830) 431; Hook. f., Fl. N.Z. i. 27, and Handbk. 28; F. Muell., 2nd Cens. Austr. Pl. 14. *E. gratioloides,* A. Cunn., Precurs. n. 610.

NORTH and SOUTH Islands; STEWART Island: on the margins of lakes and streams in lowland districts from Mongonui and Hokianga southward, but often local.

The New Zealand plant differs from the North American species with which it is usually united in the constantly trimerous symmetry and larger number of seeds.

ORDER IX.—HYPERICINEAE.

Sepals 4 or 5, imbricate. Petals 4 or 5, hypogynous, imbricate. Stamens many or few, free or united at the base into 3 or 5 sets or bundles. Ovary free, 3–7-celled by the union of the placentas in the axis, or 1-celled with 2–5 parietal placentas. Styles as many as carpels, persistent, rarely united at the base. Ovules numerous in each cell or on each placenta. Fruit a capsule or rarely an indehiscent berry. Seeds without endosperm. Embryo straight or rarely curved. Herbs, rarely shrubs or trees, with opposite exstipulate simple leaves and terminal or rarely axillary flowers.

GENERA, 8. SPECIES, 225. Widely distributed in the temperate and warm regions of the earth.

1. HYPERICUM, Linn.

Sepals 5. Petals 4. Stamens numerous, all free, or coherent in 3 or 5 sets or bundles. Ovary 1–3- or 5-celled. Fruit a capsule or berry. Seeds not winged. Embryo straight. Herbs, often suffruticose, with usually thin leaves, entire or rarely minutely toothed.

SPECIES, about 160; widely distributed. Only two species are indigenous in New Zealand.

* *Sepals equal. Fruit a capsule.*

Erect. Leaves subcordate .. .. 1. *H. gramineum.*
Procumbent, soft. Leaves oblong or broadly obovate. Petals scarcely exceeding sepals .. .. 2. *H. japonicum.*
Procumbent, wiry. Petals longer than the unequal sepals .. .. * *H. humifusum.*
Stem erect. Leaves oblong, with pellucid glands .. .. .. * *H. perforatum.*

** *Sepals unequal. Fruit a berry.*

Shrubby, 2ft.–4ft. high .. .. * *H. Androsaemum.*

1. **H. gramineum,** *Forst., Prod. n.* 281. Glabrous, strict, branched from the base. Stems angular, 6in.–12in. high, slender. Leaves rather distant, ½in.–1in. long, stem-clasping, cordate, ovate, oblong or oblong-lanceolate, obtuse, margins revolute. Flowers in terminal trichotomous cymes, with a pair of bracts at the base of each fork. Sepals lanceolate or ovate-lanceolate, acute or obtuse. Petals exceeding the sepals. Stamens usually free. Styles 3. Capsule 1-celled, 3-valved, usually longer than the calyx.—DC., Prod. i. 548; Hook. f., Fl. N.Z. i. 36, Fl. Tasm. i. 53, and Handbk. 29; Benth., Fl. Austr. i. 182. *H. japonicum*, F. Muell. in 2nd Cens. Austr. Pl. 14. *Brathys Forsteri*, Spach. in Ann. Sc. Nat. (ser. ii.) 5, 367.

NORTH and SOUTH Islands: from Poverty Bay southwards to Otago, but often local. Ascends to 1,800ft.

2. **H. japonicum,** *Thunb., Fl. Jap.* 295, *t.* 31. Stems very slender, procumbent, much branched, ascending at the tips. Leaves much smaller, broadly

oblong or ovate-oblong, obtuse, sessile, flat, sometimes glaucous. Flowers smaller, sometimes solitary or forming few-flowered cymes near the tips of the branches. Pedicels short, very slender. Sepals ovate-oblong or oblong, obtuse or subacute. Petals scarcely exceeding the sepals. Stamens free.—DC., Prod. i. 548; Hook. f., Fl. N.Z. i. 37, Fl. Tasm. i. 53, and Handbk. 29; Benth., Fl. Austr. i. 182. *Ascyron humifusum*, Labill, Pl. Nov. Holl. ii. 33, t. 175. *Brathys humifusum*, Spach. in Ann. Sc. Nat. (ser. ii.) v. 367.

NORTH and SOUTH Islands; North Cape to Dunedin. Ascends to 4,000ft. Often local.

* **H. humifusum,** *L., Sp. Pl.* 785. Stems procumbent, much branched from the base; tips ascending, very slender, 3in.–10in. long, wiry. Leaves ovate-oblong, obtuse, without pellucid glands. Flowers in few-flowered terminal cymes or solitary. Sepals connate at the base, unequal, 3 oblong, obtuse, 2 lanceolate and mucronate. Stamens adhering at the base into 3 bundles. Capsules 3-celled.

NORTH and SOUTH Islands: naturalised in many localities. Dec., Jan. Europe; Azores and Canary Islands.

* **H. perforatum,** *L., Sp. Pl.* 785. Stems slender, erect, giving off numerous runners from the base, much branched above, 1ft.–1½ft. high, 2-edged. Leaves linear-oblong or oblong, with pellucid dots. Flowers in open leafy cymes. Sepals lanceolate, acute. Petals much longer than the erect sepals. Stamens cohering at the base into 3 bundles. Capsule 3-celled; valves transversely wrinkled.

NORTH and SOUTH Islands: naturalised in many districts. *St. John's wort.* Dec., Jan. Europe.

* **H. Androsaemum,** *L., Sp. Pl.* 784. Erect, sparingly branched, 2ft.–4ft. high, woody at the base. Leaves broadly subcordate, sessile. Flowers in trichotomous terminal cymes, few-flowered. Sepals broad, unequal, persistent, and spreading in fruit. Petals as long as the sepals. Stamens cohering at the base into 5 bundles, exceeding the styles. Fruit a globose berry, imperfectly 3-celled, black.

NORTH and SOUTH Islands: plentiful in many localities. *Tutsan.* Central and South Europe.

ORDER X.—MALVACEAE.

Sepals 5, rarely fewer, more or less coherent, usually valvate. Petals 5, hypogynous, imbricate, more or less united at the base, and adnate to the base of the staminal column, rarely 0. Stamens ∞, hypogynous, usually united into a column or tube divided into filaments at the top. Anthers often reniform, 1-celled. Torus sometimes intruded into the centre of the ovary. Ovary 2–∞-celled, rarely reduced to a single carpel, the carpels verticillate round the torus. Style divided into as many or twice as many stigmas as there are cells in the ovary, rarely entire. Ovules 1 or more in each cell. Fruit usually of 1 or more indehiscent or 2-valved cocci or a capsule. Seeds with little endosperm. Cotyledons usually folded; radicle straight or curved. Herbs, shrubs, or trees with alternate stipulate leaves and hermaphrodite regular flowers. Bracteoles 3 or more, forming an involucre, free or adnate to the calyx or 0.

Tribe I. MALVEAE.—Staminal tube bearing filaments to the apex. Style-branches as many as the ovary-cells. Ripe carpels separating from the axis.

Herbs.

* LAVATERA. Bracteoles 3–6, united at the base. Stigma linear. Carpels 1-seeded.

* MALVA. Bracteoles 3, free. Stigma linear. Carpels 1-seeded.

Shrubs or trees.

1. PLAGIANTHUS. Flowers unisexual or polygamous. Fruit a naked 1-seeded capsule.
2. HOHERIA. Flowers perfect. Stigma capitate. Carpels indehiscent, winged.
3. GAYA. Flowers perfect. Stigma obliquely capitate. Carpels whorled, 2-valved, 1-seeded, wingless.

Herb.

* MODIOLA. Bracteoles 3, free. Stigmas capitate. Carpels 2-valved, 2-seeded.

Tribe II. HIBISCEAE.—Staminal column 5-toothed or truncate at the apex, bearing anthers on the outside or rarely at the apex also. Fruit a 2–10-celled capsule.

4. HIBISCUS. Ovary 5-celled. Seeds 2 or more in each cell.

*LAVATERA, Linn.

Involucre of 3–6 bracts, connate at the base. Calyx 5-lobed. Staminal column long, filaments free at the apex. Ovary several-celled; ovules 1 in each cell; style-branches as many as cells, stigmatose along the inner surface. Fruit-carpels forming a whorl round the axis, indehiscent, 1-seeded, depressed. Seed ascending. Herbs, shrubs, or trees, pubescent or hairy, with angular or lobed leaves and racemose or cymose inflorescence.

***L. arborea**, *L.*, *Sp. Pl.* 690. A stout biennial herb, 5ft.–9ft. high, rarely branched, woody at the base. Leaves on long spreading petioles, orbicular-reniform, 5–7-angled, velvety. Flowers in axillary clusters. Peduncles shorter than the petioles. Involucre with 3 rounded spreading lobes. Flowers 1½in. in diameter.

NORTH and SOUTH Islands: naturalised in waste places, especially near the sea. *Sea-mallow.* Nov. to Jan. South Europe.

*MALVA, Linn.

Involucre of 3 bracts united to the calyx. Calyx 5-fid. Petals obcordate. Stamens numerous, coherent into a tube for the greater part of their length. Anthers reniform, 1-celled, dehiscing along the convex side. Ovary many-celled. Styles many. Fruit rounded, depressed, separating into as many indehiscent 1-seeded kidney-shaped carpels as there are styles. Seed ascending; endosperm scanty, mucilaginous. Hairy or glabrous herbs with axillary flowers.

Flowers on long peduncles.

Stems erect. Flowers large. Carpels reticulated .. .. .. * *M. sylvestris.*
Stems prostrate. Flowers small. Carpels rounded at the margin .. * *M. rotundifolia.*
Stems prostrate. Flowers small. Carpels ridged at the margin .. * *M. parviflora.*

Flowers sessile or nearly so.

Stems erect. Flowers small. Carpels ridged .. .. .. * *M. verticillata.*
Stems erect. Leaves crisped at the margin .. .. * *M. crispa.*

***M. sylvestris**, *L.*, *Sp. Pl.* 689. Biennial, pilose. Stem erect, 2ft.–3ft. high, with spreading branches. Leaves palmate, sharply 5–7-lobed; stipules erect. Segments of involucre lanceolate; calyx segments ovate-lanceolate. Flowers fascicled on slender peduncles, large, 1in.–1¼in. in diameter, purple. Carpels reticulate.

NORTH and SOUTH Islands: chiefly on the borders of fields, waste places, &c. *Purple mallow.* Nov., Dec. Europe.

***M. rotundifolia**, *L.*, *Sp. Pl.* 688. Biennial. Stems mostly decumbent, pubescent. Leaves on long petioles, reniform-crenate or crenate-dentate; lobes obscure. Flowers in few-flowered fascicles. Segments of epicalyx narrow-linear, acute, fugacious; calyx segments ovate-acuminate. Corolla fully twice as long as the calyx. Carpels with rounded edges.

NORTH and SOUTH Islands: naturalised. *Dwarf mallow.* Nov. to Jan. Europe.

***M. parviflora,** *L., Diss. Dem. Pl. Nov. Amoen. Acad.* iii. 416. Biennial. Stems decumbent at the base or suberect, rather stout, furrowed, scabrid. Leaves on long petioles, 5–7-lobed; lobes acute. Flowers in 3–6-flowered fascicles, small. Peduncles very short. Segments of epicalyx oblong-ovate, ciliate; calyx segments broadly ovate, ciliate, twice as large as the calyx. Carpels when mature forming a ridge at the line of contact, transversely rugose.

NORTH and SOUTH Islands: naturalised in waste places, &c. Nov. to Jan. Europe.

***M. verticillata,** *L., Sp. Pl.* 689. Biennial, erect, larger than *M. parviflora.* Leaves on longer petioles, 5–7-lobed. Flowers small in axillary clusters, sessile. Carpels transversely rugose, forming a ridge at the line of contact.

NORTH and SOUTH Islands: naturalised by roadsides, waste places. Nov. to Jan. Europe.

***M. crispa,** *L., Syst. ed.* x. 1147. Annual, erect. Leaves on long petioles, lobed and toothed, crisped at the margins. Flowers very small, sessile, crowded, axillary.

NORTH Island: naturalised at Port Waikato. *Curled mallow.* Nov., Dec. Europe.

1. PLAGIANTHUS, Forst.

Bracteoles 0. Sepals 5-toothed or -fid. Staminal tube divided from the apex into numerous filaments. Ovary 2–5- or rarely 1-celled. Style-branches as many as the cells, united below, stigmatose along the inner face. Fruit of 1 or more separable indehiscent carpels whorled round an axis. Seeds pendulous; raphe dorsal. Herbs, shrubs, or trees, with tough inner bark. Leaves usually entire. Flowers hermaphrodite or unisexual, axillary or terminal.

SPECIES, about 10 or 12, restricted to Australia and New Zealand. Endemic in their respective countries.

ETYM. From the Greek—*oblique*, in reference to the unequal-sided petals.

1. **P. divaricatus,** *Forst., Char. Gen.* 86. A glabrous bush or shrub, excessively branched. Branches slender, divaricating, very tough. Leaves in the young state alternate or fascicled, on long slender petioles, linear-oblong, 1in. long, sinuate; in the mature state fascicled, ¼in.–¾in. long, narrow linear-obovate or cuneate, obtuse, 1-nerved. Flowers on very short slender axillary 1-flowered peduncles, fascicled or solitary, perfect or unisexual. Calyx 5-toothed, obtuse. Ovary pubescent, 1-celled. Style 1. Stigma clavate, flattened. Fruit a 1- rarely 2-seeded spherical capsule, as large as a peppercorn, globose, downy.—A. Rich., Fl. N.Z. 299; Hook. f. in Bot. Mag. t. 3271; A. Cunn., Precurs. n. 604; Hook. f., Fl. N.Z. i. 29, and Handbk. 30; A. Gray, Bot. U.S. Expl. Exped. 181; Buch. in Trans. N.Z.I. xvi. (1883) t. 34, fig. 2.

NORTH and SOUTH Islands: North Cape to Foveaux Strait. CHATHAM Islands. Littoral. Sept., Oct.

2. **P. cymosus,** *n. s.* A shrub or small tree, with slender almost divaricating branches. Leaves alternate or fascicled, distant, linear-oblong or oblong, obtuse, sinuate with one or two obtuse teeth on each side, ½in.–¾in. long; petioles very slender. Flowers axillary, solitary, or in 2–5-flowered cymes, sometimes terminal on arrested branchlets. Peduncles very short. Calyx tubular, teeth obtuse. Petals very short. Stigma club-shaped, flattened. Ovary 2-celled; cells 1-seeded. Capsule not seen.

SOUTH Island: Upper Waimakariri, *J. D. Enys!* 2,800ft. Dunedin, *Petrie!* Apparently local.

The material at my disposal is very limited, and in poor condition; but there can be no doubt as to the specific validity of the plant.

3. **P. betulinus,** *A. Cunn., Precurs. n.* 605. A shrub or tree, 30ft.–60ft. high, with tough inner bark, the young state with interlaced tortuous branches. Trunk 1ft.–3ft. in diameter. Leaves and stipules pubescent in young plants, alternate, membranous, 1in.–3in. long, ovate or ovate-lanceolate, acuminate, entire or variously lobed or coarsely serrate, or crenate or doubly serrate; petioles slender. Flowers very numerous in slender terminal or axillary panicles, unisexual or polygamous. Calyx 5-toothed. Petals small, rounded at the tips, white, adnate with the staminal tube. Filaments short. Fruit a downy ovoid capsule, 1-celled, 1-seeded. Pericarp thin, splitting along one side.—Hook. f., Fl. N.Z. i. 29, and Handbk. 30. *P. urticinus*, A. Cunn., Precurs. n. 606. *P regium*, Poit. in Ann. Sc. Nat. (ser. 2) viii. 183, t. 3.

NORTH and SOUTH Islands; STEWART Island; CHATHAM Islands: from Mongonui southwards. Ascends to 1,500ft. *Ribbon-wood. Hohere* on Chatham Islands; *manatu* in East Cape district.

[*P. linariifolia*, Buch. in Trans. N.Z.I. xvi. (1883) 394, t. 34, f. 1, appears to be an undescribed *Coprosma*, but the material is not sufficient to allow of determination. As the ovary is clearly inferior, the plant cannot be referred to *Plagianthus*.]

2. HOHERIA, A. Cunn.

Bracteoles 0. Calyx broadly cup-shaped, 5-toothed. Petals oblique, notched near the apex, obtuse. Stamens ∞, forming 5 bundles at the apex of the column. Ovary 5-celled; cells 1-seeded. Styles 5 or rarely 6, filiform; stigmas capitate. Fruit of 5 separable indehiscent carpels, whorled round a slender axis, crested with an oblique dorsal wing. Seeds pendulous. Shrubs or trees with alternate petioled serrate leaves and axillary flowers on jointed peduncles.

An endemic genus, consisting perhaps of a single species, varying to a remarkable extent in the form and toothing of the leaves.

ETYM. A modification of the Maori name.

1. **H. populnea,** *A. Cunn., Precurs. n.* 600. A shrub or tree, 10ft.–30ft. high, with tough bark. Branchlets slender, hoary when young. Leaves varying greatly, especially in the young state, ovate-oblong, lanceolate or linear, usually entire, sharply or spinnulose-toothed or rarely obtusely-serrate; petioles slender. Flowers white, produced in great profusion in axillary fascicles on jointed 1-flowered peduncles. Peduncles pubescent.—Hook. f., Fl. N.Z. i. 30; Handbk. 31.

Subsp. **vulgaris.** Leaves coriaceous, ovate, acute or acuminate, with large acute teeth; petioles 1in.–2in. long; blade 3in.–6in. Leaves of the young state differing in size only. Fascicles 5–9-flowered. Peduncles shorter than the petioles, pubescent. Flowers ½in.–¾in. in diameter.—Hook. f. in Hook. Ic. t. 565, 566; T. Kirk, Forest Fl. N.Z. t. 53.

Var. **Sinclairii.** Leaves broadly ovate, acute and more coriaceous; petiole shorter; margins with close short obtuse serratures. Fascicles 2–3-flowered. Peduncles very short. Fruits not seen.—*H. Sinclairii*, Hook. f., Handbk. 31.

NORTH Island: Mongonui to Lower Waikato. *Houhere. Lace-bark.* March, April.

Subsp. **lanceolata.** Leaves in young state ovate, rounded-ovate, or elliptic-ovate, toothed, lobed, or deeply cut; mature state coriaceous, ovate-lanceolate or lanceolate, acute or acuminate, sharply toothed. Petioles shorter.—T. Kirk, Forest Fl. N.Z. t. 54, f. 2; t. 55, f. A.

Var. **dentata.** Leaves lanceolate or elliptic-lanceolate, sharply and doubly serrate or dentate. Fascicles 3-4-flowered.—T. Kirk, Forest Fl. N.Z. t. 54A, f. 1 and 2.

NORTH and SOUTH Islands: from Whangarei to Canterbury; local in the north. *Thousand-jacket. Houhere.* Feb. to April.

Subsp. **obtusifolia.** Leaves membranous, elliptic-lanceolate or oblong, subacute or obtuse, sharply toothed or spinulose-toothed; petioles short. Fascicles 3-4 flowered.—T. Kirk, Forest Fl. N.Z. t. 54, f. 1, t. 54A, f. 3.

NORTH Island: chiefly in the Whanganui district. Dec., Jan.

Subsp. **angustifolia** (*sp.*) Raoul in Ann. Sc. Nat. (ser. i.) iii. (1844) 122; Choix de Pl. t. 26. Leaves in young state solitary or fascicled on long slender flexuous shoots, suborbicular, cuneate at base, 3-5-toothed; mature leaves linear-oblong, obtuse, cuneate at the base, spinulose-toothed; petioles very short. Fascicles 2-3-flowered. Petioles, peduncles, leaves, and branchlets often hoary or pubescent.—T. Kirk, Forest Fl. N.Z. t. 54B, f. 2.

Var. **acutifolia.** Leaves on mature plants usually soft, linear-lanceolate, acuminate, 1in.–1½in. long, with strong spinulose teeth; petioles very short. Pedicels very slender.—T. Kirk, Forest Fl. N.Z. t. 55, f. 1 and 2.

NORTH and SOUTH Islands: East Cape to Otago. Jan., Feb.—*H. sexstylosa*, Colenso in Trans. N.Z.I. xvii. (1884) 238, appears to be a form of this, characterized by "6 or 7 styles, the leaves acuminate, sometimes truncate at the tips." Dec., Jan.

All the forms are confined to lowland districts, rarely exceeding 1,000ft. above sea-level. They pass into each other by insensible gradations, so that it is impossible to draw well-defined lines of separation.

3. GAYA, Humb., Bonp., and Kunth.

Bracteoles 0. Calyx 5-fid. Staminal tube divided at the apex into ∞ filaments. Ovary ∞-celled; cells 1-seeded. Branches of the style as many as the cells, filiform. Stigma capitate or truncate. Mature carpels membranous, connivent at the apex, separable from the axis, dehiscing at the back into 2 valves, leaving the internal dorsal ligament ascending from the base round the seed. Seeds pendulous or horizontal. Herbs or shrubs, rarely small trees, often tomentose. Leaves alternate, entire. Flowers pedunculate, axillary or terminal.

SPECIES, about 7, ranging from Mexico to Brazil. One species endemic in New Zealand.

1. **G. Lyallii,** *J. E. Baker in Journ. Bot.* xxx. (1892) 37. A shrub or small tree, 10ft.–30ft. high, clothed with stellate white pubescence. Stipules large, deciduous. Leaves ovate, acute or acuminate, cordate or truncate at base, 2in.–4in. long, deeply or doubly crenate or toothed or lobed, glabrous above when mature; petioles slender, 1in.–1½in. long. Flowers in ebracteolate fascicles, rarely solitary, ¾in. in diameter. Peduncles 1in.–2in. long, 1-flowered. Calyx campanulate, teeth deltoid. Petals oblique, notched on one side near the apex, white. Staminal tube short. Ovary 10–12-celled. Style divided near the top into short arms. Stigma obliquely capitate. Fruit discoid, depressed, separating into as many carpels as stigmas. Carpels 2-valved, the valves separating from the internal dorsal ligament, 1-seeded. Seed pendulous, much compressed.—*Plagianthus Lyallii*, A. Gray *ex* Hook. f., Fl. N.Z. ii. 326, and Handbk. 30; T. Kirk, Forest Fl. N.Z. t. 134. *Sida Lyallii*, F. Muell. in Veg. Chath. Isds. *Hoheria Lyallii*, Hook. f., Fl. N.Z. i. 31, t. 11.

Var. **ribifolia.** Leaves deeply lobed or partite.

SOUTH Island: margins of subalpine forest from Nelson and Marlborough to Otago. Ascends to 3,000ft. Usually deciduous at high levels. *Lacebark.* Dec., Jan.

* **MODIOLA,** Moench.

Involucre of 3 bracts, free. Calyx 5-fid. Staminal column divided at the apex into numerous filaments. Ovary ∞-celled. Style-branches as many as cells. Carpels with 2 stout dorsal bristles, ∞, 2-valved, transversely septate. Herbs with prostrate stems, divided leaves, and axillary flowers.

* **M. multifida,** *Moench., Meth.* 620. A much-branched perennial with prostrate stems. Leaves on slender petioles, palmate or ovate-cordate, 3–5-lobed; segments coarsely toothed or incised, glabrate. Flowers axillary, on slender 1-flowered pedicels. Segments of involucre ovate-lanceolate, short. Calyx segments ovate, hirsute. Petals slightly exceeding the calyx, red. Carpels 15–20, hairy at the top, each valve crested with hairs. Seeds reniform.

NORTH Island: naturalised in many localities. SOUTH Island: Nelson. North and South America.

4. HIBISCUS, Linn.

Bracteoles numerous or rarely few, usually narrow, free or coherent. Calyx 5-lobed or toothed. Staminal column with the filaments inserted below the 5-toothed apex. Ovary 5-celled; ovules 2 or more in each cell; styles 5, spreading; stigmas capitate. Capsule 5-valved, loculicidal. Seeds glabrous or hispid. Herbs, shrubs, or trees, with alternate entire or divided leaves. Stipules fugacious.

SPECIES, 150. Widely distributed in tropical regions; rare in temperate countries.

ETYM. From the Greek, but of doubtful meaning.

Flowers axillary .. .. .. 1. *H. Trionum.*
Flowers in terminal racemes .. 2. *H. diversifolius.*

1. **H. Trionum,** *L., Sp. Pl.* 697. Annual or biennial. Stems nearly simple or with spreading branches, almost woody below, 1ft.–2ft. high, hispid or scabrid, pubescent. Leaves shortly petioled, 1in.–3in. long, palmately deeply 3–5-lobed; lobes oblong or lanceolate, irregularly toothed or sinuate. Flowers large, 1in.–1¼in. in diameter, axillary on short peduncles. Bracteoles 7–12, narrow-linear, acute, hispid. Calyx membranous, inflated, with raised veins. Capsule ovoid, hispid, enclosed in the calyx. Seeds wrinkled, glabrous.—DC., Prod. i. 453; Bot. Mag. t. 209; Hook. f., Fl. N.Z. i. 28, Handbk. 31; Benth., Fl. Austr. i. 210. *H. vesicarius,* Cav., Diss. iii. 171, t. 62, f. 2; A. Cunn., Precurs. n. 607. *H. tridactylites,* Lindl. in Mitch. Three Exped. i. 85.

NORTH Island: chiefly from North Cape to Whangarei and Kaihu, but local. Great Barrier Island, *T. K.* SOUTH Island: South Whanganui, *Lyall.*

Flowers handsome, straw-coloured, with a dark eye. I fully agree with Mr. Colenso in considering this species indigenous in the North. It is common in Australia, South Africa, southern Asia, and China.

2. **H. diversifolius,** *Jacq., Icon. Pl. rar.* 3, *t.* 551. A stout rigid perennial herb, often woody at the base. Branches, petioles, and rarely the principal ribs of the leaf clothed with small prickles mixed with setae. Leaves on stout petioles 2in.–3in. long; blade 2in.–4in., broadly cordate or rounded-cordate, obscurely 3–5-lobed, coarsely irregularly toothed. Flowers large, handsome, 2in. in diameter or more, in terminal elongated racemes. Floral leaves small, 3-fid or lanceolate. Pedicels very short or rarely 0, solitary or in pairs.

10

Bracteoles linear. Calyx-lobes lanceolate, hispid or bristly. Capsule acuminate, densely hispid or strigose. Seeds glabrous.—DC., Prod. i. 449; Bot. Reg. t. 381; Benth., Fl. Austr. i. 213; T. Kirk, Trans. N.Z.I. iii. (1870) 163. *H. Beckleri*, F. Muell., Fragm. ii. 117. *H. Taylori*, Buch. in Trans. N.Z.I. ii. (1869) 173.

NORTH Island: North Cape to Maunganui Bluff and Bay of Islands; local. Also in Australia, South Africa, Mauritius, Madagascar, and South Pacific islands.

ORDER XI.—TILIACEAE.

Sepals 5, rarely 4 or 3, free or connate at the base, usually valvate. Petals as many as the sepals, lobed or cut, inserted round the base of the torus, rarely 0. Stamens ∞, free or the filaments coherent at the base. Anthers 2-celled, opening by slits or by apical pores. Ovary sessile on the torus, free, 2–10-celled; style entire or the number of stigmatic lobes equalling the number of cells; ovules 1 or 2 or more attached to the axis of each cell, erect, pendulous or horizontal. Fruit a capsule, drupe, or berry. Seeds usually with fleshy endosperm. Radicle next the hilum. Trees or shrubs, rarely herbs, with alternate or rarely opposite entire leaves. Stipules free or rarely 0. Flowers hermaphrodite or unisexual, axillary or terminal.

A large order, chiefly distributed through tropical or subtropical countries. *Tilia* extends into temperate and cold regions in the Northern Hemisphere, and *Aristotelia* has a similar but more restricted distribution in the Southern Hemisphere. *Entelea* is endemic in New Zealand. GENERA, 40. SPECIES, 340.

1. ENTELEA. Leaves alternate. Fruit a spinous capsule.

2. ARISTOTELIA. Leaves opposite. Fruit a berry.

3. ELAEOCARPUS. Leaves alternate. Fruit a drupe.

1. ENTELEA, R. Br.

Sepals 4 or 5, free. Petals 4 or 5, undulate. Stamens ∞, inserted on a low torus; anthers versatile. Ovary 4–6-celled, many-ovuled; style simple, stigma terminal. Capsule globose, 4–6-valved, echinate, loculicidal. Seeds ∞. A shrub or small tree, with alternate palmatinerved cordate toothed leaves and umbellate cymes of large white flowers.

A monotypic genus, restricted to the northern part of the colony.

ETYM. From the Greek, signifying *perfect*, referring to the stamens as contrasted with those of *Sparmanniae*, many of which are abortive.

1. **E. arborescens**, *R. Br. in Bot. Mag. t.* 2480. A shrub or small tree, sometimes 25ft. high. Young branchlets, peduncles, and leaves beneath, &c., clothed with white stellate down. Leaves broadly cordate, 4in.–9in. long or more, acuminate, oblique, sometimes lobed, doubly or trebly crenate or serrate. Stipules linear, persistent. Flowers in erect axillary or terminal umbellate cymes. Capsules 4–7-celled, echinate; bristles stout, 1in. long or more, mixed with hairs at the base. Seeds numerous in two rows; endosperm oily. Radicle very short.—A. Cunn., Precurs. n. 601; Hook. f., Fl. N.Z. i. 33, Handbk. 32; T. Kirk, Forest Fl. N.Z. t. 33. *Apeiba australis*, A. Rich., Fl. N.Z. t. 34. *Corchorus sloanoides*, Banks and Sol., Ic. *et* MSS.

NORTH Island: North Cape to Cook Strait; Great and Little Barrier Islands, &c. Chiefly littoral, and very rare outside the Auckland District. Cape Palliser, Paikakariki, &c., rare. SOUTH Island: Nelson: Collingwood, and islands near Cape Farewell, *Hector, Kingsley!* *Whau. Hau-ama. Corkwood.* Oct. to Jan.

The wood is very light, and is used by the Maoris as floats for their fishing-nets.

2. ARISTOTELIA, L'Hérit.

Sepals 4 or 5, valvate. Petals 4 or 5, imbricate, 3-lobed or toothed, rarely entire or minute, inserted around the base of the thickened torus. Stamens 4 or 5 or ∞, inserted on the torus; anthers linear, with short apical slits. Ovary 2–4-celled; ovules 2 in each cell; style subulate. Fruit a berry. Seeds ascending or pendulous, often fleshy outside the hard testa. Embryo straight. Shrubs or small trees, usually with opposite entire or toothed leaves and unisexual often polygamous flowers.

Besides the 2 or 3 species endemic in New Zealand, 3 others are found in Australia, 1 in the New Hebrides, 1 in Chili, and 1 in Central America.

ETYM. In memory of the Macedonian philosopher Aristotle.

Leaves on long petioles. Flowers in large axillary panicles .. .. .. 1. *A. racemosa.*
· · Leaves on short petioles. Flowers in small panicles or simple racemes .. .. 2. *A. Colensoi.*
Flowers solitary or in small cymes or racemes .. .. 3. *A. fruticosa.*

1. **A. racemosa,** *Hook. f., Fl. N.Z.* i. 33. A shrub or small tree, 6ft.–30ft. high. Bark of young branches red. Branchlets, petioles, young leaves, and panicles pubescent. Leaves on slender petioles, opposite or sub-opposite, membranous, ovate-cordate or ovate-acuminate, irregularly and sharply serrate. Flowers in much-branched slender axillary panicles. Sepals 4, free. Petals 4-lobed at the tips. Stamens inserted on a glandular disk; anthers equalling or exceeding the filaments. Female flowers: ovary 3- or 4-celled; styles 3 or 4, short, straight. Fruit a 3- or 4-celled berry, about the size of a pea, red. Seeds 8, angular.—Handbk. 33; T. Kirk, Forest Fl. N.Z. t. 113. *Friesia racemosa,* A. Cunn., Precurs. n. 603; Hook. f. in Hook. Ic. t. 601. *Elaeocarpus Dicera,* Vahl., Symb. iii. 67. *Triphalia rubicunda,* Banks and Sol. MSS.

NORTH and SOUTH Islands; STEWART Island: chiefly in lowland districts, but ascends to fully 2,000ft. Flowers very attractive. *Makomako. Wineberry.* Nov., Dec.

The wood is converted into charcoal for the manufacture of gunpowder. The berries seem likely to be used for colouring wines.

2. **A. Colensoi,** *Hook. f., Handbk.* 33. A shrub, rarely exceeding 6ft. in height, with opposite branches. Leaves 1in.–1½in. long, ovate-lanceolate or ovate, serrate, membranous. Racemes axillary. Female 5–8-flowered; pedicels very slender. Berry the size of a peppercorn. Seeds 4-angled.

NORTH Island: Wairarapa Valley. SOUTH Island: Otago: chiefly in subalpine districts. Perhaps a variety of the preceding.

3. **A. fruticosa,** *Hook. f., Fl. N.Z.* i. 34. A much-branched shrub, 3ft.–6ft. high, erect, suberect, or decumbent. Branchlets red. Leaves in young state narrow-linear-lanceolate, acute or acuminate, deeply toothed, lobed or pinnatifid, ¾in.–1½in. long. Leaves of mature state on very short petioles,

coriaceous or rarely membranous, ¼in.–1in. long, ovate-obovate, linear-oblong or elliptic-oblong, rounded at the tip, quite entire or crenate or serrate. Flowers solitary or in axillary 3–6-flowered cymes or racemes. Pedicels pubescent. Sepals 4, oblong, obtuse, pubescent or ciliate. Petals 4, equalling or slightly exceeding the sepals, with 1–4 crenatures at the apex. Stamens 6–8; anthers much longer than the filaments. Seeds 4, angled.—Handbk. 33. *Myrsine brachyclados,* Col. in Trans. N.Z.I. xxii. (1889) 478.

Var. **erecta** (*sp.*), Buch. in Trans. N.Z.I. iii. (1870) 209. Branches rigid, usually opposite, pubescent in the young state. Leaves oblong or elliptic-oblong or oblong-lanceolate, acutely serrate. Cymes or racemes ⅔in. long, axillary. Flowers as in the typical form.

NORTH and SOUTH Islands; STEWART Island: from the Thames southwards. Ascends to upwards of 3,000ft., but varies to a remarkable degree in habit and foliage.

3. ELAEOCARPUS, Linn.

Sepals 4 or 5, valvate. Petals 4 or 5, valvate, lobed or laciniate, inserted at the base of the torus. Stamens ∞, inserted on the glandular torus. Anthers long, awned, opening by a short terminal slit. Ovary 2–5-celled, with 2 or more pendulous ovules in each cell; style subulate. Fruit a drupe, with a hard or bony putamen, 2–5- rarely 1-celled, furrowed or rugose. Seed solitary in each cell, pendulous; testa hard. Trees, with usually alternate entire or serrate exstipulate leaves. Flowers usually hermaphrodite, rarely polygamous, racemose.

SPECIES, about 55. Most plentiful in tropical Asia, extending to the Pacific Islands, New Caledonia, and Australia; but the New Zealand species are endemic.

ETYM. From the Greek, signifying *olive* and *fruit*, the drupe often closely resembling an olive.

Leaves oblong-obovate, whitish beneath, margin recurved .. 1. *E. dentatus.*
Leaves linear-oblong or lanceolate, margins flat .. .. 2. *E. Hookerianus.*

1. **E. dentatus,** *Vahl., Symb. Bot.* iii. 66. A round-headed tree. Trunk 1ft.–3ft. in diameter. Branchlets often naked except at the tips, silky when young. Leaves alternate, in the young state 3in.–6in. long or more, subcoriaceous, elliptic, acute, sinuate; mature state 2in.–3in. long, very coriaceous, linear-oblong or obovate, acute, white with silky appressed down beneath, serrate or sinuate-serrate, margins recurved; petioles short. Racemes numerous, silky, 8–10-flowered. Flowers ⅓in. in diameter, drooping. Petals broadly obovate, 4–5-lobed above, lobes rounded. Filaments 10–12, very short; anthers tetragonous, with a flat recurved tip. Ovary silky, 2-celled; ovules 4 in each cell. Drupe ½in.–¾in. long; stone rugose, 1-celled, 1-seeded.—Hook. f. in Hook. Ic. Pl. t. 602; Fl. N.Z. i. 32; Handbk. 31; T. Kirk, Forest Fl. N.Z. t. 11. *E. hinau,* A. Cunn., Precurs. n. 602. *E. Cunninghamii,* Raoul, Choix de Pl. 25. *Dicera dentata* and *D. serrata,* Forst., Char. Gen. 80.

NORTH and SOUTH Islands: from the North Cape to Catlin's River. Ascends to 2,000ft. *Hinau.* Oct., Nov.

2. **E. Hookerianus,** *Raoul, Choix de Pl.* 26, *t.* 25. A glabrous tree. Trunk 1ft.–3ft. in diameter. Bark white. Branchlets tortuous and interlaced in young state, with narrow-linear leaves ¾in.–1½in. long, irregularly toothed or

lobed, sometimes shortly obovate or almost orbicular; in mature state linear-oblong, narrowed at both ends or almost lanceolate, very coriaceous, flat, obtuse, 1½in.–2½in. long, sinuate or serrate. Racemes very slender, shorter than the leaves, 10–14-flowered. Flowers drooping, small. Petals 4–5-lobed at the tip; lobes obtuse, unequal. Drupe smaller, rugose.—Hook. f., Fl. N.Z. i. 32, Handbk. 34; T. Kirk, Forest Fl. N.Z. t. 12, 13. *E. serratus*, Banks and Sol. MSS.

NORTH and SOUTH Islands; STEWART Island: rare and local north of the Auckland Isthmus; more plentiful in the South. Ascends to 3,000ft. *Pokaka. Mahimahi.* Nov. to Jan.

ORDER XII.—LINEAE.

Sepals 5, rarely 4, free or coherent at the base, imbricate or rarely valvate. Petals 5, rarely 4, hypogynous or perigynous, contorted. Stamens as many as the petals or twice or thrice as many, monadelphous below, with 5 minute glands at the base of the tube. Ovary 3–5-celled, free; styles 3–5, free or coherent below; stigmas terminal; cells 1–2-ovuled. Fruit a capsule splitting septicidally into 1–2-seeded cocci, or a drupe with 1 or more pyrenes. Seeds with scanty endosperm. Embryo usually straight. Radicle superior. Herbs, shrubs, rarely trees, with simple leaves and hermaphrodite flowers.

GENERA, 14. SPECIES, 140. Most of the genera are restricted to tropical countries, but *Linum*, the largest genus, is chiefly found in temperate and extra-tropical regions.

1. LINUM, Linn.

Sepals 5, entire, persistent. Petals 5, contracted. Stamens 5, alternating with staminodia. Ovary 5-celled; cells 2-ovuled, imperfectly divided by a longitudinal septum. Cocci 5, 2-seeded; or 10, 1-seeded, by each coccus splitting along the septum. Glabrous herbs, with narrow entire simple leaves, fibrous cortex, and terminal flowers.

SPECIES, 85. Generally distributed through the temperate and extra-tropical regions.

ETYM. The Latin word for *flax*.

The flax of common use is afforded by *L. usitatissimum*. The New Zealand species is endemic, but several exotic species are naturalised in the colony.

Leaves alternate, linear.

Perennial, erect. Flowers large, white. Sepals ovate-lanceolate	1. *L. monogynum.*
Perennial. Stems spreading. Flowers blue. Sepals ovate, acute, with scarious margins	* *L. marginale.*
Annual, erect. Flowers blue. Sepals ovate, acute ..	* *L. usitatissimum.*
Annual, erect. Sepals lanceolate-acuminate	* *L. gallicum.*

Leaves opposite.

Annual. Leaves obovate. Flowers nodding in bud, white ..	* *L. catharticum.*

1. **L. monogynum,** *Forst., Prod. n.* 145. A perennial herb, woody at the base, simple or branched, 6in.–24in. high, usually erect. Leaves numerous, scattered, ascending, narrow-linear-oblong or lanceolate or linear-subulate, ¼in.–1in. long, 1–3-nerved. Flowers in terminal corymbs, usually white, rarely pale-blue. Sepals ovate or ovate-lanceolate, acute. Styles united at the base, recurved above. Capsule large, rounded-ovate, divided into 10 1-seeded cocci,

—DC., Prod. i. 428; A. Rich., Fl. N.Z. 317; A. Cunn., Precurs. n. 608; Hook., Bot. Mag. t. 3574; Hook. f., Fl. N.Z. i. 28, Handbk. 35.

NORTH and SOUTH Islands; STEWART Island; CHATHAM Islands: extends northward to the Three Kings Islands. Chiefly littoral, but is occasionally found inland, and ascends to 2,000ft.

A beautiful plant, varying greatly in habit according to situation and exposure. *Rauhuia*. Oct. to Jan.

* **L. marginale,** *A. Cunn. in Hook. Lond. Journ. Bot.* vii. (1848) 169. A glabrous perennial, usually much branched from the base. Stems slender, spreading or ascending, 1ft.–2ft. high. Leaves scattered, ½in.–¾in. long, linear or narrow-linear-lanceolate, acute. Flowers in a lax irregular corymb, blue. Sepals ovate or ovate-lanceolate, acute or acuminate, with a narrow membranous margin and strong median nerve. Styles united below. Capsule smaller than in *L. monogynum*, dividing into 10 1-seeded cocci.

NORTH and SOUTH Islands: abundantly naturalised in Auckland; less frequent in the southern provinces.

I formerly supposed this species to be indigenous in the Auckland District, but the balance of evidence shows that it must have been introduced. Nov. to Jan. Australia.

* **L. usitatissimum,** *L., Sp. Pl.* 277. A glabrous annual, or rarely perennial. Stems erect, 12in.–18in. high. Leaves linear-lanceolate, ¾in.–1¼in. long. Sepals ovate, acute, ciliated when young, 3-nerved, lateral often obscure. Capsule glabrous within.

NORTH and SOUTH Islands: frequent, but scarcely naturalised. *Common flax*. Dec., Jan. Europe, temperate Asia, &c.

* **L. gallicum,** *L., Sp. Pl. ed.* ii. 401. A slender annual or rarely perennial herb, sometimes woody at the base. Leaves narrow-linear-lanceolate or linear-subulate, slightly scabrid at the margins. Flowers numerous, small, yellow, forming lax irregular corymbs. Pedicels often capillary. Sepals lanceolate-acuminate, glandular, ciliate or slightly scabrid when young. Capsule small, 10-seeded.

NORTH Island: naturalised from the Waitemata to Middle Waikato. Nov. to Feb. Mediterranean, Abyssinia, &c.

* **L. catharticum,** *L., Sp. Pl.* 281. A slender glabrous annual, 3in.–6in. high. Flower-buds nodding. Lower leaves opposite, linear-oblong or obovate, obtuse; upper leaves alternate, lanceolate. Panicles forked. Branches short, spreading. Flowers small, white. Sepals minutely serrate when young. Capsule small, 10-seeded.

NORTH and SOUTH Islands: naturalised in several districts, but local at present, or possibly overlooked. *Heath-flax*. Nov., Dec.

ORDER XIII.—GERANIACEAE.

Sepals 5 or fewer, free, imbricate or rarely valvate; one sometimes spurred. Petals 5 or fewer, hypogynous or slightly perigynous, imbricate. Torus more or less expanded into an inconspicuous disk often bearing glands alternating with the petals. Stamens 10, hypogynous, 5 of them often imperfect; filaments often united below. Ovary with as many lobes or cells as sepals; ovules 1 or more in each cell. Fruit of 5 carpels, united round a long beaked axis, each terminated by an elastic cartilaginous style, which ultimately separates from the axis and becomes rolled or spirally twisted upwards, carrying the seed with it, or the lobes open loculicidally, or the fruit separates into cocci; rarely an indehiscent drupe. Endosperm usually scanty or 0. Embryo straight or curved. Radicle sometimes long and folded over the cotyledons. Herbs, shrubs, or trees, with opposite or alternate rarely entire stipulate leaves and axillary peduncles.

GENERA, 20. SPECIES, 750. Chiefly inhabiting temperate regions; plentiful in South Africa. A few species are extra-tropical or tropical. The New Zealand genera have a wide distribution.

Capsule beaked, the lobes 1-seeded and elastically carried upwards to the apex of the beak. Leaves toothed or divided.

1. GERANIUM. Flowers regular. Anthers usually 10. Styles combined.

Beaks of the carpels revolute.

* ERODIUM. Flowers regular. Styles combined. Anthers 5.

Beaks of the carpels spirally twisted.

2. PELARGONIUM. Flowers irregular, with a short spur adnate to the pedicel. Anthers 5-7.

Capsule separating into 3 indehiscent cocci.

* TROPAEOLUM. Flowers irregular, with a spur. Leaves peltate.

Capsule opening loculicidally.

4. OXALIS. Flowers regular. Leaves 3-foliolate.

1. GERANIUM, Linn.

Sepals 5. Petals 5, alternating with 5 glands. Stamens 10, all with perfect anthers, free or monadelphous. Ovary 5-celled, beaked, the beak terminated by the 5-lobed stigma; cells 2-ovuled. Lobes of the capsule separating from the axis and curved upwards on a long awn parted from the beak and glabrous inside. Radicle turned back on the plicate or convolute cotyledons. Herbs, rarely woody at the base, with stipulate lobed or divided leaves and axillary bracteolate 1–2-flowered peduncles.

SPECIES, about 100. Widely distributed over the globe, but most abundant in the Northern Hemisphere; rare within the tropics. Two of the New Zealand species are endemic.

ETYM. An old Greek name, from *crane*, in reference to the beak-like carpels.

Leaves orbicular or suborbicular.

Stems erect. Leaves much divided. Peduncles 2-flowered. Sepals awned .. 1. *G. dissectum.*

Stems prostrate, stout. Leaves with broader lobes .. .. .. var. *glabratum.*

Stems prostrate, weak, downy. Peduncles 1-flowered. Sepals scarcely awned .. .. 2. *G. microphyllum.*

Stemless. Rootstock thick. Peduncles 1-flowered. Sepals awned .. 3. *G. sessiliflorum.*

Stems prostrate, stout. Leaves hoary. Flowers large. Peduncles 1-flowered. Sepals awnless .. .. 4. *G. Traversii.*

Stems prostrate. Leaves soft. Peduncles 2-flowered. Sepals awnless. Carpels wrinkled .. .. * *G. molle.*

Leaves 1- or 2-pinnatifid.

Stems erect or suberect. Sepals with long awns .. .. .. * *G. Robertianum.*

1. **G. dissectum**, *L., Cent.* i. 21, *var.* **australe**. Annual or perennial, 1ft.–2ft. high, erect, decumbent or prostrate, usually clothed with soft spreading or retrorse hairs, rarely downy or glabrate. Leaves on long petioles, nearly orbicular, cut to the base into 5–7 lobes; lobes pinnatifid; segments broad or narrow, obtuse. Peduncles 2-flowered. Sepals ovate or ovate-lanceolate, or oblong with a long awn, hairy. Petals notched, about as long as the sepals or

exceeding them, red-purple. Carpels with long erect hairs; beaks hairy or downy. Seeds reticulated or pitted.—Benth., Fl. Austr. i. 296. Var. *carolinianum*, Hook. f., Fl. N.Z. i. 37, Handbk. 36.

Var. **pilosum**. Stems erect or spreading, clothed with spreading hairs. Seeds pitted.—*G. patagonicum*, Hook. f., Fl. Antarc. ii. 252.

Var. **patulum**. Usually spreading, clothed with retrorse or spreading hairs. Seeds reticulated.—*G. retrorsum*, DC., Prod. i. *G. patulum*, G. Forst., Prod. n. 531. Perennial.

Var. **glabratum**. Stems prostrate, stout, glabrate or nearly glabrous. Leaves 3–5-partite to below the middle; segments shortly pinnatifid or lobed; tips subacute. Sepals broadly oblong, awned. Carpels glabrate or pubescent. Seeds reticulated; beak glabrous or downy.

NORTH and SOUTH Islands; KERMADEC Islands: CHATHAM Islands: common in the North, less frequent in the South. Also in North and South America, Australia.

Var. *glabratum* appears to be confined to the North Island. It looks very different from the forms with much-divided leaves, but the characters are not sufficiently distinctive to warrant its separation. It is certainly perennial, but the other forms are often annual.

2. **G. microphyllum**, *Hook. f., Fl. Antarc.* i. 8, *t.* 5. A prostrate straggling perennial with numerous slender branches, 6in.–18in. long, pubescent or clothed with short spreading hairs. Leaves on long petioles, orbicular, cut to below the middle into 3–7 broad or narrow lobes, more or less cuneate at the base; lobes toothed but not pinnatifid. Peduncles 1- rarely 2-flowered. Sepals ovate-lanceolate, with short awns, pubescent. Petals white, exceeding the sepals, entire or retuse. Carpels smooth, unequally pubescent. Seeds minutely longitudinally striated, scarcely reticulated or dotted.—*G. potentilloides*, L'Hérit. *ex* DC. Prod. i. 639; Hook. f., Fl. N.Z. i. 40. *G. dissectum*, var., Benth., Fl. Austr. i. 296. *G. australe*, Nies in Pl. Preiss. i. 162.

From the KERMADEC to the AUCKLAND Islands: common. Ascends to 2,800ft. Also in Australia, &c.

This is united with *G. dissectum* by Bentham and Mueller, but, as it seems to me, without cause. Certainly the seeds differ in shape and in the narrow reticulations from any form of that plant in New Zealand.

3. **G. sessiliflorum**, *Cav., Diss.* 198, *t.* 77, *f.* 2. Perennial. More or less silky in all its parts. Rootstock stout and woody. Leaves ½in.–¾in. in diameter on slender petioles, crowded, 3–7-partite; segments broad, lobed; stipules broad. Flowering stems very short or 0. Peduncles short, 1- rarely 2-flowered. Sepals broadly oblong, shortly awned, silky. Petals exceeding the sepals, retuse, white. Carpels hairy. Seeds smooth. *G. brevicaule*, Hook., Journ. Bot. i. (1834) 252; Hook. f., Fl. N.Z. i. 40.

NORTH and SOUTH Islands: STEWART Island: from Lower Waikato southwards. Ascends to 3,000ft. Nov. to Jan. Also in Fuegia, Chili, and Australia.

4. **G. Traversii**, *Hook. f., Handbk.* 726. Perennial. Stems prostrate, 6in.–18in. long or more. Hoary with dense silvery pubescence. Radical leaves on long petioles, orbicular, 1in.–2in. in diameter, 5–7-lobed to the middle; lobes toothed, silky on both surfaces. Stipules large, ovate-acuminate. Cauline leaves smaller. Peduncles 1-flowered, with 2 linear-acuminate bracts about the middle. Flowers ¾in. in diameter, handsome. Sepals broadly ovate, acute, scarcely awned. Petals broadly obovate, entire, whitish-red. Carpels silky. Seeds finely reticulated.

CHATHAM Islands. *H. H. Travers!* Nov., Dec.

A beautiful plant, with larger flowers than any other New Zealand species.

* **G. molle,** *L., Sp. Pl.* 682. Annual or perennial. Stems procumbent or ascending, 6in.–12in. long, more or less pilose, slender or rather stout. Leaves on long petioles, orbicular, 5–9-lobed; segments partite, lobed or obtusely toothed. Peduncles 2-flowered, axillary; bracts membranous, usually small. Sepals oblong or oblong-lanceolate, mucronate. Petals shortly exceeding the sepals, reddish or white, 2-fid; claw very short, hairy. Carpels transversely wrinkled or nearly smooth, glabrous.—Hook. f., Fl. N.Z. i. 40, and Handbk. 37.

NORTH and SOUTH Islands; STEWART Island; CHATHAM Islands: abundantly naturalised. Common in lowland situations. Dec., Jan.

Some of the carpels are quite as much wrinkled as in European specimens, while others are perfectly smooth; but all intermediate grades may be found. Nov. to Jan. Europe, North Africa, West Asia. This was first observed by Dr. Lyall and Mr. Colenso.

* **G. Robertianum,** *L., Sp. Pl.* 681. Annual or rarely perennial. Stems branched, spreading, reddish, 6in.–15in. long, sparsely clothed with spreading hairs. Leaves on long petioles, 3- or rarely 5-foliolate; leaflets petiolulate, 2-pinnatifid; segments lobed. Flowers in pairs on long axillary peduncles. Pedicels divergent. Sepals with long awns, deeply furrowed, hairy. Petals with the blade equalling the glabrous claw, red. Carpels more or less wrinkled, hairy or glabrous.

NORTH and SOUTH Islands: naturalised in cool, damp situations, but rather local. Abundant about Wellington. *Herb-Robert.* Dec., Jan. Europe, North Africa, South Africa, Siberia, West Asia.

* ERODIUM, L'Hérit.

Sepals 5, imbricate. Petals 5, rarely 3 by suppression, hypogynous. Stamens monadelphous, 5 fertile with glands at their base, 5 reduced to staminodes. Carpels with 2 apical pits, the long beaks spirally twisted and bearded on the inner surface. Herbs, with more or less pubescent leaves and membranous stipules. Flowers solitary or in umbels on naked axillary peduncles.

Leaves entire or lobed, usually cordate .. .. * *E. malachoides.*
Leaves pinnate. Apical pits glandular .. .. * *E. moschatum.*
Leaves pinnate or 2-pinnate. Apical pits glandular .. .. .. * *E. cicutarium.*

* **E. malachoides,** *Willd., Phyt.* 10. Stems suberect, spreading or prostrate, 3in.–18in. high. Whole plant glandular-pubescent. Leaves petiolate, ovate or broadly ovate, usually cordate, obtuse, entirely or deeply lobed or coarsely serrate; blade ¾in.–2in. long. Peduncles long or short; umbels 3–7-flowered. Sepals strongly awned, 3–5-nerved, with membranous margins. Filaments all glabrous, very slightly coherent at base. Carpels with spreading hairs. Apical pits glandular, with deep furrow beneath. In dry seasons the leaves are small, narrow-ovate, and rounded at the base.

NORTH Island: naturalised, Bay of Islands and Wellington. Oct. to April. South Europe, North Africa, &c.

* **E. moschatum,** *L'Hérit.* ex *Ait. Hort. Kew. ed.* 1, ii. 414. Stems 6in.–24in. high, stout, glabrate or clothed with spreading hairs. Leaves 6in.–9in. long, pinnate; leaflets sessile, unequally ovate, obtuse; margins inciso-serrate. Stipules ovate. Umbels on long many-flowered peduncles. Filaments of perfect stamens toothed at the base. Carpels hairy; apical pit glandular, with a concentric furrow beneath. Often diffusing a musky odour when bruised.

NORTH and SOUTH Islands: naturalised in many places, but rather local. *Musky stork's-bill.* Oct. to Feb. Europe, North Africa, West Asia.

* **E. cicutarium,** *L'Hérit.* ex *Ait. Hort. Kew. ed.* 1, ii. 414. Stems 6in.-24in. high, procumbent or suberect; whole plant more or less hairy. Leaves pinnate or 2-pinnate; leaflets sessile, deeply toothed or lobed or cut into narrow segments. Stipules lanceolate. Flowers in 3–5-flowered umbels. Filaments of perfect anthers, not toothed below. Carpels hairy; apical pit eglandular, with a concentric furrow beneath.

Var. **pilosum.** Peduncles greatly exceeding the leaves, pilose. Leaflets deeply pinnatifid or almost pinnate. Upper petals not spotted. Carpels not furrowed.

NORTH and SOUTH Islands; STEWART Island; CHATHAM Islands: abundantly naturalised. *Stork's-bill.* Sept. to March. Europe, North and East Africa, &c.

2. PELARGONIUM, L'Hérit.

Flowers irregular. Sepals 5, slightly coherent at the base and produced into a spur adnate with the pedicel. Petals 5 or fewer, the two upper larger than the lower. Disk with 5 glands. Stamens 10, of which 3 or 5 only are fertile, slightly coherent at the base. Ovary and fruit as in *Erodium.* Herbs or under-shrubs, with opposite or alternate simple or divided stipulate leaves. Flowers in umbels on naked axillary peduncles.

SPECIES, about 180. All restricted to South Africa except 3 in North Africa and the Levant, and 2 others in Australia and New Zealand.

Leaves cordate or ovate-cordate .. .. 1. *P. australe.*
Leaves palmatifidly divided. Segments pinnatifid or toothed .. * *P. quercifolium.*
Leaves orbicular-reniform, obscurely lobed .. .. .. * *P. zonale.*

1. **P. australe,** *Jacq., Eclog. Pl. t.* 100. Rootstock stout, stems prostrate or erect, simple or branched, 6in.–18in. high, more or less hairy in all its parts. Leaves on slender petioles much longer than the blade, orbicular-cordate or ovate-cordate, 3–5-lobed, crenate or serrate. Stipules ovate, acute. Peduncles exceeding the leaves, slender. Umbels 10–12-flowered. Pedicels short. Flowers small. Sepals acute, hairy, spur short. Petals about one-third or one-half longer than the sepals, spathulate. Fertile stamens 5, alternating with scale-like staminodes. Carpels hairy, the beaks bearded on the inner face. Seeds minutely punctulate.—Willd., Sp. Pl. iii. 675; Sweet, Geran. t. 68; DC., Prod. i. 654; Hook. f., Fl. Tasm. i. 57; Benth., Fl. Austr. i. 298; F. Muell., Pl. Vict. i. 170. *P. clandestinum,* L'Hérit. *ex* A. Cunn., Precurs. n. 595; Hook. f., Fl. N.Z. i. 41. *P. australe,* var. *clandestinum,* Hook. f., Handbk. 298. *P. glomeratum,* Jacq. DC., Prod. i. 659. *P. Acugnaticum,* Thou., Fl. Tristan d'Acugn. 44, t. 13. *Geranium amaenum,* Banks and Sol., MSS.

NORTH and SOUTH Islands; STEWART Island; CHATHAM Islands: common. Also in Tristan d'Acunha and Australia. *Kopata.* Nov. to Feb.

P. grossularioides, Ait., of South Africa is doubtless identical, but the leaves are often deeply divided. A decoction of the leaves is used as a lotion for burns, scalds, &c.

* **P. zonale,** *L'Hérit. in Ait. Hort. Kew. ed.* 1, ii. 424. Suffruticose. Stems robust, pubescent or hairy, 2ft.-3ft. high or more. Leaves on long petioles, orbicular-reniform, 3–5-lobed, crenate. Stipules large, apiculate. Peduncles exceeding the leaves. Umbels many-flowered. Petals cuneate, crimson, red, or pink. The leaf-blade has a curved dark-brown band midway between the base and margin.

NORTH Island: on the site of deserted gardens, &c.; not infrequent, and persistent for years, but not naturalised, although it produces seeds freely. *Horseshoe-leaved geranium.* Dec. to March. South Africa.

* **P. quercifolium,** *L'Hérit., Geran. t.* 14. Suffruticose, much-branched, 1ft.–2ft. high, more or less hispid in all its parts. Leaves palmatifid to below the middle, with rounded sinuses; the segments pinnatifid, lobed or toothed, obtuse or subacute, crenate. Peduncles short, many-flowered. Pedicels short. Sepals hispid.

NORTH Island: in similar situations to the last, but does not produce seed so freely. Nov. to March. South Africa.

* TROPAEOLUM, Linn.

Flowers irregular; receptacle concave, produced backwards into a nectariferous spur. Sepals 5, hypogynous. Petals 5. Stamens 8 in two series, perigynous; filaments free. Ovary 3-celled. Carpels 3, indehiscent, furrowed, 1-seeded, separating when mature. Herbs, usually glabrous, with alternate simple or divided exstipulate leaves, and climbing or prostrate stems.

* **T. majus,** *L., Sp. Pl.* 345. A glabrous subscandent or trailing herb. Stems 2ft.–6ft. long. Leaves on long petioles, peltate, obscurely 5-lobed or angled. Flowers solitary, axillary on long peduncles. Lower petals on long claws, fringed at the base. Style long, 3-fid at the apex.

NORTH Island: Auckland and New Plymouth: scarcely naturalised. *Indian cress.* Jan. to March. Peru.

3. OXALIS, Linn.

Flowers regular. Sepals 5, persistent, imbricate. Petals 5, contorted, sometimes cohering. Stamens 10 in 2 series, all fertile, free or cohering at the base. Ovary 5-lobed, 5-celled; styles 5, with terminal capitate or lobed stigmas; ovules 1 or more in each cell. Capsule 5-celled, opening loculicidally, the valves partially cohering and persistent. Seeds with an outer fleshy coat that opens elastically; endosperm fleshy; embryo straight. Herbs, or rarely small shrubs, with simple or much-branched stems, alternate or radical 3-foliolate or rarely pinnate leaves. Stipules usually minute or 0. Peduncles axillary or radical, 1- or many-flowered.

SPECIES, about 240, of which 3 or 4 are widely distributed, the others being restricted to South America and extra-tropical South Africa. Cleistogamous flowers are produced by some species, and the flowers of others are dimorphic or trimorphic.

ETYM. From the Greek, in allusion to the acidity of many species.

1. **O. corniculata,** *L., Sp. Pl.* 435. Perennial. Stems erect, prostrate, or decumbent, often matted, 1in.–10in. long. Leaves on long or short petioles, 3-foliolate, glabrous or pubescent. Petioles ½in.–4in. long. Leaflets deeply obcordate. Stipules minute, adnate to the petiole or 0. Peduncles axillary, 1- 2- 3- 6-flowered on reflexed pedicels. Flowers ⅓in.–½in. in diameter. Sepals acute or obtuse. Petals yellow, notched. Capsule oblong, linear or ovate, with few or several seeds in each cell.—Hook. f., Fl. N.Z. i. 43, Fl. Tasm. i. 59, Handbk. 38; Benth., Fl. Austr. i. 301. *O. ambigua,* A. Rich., Fl. N.Z. 296. *O. flaccida,* Banks and Sol., MSS.

KERMADEC Islands; NORTH and SOUTH Islands: common in lowland situations, cultivated and waste ground, &c. Extremely variable.

Var. **corniculata.** Stems decumbent. Leaves stipulate. Pod ½in.–¾in. long or more.

Var. **stricta.** Erect or suberect. Leaves exstipulate. Flowers small. Capsule large.—*O. stricta*, L.; *O. Urvillei*, *O. propinqua*, *O. divergens*, *O. lacicola*, A. Cunn., Precurs. n. 584, 586, 588, 590. In warm situations.

Var. **microphylla.** Stems procumbent, slender. Leaves stipulate. Leaflets usually small. Flowers large. Sepals narrow-linear, obtuse.—*O. microphylla*, *O. exilis*, A. Cunn.

Var. **ciliifera.** Stems procumbent, filiform, often matted. Leaflets ciliated. Peduncles 1-flowered. Flowers small. Sepals obtuse. Capsule broadly ovate. *O. tenuicaulis*, *O. ciliifera*, A. Cunn., Precurs. n. 589, 591.

Var. **crassifolia.** Stems rigid, matted. Leaves small, thick, pilose. *O. crassifolia*, A. Cunn., Precurs. n. 592.

2. **O. magellanica,** *Forst. in Comm. Gotting.* ix. (1789) 33. Glabrous or pubescent, 2in.–4in. high. Rhizome slender, simple, creeping. Petioles more or less hairy. Stipules large, often persistent on the rhizome after the leaves have fallen. Leaflets obcordate, glaucous beneath. Peduncles radical, longer or shorter than the leaves, 2-bracteolate, above the middle, 1-flowered. Sepals ovate, blunt, silky. Petals white, oblique, often lobed and ciliated. Capsule globose.—Hook. f., Fl. Antarc. ii. 253, Fl. N.Z. i. 42, t. xiii., Fl. Tasm. i. 39; Handbk. 38; F. Muell., Pl. Vict. i. 76; Benth., Fl. Austr. i. 300. *O. cataractae*, A. Cunn., Precurs. n. 585; Hook., Ic. Pl. t. 418.

NORTH and SOUTH Islands: Bay of Islands to Otago, but often local. Sea-level to 3,500ft. Also in Victoria, Tasmania, South Chili, and Fuegia. Aug. to Oct.

This species is closely related to *O. acetosella.*, L., of the Northern Hemisphere.

* **O. variabilis,** *Jacq., Oxal.* 89. Stemless, tuberous. Leaves all radical, 1in.–3in. long. Petioles hairy. Leaflets rounded at the apex, cuneate below, hairy, often ciliated. Peduncles shorter or longer than the petioles, hairy. Sepals linear, acute, glabrate or silky. Flowers large. Petals crimson, red, or white.

NORTH Island: Auckland; New Plymouth: a garden escape, scarcely naturalised. Aug. to Oct. South Africa.

* **O. hirta,** *L., Sp. Pl.* 434. Rhizome tuberous, creeping below, branches spreading above, hairy. Petioles extremely short and broad or rarely 0. Leaflets very shortly petiolulate, linear-obovate or spathulate, retuse. Flowers solitary on axillary peduncles, exceeding the leaves. Calyx silky.

NORTH Island: Auckland: a garden escape, scarcely naturalised. Sept. South Africa.

* **O. cernua,** *Thunb., Diss. Oxal.* 14. Stemless or nearly so, glabrous. Petioles radical, 5in.–9in. long. Leaflets broadly cuneate, deeply obcordate. Peduncles 6in.–12in. high. Flowers in terminal umbels, drooping in bud. Pedicels slender, pubescent. Sepals linear, blunt, with purple tips. Flowers large, yellow.

NORTH Island: Auckland, Wellington, &c.: in cultivated ground and on shaded banks. Scarcely naturalised. South Africa.

ORDER XIV.—RUTACEAE.

Flowers regular, perfect or rarely unisexual. Calyx 4- or 5-lobed or divided, imbricate, rarely valvate. Petals 4 or 5, free or rarely cohering, hypogynous or slightly perigynous, imbricate or valvate. Stamens as many or twice as many as the petals, rarely more, inserted at the base of a swollen disk; anthers versatile. Ovary of 2–5 coherent 1-celled carpels; styles united from their base or by the capitate stigma only; ovules 2 in each cell. Fruit capsular, separating into 2-valved or rarely indehiscent cocci, the outer coat separating from the inner; rarely a berry or drupe. Seeds 1 in each cell; testa usually crustaceous and often shining; endosperm copious or 0; embryo large, radicle superior. Shrubs or trees, rarely herbs, with simple or compound exstipulate, pellucid-dotted leaves, affording a pungent aromatic volatile oil.

A large order, distributed through the tropical and warm temperate regions; especially plentiful in South Africa and Australia. GENERA, 78. SPECIES, about 700.

1. PHEBALIUM. Flowers 5-merous. Leaves simple; petiole terete.

2. MELICOPE. Flowers 4-merous. Leaves 3-foliolate, or, if simple, with winged petioles.

1. PHEBALIUM, Vent.

Calyx small, 5-lobed or -parted. Petals 5, valvate or laterally imbricate, with valvate inflexed tips. Stamens 10, longer or shorter than the petals; filaments filiform, glabrous. Carpels 2–5, separating nearly to the base. Style simple, springing from the centre of the lobes; stigma small, capitate. Ovules 2 in each cell, superposed. Cocci 2-valved, 1-seeded, usually more or less beaked or truncate. Endocarp cartilaginous and separating elastically. Testa black, shining. Shrubs, with alternate simple entire or slightly toothed leaves, usually with axillary or terminal corymbs.

The genus comprises 28 species restricted to Australia, and 1 in New Zealand.

1. **P. nudum,** *Hook., Ic. Pl. t.* 568. A glabrous much-branched shrub, 4ft.–10ft. high, with slender branchlets, red bark, and alternate spreading or ascending leaves. Leaves 1in.–1½in. long, linear-oblong or narrow-oblong-lanceolate, blunt, obscurely crenate, coriaceous, dotted beneath. Flowers whitish, in terminal much-branched corymbs; pedicels short. Calyx small, 5-toothed. Petals 5. Cocci compressed, wrinkled, splitting into 2 valves.—Raoul, Enum. Pl. Nov. Zel. 48; Hook. f., Fl. N.Z. i. 44; Handbk. 39.

NORTH Island: Kaitaia southward to the Thames Goldfield; Great Barrier Island. Sea-level to 2,500ft. *Mairehau.* Nov., Dec.

Endemic. Closely related to *P. elatius*, F. Mueller, of Queensland.

2. MELICOPE, Forst.

Flowers regular, perfect or unisexual. Sepals 4. Petals 4, valvate or slightly imbricate with inflexed tips. Stamens 8; filaments subulate. Ovary of 4 carpels separating nearly to the base, 2-ovuled; styles (in the New Zealand species) simple, capitate, 4-lobed. Cocci 1–4, 1-seeded, spreading, free, 2-valved, the outer coat separating from the inner. Testa crustaceous, shining; endosperm fleshy; embryo straight or slightly curved. Shrubs, or rarely trees, with opposite 3-foliolate or simple pellucid dotted leaves and terminal axillary cymes of rather small flowers.

ETYM. From the Greek, in reference to the lobed glands around the ovary.

SPECIES, about 15, of which 2 are endemic in New Zealand, 3 in Australia, and the remainder in the Pacific islands.

Leaves 3-foliolate; petioles terete; leaflets acute .. .. 1. *M. ternata*.
Leaves 3- or 1-foliolate; leaflets obovate, rounded .. .. var. *Mantellii*.
Leaves 1-foliolate, orbicular; petioles winged .. .. 2. *M. simplex*.

1. **M. ternata,** *Forst., Char. Gen.* 56. A glabrous shrub or small tree, 6ft.–15ft. high. Leaves opposite, 3-foliolate; leaflets 2in.–4in. long, exceeding the petiole, linear-oblong or ovate, narrowed below, acute, entire, minutely pellucid-dotted. Flowers in axillary trichotomous panicles, more or less unisexual. Pedicels short. Petals ovate, dotted. Style very short and stout; stigma minutely 4-lobed. Carpels coriaceous, wrinkled and punctate, coherent at the base only. Seed black, shining, attached by a slender funiculus, and often projecting from the opening valves.—Forst., Prod. n. 166; DC., Prod. i. 723; A. Rich., Fl. N.Z. p. 293; A. Cunn., Precurs. n. 582; Raoul, Enum. Pl. N.Z. 48; Hook., Ic. Pl. t. 603; Hook. f., Fl. N.Z. i. 43; Handbk. 40; A. Gray, U.S. Expl. Exped. Bot. i. 350; T. Kirk, Forest Fl. N.Z. t. 66. *Entoganum laevigatum*, Sol. *ex* Gaertn. Fruit, i. 331, t. 68.

KERMADEC Islands; NORTH Island: sea-level to fully 1,000ft. Common. SOUTH Island: coast of Nelson and Marlborough, but local; D'Urville Island. *Wharangi*. Sept., Oct.

Var. **grandis**, Cheesem. in Trans. N.Z.I. xx. (1887) 166. Leaves and fruit much larger. Leaflets obtuse or rounded at the apex. KERMADEC Islands.

Var. **Mantellii** (*sp.*), Buch. in Trans. N.Z.I. iii. (1870) 212! Excessively branched; twigs crowded, strict. Bark red. Leaves 3- rarely 1-foliolate, varying greatly in size; leaflets usually rounded at the apex. Flowers in axillary 3–6-flowered cymes. Petals linear-oblong. Ovary silky. Carpels wrinkled, punctate.—T. Kirk, Forest Fl. N.Z. t. 67.

NORTH Island: Mongonui to Wellington, but very local. A hybrid between *M. ternata* and *M. simplex*. Trunk sometimes 1ft. in diameter.

2. **M. simplex,** *A. Cunn., Precurs. n.* 583. A glabrous shrub or small tree, 3ft.–12ft. high. Leaves alternate or fascicled, in young plants 3-foliolate; petiole not winged: in mature plants, blade orbicular-obovate or ovate, ½in.–¾in. long, obtuse, obscurely or doubly crenate; petiole winged. Flowers whitish, in axillary 2–4-flowered cymes or fascicles. Pedicels very slender, exceeding the petioles. Petals linear-oblong, reflexed, exceeding the stamens. Style very short, obscurely 4-lobed. Fruit smaller than in *M. ternata*.—Hook., Ic. Pl. t. 584; Raoul, Enum. Pl. N.Z. 48; Hook. f., Fl. N.Z. i. 43; Handbk. 40; T. Kirk, Forest Fl. N.Z. t. 68. *M. parvula*, Buch. in Trans. N.Z.I. xx. (1887) 255. *Astorganthus Huegelii*, Endlich., Cat. Hort. Vindob. ii. 196.

North Cape to Southland. Sea-level to 2,000ft. Not uncommon. Sept. to Nov.

The blade is invariably jointed to the petiole, which is grooved on the upper surface. A form with cleistogamous flowers was discovered by G. M. Thomson, on Pigeon Island, Lake Wanaka.

ORDER XV.—MELIACEAE.

Flowers regular, usually perfect. Calyx small, 4–5-lobed or divided into free sepals, imbricate. Petals 4–5, rarely 3 or more than 5, free or adnate with the staminal tube, imbricated, rarely contorted or valvate. Disk free, annular or tubular within the staminal tube. Stamens 5–20, more or less united into a tube inserted with the petals outside the base of the disk. Ovary free, 3–5-

celled; style simple; cells 2- rarely 4-ovuled, the radicle superior. Fruit a drupe, berry, or capsule, indehiscent or loculicidally 3-valved. Seeds 1 rarely 2 or more in each cell, usually arillate. Endosperm 0 or fleshy. Shrubs or trees, with alternate exstipulate leaves and paniculate flowers.

This order comprises numerous tropical forest-trees, as the mahogany, *Swietenia Mahagani*; the satin-wood, *Chloroxylon Swietenia*; the Australian cedar, *Cedrela australis*, &c. It is most abundantly represented in tropical Asia and America, but is rare in Africa. GENERA, 37. SPECIES, 275.

1. DYSOXYLUM, Blume.

Calyx small, 4–5-toothed -lobed or -parted, imbricate. Petals 4–5, linear-oblong, valvate, free or adnate with the staminal tube, which is cylindrical and truncate or toothed at the mouth. Anthers 8–10, included. Disk shortly tubular, sheathing the ovary. Ovary 3–5-celled; ovules 2 or rarely 1 in each cell. Capsule coriaceous, 2–5-valved, 1–5-celled. Seeds usually arillate, oblong; hilum large, coriaceous, shining. Cotyledons large. Endosperm 0. Trees, often foetid, with alternate pinnate or opposite leaves and axillary paniculate flowers.

A genus of large forest-trees, distributed through tropical Asia, the Pacific Islands, and Australia. SPECIES, about 30. One is endemic in New Zealand, and another in Norfolk Island.

1. **D. spectabile,** *Hook. f., Handbk. N.Z. Fl.* 41. A round-headed tree, 20ft.–50ft. high; trunk 1ft.–3ft. in diameter. Leaves glabrous, unequally pinnate, 9in.–18in. long; leaflets in 2–4 pairs, petioled, broadly-oblong to oblong-obovate, entire, the upper leaflets often narrowed into the petiole. Panicles 6in.–18in. long, pendulous, given off from the trunk or branches, rarely axillary, shortly pedicelled. Sepals short, pubescent. Petals 5, spreading, linear, free or nearly so. Staminal tube toothed. Disk forming a fluted cup around the silky 3–4-celled ovary. Style exceeding the staminal tube; stigma broadly discoid, with a curious circular lip or rim. Capsule long, pyriform or obovate, 3–4-celled; valves coriaceous, splitting down the middle of the cells. Cells 2-seeded, enveloped in a scarlet aril.—T. Kirk, Forest Fl. N.Z. tt. 64, 65. *Hartighsea spectabilis*, A. Juss. in Mem. Mus. Par. xix. (1830) 228; A. Cunn., Precurs. n. 597; Hook., Ic. Pl. t. 616, 617; Hook. f., Fl. N.Z. i. 39. *Trichilia spectabilis*, G. Forst., Prod. n. 188; DC., Prod. i. 623; A. Rich., Fl. N.Z. 306; *T. cauliflora*, Banks and Sol. MSS.

NORTH and SOUTH Islands: Mongonui to Nelson and Marlborough; D'Urville Island. Sea-level to 1,500ft. *Kohekohe*. Aug., Sept.

The leaves are bitter, and are said to have been used as a substitute for hops. A spirituous infusion has been used as a stomachic.

ORDER XVI.—OLACINEAE.

Flowers regular, hermaphrodite or rarely unisexual. Calyx small, 4–5- or rarely 6-toothed -lobed or -divided, free or adnate with the disk. Petals as many as the sepals, valvate, free or united into a tube. Stamens as many or twice as many as the petals, hypogynous or adnate to the base of the petals. Ovary free or invested by the disk, 1-celled or imperfectly 2–3-celled; style simple; stigma entire or lobed; ovules 1–3, pendulous from a central placenta, or from the side or apex of the cell. Fruit an indehiscent drupe, superior or rarely inferior, 1-celled, 1-seeded. Seed pendulous; testa thin; endosperm fleshy or rarely 0; embryo minute; radicle superior. Trees or shrubs, with exstipulate simple alternate or rarely opposite leaves.

GENERA, about 36. SPECIES, 175. Widely distributed through the tropical and subtropical regions of the earth. The order has strong affinities with *Santalaceae* and *Loranthaceae*.

1. PENNANTIA, Forst.

Flowers dioecious or polygamous. Calyx minute. Petals 5, valvate. Stamens 5, hypogynous. Anthers versatile or basifixed. Ovary oblong, 1-celled; stigma sessile or nearly so; ovule solitary, suspended from near the apex of the cell. Drupe small, fleshy or coriaceous; stone obliquely trigonous, spuriously erect, grooved at the back to receive a flattened cord, which enters a perforation near the apex, and carries the pendulous seed at its tip. Shrubs or trees, with alternate leaves and terminal many-flowered cymes or panicles.

The genus comprises 3 species: the present, another in Norfolk Island, and a third in New South Wales.

Named in honour of *Thomas Pennant*, a Scottish naturalist.

1. **P. corymbosa**, *Forst., Char. Gen.* 134. A shrub or small tree, 10ft.–40ft. high. In the young state the branchlets are slender, tortuous, and interlaced, bearing distant cuneate 3-lobed leaves ½in. long. Mature leaves shortly petioled, 1in.–4in. long, ovate, oblong-ovate or obovate, obtuse, entire or sinuate, lobed or with large coarse teeth. Panicles terminal, the rhachis and branches white and pubescent. Pedicels jointed. Male: ovary 0; petals waxy; anthers sagittate, versatile, on long pendulous filaments. Female: smaller; anthers oblong, basifixed on short erect filaments; ovary oblong; stigma 3-lobed. Berries black, fleshy.—Forst., Prod. n. 396; Willd., Sp. Pl. iv. 1122; Rich., Fl. N.Z. 368; A. Cunn., Precurs. n. 576; Hook. f., Fl. N.Z. i. 35, t. 12; Handbk. 41; T. Kirk, Forest Fl. N.Z. t. 77, 78. *Fagoides triloba* and *Meristoides paniculata*, Banks and Sol. MSS.

NORTH and SOUTH Islands: Kaipara to Southland. Sea-level to 2,000ft. *Kaikomako. Kahikomako.* Nov., Dec.

Flowers fragrant. Young shoots and leaves pubescent. Wood white, brittle; used for obtaining fire by friction.

ORDER XVII.—CELASTRINEAE.

Calyx small, persistent, 4–5-cleft, rarely 3–6-cleft. Petals as many as the sepals, both imbricated in the bud or rarely valvate. Stamens as many as the petals, alternating with them, inserted on or at the margin of the fleshy disk which fills the bottom of the calyx and sometimes covers the ovary. Ovary sessile, 2–5-celled; ovules 1 or 2 in each cell, anatropous; style 1, short; stigma entire or lobed. Fruit various, dehiscent or indehiscent, 2–5-celled, superior. Seeds usually arillate, erect, rarely winged; endosperm usually copious and fleshy; embryo rather large. Shrubs or trees, rarely spinous or climbing. Leaves opposite or alternate, entire or toothed. Stipules, when present, minute, fugacious. Flowers small, in axillary cymes or racemes or terminal panicles, regular hermaphrodite or polygamous.

The order must be considered of doubtful occurrence in New Zealand. I am indebted to Sir Joseph Hooker for descriptions of three species of *Elaeodendron*, which he has kindly copied and forwarded, nothing being known of the plants in the colony.

1. ELAEODENDRON, Jacq. f.

Calyx 4–5- rarely 3-cleft. Petals as many as sepals, spreading. Stamens inserted under the margin of a thickened disk; filaments short. Ovary conical, usually 3-celled, rarely 2- 4- or 5-celled; style very short; cells 2-ovuled. Drupe succulent or nearly dry, the putamen hard, 1–3-celled. Seeds exarillate, usually solitary; testa membranous or porous; endosperm scanty or copious. Glabrous shrubs or trees, with opposite or alternate entire or crenate leaves, and small flowers in dichotomous axillary cymes, often clustered.

The genus comprises about 40 species, distributed through East India, South Africa, and Madagascar, &c. Two species are found in Australia, one in Norfolk Island, and a few others in New Caledonia, California, China, and Japan.

Leaves oblong or obovate, cuneate at base .. 1. *E. Novae-Zelandiae.*
Leaves ovate or ovate-oblong, fleshy .. .. 2. *E. carnosum.*
Leaves ovate, scarcely attenuate at base .. 3. *E. punctulatum.*

1. **E. Novae-Zelandiae,** *Turcz. in Bull. Soc. Nat. Mosc.* xxxvi. (1863) i. 602. Branches somewhat terete or trigonous. Leaves alternate, oblong or obovate, cuneate at the base, acuminate, coriaceous, glabrous; blade much longer than the petiole. Flowers polygamous. Male flowers: anthers subsessile, dehiscing longitudinally; style 0. Female with staminodia; style angular; stigma 5-lobed. Fruit unknown.

New Zealand. *A. Cunningham*, No. 67, Herb., Kew.

2. **E. carnosum,** *Turcz., l.c.* Branches terete. Leaves alternate, ovate or ovate-oblong, rounded or attenuate at the base; blade three times longer than the petiole, obtuse, entire, fleshy; veins not reticulate. Racemes axillary, longer than the petiole, many-flowered. Filaments short; anthers subrotund, dehiscing horizontally.

New Zealand. Collector's name not stated.

3. **E. punctulatum,** *Turcz., l.c.* Leaves alternate, ovate, entire, scarcely attenuate at the base, shortly and obtusely acuminate; blade three times longer than the petiole, inferior surface rough with minute raised points. Peduncles 1–2-flowered, axillary, forming small umbels. Drupe 1-seeded.

New Zealand. Collector's name unknown.

ORDER XVIII.—STACKHOUSIEAE.

Flowers regular, hermaphrodite. Calyx 4–5-toothed or -lobed, imbricate. Petals 5, perigynous, clawed, free at the base but more or less united above, forming a tubular corolla rarely quite free, with spreading or reflexed lobes, imbricate in bud. Stamens 5, free, inserted on the margin of the disk; filaments slender. Ovary free, 2–5-lobed, 2–5-celled; styles 2–5-lobed or simple; ovules solitary, erect, anatropous. Fruit of 2–5 indehiscent angular rounded or winged cocci attached to the axis. Seed solitary; testa membranous; endosperm fleshy; embryo straight; radicle inferior. Perennial herbs of lowly growth, with alternate almost exstipulate leaves. Flowers in short terminal spikes or rarely solitary.

The order is limited to 2 genera, containing about 12 species, of which 1 is endemic in New Zealand. The others are restricted to Australia, except a single species which extends to the Philippine Islands.

1. STACKHOUSIA, Smith.

Calyx 5-lobed. Petals free at the base, more or less united above. Ovary 3-lobed; style simple; stigma 3-lobed.

Named in honour of *J. Stackhouse*, an English algologist.

1. **S. minima**, *Hook. f., Fl. N.Z.* i. 47. A very small slender glabrous herb. Rhizomes filiform, often matted, white. Stems erect, leafy, ¼in.–2in. high, simple or branched. Leaves slightly fleshy, linear-acute or obovate, ⅙in.–¼in. long, close-set or distant. "Flowers very minute, solitary or few together towards the tops of the stems." Calyx-teeth acute. Petals united above the middle; tips free, recurved, acute. "Anthers pubescent." Ovary 2–3-lobed; stigmas 3-lobed. Only a single carpel ripens.—Handbk. 42.

Var. **uniflora** (*sp.*), Colenso in Trans. N.Z.I. xviii. (1885) 258. Flowers all solitary, terminal, shortly peduncled.

NORTH Island: Hawke's Bay: open downs on the east coast, *Colenso*. Waipawa County, *H. Hill!* SOUTH Island: Nelson: Mount Arthur Plateau, *Cheeseman!* Spencer Mountains, *T. K.* Canterbury: Broken River Basin, *Enys!* Burnham, *T. K.* Central Otago, *Petrie!* Descends to 100ft. above sea-level near Burnham. Ascends to 4,000ft. in the Spencer Mountains. Dec., Jan.

I have copied Sir Joseph Hooker's original description of the arrangement of the flowers, but have not been so fortunate as to find specimens with "few-flowered spikes" or "pubescent anthers." Possibly the original description may be inaccurate on these points. In Mr. Cheeseman's Mount Arthur specimens the petals are usually free to just below the limb, and in one or two flowers for their entire length.

ORDER XIX.—RHAMNEAE.

Flowers regular, hermaphrodite, rarely polygamous. Calyx 4–5-toothed or -lobed; lobes valvate, mostly with a raised longitudinal line on the inner surface. Petals 4–5 or 0, concave, minute, alternating with the base of the calyx-lobes. Stamens 4–5, opposite the petals; anthers often enclosed in the petals; filaments short. Disk fleshy, rarely 0, hypogynous or epigynous. Ovary sessile on the disk or closely invested by it or more or less inferior, usually 3-celled; style simple; stigma capitate or with 4, 3, or 2 lobes. Ovules solitary, erect, anatropous, usually with a dorsal raphe. Fruit a drupe or capsule, the latter separating into as many cocci as cells. Seeds solitary, erect; testa crustaceous or coriaceous; endosperm usually fleshy, often scanty; embryo straight; radicle inferior. Usually shrubs or trees, with alternate or rarely opposite undivided stipulate leaves. Stipules deciduous, rarely spinous. Flowers in axillary or terminal cymes or panicles.

GENERA, about 37. SPECIES, 440. Widely distributed through the temperate and tropical regions of both hemispheres.

1. POMADERRIS. Ovary inferior. Tomentose leafy shrubs.

2. DISCARIA. Ovary superior. Spinous shrub. Leaves few or 0.

1. POMADERRIS, Labill.

Calyx-tube adnate with the ovary, the limb divided into 5 lobes, deciduous or reflexed. Petals 5 or 0. Stamens 5; filaments exceeding the petals; anthers free. Disk surrounding the ovary at the margin of the calyx-tube, inconspicuous. Ovulary more or less inferior. Style 3-cleft. Capsule small, separating into 3 cocci, which open by an oblong lid at the base or by splitting down the inner face. Seed carried on a short thick funiculus. Shrubs, more or less clothed with white hoary or ferruginous stellate tomentum, often mixed with silky hairs. Leaves alternate. Stipules deciduous. Flowers in small cymes, forming terminal or axillary panicles or corymbs.

A small genus, comprising about 22 species, restricted to Australia, New Zealand, and New Caledonia.

ETYM. From the Greek, signifying *a covering* and *the skin*, in reference to the capsule being loosely invested by the calyx.

Petals on slender claws.

Leaves elliptic-oblong, smooth, white beneath .. .. .. 1. *P. elliptica.*

Petals 0.

Leaves rugose, linear-oblong, ferruginous .. .. 2. *P. Edgerleyi.*
Leaves very rugose, broadly oblong, white beneath .. .. 3. *P. apetala.*
Leaves ½in.-¾in. long; margins revolute .. .. 4. *P. phylicaefolia.*

1. **P. elliptica,** *Labill, Nov. Holl. Pl.* i. 61, *t.* 86. A shrub with slender branches, 2ft.–10ft. high. Young branches, petioles, and pedicels clothed with fine stellate pubescence. Leaves 2in.–3in. long, petiolate, glabrous above, white beneath with close tomentum, entire, elliptic-oblong or ovate-oblong, narrowed at the base, somewhat membranous. Flowers in much-branched terminal cymose corymbs or panicles. Calyx white with close tomentum mixed with a few silky hairs. Petals on slender claws; blade slightly waved. Style divided for less than half its length; stigmas capitate. Capsule small, the free portion smaller than the calyx-tube. Carpels small, opening by an oblong lid on the inner face.—Bot. Mag. t. 1510; DC., Prod. ii. 33; Hook. f., Fl. N.Z. i. 46; Handbk. 43; Fl. Tasm. i. 76; Benth., Fl. Austr. i. 417. *P. Kumeraho,* A Cunn., Precurs. n. 248. *P. intermedia,* Sieb. *ex* DC. Prod. ii. 33.

NORTH Island: North Cape to Katikati, Tauranga; Great Barrier Island. In open country. *Kumarahou.* Sept.

Also in south-east Australia and Tasmania.

2. **P. Edgerleyi,** *Hook. f., Handbk. N.Z. Fl.* 43. A low spreading shrub, 1ft.–2ft. high, or rarely strict and erect, 2ft.–6ft. Young shoots, cymes, and lower surface of leaves covered with loose rusty tomentum and whitish stellate pubescence. Leaves ¾in.–2in. long, petiolate, elliptic-oblong or linear-oblong, rounded at the base, obtuse, glabrous above, veins prominent below. Cymes axillary or terminal, ½in.–3in., rarely forming racemes or narrow

panicles. Pedicels often very short or 0. Calyx deeply 5-lobed, turbinate below; lobes large, acute, densely tomentose. Petals apparently 0. Ovary wholly inferior, sunk in the calyx-tube. Cocci opening by a lid.—*Pomaderris*, n. s., Hook. f., Fl. N.Z. i. 46.

NORTH Island: Auckland: North Cape and Spirits Bay to Kaipara and Waipu, &c.; between Orewa and the Wade; Cape Colville to Tapu Creek, Thames; Mercury Bay, &c. Sea level to 1,200ft. Flowers not seen. Endemic.

3. **P. apetala,** *Labill., Pl. Nov. Holl.* i. 62, *t.* 87. A shrub or small tree, 6ft.–20ft. high; trunk 5in.–6in. in diameter. Branchlets, leaves below, and inflorescence densely clothed with short white stellate pubescence and floccose tomentum. Leaves petioled, oblong-lanceolate or broadly oblong, 2in.–4in. long, obtuse or subacute, very rugose above; margins crenulate or minutely erose, glabrous above; veins prominent beneath. Flowers very numerous, in terminal or axillary cymes forming loose panicles 4in.–7in. long. Calyx-tube short. Petals 0. Anthers usually with a minute gland at the apex. Styles rather short, divided to the middle. Ovary half-superior; upper portion of capsule broad, more or less clothed with minute patches of stellate pubescence; cocci broad, opening by a lid on the inner face.—Hook. f., Fl. Tasm. i. 77; F. Muell., 2nd Cens. Austr. Pl. 103; Benth., Fl. Austr. i. 419; T. Kirk, Forest Fl. N.Z. t. 8. *P. aspera*, Sieb. in DC., Prod. ii. 33. *P Tainui*, Hector in Trans. N.Z.I. xi. (1870) 428.

NORTH Island: Auckland: formerly abundant at Kawhia, but has been destroyed by goats. Between Kawhia and Mokau, *Gilbert*. Taranaki: between the Mokau and Mohakatina Rivers, *Hector*, *T. K.* Chatham Islands, *Cox!* *Tainui*. Oct., Nov.

Var. **mollis** (*sp.*), Colenso in Trans. N.Z.I. xxv. (1892) 327. A form with leaves less rugose, obtuse or acute, whiter beneath. Flowers smaller, on more slender pedicels. NORTH Island: Puketapu, Hawke's Bay, *Colenso*, *Petrie!* SOUTH Island: Geraldine, Canterbury, *L. Mathias!* Planted. Naturalised in the southern locality. A common Australian form.

The northern and Chatham Island natives assert that the plant was introduced in the Tainui canoe. Also in Australia.

4. **P. phylicaefolia,** *Lodd., Bot. Cab.* i. 120A. A small heath-like shrub, with spreading or erect fastigiate villous branches, 1ft.–3ft. high. Leaves on young plants $\frac{1}{2}$in.–$\frac{3}{4}$in. long, petiolate, membranous, ovate, flat, hairy on both surfaces; on mature plants small, $\frac{1}{8}$in.–$\frac{1}{4}$in. long, narrow-oblong, obtuse, with scabrid hairs on both surfaces, very shortly petioled or sessile, coriaceous, margins broadly revolute, deeply grooved above. Flowers in small axillary cymes, slightly exceeding the leaves. Calyx very small, globose, villous. Petals 0. Ovary half-superior; style very short. Capsule small, globose, silky. Cocci opening by the entire inner face falling away.—DC., Prod. ii. 34; Hook. f., Handbk. 43; Benth., Fl. Austr. i. 422; F. Muell., 2nd Cens. Austr. Pl. 103. *P ericifolia*, Hook., Lond. Journ. Bot. i. 257; A. Cunn., Precurs. n. 578; Hook. f., Fl. N.Z. i. 46; Fl. Tasm. i. 78. *P. polifolia*, Reissek in Linnaea xx. 269. *P. amaena*, Col. in Trans. N.Z.I. xviii. (1885) 258. *Stiphrum ledifolium*, Banks and Sol. MSS.

NORTH Island: North Cape to Otaki and Cape Palliser. Abundant in open country. Sea-level to 2,000ft. *Tauhinu* Nov., Dec.

Also in Victoria and Tasmania.

2. DISCARIA, Hook.

Flowers hermaphrodite. Calyx membranous, campanulate, free or adnate with the base of the ovary; lobes 4–5, recurved. Petals 4–5 or 0. Stamens 4–5; filaments short. Disk annular, adnate with the base of the calyx-tube. Ovary more or less sunk in the disk or free, 3-lobed, 3-celled; style divided near the apex; stigmas 3, capitate or clavate. Fruit a drupe, finally coriaceous or capsular, the endocarp separating into three 2-valved crustaceous cocci. Seeds plano-convex; testa coriaceous; endosperm fleshy. Much-branched rigid shrubs or small trees, with opposite green branches, articulate at the nodes, often spinous. Leaves opposite or fascicled or 0. Stipules small. Flowers in axillary fascicles or solitary.

SPECIES, about 14, chiefly in extra-tropical and alpine South America; 2 species in the Galapagos Islands, 1 in Australia, and another in New Zealand.

ETYM. From the Greek, signifying *a disk*, the ovary being seated in a broad disk.

1. **D. Toumatou,** *Raoul, Ann. Sc. Nat.* ii. (1844) 123. A thorny bush or small tree, 1ft.–20ft. high, usually glabrous. Branches grooved and articulated at the nodes; branchlets reduced to opposite woody spines, 1½in.–2in. long. Leaves small, fascicled or solitary in the axils of the spines or below them, ½in.–¾in. long, shortly petioled, oblong or narrow-obovate-oblong, obtuse or retuse, rarely 0. Flowers in few-flowered fascicles with the leaves. Pedicels shorter than the leaves, finely pubescent. Calyx-lobes 4–5, white, reflexed, deciduous. Base of ovary adnate with the disk, 3-lobed. Disk broad, with a thickened margin.—Hook. f., Fl. Tasm. i. 69; Handbk. 44; Benth., Fl. Austr. i. 445; T. Kirk., Forest Fl. N.Z. t. 136. *D. australis*, Hook., var. *apetala*, Hook. f., Fl. N.Z. i. 47.

NORTH and SOUTH Islands: south head of Manukau Harbour to Southland. Sea-level to 3,000ft. *Tumatakuru.* Dec., Jan.

Scarcely differs from *D. australis*, Hook., except in the absence of petals.

ORDER—* AMPELIDEAE.

Flowers hermaphrodite or unisexual. Calyx minute, 4–5-toothed or entire. Petals 4–5, free or united, valvate in the bud, inserted at the outer base of the disk, united by their tips. Stamens opposite the petals, 4 or 5, free or slightly coherent. Disk adnate to the ovary or free. Ovary free, usually 2-celled with two erect ovules in each cell, rarely 3–6-celled with one anatropous ovule in each cell; style short or 0; stigma capitate. Fruit a berry. Seeds 1–6; testa hard or bony; endosperm scanty; radicle short, inferior. Woody climbers or erect shrubs. Leaves simple or compound, usually alternate, opposite the peduncles, which are often changed into simple or branched tendrils. Stipule petiolar. Flowers greenish, in paniculate cymes or racemes.

* VITIS, Tourn.

Flowers hermaphrodite, dioecious or polygamous. Petals calyptrate, falling away from the receptacle before separating from each other. Disk 5-lobed. Ovary 2-celled. Fruit a berry. Seeds pyriform. Woody climbers, with tendrils opposite the leaves. Flowers in simple or compound thyrsoid panicles.

* **V. vinifera,** *L.*, *Sp. Pl.* 202. A robust woody climber, the old bark separating in long fibrous ribbons. Leaves palmate, cordate at base, 5-lobed, lobes acute or sub-acute, margins coarsely serrate. Calyx forming a raised margin; teeth obsolete.

NORTH Island: on the sites of deserted gardens and Maori cultivations, &c., where it maintains itself for years, although not naturalised. *Vine.*

ORDER XX.—SAPINDACEAE.

Flowers regular or irregular, unisexual or polygamous. Sepals 4 or 5, free or united in a toothed or lobed calyx, imbricate or rarely valvate. Petals equalling the sepals or fewer or 0, sometimes with a scale on the inner face, usually small. Disk rarely 0. Stamens 5–8, inserted within the disk or hypogynous. Ovary entire or lobed, usually 3-celled, rarely 1–4-celled; style simple; stigma more or less lobed or divided; ovules usually 1–2 in each cell, ascending or horizontal. Fruit dry or succulent, dehiscent or indehiscent, entire or separating into cocci. Seeds often arillate, usually without endosperm; cotyledons often unequal, large, thick, spirally coiled or superposed; radicle short, incurved. Trees or rarely shrubs or climbers, with alternate rarely opposite entire or divided exstipulate leaves and axillary flowers.

A large order, most abundant in the tropical and warm regions of the Southern Hemisphere; rare in the temperate regions and in the Northern Hemisphere. GENERA, about 75. SPECIES, about 670.

1. DODONAEA. Leaves entire (in the New Zealand species).
2. ALECTRYON. Leaves pinnate, green. Flowers regular.

* MELIANTHUS. Leaves pinnate. Flowers irregular.

1. DODONAEA, Linn.

Flowers unisexual or polygamous, regular. Sepals 3–5. Petals 0. Male flowers: disk small or 0; stamens 5–8, on very short filaments; anthers usually linear-oblong; ovary 0. Female flowers: ovary sessile, 3–6-celled, 3–6-angled; ovules 2 in each cell. Fruit a septicidal membranous or coriaceous capsule with as many valves as cells; valves broadly winged. Seeds 1 or 2, globular or compressed, with a thickened funicle; cotyledons spiral; endosperm 0; arillus 0. Shrubs or small trees, often viscid. Leaves simple or pinnate. Flowers terminal or axillary.

SPECIES, about 50, of which nearly 40 are endemic in Australia, the others being distributed through the tropical and warm regions of the earth.

Named in honour of *Rambert Dodoens*, a German botanist.

1. **D. viscosa,** *Jacq.*, *Enum. Pl. Carib.* 19. A shrub or small tree, 1ft.–30ft. high, with loose brown bark and viscid shoots, more or less compressed. Leaves 1in.–3in. long, shortly petiolate, linear-oblong-obovate or linear-spathulate, entire, obtuse or retuse. Flowers in terminal or axillary cymes or racemes, or small panicles. Male: sepals 4, free, membranous, orbicular, ovate; stamens usually 10, rarely 12; filaments extremely short; anthers apiculate. Female: calyx deeply 4-cleft, coriaceous; style stout, 2-fid. Fruit orbicular, flat, with 2–3 broad membranous wings. Seeds orbicular, compressed, brown.—Linn., Mant. 228; G. Forst., Prod. n. 164; DC., Prod. i. 616;

Hook. f., Fl. N.Z. i. 38; Handbk. 45; Benth., Fl. Austr. i. 439; T. Kirk, Forest Fl. N.Z. t. 17. *D. viscosa*, var. *oblongata*, Banks and Sol. MSS. *D. spatulata*, Sm. in Rees. Cycl. xii. n. 2; A. Rich., Fl. N.Z. 308; A. Cunn., Precurs. n. 599.

NORTH and SOUTH Islands: North Cape to Banks Peninsula and Westland. Also in most tropical and warm temperate regions. Ascends to 1,800ft. *Akeake.* Oct., Nov.

2. ALECTRYON, Gaertn.

Flowers small, regular, hermaphrodite or unisexual. Calyx 4–5-lobed; lobes almost equal, villous on the inner surface, imbricate. Petals 0. Disk small, 8-lobed. Stamens 5–8, inserted in the lobes of the disk; filaments at length exceeding the anthers. Ovary obliquely obcordate, 1-celled, villous, compressed; stigma simple or 2–3-lobed; ovule solitary. Fruit coriaceous or almost woody, tumid, gibbous, with a compressed ridge on one side. Seed subglobose, arillate; cotyledons spirally coiled; endosperm 0. A handsome tree, with pinnate exstipulate leaves and axillary panicles of hermaphrodite or polygamous flowers.

Etym. From the Greek, in reference to the scarlet arillus, which resembles a cock's comb.

A genus usually restricted to the following species, but, according to some botanists, including about a dozen plants from Australia, New Caledonia, Timor, &c.

1. **A. excelsa,** *Gaertn., Fruct.* i. 216, *t.* 46. An erect tree, 40ft.–60ft. high, with black bark. Branches, leaves below, inflorescence, and fruit clothed with rusty pubescence. Leaves unequally pinnate, 4in.–18in. long; leaflets 2in.–4in., shortly petioled, oblique at base, ovate-lanceolate, acute or acuminate, entire or obscurely crenate or with distant teeth. Panicles 4in.–10in. long; flowers pedicellate. Fruit ⅓in. long, with a flattened crest on one side terminating in a spur. Seed jet-black, shining, imbedded in a fleshy scarlet arillus. —DC., Prod. i. 617; A. Cunn., Precurs. n. 598; Hook., Ic. Pl. t. 570; Hook. f., Fl. N.Z. i. 38; Handbk. 45; T. Kirk, Forest Fl. N.Z. i. 92, 93. *Euonymoides excelsa,* Banks and Sol. MSS.

NORTH and SOUTH Islands: North Cape to Banks Peninsula and Ross. Ascends to 2,000ft. *Titoki. Titongi. Tokitoki.* Nov., Dec.

A beautiful tree, which affords a tough and elastic timber, termed by the settlers "the New Zealand ash." The leaves of the young plant are deeply lobed and toothed.

Var. **grandis.** Leaves equally pinnate; leaflets broadly ovate, oblong, 4in.–5in. long, 3in. broad, with 2 or 3 coarse teeth. THREE KINGS Islands: *Cheeseman!*

* MELIANTHUS, Linn.

Flowers hermaphrodite, unequal. Calyx persistent, compressed, deeply 5-lobed, the lowest inflated and smaller than the others. Petals 5, perigynous, the lowest very small or 0. Disk unilateral, lining the bottom of the calyx. Stamens 4, unequal, hypogynous. Ovary oblong, 4-lobed, 4-celled; style central; ovules 2–4 in each cell. Capsule papyraceous, deeply 4-lobed, 4-celled; cells 1-seeded. Seeds subglobose, with copious endosperm; arillus 0; testa crustaceous; embryo small. Shrubs or undershrubs, with pinnate stipulate leaves and terminal axillary or racemose inflorescence.

* **M. major,** *L., Sp. Pl.* 639. An erect glabrous shrub, 3ft.–4ft. high. Leaves pinnate, glaucous; stipules large, connate for their entire length; leaflets sessile, deeply and acutely serrate. Flowers in erect terminal racemes. Pedicels short. Sepals pubescent. Disk excessively nectariferous. Ovules ascending.

NORTH Island: a garden escape, thoroughly established in many localities between Ahipara, Auckland, and Wellington. Aug. to Oct. South Africa.

Order XXI.—ANACARDIACEAE.

Flowers usually regular, hermaphrodite, unisexual or polygamous. Calyx inferior, 3–5-lobed or -divided. Petals perigynous, 3–7 or 0. Disk annular or broad. Stamens as many as the petals or twice as many, inserted on the disk or at its base; filaments free; anthers versatile. Ovary free, 1-celled, with 1–3 styles, or 2–5-celled, or rarely of 2–5 free carpels; ovules solitary, pendulous from a basal funicle or adnate to the wall of the cavity, rarely erect; micropyle inferior. Fruit superior, 1–5-celled, usually drupaceous and indehiscent. Seed pendulous, horizontal or erect; endosperm 0; cotyledons thick, fleshy; radicle short. Trees or shrubs, with alternate simple or compound exstipulate leaves. Bark often exuding a balsamic gum.

Genera, about 45. Species, 450. Common in tropical regions; less frequent in warm temperate regions.

1. CORYNOCARPUS, Forst.

Flowers hermaphrodite. Calyx deeply 5-partite, lobes imbricate. Petals 5, perignyous, imbricate, erose. Disk 5-lobed, fleshy. Stamens 5, inserted between the lobes of the disk, alternating with 5 petaloid staminodia. Ovary superior, sessile, 1-celled; style erect; stigma capitate; ovule pendulous from the upper wall of the cell. Drupe large, obovoid, obtuse, pulpy when ripe; endocarp coriaceous, fibrous, forming a network outside the membranous testa of the pendulous seed; cotyledons oblong, plano-convex, fleshy; radicle minute, superior. A tree, with alternate exstipulate leaves and terminal panicles of small green flowers.

1. **C. laevigata,** *Forst., Char. Gen.* 31, *t.* 16. A glabrous evergreen tree, 20ft.–50ft. high; trunk 1ft.–2½ft. in diameter. Leaves 3in.–7in. long, oblong or broadly lanceolate, narrowed into short stout petioles, glossy. Flowers in erect rigid panicles; pedicels stout. Petals as long as the calyx-lobes, concave. Disk fleshy. Ovary small. Drupe 1in.–1½in. long, orange-coloured; sarcocarp fleshy; endocarp reticulate, coriaceous; testa membranous, veined. —A. Cunn., Precurs. n. 638; Hook., Bot. Mag. t. 4379; Hook. f., Fl. N.Z. i. 48; Handbk. 46; T. Kirk, Forest Fl. N.Z. t. 88. *Merretia lucida*, Banks and Sol. MSS.

NORTH and SOUTH Islands: Kermadec Islands to Ross and Banks Peninsula. CHATHAM Islands. Chiefly in littoral situations. *Karaka. New Zealand laurel.* Aug. to Nov., Dec.

The sarcocarp is edible, but the seed is highly poisonous until deprived of its injurious properties by boiling or maceration, when it becomes nutritious. The wood although white and perishable is used for canoes.

ORDER XXII.—CORIARIEAE.

Flowers regular, hermaphrodite or polygamous. Sepals 5, ovate-triangular, imbricate in bud, persistent. Petals 5, shorter than the sepals, at length accrescent and appressed to the carpels, keeled within. Stamens 10 in two series, all free or sometimes 5 adnate with the keel of the petals; filaments short, elongating after fertilisation. Carpels 5 or 10, inserted at the summit of the depressed conical receptacle, 1-celled. Styles as many as the carpels, free, flexuous, stigmatiferous over the entire surface; ovules solitary, pendulous; micropyle superior. Fruit 5–10 compressed achenes, enclosed in the fleshy petals, keeled on the back and sides. Seeds compressed; testa membranous; endosperm scanty, thin; embryo compressed, fleshy; cotyledons plano-convex; radicle short, superior. Woody or suffruticose shrubs, or rarely small trees or herbs, with angular branches and opposite or rarely 3-nate entire exstipulate leaves. Flowers axillary, rarely terminal, solitary or racemed.

A small monotypic order, comprising about 12 species, natives of Europe, China, Japan, India, Peru, and New Zealand. All the New Zealand forms are perennial; they vary to a remarkable extent, and should probably be referred to a single species.

1. CORIARIA, Linn.

CHARACTERS OF THE ORDER.

Stems branched, woody. Racemes drooping .. .. .. 1. *C. ruscifolia.*
Suffruticose, herbaceous, branched. Racemes erect .. .. 2. *C. thymifolia.*
Herbaceous. Stems simple, strict. Racemes erect .. .. 3. *C. lurida.*
Herbaceous, much branched, slender .. .. 4. *C. angustissima.*

1. **C. ruscifolia,** *L., Sp. Pl.* 1037. A shrub or small tree, with trunk rarely 12in. in diameter. Branches spreading. Leaves ovate or oblong-ovate, acute or acuminate, sessile or very shortly petioled, 5-nerved, the outer nerves often obscure. Racemes drooping, many-flowered, 6in.–12in. long or more, pubescent or glabrous. Pedicels slender, bracteolate at the base. Flowers proterogynous. Sepals broadly obtuse or subacute. Filaments elongating after fertilisation. In the fruiting stage the petals are fleshy and full of purple juice.—Hook. f., Fl. N.Z. i. 45; Handbk. 46; T. Kirk, Forest Fl. N.Z. t. 139. *C. sarmentosa,* G. Forst., Prod. n. 377; DC., Prod. i. 739; Hook, Bot. Mag. t. 2470; A. Cunn., Precurs. n. 581. *C. arborea* and *C. Tutu,* Lindsay, Contrib. to N.Z. Bot. 84. *C. hermaphrodita,* Banks and Sol. MSS.

KERMADEC Islands to Southland; CHATHAM Islands; STEWART Island. Sea-level to 3,000ft. Abundant in lowland districts. Also in Chili and Peru. *Puhou. Tutu. Tupakihi.*

The juice of the leaves, &c., in the spring and the seeds in the autumn are highly poisonous. The fleshy petals afford a pleasant wine. In the "Index Kewensis" the New Zealand plant is considered distinct from the Peruvian.

C. ruscifolia often develops very robust shoots from the rootstock, which attain the height of from 6ft.–10ft. in a single season.

2. **C. thymifolia,** *Humb. and Bonp.* ex *Willd., Sp. Pl.* iv. 819. Suffruticose or herbaceous, 6in.–3ft. high. Branches slender. Leaves ⅛in.–1in. long, ovate, lanceolate or ovate-lanceolate, flat, sessile or very shortly petioled, often with short stout rhizomes. Racemes slender, spreading, but not drooping, 1in.–3in.

13

long. Pedicels very slender. Flowers as in *C. ruscifolia*, but smaller.—DC., Prod. i. 470; Hook. f., Fl. N.Z. i. 45; Handbk. 47; Lindsay, Contrib. to N.Z. Bot. 87.

NORTH and SOUTH Islands: from the East Cape to Southland. 1,000ft. to 5,000ft. Said to grow on Stewart Island, but I have not seen specimens. *Tutupapa. Tutuheuheu.*

Large forms are very close to the preceding; alpine forms are easily distinguishable.

3. **C. lurida,** *n. s.* Whole plant lurid-purple; herbaceous. Stems developed from the subterranean bases of old branches, erect, strict; angles winged. Branches 3in.–4in. long, weak, simple, opposite or 3-nate, ascending. Leaves sessile or scarcely petioled, ovate-lanceolate or lanceolate, acuminate, ½in. long, 3-nerved, the intramarginal nerves often indistinct. Racemes 3in.–4in. long, erect, very slender, many-flowered; rhachis and pedicels pubescent. Flowers hermaphrodite or polygamous. Stamens in the female flowers represented by thread-like staminodia. Styles 1–5.

SOUTH Island: mountains of Canterbury and Otago, from 1,200ft. to 3,500ft. Nov., Dec.

Easily distinguished by its lurid hue and strict habit, but presenting no structural points of difference. The flowers are usually restricted to the middle portion of the stem. The subterranean portions of the stem are often matted.

4. **C. angustissima,** *Hook. f., Handbk. N.Z. Fl.* 47. Herbaceous, usually forming broad patches. Stems springing from the bases of subterranean branches, flexuous. Branches and ultimate branchlets filiform or almost capillary, drooping or suberect. Leaves very numerous, opposite or subopposite, ⅛in.–⅓in. long, $\frac{1}{30}$in.–$\frac{1}{20}$in. broad, narrow-linear, subulate or acuminate, very shortly petioled, margins usually involute. Racemes few, rarely branches. Bracteoles nearly as long as the leaves, glabrous or puberulous. Flowers very small, hermaphrodite or polygamous. Sepals oblong-ovate or broadly ovate, acute. Styles variable in number. Staminodia thread-like, thickened at the tips. Fruits large, black, glossy.

NORTH Island: Mount Egmont (?), *Handbook.* SOUTH Island: mountains of Canterbury and Otago, but often local. 2,000ft. to 4,000ft. Dec., Jan.

A beautiful plant, of plumose habit. Some varieties make a close approach to small forms of *C. thymifolia.* The Mount Egmont habitat is probably erroneous.

Order XXIII.—LEGUMINOSAE.

Calyx 5- rarely 4-toothed or the sepals free. Corolla of 5 petals or fewer, perigynous, irregular or rarely regular, usually free, rarely coherent. Stamens 10, rarely fewer or indefinite, usually 9 with the filaments united and 1 free, or all united, or rarely all free, perigynous. Ovary of a single carpel, sessile or stipitate; ovules several, attached along the upper angle, abruptly narrowed into a straight or oblique style. Fruit a 2-valved pod (legume), rarely indehiscent or spirally coiled, 1 or more seeded. Seeds usually without endosperm. Herbs, shrubs, trees, or climbers. Leaves alternate, rarely opposite, usually stipulate, compound or rarely simple. Flowers in axillary or terminal racemes, spikes, fascicles, or corymbs.

One of the largest natural orders. Genera, 400. Species, 6,600. Distributed over the temperate and tropical regions of the earth, but less developed in New Zealand than in any other temperate country, only 8 genera being represented, of which *Carmichaelia*, *Corallospartium*, *Huttonella*, and *Notospartium* are endemic, except one species of the first, which is restricted to Lord Howe's Island, while *Swainsona* and *Clianthus* are restricted to Australia and New Zealand. Representatives of nearly 20 exotic genera are more or less naturalised, and often occur in great abundance. The order is divided into three suborders. All the indigenous species belong to the first.

Suborder I.—PAPILIONACEAE.

Corolla irregular or rarely nearly regular, imbricate, the upper petal or standard broadest, outside the others in bud and in most cases reflexed; the two equal lateral petals or wings and two lower petals parallel with the wings are often coherent by their lower margins, forming the keel. Stamens usually 10, the uppermost free, 9 united by the filaments forming a sheath around the ovary, open above, or rarely all the filaments united or all free.

This suborder includes a large number of economic plants—the pea, bean, kidney-bean, and other legumes; various clovers, medicks, and other forage-plants. Many species yield drugs and timbers of great value.

Tribe I. GENISTEAE.*—Herbs or shrubs, with simple or 3-foliolate leaves. Stamens all united by their filaments into a sheath.

- * Lupinus. Calyx deeply bilabiate. Pod compressed. Leaflets 3–12, digitate.
- * Ulex. Calyx 2-lipped. Branches spinous.
- * Cytisus. Calyx 2-lipped, minutely toothed. Leaves 1–3-foliolate. Branches furrowed.

II. TRIFOLIEAE.*—Herbs or shrubs. Leaves pinnately or digitately 3-foliolate; veinlets usually produced into minute teeth. Racemes or flower-heads axillary. Upper stamens free.

- * Medicago. Racemes short. Pod spiral, rarely small and 1-seeded.
- * Melilotus. Racemes elongated. Pods short, thick, indehiscent.
- * Trifolium. Flower-heads terminal or axillary. Petals with their claws adherent with the staminal tube. Pods mostly included in the calyx.

III. LOTEAE.*—Herbs or small shrubs. Leaves pinnate; leaflets entire. Flowers capitate or umbellate. Stamens mon- or diadelphous. Pod 2-valved, dehiscent or indehiscent.

- * Anthyllis. Pod included in the inflated calyx, 1- or many-seeded.
- * Lotus. Pod exserted, linear, dehiscent, many-seeded.

IV. GALEGEAE.—Herbs, shrubs, or trees, rarely climbers. Leaves pinnately 3- or many-foliolate. Upper stamen free, or rarely all united. Ovules 2 or more, rarely 1. Pod 2-valved, turgid or flat, not articulate.

- * Indigofera. Anthers tipped with a small gland. Pod 4-angled. Leaflets with obscure veins.
- * Robinia. Leaflets stipellate. Flowers large, in pendulous racemes. Pod flat.

1. Corallospartium. Stem yellow, stout, leafless. Pods compressed, silky, dehiscing along the suture.
2. Carmichaelia. Leafless or leaves very deciduous. Pods short, turgid or compressed; valves falling away from the persistent consolidated margins.
3. Huttonella. Branchlets terete or compressed. Pod obovate, with an upturned beak, broader than deep, sutural but indehiscent.
4. Notospartium. Branchlets pendulous, leafless. Pods shortly stipitate, linear, torulose or constricted, many-seeded, the seeds solitary in the cells.
5. Clianthus. Racemes pendulous. Flowers large, crimson. Petals acute. Style bearded beneath the stigma. Pod terete, many-seeded.
6. Swainsona. Herb. Racemes axillary, peduncled, erect. Petals obtuse. Pod membranous, inflated.

Tribe V. HEDYSAREAE.*—Herbs or shrubs. Leaves pinnate; leaflets 3 or many. Upper stamen free. Pod indehiscent, articulations few or many, 1-seeded, separating transversely.

* ONOBRYCHIS. Pod 1-celled, 1-seeded.

VI. VICIEAE.*—Herbs, often climbing. Leaves pinnate; rhachis ending in a simple or branched tendril or a bristle; leaflets often toothed. Upper stamen free. Pods 2-valved, 1-celled.

* VICIA. Style filiform, hairy below or all round.

* LENS. Style filiform or slightly flattened, hairy on the upper surface only.

* LATHYRUS. Style flattened, hairy on the upper surface only.

VII. PHASEOLEAE.—Twining or prostrate herbs, rarely suffruticose. Leaves 1–7-foliolate; leaflets stipellate. Upper filament quite free or at the base only. Pod 2-valved, several-seeded.

7. CANAVALIA. Two upper lobes of the calyx coherent, three lower lobes minute, free. Style beardless. Pod straight.

* DOLICHOS. Two upper lobes of the calyx coherent. Wings adnate to the keel. Pod falcate.

VIII. SOPHOREAE.—Shrubs or trees. Leaves pinnate. Stipellae 0. Stamens all free or slightly united at the base.

8. SOPHORA. Corolla papilionaceous; standard short. Stamens exserted. Pod terete or 4-angled or winged, moniliform.

SUB-ORDER II.—MIMOSEAE.

Flowers small, regular, sessile in spikes or heads, rarely pedicellate. Sepals and petals 5 or 4, rarely 3 or 6, valvate, rarely imbricate, free or united. Stamens equalling the petals or indefinite, free or united. Radicle straight.

* ACACIA. Stamens indefinite, all free.

ALBIZZIA. Stamens indefinite, monadelphous.

*LUPINUS, Tourn.

Calyx unequally 2-lipped; lobes toothed. Standard reflexed at the sides; keel arcuate. Stamens monadelphous; anthers oblong or roundish. Pod oblong, flattened, often constricted between the seeds. Cotyledons fleshy. Herbs, rarely shrubs. Leaves palmately 1–many-foliolate. Flowers in terminal racemes or spikes.

***L. arboreus**, *Sims., Bot. Mag. t.* 682. A shrub, 3ft.–6ft. high, with spreading branches. Leaves and pedicels silky. Stipules adnate to the petiole. Leaflets 6–9, linear-oblong-spathulate. Racemes terminal. Flowers verticillate, pale-yellow. Calyx deeply 2-fid, unequal, silky; upper lobe linear-ovate; lower broadly ovate, shorter. Pod 6-seeded, silky.

NORTH and SOUTH Islands: Turakina, &c.; Paikakariki; Kaikoura, &c. Scarcely naturalised, but persistent and increasing. Dec., Jan. California.

*ULEX, Linn.

Calyx membranous, 2-fid, the upper part with 2 minute teeth, the lower with 3; bracts 2, minute. Stamens monadelphous. Pod 2-valved; ovules many. Much-branched spinous shrubs.

***U. Europaeus**, *L., Sp. Pl.* 241. Stems furrowed, 1ft.–6ft. high. Leaves exstipulate, 1–3-foliolate on seedling plants, ultimately reduced to spines or small scales; spines branching at the base. Flowers lateral, solitary or racemose, bright-yellow. Calyx hairy. Wings exceeding the keel, obtuse.

Naturalised throughout the colony. *Furze. Whin. Gorse.* Aug. to April. Europe: Denmark to Italy, &c.

* CYTISUS, Linn.

Calyx 2-lipped, upper lip often 2–3-toothed. Wings oblong, deflexed after flowering; keel obtuse. Stamens monadelphous. Stigma terminal, capitate; ovules many. Pod flat. Funicle tumid. Shrubs, with 1–3-foliolate leaves.

Branchlets angled and furrowed. Flowers yellow. Pods many-seeded .. .. * *C. scoparius.*
Branchlets slender. Flowers white. Pods 2-seeded .. .. * *C. albus.*
Branchlets leafy. Flowers yellow. Pods linear, 5-seeded .. .. .. * *C. candicans.*

* **C. scoparius,** *Link., Enum. Hort. Berol.* ii. 241. Much-branched, 1ft.–3ft. high; branchlets angular or furrowed. Leaves 1–3-foliolate; leaflets obovate, silky. Flowers bright-yellow. Pedicels short. Styles spiral. Pod 1in.–2in. long, many-seeded; valves twisted after dehiscence.

Naturalised from Mangonui to Southland. *Broom.* Nov., Dec. Europe, Northern Asia, &c.

* **C. albus,** *Link., Enum. Hort. Berol.* ii. 241. Erect, branched, 1ft.–6ft. high. Branchlets very slender, furrowed. Leaves 1–3-foliolate, deciduous. Flowers in fascicles or short racemes, white. Pod 2-seeded, compressed, villous.

SOUTH Island: near Lake Ellesmere. Apparently established. *White* or *Portugal broom.* Sept., Oct. Italy, Spain, Portugal, &c.

* **C. candicans,** *Lam., Fl. Fr.* ii. 623. A slender leafy shrub, 2ft.–4ft. high, more or less clothed with white pubescence. Leaves 3-foliolate; leaflets obovate, obtuse. Flowers yellow, on short peduncles in few-flowered terminal racemes or fascicles. Calyx silky, 2-lipped; 2 upper sepals entire, acute; lower sepal with 3 short teeth. Standard exceeding the wings, not reflexed; wings equalling the keel. Pods linear, imperfectly septate, 4–5-seeded, acute, villous; seeds black.

NORTH and SOUTH Islands: naturalised in many localities, and increasing. Sept. South Europe.

* MEDICAGO, Linn.

Calyx 5-toothed. Wings exceeding the obtuse keel. Stamens diadelphous, the upper one free. Ovary curved; stigma sub-capitate; ovules few or many. Pod 1-celled, hooded, curved or spirally twisted, sometimes spinous, rarely dehiscent. Herbs, with pinnately 3-foliolate leaves and stipules adnate with the petiole.

Stems erect. Pods loosely spiral .. .. * *M. sativa.*
Pods small, compressed, reniform .. .. * *M. lupulina.*
Pods spirally coiled, spinous. Stipules laciniate .. .. .. * *M. denticulata.*
Pods spirally coiled, spinous. Stipules toothed .. .. .. * *M. maculata.*

* **M. sativa,** *L., Sp. Pl.* 778. Stem erect, 6in.–24in. high. Leaflets narrow-linear or obovate-oblong, dentate, emarginate, mucronate at the apex. Stipules large, subulate. Flowers on long peduncles, racemose or subcorymbose, purple; pedicels short. Pods laxly twisted or coiled, downy.

NORTH and SOUTH Islands: in many localities; a denizen only. *Lucerne. Alfalfa.* Sept. to Dec.

* **M. lupulina,** *L., Sp. Pl.* 779. Annual. Stems procumbent or suberect, spreading, 6in.–18in. long, pubescent. Leaflets cuneate-obovate, toothed, apiculate. Peduncles slender, exceeding the leaves. Pedicels very short, or flowers sessile. Flowers small, yellow. Pods compressed, kidney-shaped, shortly toothed, black, indehiscent, 1-seeded.

NORTH and SOUTH Islands: generally naturalised. *Black medick.* Sept. to Feb. Europe, &c.

* **M. denticulata,** *Willd.*, *Sp. Pl.* iii. 1414. Annual. Stems 6in.–18in. long, prostrate or ascending, almost glabrous. Stipules laciniate. Leaflets broadly obcordate. Peduncles 2–10-flowered. Pedicels very short. Flowers small, yellow. Pod spirally coiled in two or three lax turns, reticulate; margins thin, with 2 rows of divergent hooked spines.

NORTH and SOUTH Islands; STEWART Island; CHATHAM Islands: generally naturalised. *Toothed medick.* Sept. to Jan. Europe, North Africa.

* **M. maculata,** *Willd.*, *Sp. Pl.* 1412. Annual, procumbent or spreading, pubescent or nearly glabrous. Stipules toothed. Leaflets triangular, obcordate, with a dark central spot, minutely toothed. Peduncles 3–5-flowered. Flowers yellow. Pods compactly spiral in 3–5 compact coils, compressed, furrowed on the thick edge and faintly reticulate; spines in 2 divergent rows, subulate, hooked.

NORTH Island: Auckland, Hawke's Bay, Wellington. Naturalised, but rather local. *Spotted medick.* Sept. to Dec. Europe, North Africa.

* MELILOTUS, Tourn.

Calyx-teeth 5, nearly equal. Petals deciduous; standard oblong; keel obtuse. Ovary straight. Pod oblong or subglobose, short but exceeding the calyx, 1–4-seeded. Annual or perennial herbs, with pinnately 3-foliolate leaves and erect racemes of small flowers.

Stipules setaceous. Pod acute, black.. .. .. .. * *M. officinalis.*
Stipules subulate. Pod obtuse, brown .. .. .. * *M. arvensis.*
Stipules linear-subulate. Pod acute, reticulate, black .. .. * *M. alba.*

* **M. officinalis,** *Lam.*, *Fl. Fr.* ii. 594. Stems erect, furrowed, 1ft.–2ft. high. Stipules setaceous. Leaflets obovate-oblong, obtuse, toothed. Racemes axillary, 2in.–4in. long, naked below. Flowers yellow, drooping. Wings, keel, and standard of equal length. Pod ovoid, acute, compressed, hairy, black.

NORTH Island: Auckland to Wellington: naturalised, but not common. SOUTH Island: near Christchurch. *Melilot.* Dec., Jan.

* **M. arvensis,** *Wallr.*, *Sched. Crit.* 391. Stems spreading, wiry. Stipules subulate. Leaflets linear-oblong, sharply toothed. Racemes naked below. Flowers small; wings and standard equal, longer than the keel. Pod ovoid, obtuse, mucronate, rounded on the back, transversely wrinkled, glabrous, brown.

NORTH and SOUTH Islands; CHATHAM Islands: naturalised in waste places, &c. Dec. to Feb. Europe, &c.

* **M. alba,** *Desr. in. Lam. Encyc.* iv. 63. Stems erect, wiry, sparingly leafy. Stipules linear-subulate. Leaflets linear-obovate or cuneate-obovate, obtuse. Flowers white. Wings and keel equal, shorter than the standard. Pods ovate, acute, apiculate, reticulate, glabrous, black.

NORTH Island: Napier. SOUTH Island: near Lincoln and Springston. Sparingly naturalised in waste places, &c. *White melilot. Bokhara clover.* Jan. to March.

* TRIFOLIUM, Linn.

Calyx-teeth 5, unequal. Corolla persistent; the claws of all the petals, or sometimes with the exception of the standard, united below with the staminal tube. Filaments slightly enlarged upwards; upper stamen free. Pod indehiscent, included within the calyx or partially extruded, 1–4-seeded. Small annual or perennial herbs, with digitate or pinnate 3-foliolate leaves, stipules adnate with the petiole, and flowers in dense heads or spikes.

1. Heads many-flowered, globose or oblong. Calyx with a ring of hairs at the throat, not inflated. Petals persistent. Pods sessile, 1-seeded.

Heads cylindrical or oblong.

Stipules with long setaceous points .. .. * *T. arvense.*
Stipules obtuse .. .. * *T. incarnatum.*

Heads ovoid or globose.

Heads ovoid, sessile. Calyx hairy .. .. * *T. pratense.*
Heads subglobose. Calyx glabrous .. .. * *T. medium.*
Heads sessile. Calyx-tube cylindrical in fruit .. .. * *T. scabrum.*

2. Heads many-flowered. Calyx not inflated; throat naked. Pods 2-4-seeded.

Calyx-teeth ovate, acute, reflexed in fruit .. .. * *T. glomeratum.*
Calyx campanulate. Stems creeping .. .. * *T. repens.*
Calyx campanulate. Stems erect .. .. * *T. hybridum.*

3. Heads many-flowered. Calyx 2-lipped, inflated; throat naked. Pods 1-2-seeded, sessile .. .. * *T. resupinatum.*

4. Heads many-flowered. Flowers ultimately drooping. Calyx not inflated; throat naked. Corolla persistent. Standard covering the pod. Stems erect .. .. * *T. agrarium.*

* *Stems procumbent or ascending, slender.*

Heads about 40-flowered .. .. * *T. procumbens.*
Heads 10–12-flowered .. .. * *T. dubium.*
Heads 3–7-flowered .. .. * *T. filiforme.*

* **T. arvense,** *L., Sp. Pl.* 769. Annual, erect or spreading, 3in.–10in. high, very slender, sparingly leafy, silky. Stipules linear-acuminate or setaceous. Leaflets narrow-oblong or linear-obovate. Heads terminal, dense, oblong or cylindrical. Flowers minute. Calyx-teeth equalling or exceeding the petals, silky, plumose.

NORTH and SOUTH Islands: naturalised in numerous localities, but rarely abundant. *Hare's-foot trefoil.* Jan. to March. Europe.

* **T. incarnatum,** *L., Sp. Pl.* 769. Annual, erect, 6in.–18in. high, villous or pubescent, hairs spreading. Stipules ovate, obtuse. Leaflets broadly obovate, retuse or emarginate. Heads terminal, ovate or cylindrical. Calyx hairy; teeth lanceolate-subulate, nearly equal, shorter than the corolla, patent in fruit.

NORTH and SOUTH Islands: an occasional escape from cultivation, scarcely permanent. *Crimson clover.* Dec. to Feb. South Europe, West Europe.

* **T. pratense,** *L., Sp. Pl.* 768. Perennial. Stems solid or fistulose, erect or ascending, pubescent, 6in.–18in. high. Stipules ovate, obtuse, abruptly bristle-pointed. Leaflets oblong, elliptic or obovate, entire or retuse, sometimes apiculate. Heads terminal, globular or ovate, dense, sessile, usually subtended by opposite almost sessile leaves. Calyx much shorter than the corolla; teeth setaceous, hairy, slender, the lowest much the longest.

Naturalised throughout the colony. *Red clover. Cow-grass.* Nov. to Feb. Europe.

* **T. medium,** *L.* ex *Huds., Fl. Ang. ed.* i. 284. Stems flexuous, erect or spreading, 6in.–18in. high, pubescent. Stipules lanceolate, acuminate. Leaflets elliptic, oblong or lanceolate, entire apiculate. Heads shortly pedunculate, sub-globose, subtended by opposite leaves. Flowers purple or white. Calyx not half as long as the corolla; teeth setaceous, hairy, spreading in fruit. Pod dehiscing longitudinally.

NORTH and SOUTH Islands: naturalised in numerous localities. *Zig-zag clover.* Dec. to Feb. North Europe.

* **T. scabrum,** *L., Sp. Pl.* 770. Annual. Stems pubescent, rigid, flexuous, 3in.–10in. high. Leaves on very short petioles. Leaflets obovate or narrow-obovate, strongly nerved, minutely toothed. Stipules ovate-cuspidate. Heads ovate, solitary, sessile, terminal or axillary. Flowers small. Calyx-tube strongly veined; teeth erect, equalling the corolla, spreading in fruit.

NORTH Island: naturalised, Auckland, Hawke's Bay, Wellington, &c. Nov. to Jan. Europe.

* **T. glomeratum,** *L., Sp. Pl.* 770. Annual. Stem 2in.–10in. long, prostrate, spreading, glabrous, obovate or obcordate. Stipules ovate-acuminate, toothed. Heads sessile, terminal and axillary. Flowers reddish-purple. Calyx-teeth ovate-acute, reflexed. Standard persistent, striate.

NORTH Island: naturalised in various localities, especially in Auckland and Wellington. Nov., Dec. Europe.

* **T. hybridum,** *L., Sp. Pl.* 776. Perennial. Stems erect, 12in.–18in. high, almost glabrous, flexuous, fistulose. Stipules ovate-lanceolate. Leaflets obovate-oblong or lanceolate. Heads axillary. Peduncles longer than the leaves. Pedicels half as long as the corolla, ultimately deflexed. Flowers white or pinkish. Calyx-tube campanulate; teeth subulate, nearly equal. Seeds 2–4.

NORTH and SOUTH Islands: naturalised in many localities. Ascends to 3,000ft. *Alsike clover.* Jan. to March. Europe.

* **T. repens,** *L., Sp. Pl.* 767. Perennial. Stems 3in.–18in. long, creeping, glabrous. Leaflets obovate or obcordate, often with a dark spot. Stipules ovate, cuspidate, or acuminate. Peduncles axillary, greatly exceeding the leaves. Flowers white. Pedicels ultimately reflexed. Calyx-tube half the length of the corolla, gibbous; teeth lanceolate, unequal. Pod 2–4-seeded.

Naturalised throughout the colony. *White* or *Dutch clover.* Oct. to March. Europe, North Africa.

* **T. resupinatum,** *L., Sp. Pl.* 771. Annual. Stems prostrate, often appressed to the ground, 2in.–10in. long. Stipules lanceolate, subulate. Leaflets obovate, minutely serrate. Peduncles short, axillary. Heads dense, globose. Flowers reversed. Calyx ventricose, netted, woolly; teeth short. Pod wholly included, 2-seeded.

NORTH Island: naturalised in many localities, and increasing. *Hare's-foot trefoil.* Oct. to Jan. Europe.

* **T. agrarium,** *L., Sp. Pl.* 772. Annual. Stems erect, 4in.–12in. high, sparingly branched, glabrate, pubescent or almost pilose. Leaves on short petioles, digitately 3-foliolate; leaflets narrow-obovate, oblong, obtuse or truncate. Stipules ovate-lanceolate, hairy. Peduncles slender, axillary. Flowers on short pedicels, yellow, drooping. Calyx-tube one-third the length of the corolla; teeth unequal, short.

SOUTH Island: naturalised, Broken River Basin, Canterbury. Dec., Jan. Central Europe.

* **T. procumbens,** *L., Sp. Pl.* 727. Annual. Stems spreading or ascending, 6in.–15in. long, pubescent. Stipules ovate, acute, hairy. Leaves pinnately 3-foliolate; leaflets narrow, cuneate-obovate, retuse or emarginate, minutely toothed. Heads on slender axillary peduncles, very dense. Flowers about 40, on short pedicels, ultimately reflexed. Calyx small, campanulate; teeth very short, unequal. Standard dilated, reflexed in front. Pod much shorter than the standard.

Naturalised throughout the colony. *Hop trefoil.* Nov., Feb. Europe.

*T. **dubium,** *Sibth., Fl. Oxon.* 231. Annual. Stems slender, wiry, prostrate or ascending, 3in.–18in. long, glabrate or hairy. Leaves pinnately 3-foliolate; leaflets obovate, emarginate, minutely serrate. Stipules ovate, acuminate. Peduncles axillary, very slender. Heads small, dense. Flowers about 12, yellow. Pedicels very short. Calyx short, half as long as the corolla; teeth linear-subulate. Standard folded, furrowed, slightly exceeding the pod.

Abundantly naturalised throughout the colony. *Little yellow suckling.* Oct. to March. Europe.

*T. **filiforme,** *L., Sp. Pl.* 773. Annual. Stems 2in.–6in. long, prostrate, very slender, pubescent. Stipules ovate, acute. Leaflets emarginate. Peduncles axillary, capillary, short. Flowers 2–6, very small, yellow. Pedicels as long as the calyx-tube, at length reflexed. Standard folded, not furrowed, emarginate, about equalling the pod.

SOUTH Island: Dunedin; Bluff Hill. Dec., Jan. Europe.

*ANTHYLLIS, Linn.

Calyx tubular, inflated; teeth 5, unequal; mouth oblique. Standard auricled at the base; wings and keel with the claws adnate to the staminal tube. Stamens monadelphous. Ovules 2–4. Pod 1–3-seeded, dehiscent or indehiscent. Herbs, with unequally pinnate leaves, and flowers in compact naked or involucrate cymes.

*A. **vulneraria,** *L., Sp. Pl.* 719. Perennial. Stems 6in.–12in. high, silky. Leaves simple or with 2–6 pairs narrow-oblong leaflets. Heads in pairs, pedunculate, with an involucre of linear leaflets. Calyx exceeding the petals, contracted at the mouth; teeth minute. Pod small, 1-seeded, upper suture curved outwards.

SOUTH Island: Nelson; Dunedin. Sparingly naturalised. *Kidney-vetch.* Dec., Jan. Europe, North Africa, West Africa.

*LOTUS, Linn.

Calyx-teeth 5, nearly equal. Petals free from the staminal tube; keel ascending, narrowed into a beak; wings connivent at their upper margin. Alternate filaments dilated upwards, upper free. Ovary sessile; ovules numerous; style kneed at the base. Pod linear, many-seeded, imperfectly septate. Annual or perennial herbs or undershrubs. Leaves pinnate or palmate.

Stems decumbent. Heads 5–10-flowered .. .. * *L. corniculatus.*
Stems erect. Heads 8–12-flowered .. .. * *L. uliginosus.*
Stems procumbent. Heads 1–3-flowered .. .. * *L. angustissimus.*

*L. **corniculatus,** *L., Sp. Pl.* 775. Stems decumbent, 6in.–16in. long, glabrous or slightly hairy. Rootstock short. Petioles very short. Stipules ovate-lanceolate. Leaflets obovate. Heads on slender axillary peduncles, depressed. Flowers 5–10, shortly pedicelled, yellow with red or crimson streaks. Calyx-teeth adpressed in the bud, points of the two upper teeth converging. Corolla twice as long as the calyx. Pod 1in. long or more.

NORTH and SOUTH Islands: sparingly naturalised, but often local. *Bird's-foot trefoil.* Nov. to Jan. Europe.

*L. **uliginosus,** *Schkukr., Handbk.* ii. 412, *t.* 211. Stems erect or ascending, glabrous or hairy, 6in.–18in. high. Rootstock elongate. Leaflets obliquely obovate. Heads on long peduncles, 8–12-flowered. Calyx-teeth spreading in bud, two upper diverging.

NORTH and SOUTH Islands: more plentiful than the preceding, but not common. Jan., Feb. Europe, North Africa.

* **L. angustissimus,** *L., Sp. Pl.* 774. Annual. Stems slender, wiry, prostrate or ascending, 6in.–12in. high, glabrate or hairy. Leaflets obovate-oblong. Stipules equalling the leaflets. Heads on short peduncles, 1–3-flowered. Calyx-teeth subulate, straight in the bud. Pod narrow-linear, many-seeded.

NORTH Island: Auckland. Local. *Cheeseman!* Jan., Feb. Europe, North Africa.

* INDIGOFERA, Linn.

Calyx small; teeth 5, equal. Standard small, silky on the inferior surface; wings coherent, slightly adnate to the keel; keel erect, spurred or gibbous at the base. Upper stamen free. Pod 1- or several-seeded, septate within. Herbs or shrubs, clothed with appressed hairs. Leaves unequally pinnate, minutely stipulate. Flowers in axillary spikes or racemes.

* **I. viscosa,** *Lam., Encyc.* iii. 247. Suffruticose or almost herbaceous, 1ft. high; branchlets, leaves, and pedicels grey with appressed silky hairs, clammy. Leaves unequally pinnate; leaflets in about six pairs, elliptical-oblong or obovate, mucronate, minutely stipellate. Racemes axillary, equalling or shorter than the leaves. Calyx campanulate, silky, pilose; teeth linear-subulate. Style very short. Pods straight, spreading, compressed, 8–10-seeded.

NORTH Island: in several localities about Auckland, but scarcely naturalised. East Indies, &c.

* ROBINIA, Linn.

Calyx short, 2-lipped, 5-toothed. Standard large, rounded, reflexed, equalling or slightly exceeding the wings and keel. Upper stamen free. Pod linear, flat, margined on the ventral suture, 2-valved, several-seeded. Trees or shrubs, with pinnate leaves and pendulous racemes.

* **R. Pseud-acacia,** *L., Sp. Pl.* 722. A tall tree, affording timber of great durability. Leaves 6in.–10in. long, unequally pinnate. Stipules converted into spines. Leaflets broadly oblong, stipellate. Base of the petiole protecting the buds of the next season. Flowers large, white.

NORTH Island: naturalised in several localities. Forming groves above Taupiri, Waikato, and affording good timber. *Cobbet's locust-tree,* or *False acacia.* United States.

1. CORALLOSPARTIUM, J. B. Armst.

Calyx turbinate, woolly, 5-toothed, the superior teeth slightly shorter and broader than the others. Standard large, sharply reflexed; wings falcate, much shorter than the keel. Upper stamen free. Ovary villous, 2–3-ovulate; style silky at the base; stigma minute. Pod deltoid, rounded and winged at the back, shortly beaked, silky or villous; valves thin, cartilaginous, dehiscence sutural, 1-seeded. Radicle with a double fold; cotyledons thin. An erect sparingly-branched shrub, with a terete deeply-grooved stem and stout compressed or subterete branches and branchlets. Leaves only developed on very young plants, 1-foliolate, fugacious. Flowers in dense fascicles or short racemes in the denticulations of the branches.—*Carmichaelia,* Hook. f., Handbk. N.Z. Fl. 48.

A singular monotypic genus, distinguished from *Carmichaelia* by the absence of the persistent consolidated replum. Mr. Armstrong did not give a technical definition.

1. **C. crassicaule,** *J. B. Armst. in Trans. N.Z.I.* xiii. (1880) 333. Stems leafless, erect, 1ft.–3ft. high, terete, yellow, ½in. in diameter, deeply

grooved, the grooves tomentose; branchlets terete or rarely compressed. Leaves only found on the young plant, 1-foliolate, articulated on short grooved petioles, orbicular or oblong, entire or emarginate, gradually reduced to sessile scales, very fugacious. Fascicles 8–15-flowered, rarely umbellate; pedicels short, woolly; bracteoles 2, linear, below the base of the calyx or adnate with the calyx-tube. Flowers ⅓in. long; keel incurved. Style exserted quickly after expansion.—*Carmichaelia crassicaulis*, Hook. f., Handbk. 48.

SOUTH Island: mountains of Canterbury and North Otago. Rare and local. Mount Torlesse, &c., Lake Lyndon, Lake Ohau, *Haast! Enys!* Lindis Pass, *Hector! Buchanan!* North Central Otago, *Petrie*. 1,500ft. to 3,000ft. *Sticks* of the shepherds. *Coral-broom*. Dec., Jan.

The robust yellow stems, cream-coloured fascicled flowers, and villous pods distinguish this from all other New Zealand plants.

Var. **racemosa.** Branchlets narrow, compressed, ⅛in. broad, flat, edges convex, deeply grooved. Flowers less than ¼in. long, similar to the typical form, solitary, or in solitary or fascicled 3–5-flowered racemes, but the pedicels shorter, and with the rhachis and calyx less woolly. Otago: near the Lindis Pass, *Buchanan!* The small solitary or racemose flowers and much-compressed branchlets easily differentiate this from the type, but are scarcely sufficient to confer specific rank. I have only seen one small and imperfect specimen.

2. CARMICHAELIA, R. Br.

Calyx campanulate or turbinate, 5-toothed; teeth nearly equal, imbricate. Standard orbicular or obovate, usually reflexed, claw short; wings auricled at the base, claw slender. Stamens diadelphous, 9 united by their filaments into a sheath open above, the upper one free. Ovary rarely pubescent, narrowed into a slender beardless style; stigma minute, terminal; ovules few or many. Pods usually small, coriaceous, compressed or turgid, 1–12-seeded; valves separating from the persistent thickened margins. Radicle with a double fold; cotyledons fleshy. Depressed or erect shrubs, always leafless after the flowers have fallen. Leaves 1- or pinnately 3–5-foliolate, in many species restricted to the young plant or reduced to scales. Flowers usually small, solitary or racemose in the denticulations of the striated branchlets.

This is perhaps the most characteristic genus in the New Zealand flora. All the species are endemic, and are variously distributed from the North Cape to Southland, ascending from the sea-level to 4,000ft. Many vary to a great extent at different stages of growth, and are often difficult of identification. Branchlets which, previous to the expansion of the flowers, are severely flattened become terete or semi-terete before the fruit is matured. As far as possible it is desirable that the student should obtain both flowers and pods from the same plant, as species with similar branches and flowers may produce dissimilar pods. My herbarium contains fruiting specimens of several forms, apparently undescribed, but, as the flowers are unknown, they must remain for future description. With the exception of a single species endemic on Lord Howe's Island, the genus is restricted to New Zealand.

I. NANA.—Depressed leafless plants with fastigiate compressed branchlets forming dense patches, 1in.–4in. high. Flowers red.

* *Branchlets linear or narrow-linear, thin.*

Flowers solitary or racemose. Pods 1-seeded .. .. 1. *C. Enysii.*
Flowers solitary on long peduncles. Pods 3-seeded .. .. 2. *C. uniflora.*
Flowers racemose. Pods 2–4-seeded .. .. 3. *C. nana.*

** *Branchlets robust.*

Flowers racemose. Pods 6–12-seeded .. .. 4. *C. Monroi.*

II. Eucarmichaelia.—Erect or spreading plants, 1ft.–6ft. high; branchlets compressed or rarely terete; leafless when in flower, except 7, 13, 15, 16, 17, 18. Flowers usually purple-streaked, rarely yellowish, white, pink, or deep purple.

* *Pods convex or turgid.*

A. Stems strict, erect.

Flowers few, 1in. long. Pods 1in. long, straight, erect .. .. .. 5. *C. Williamsii.*
Flowers numerous, small. Pod oblong. Seeds red .. .. 6. *C. australis.*
Leafy. Branchlets flat or terete, $\frac{1}{12}$in.–$\frac{1}{4}$in. broad. Pods erect .. .. 7. *C. grandiflora.*
Branchlets stout, plano-convex. Pods elliptic-oblong. Seeds 3–6 .. .. 8. *C. robusta.*
Branchlets spreading, terete. Pods broadly oblong. Seeds 1–2 .. .. 9. *C. Petriei.*
Branchlets terete, deeply grooved. Flowers violet. Seeds 2–4 .. .. 10. *C. violacea.*
Branchlets almost filiform. Pod $\frac{1}{8}$in.–$\frac{1}{4}$in. long, 1-seeded .. .. .. 11. *C. diffusa.*
Branchlets terete or plano-convex. Pods drooping. Seeds 1–2 .. .. 12. *C. virgata.*
Branchlets compressed or plano-convex. Pods subulate .. .. .. 13. *C. subulata.*

B. Stems slender, flexuous, very narrow.

Leafy. Racemes short. Pod with a straight subulate beak .. .. .. 14. *C. Kirkii.*

** *Pods compressed.*

† Leafy when in flower.

Branchlets numerous, compressed, $\frac{1}{20}$in.–$\frac{3}{16}$in. broad. Pods in erect racemes, or apparently pendulous when growing on drooping branchlets, 2-seeded 15. *C. odorata.*
Branchlets few, compressed or terete. Pods numerous, on spreading racemes, 2-seeded .. .. 16. *C. angustata.*
Branchlets numerous, fastigiate, compressed, $\frac{1}{20}$in.–$\frac{1}{10}$in. broad. Seeds 2–4 .. 17. *C. flagelliformis.*

†† Leafless, or nearly so, when in flower.

Branchlets compressed, $\frac{1}{12}$in. broad. Pod abruptly acuminate .. .. 18. *C. acuminata.*
Branchlets compressed, $\frac{1}{16}$in.–$\frac{1}{12}$in. broad. Pod ovate, oblong, small .. .. 19. *C. Hookeri.*

1. **C. Enysii,** *T. Kirk in Trans. N.Z.I.* xvi. (1883) 379, *t.* 30. Forming dense patches 1in.–2in. above the surface of the soil. Root and lower branches stout; secondary branches 1in. long; branchlets fastigiate, $\frac{1}{2}$in.–$\frac{3}{4}$in. long, $\frac{1}{20}$in.–$\frac{1}{12}$in. broad, compressed, glabrous. Leaves only found on very young plants, small, orbicular, emarginate, shortly petiolate. Flowers solitary or in 3–5-flowered fascicles or racemes, red or pink. Pedicels slender. Calyx campanulate; teeth short, acute. Standard rather narrow, reflexed; wings equalling the keel. Ovules 2–5. Pod $\frac{3}{10}$in. long; pedicels erect or recurved, shorter than the pod, ovate-orbicular, with a short recurved beak, at first turgid, compressed when mature, 1- rarely 2–3-seeded. Seeds reniform, black. Replum incomplete. In dehiscence one valve becomes partially separated from the replum, remaining attached near the apex, and both valves become contorted in such a manner as to give the pod a curious deltoid appearance.

NORTH and SOUTH Islands: rare and local. Near the Wangaehu River, on the south-east side of Ruapehu, *T. K.* Broken River Basin, *Enys!* Ashburton Mountains, *T. H. Potts!* Eweburn Creek, Naseby, &c., *Petrie!* 1,500ft. to 3,000ft. Dec.

Var. **orbiculata** (*sp.*), Colenso in Trans. N.Z.I. xxii. (1889) 459. Larger and more robust than the type, 2in.–4in. high; branchlets $\frac{1}{8}$in. broad. Flowers solitary or in 3–6-flowered umbels or racemes. Valves rugulose. NORTH Island: Rangipo Plain, *T. K.* SOUTH Island: Mount Ida Range, &c., *Petrie!*

2. **C. uniflora,** *T. Kirk in Gard. Chron.* (1884) i. 512. Forming large compact patches 1in.–2in. high. Stems subterranean, giving off distant slender branches. Branchlets compressed, glabrous, $\frac{3}{4}$in.–1in. long, $\frac{1}{25}$in.–$\frac{1}{20}$in. broad; notches few, distant. Leaves not seen. Flowers solitary, $\frac{1}{4}$in. long, on long capillary puberulous peduncles, jointed about the middle, minutely bracteolate. Calyx turbinate or almost campanulate; teeth very short, acute. Standard large for the size of the flower, slightly reflexed; wings shorter than the keel. Pod pendulous, about $\frac{1}{2}$in. long including the oblique beak, oblong; valves slightly corrugated. Seeds usually 3.—Trans. N.Z.I. xvi. (1884) 379, t. 31; Buch. *l.c.* 394. *C. Suteri,* Colenso in Trans. N.Z.I. xxiii. (1890) 383.

SOUTH Island: mountains of Nelson, Canterbury, and North Otago. Cheviot, Lake Grasmere, Lochnavar, Poulter River, *Enys!* Mount Cook, *H. Suter!* Waitaki Valley, *Buchanan!* Head of Lake Hawea, *Petrie!* 1,800ft. to 3,000ft. Dec.

Probably not infrequent. Best distinguished from *C. nana* by the narrow branchlets, long slender peduncles and solitary flowers. Mr. Colenso described the Mount Cook plant as 7-seeded, but I could only find 3 seeds in pods from that locality.

3. **C. nana,** *Col.* ex *Hook. f., Handbk. N.Z. Fl.* 49. Forming broad patches. Branchlets 2in.–4in. high, thin, much compressed, strict, erect, $\frac{1}{12}$in.–$\frac{1}{6}$in. broad. Leaves not seen. Racemes 2–5-flowered. Pedicels long, slender. Flowers $\frac{1}{3}$in.–$\frac{1}{2}$in. long, red. Calyx campanulate; teeth broadly deltoid, subacute. Standard with a short broad claw, broadly oblong, deeply emarginate; wings broad, shorter than the keel. Ovary glabrous. Pod $\frac{1}{3}$in. in length, linear-oblong, as long as the expanded flower, narrowed at the base; beak short, straight. Seeds 2–4, black.—*C. australis β nana,* Benth. in Hook. f., Fl. N.Z. i. 50.

NORTH Island: open country, Lake Taupo, Ngauruhoe, Tongariro, Ruapehu, &c. SOUTH Island: abundant, Nelson to central Otago. Descends to within 100ft. of sea-level; ascends to 2,500ft. Dec.

Distinguished from *C. uniflora* by the broader more obtuse branchlets and large racemose flowers.

4. **C. Monroi,** *Hook. f., Handbk. N.Z. Fl.* 49. Excessively branched, forming compact hemispherical masses, 3in.–6in. high. Branchlets crowded, very stout, $\frac{1}{8}$in.–$\frac{1}{4}$in. broad, grooved. Leaves only seen on very young plants, cuneate or obovate, emarginate, silky, fugacious. Racemes 2–3-flowered, but sometimes fascicled, rarely forming a lax corymb. Pedicels long and slender. Calyx funnel-shaped; teeth linear-subulate, acute, nearly equalling the tube, ciliated. Standard exceeding the keel, deeply emarginate, claw narrow; wings shorter than the keel. Pod falcate or nearly straight, $\frac{1}{3}$in.–$\frac{5}{8}$in. long, often narrowed at the base, remarkably turgid; valves strongly corrugated when mature; teeth very short, straight or oblique. Seeds 4–12, mottled.—*C. corrugata,* Col. in Trans. N.Z.I. xv. (1882) 320.

SOUTH Island: Marlborough: Wairau and Awatere Valleys. Canterbury: Broken River Basin, Upper Waimakariri, Poulter Valley, &c. Otago: Kurow, Mount Ida &c., mountains of Lake district &c. 200ft. to 4,500ft. Dec., Jan.

Distinguished from all the species of this section by the stout branchlets, lax racemes, and linear-subulate calyx-teeth. The flowers are smaller than those of *C. nana*, and red, as in all the species of this section.

5. **C. Williamsii,** *T. Kirk in Trans. N.Z.I.* xii. (1879) 394. Erect, 3ft.–6ft. high, much branched. Branchlets thin, excessively compressed, $\frac{3}{8}$in.–$\frac{5}{8}$in. broad, closely striated or almost grooved, minutely pubescent when young; denticulations alternate, distant. Leaves 1–3-foliolate; leaflets obovate, obcordate or emarginate. Flowers large, $\frac{3}{4}$in.–1in. long, pendulous, solitary or in 2–6-flowered fascicles or very short racemes. Pedicels slender, silky. Calyx tubular, pubescent; teeth linear-subulate, acute. Corolla cuneate, lurid-yellow, veined; standard equalling the keel, sharply reflexed, forming a right angle with its base; wings narrow-falcate, shorter than the keel. Pods on stout erect pedicels, oblong or obliquely oblong, turgid, 1in.–1$\frac{1}{4}$in. long; beak stout, oblique. Seeds 9–10, red mottled with black.

NORTH Island: Auckland, rare and local. Te Kaha Bay and Raukokore Bay, Bay of Plenty, Hicks Bay, East Cape district, *Bishop Williams! H. B. Kirk!* Nov., Dec.

Distinguished from all other species by the broad but thin branchlets, large yellowish-red flowers, and stout erect pods. The sinus between two teeth of the calyx is rounded, not acute.

6. **C. australis,** *R. Br. in Bot. Reg.* xi. (1825) *t.* 242. An erect much-branched glabrous usually leafless shrub, 3ft.–9ft. high. Branchlets flat, straight, often much elongated, and occasionally distichous, $\frac{1}{15}$in.–$\frac{1}{3}$in. broad, closely striated; denticulations close. Leaves confined to young plants, $\frac{3}{4}$in.–2in. long, deciduous, unifoliolate or pinnately 3–5-foliolate; leaflets membranous, sessile, obovate-cuneate or obovate, emarginate, rarely lobed. Racemes often fascicled, 3–12-flowered. Flowers small, purplish, dense or rarely spreading. Rhachis and pedicels puberulous or glabrous. Calyx glabrous, campanulate; teeth minute. Standard much broader than long, rounded at the apex, reflexed; wings broadly oblong, shorter than the keel, claw minute. Pods numerous, spreading, oblong, abruptly narrowed to a short beak, which is sometimes reduced to a mere point. Replum persistent long after the valves have fallen. Seeds 1–4, red.—A. Cunn., Precurs. n. 574; Hook. f., Fl. N.Z. i. 50; Handbk. 50. *C. Cunninghamii,* Raoul, Choix. t. 28B. *Bossiaea scolopendra,* A. Rich., Fl. N.Z. 345 (not of R. Br.). *Genista compressa,* Banks and Sol. MSS.

NORTH Island: common from the North Cape to the northern portion of the Wellington District. SOUTH Island: extremely rare and local. Queen Charlotte Sound, *J. Rutland!* Reported from Banks Peninsula, but requires confirmation. Sea-level to 3,000ft. *Maukoro. Makaka.* Nov., Dec.

Varying greatly in the width of the branchlets and the length of the racemes. It cannot be the *Lotus*(?) *arboreus*, Forst., Prod. 278. Of *C. stricta*, Lehm., usually referred to this species, nothing certain is known.

Var. **alata.** Branchlets $\frac{1}{2}$in. wide. Racemes lax. Pedicels longer than the type; wings broader. Keel abruptly expanded above, with a large lobe or tooth at the base of the expanded portion. NORTH Island: Hokianga.

Var. **strictissima.** Branchlets $\frac{1}{4}$in.–$\frac{3}{8}$in. broad. Racemes strict, very dense. Pedicels very short. Pods not seen. White Cliffs, Taranaki, *Cheeseman!*

7. **C. grandiflora,** *Hook. f., Handbk. N.Z. Fl.* 49. Erect, glabrous, 1ft.–3ft. high, much branched. Branchlets compressed, spreading, rarely subterete or fastigiate, $\frac{1}{12}$in.–$\frac{1}{8}$in. broad, deeply grooved. Leaves numerous, unequally 3–5-foliolate; leaflets very shortly petiolulate, broadly obcordate-cuneate, glabrous. Racemes $\frac{1}{2}$in.–1in. long, pedunculate, 5–9-flowered. Pedicels

short. Calyx campanulate or almost turbinate, large; teeth acute, glabrous or ciliolate. Standard broader than long, retuse, exceeding the keel. Ovary glabrous, many-ovuled. Fruiting racemes erect, 1in. long. Pods crowded, not compressed, oblong, somewhat turgid, narrowed into a short subulate beak. Seeds 2 or 3.—*C. australis*, var. γ *grandiflora*, Benth. in Hook. f., Fl. N.Z. i. 50.

SOUTH Island: Nelson to Otago; chiefly on the banks of rivers, &c.; local. Sea-level to 2,300ft. Dec., Jan.

Distinguished from *C. odorata* by the wider branchlets, more robust habit, larger flowers and longer pods.

Var. **alba.** Branchlets more robust, compressed, deeply grooved, fastigiate or nearly so. Flowers as in the typical form, but white. Ripe pods not seen. Smells disgustingly of mice. Near the Waimakariri glaciers, *Enys* and *Kirk*. Jan. Possibly a distinct species.

Var. **dumosa.** Stems creeping below the surface. Branches excessively numerous, 4in.–5in. high. Flowers as in the type, but smaller. Pods not seen. Broken River Basin. Forming compact patches 2ft.–4ft. in diameter.

Var. **divaricata.** A spreading shrub. Branches and branchlets divaricating at right angles, curved, subterete or terete, compressed at the tips, flexuous, rigid, grooved. Leaves few or many, 3–5-foliolate. Racemes 5–10-flowered, slender. Flowers small. Pods elliptic-oblong, narrowed at both ends; beak very short. SOUTH Island: Mount White and valley of the Poulter, 2,300ft., *Enys!* Near Greymouth, Westland, *Helms!* Jan. For the present this is retained as a variety of *C. grandiflora*, but better specimens will probably show that it has good claims to specific rank.

8. **C. robusta,** *n. s.* A stout spreading species, 12in.–18in. high, with distant terete or subterete branches, which are sometimes distichous. Branchlets often curved, stout, thick, compressed, plano-convex or almost terete in the autumn, strongly grooved. Leaves not seen. Racemes 3–8-flowered, lax. Rhachis and pedicels slender, puberulous or pubescent; pedicels ebracteolate. Calyx funnel-shaped; teeth linear, acute. Standard much broader than long, retuse, and with the wings wholly enclosing the keel; wings very broad, obtuse, with a very small auricle. Ovules numerous. Pod elliptic-oblong, ⅓in.–½in. long, slightly narrowed at the base, somewhat turgid; beak reduced to a very short subacute point. Seeds 3–6, mottled.

SOUTH Island: Broken River Basin, 2,000ft. to 2,800ft., *Enys* and *T. K.* Dec., Jan.

Easily distinguished from all other species by the robust habit: most nearly related to *C. Petriei*, from which it is separated by the elliptic-oblong many-seeded pod. The leaves of young plants supposed to belong to this species are 1–3-foliolate, with obcordate leaflets; the terminal leaflet much the largest.

9. **C. Petriei,** *n. s.* Stout, sparingly branched. Branchlets spreading, terete or plano-convex, finely striated, $\frac{1}{16}$in.–⅛in. in diameter. Leaves not seen. Racemes 3–8-flowered, solitary or excessively crowded. Rhachis and pedicels silky or almost villous; pedicels equalling or much longer than the flowers. Calyx broadly campanulate, with very short subacute or acute teeth, silky or villous. Standard rather longer than broad, exceeding the keel; wings very short, broad, rounded at the tips. Ovary glabrous, 2–4-ovulate. Pods broadly oblong, turgid, oblique at the apex; beak minute, subulate, often reduced to a mere point; valves thick, strongly reticulated. Seeds 1, rarely 2, large, mottled. Radicle very short and stout.

SOUTH Island: Otago: Clutha Valley; north of Clyde; valleys and terraces on east and west of the Dunstan mountains; *Petrie!*

A very distinct species, which I have much pleasure in dedicating to its energetic discoverer.

10. **C. violacea,** *n. s.* Erect, 1ft.–2ft. high, distichously branched, glabrous. Branches terete, strongly grooved; branchlets terete, except at the compressed striated tips, or when very young. Leaves not seen. Racemes $\frac{1}{4}$in.–$\frac{1}{2}$in. long, 5–8-flowered, solitary or in dense fascicles. Rhachis and pedicels puberulous or silky; bracteoles minute. Flowers $\frac{3}{16}$in. long or less (without pedicels). Calyx glabrate or pubescent; teeth short, acute; sinus rounded. Standard and wings exceeding the keel. Standard obovate, claw broad; wings obovate, broad, rounded at the tip; keel toothed, not auricled. Ovary glabrous; ovules 8–9. Pods about $\frac{1}{4}$in. long or more; beak short, stout. Seeds 2–4(?), mottled.

SOUTH Island: Coleridge Pass, 2,500ft. to 3,000ft., *Enys* and *T. K.* Jan.

The stout terete deeply-grooved branches and vast profusion of minute violet-coloured flowers enable this plant to be easily recognised. The pods on my specimen are too old for a good description, only the naked replum with its attached seeds remaining. The only leaf observed was digitately 3-foliolate. The tooth at the base of the keel-petals is unusual.

11. **C. diffusa,** *Petrie in Trans. N.Z.I.* xxv. (1892) 272. Erect, 1ft.–2ft. high, slender, glabrous, much branched. Branchlets sometimes distichous, compressed or plano-convex or almost terete or filiform, $\frac{1}{20}$in.–$\frac{1}{15}$in. wide, striated. Leaves not seen. Racemes numerous, 3–6-flowered. Pedicels shorter than the flowers, glabrous, puberulous or silky. "Calyx glabrous, undulate and jagged or ciliate at the margin, which is hardly toothed. Ovary glabrous." Pods very small, $\frac{1}{6}$in.–$\frac{1}{5}$in. long, obliquely oblong, slightly narrowed below but not obovate, turgid; beak short, subulate. Seeds 1, rarely 2, mottled.

NORTH Island: Wellington: Dry River (pods only), *T. K.* SOUTH Island: Akaroa (pods only), *T. K.* Otago, *Buchanan!* Otepopo River, *Petrie!*

A very slender species, varying greatly in habit; rarely prostrate. When growing on the margins of forests it attains 5ft. to 6ft. in height, with drooping almost filiform branches and few pods. I have not seen flowering specimens. Mr. Petrie states that the pod resembles that of "some states of *C. flagelliformis*"; but that species has the pod compressed, not turgid, when mature. I have not seen flowers. Much better specimens of this plant are required before a satisfactory diagnosis can be drawn.

12. **C. virgata,** *n. s.* An erect rigid glabrous shrub, 3ft.–4ft. high, spreading from the base. Branchlets numerous, terete or rarely plano-convex, grooved. Leaves not seen. Racemes few, 3–5-flowered, lax, spreading. Rhachis and pedicels glabrous, puberulous. Calyx glabrous, broadly turbinate; teeth short, acute. Bracteoles 0. Standard broader than long, rounded, and with the wings exceeding the keel, claw very short and broad; keel with a very long narrow claw. Ovules 5–6. Pods (immature) on slender spreading or drooping pedicels, oblong, but sharply narrowed below; beak very short, straight, subulate. Seeds 1 or 2.

SOUTH Island: Otago, *Petrie!* Makarewa and Orepuko, Southland, *T. K.* Dec., Jan.

Distinguished by the paucity of its racemes, small whitish flowers, and oblong pod narrowed at both ends.

13. **C. subulata,** *n. s.* Erect, slender, 1ft.–2ft. high, leafy or leafless. Branches terete; branchlets compressed or plano-convex, $\frac{1}{20}$in.–$\frac{1}{12}$in. broad, rigid, grooved or striated, strict, often elongated, rarely twisted. Leaves

3-foliolate; leaflets sessile, obcordate. Racemes solitary or fascicled, forming loose corymbs. Rhachis and pedicels puberulous or almost silky; pedicels shorter than the flowers. Calyx campanulate; teeth short, acute. Standard broader than long, and with the wings equalling the keel. Ovary glabrous; ovules 3–4. Pod ¼in.–⅜in. long, subulate, acuminate; beak short, straight, stout. Seeds usually 2.

SOUTH Island: Marlborough: Whakamarina, Blenheim, *T. K.* Nelson: Cheviot, *Haast!* Canterbury: Burnham, Lincoln, Akaroa, &c., *T. K.* Broken River, *Enys* and *T. K.* Otago, *Buchanan!* Sea-level to 2,000ft. Dec., Jan.

Distinguished from all other species by the slender, strict, almost filiform branchlets and subulate pods.

14. **C. Kirkii,** *Hook. f. in Hook. Ic. Pl. t.* 1332. Branches elongated, terete, very slender and flexuous, 2ft.–3ft. long or more, often subscandent or scrambling over other shrubs, finely grooved, often interlaced. Leaves pinnately 3–5-foliolate, ½in.–1in. long; leaflets sessile, orbicular, emarginate, ¼in. long, glabrous. Racemes laxly 3–5-flowered. Rhachis and pedicels very slender; bracteoles ciliate. Calyx with long narrow deltoid teeth, villous within. Standard broad, 2-lobed, exceeding the keel; wings equalling the keel. Pod pendulous, 1ft. long, ellipsoidal, with a long straight subulate beak. Seeds 2. Replum stout, broad.—*C. gracilis,* J. B. Armst. in Trans. N.Z.I. xiii. (1880) 336 (in part).

SOUTH Island: Canterbury: by the Avon, Christchurch, *Haast! Cockayne!* Otago: rare and local. Cardrona Valley, *T. K.* Otepopo; valleys east and west of Rock and Pillar Range, Sowburn, *Petrie!* Nov., Dec.

Best distinguished by the slender flexuous leafy stems and 2-seeded pod with straight subulate beak.

15. **C. odorata,** *Col.* ex *Hook. f., Fl. N.Z.* i. 50. Erect, 3ft.–8ft. high. Branchlets leafy, distichous, drooping, compressed or plano-convex, 1/20in.–3/16in. broad, grooved, pubescent towards the tips. Leaves small, pinnately 3–7-foliolate, pubescent or almost silky, ⅛in.–¾in. long; leaflets narrow, cuneate-oblong or cuneate-obcordate, deciduous. Racemes strict, erect, 10–20-flowered, pubescent or silky in bud. Bracteoles at the base of the calyx. Calyx-teeth acute, short, ciliated. Standard much broader than long, equalling or slightly exceeding the keel, widely retuse. Ovary usually glabrous. Pods in erect racemes, compressed, obliquely ovate, narrowed into a short straight subulate beak, 2-seeded. —Handbk. 50.

NORTH Island: from the Ruahine Range to the Pahau River. SOUTH Island: Pelorus Sound, *T. K.* Nelson, *Monro, Travers.*

A beautiful species—perhaps the tallest of the genus. Much confusion has been caused from the racemes having been described as pendulous: they are invariably erect, but appear pendulous in herbarium specimens owing to their being carried on drooping branchlets. The pods are shorter and broader than in *C. grandiflora.*

Var. **pilosa** (*sp.*), Col. *ex* Hook. f., Fl. N.Z. i. 50. Branchlets 1/12in.–3/16in. wide, much compressed, and grooved on both surfaces. Ovary silky. *C. odorata,* var. *pauciflora,* Benth. in Herb. Kew. NORTH Island: east coast, *Colenso,* Handbk.

A plant in immature fruit collected by me on the Pahau River (n. 902) was authenticated at Kew as *C. pilosa,* but the young pods were glabrous. I have not seen the flowers, and consider the plant to be a variety of *C. odorata.* So far as known to me, *C. pilosa* has only been collected by Mr. Colenso. The ovary of several species is *occasionally* silky.

16. **C. angustata,** *n. s.* Erect, 1ft.–3ft. high, glabrous, leafy. Branches spreading, terete; branchlets slender, filiform or compressed above, $\frac{1}{20}$in.–$\frac{1}{12}$in. broad. Leaves deciduous, glaucous beneath, $\frac{3}{4}$in.–1$\frac{1}{2}$in. long, 3–5-foliolate; leaflets obcordate-cuneate. Flowers not seen. Fruiting racemes numerous, erect, 1in.–1$\frac{1}{2}$in. long, slender. Pedicels spreading. Pods 20–40, obliquely oblong, compressed, abruptly narrowed into a short straight or oblique subulate beak pointing outwards. Seeds 1 or 2, oblong, reniform, much compressed.

SOUTH Island: Nelson: valley of the Buller; especially plentiful near the Lyall Junction; *T. K.*

I regret my inability to obtain flowers of this interesting species, which differs from *C. grandiflora* in the slender habit, narrow or terete branchlets, large leaves, and many-podded racemes, the last character alone differentiating it from all other species. Usually the leaves are much larger than those of *C. grandiflora*, while the pods approach *C. odorata*, but in the latter species the branchlets are always much compressed.

17. **C. flagelliformis,** *Col.* ex *Hook. Fl. N.Z.* i. 51. Erect, glabrous, 1ft.–4ft. high. Branches spreading; branchlets very numerous, often fastigiate, slender, compressed or plano-convex, grooved or striated, $\frac{1}{20}$in.–$\frac{1}{10}$in. broad. Leaves, on the elongated whip-like shoots produced by young plants, pinnately 3–5-foliolate, about 1in. long or more; leaflets sessile, hardly oblong-obovate, cuneate below, the notch usually with a rounded sinus; on mature branchlets $\frac{1}{4}$in.–$\frac{1}{3}$in. long, 1–3-foliolate. Racemes 3–7-flowered, often reduced to fascicles. Pedicels short. Calyx campanulate; teeth deltoid, acute, ciliated. Standard very broad, retuse, and with the wings exceeding the keel. Pod often solitary, erect, compressed, oblong or obliquely orbicular, narrowed into a short stout subulate beak. Seeds 2–4, orbicular, much compressed.—Handbk. 50. *C. australis,* Raoul, Choix. t. 28A. *C. multicaulis,* Col. in Trans. N.Z.I. xxv. (1890) 329. *C. micrantha,* Col. *l.c.* xxvi. 313. *Lotus arboreus,* Forst., Prod. n. 258.

NORTH Island: local. Patetere Plateau, *Cheeseman!* Tuhoe Country, *Best!* Hawke's Bay, &c. SOUTH Island: Marlborough, Nelson, Westland, Canterbury, and Otago. Sea-level to 2,300ft. *Makaka.* Dec., Jan.

In Mr. Cheeseman's Patetere specimens the wings are shorter than the keel. Forster's drawing conclusively shows this species to be his *Lotus arboreus*, but the pods represented are immature and turgid instead of compressed. I have not seen authentic specimens of *C. multicaulis* and *C. micrantha*, Col.

Var. **corymbosa** (*sp.*), Col. in Trans. N.Z.I. xxi. (1889) 80. Branchlets very slender, spreading or drooping, compressed or subterete, striated. Racemes 3–5-flowered, solitary or aggregated in corymbose fascicles. Pedicels very short. Pod broadly oblong, compressed, oblique, dimidiate. Seeds 1 rarely 2, mottled. NORTH Island: near Dannevirke; Waipawa, *Colenso!*

I am indebted to Mr. Colenso for the only specimens I have seen of this very slender plant. Mr. N. E. Brown, of Kew, considers it identical with *C. flagelliformis*, but it cannot be refused varietal rank.

18. **C. acuminata,** *n. s.* Erect, much branched, glabrous. Branchlets much compressed or plano-convex, $\frac{1}{12}$in. broad, finely striated or almost grooved. Leaves and flowers not seen. Fruiting racemes of 2–6-pods. Rhachis and pedicels very short, apparently glabrous. Pods spreading, $\frac{1}{4}$in.–$\frac{3}{8}$in. long, compressed, obpyriform, falcate, abruptly acuminate; beak straight, oblique. Seeds 2 or 1.

NORTH Island: White Rock, East Coast, *T. K.*

The pods cannot be mistaken for those of any other species. The valves detached from the replum have the general form of the mussel-scale (*Mytilaspis pomorum*), but are relatively much broader at the base.

19. **C. Hookeri,** *n. s.* Erect, 2ft.–3ft. high, glabrous, much branched. Branchlets numerous, fastigiate or spreading, striated, often elongated, very slender, $\frac{1}{16}$in.–$\frac{1}{12}$in. broad. Leaves few, 1–3-foliolate. Racemes numerous, 5–8-flowered, often fasciculate, or the internodes greatly reduced, forming compact many-flowered corymbs. Rhachis and pedicels puberulous or pubescent. Calyx turbinate, glabrous; teeth reduced to mere points, ciliolate. Standard very broad, sharply reflexed, slightly exceeding the keel; wings equalling the keel, obtuse. Pods $\frac{1}{4}$in.–$\frac{5}{16}$in. long, compressed, ovate-oblong, narrowed into a short straight oblique beak. Seeds 2, rarely 1 or 3, obscurely mottled, dark.

NORTH Island: east coast, from the Akiteo River southwards; South Makara Stream; Plimmerton; Pencarrow, Wellington, *T. K.* Dec., Jan.

In general appearance this species approaches *C. flagelliformis*, but the leaves are very few, the flowers are larger, and the short pod is less obviously compressed. A beautiful shrub, flowering in vast profusion.

3. HUTTONELLA, n. g.

General characters of *Carmichaelia*, but pods indehiscent, very small, turgid or almost inflated, the breadth exceeding the depth; beak short, turned upwards. Seeds 1–3. Branchlets terete or compressed. Leaves 1–3-foliolate. Flowers minute. Radicle with a single fold. Cotyledons oblong or obovate, fleshy.

The pods often remain on the branches until the spring, when they fall to the ground and decay, the valves being inseparable from the replum, which is usually imperfect.

Named in honour of *Professor F. W. Hutton, F.R.S., &c.*

**Leafless when in flower.*

Branchlets numerous, erect. Racemes lax .. .. 1. *H. compacta.*

† *Racemes compact.*

Erect. Branchlets few, terete. Pods broadly oblong, 2–3-seeded .. 2. *H. curta.*
Erect or prostrate. Branchlets terete or compressed, 1–3-seeded .. 3. *H. juncea.*

** *Leafy.*

Prostrate. Branchlets compressed. Pods 1-seeded .. .. 4. *H. prona.*

1. **H. compacta.** Erect, 2ft.–4ft. high, much branched above. Branchlets terete, slender, naked, striated, strict or somewhat flexuous. Leaves not seen. Racemes numerous, lax, $\frac{1}{2}$in.–$\frac{3}{4}$in. long, 3–6-flowered. Rhachis naked below; pedicels longer than the flowers, very slender, spreading. Calyx campanulate, slightly inflated; teeth small, deltoid, acute. Standard much broader than long, 2-lobed, and, like the wings, exceeding the keel; keel with a long slender claw. Petals free at the apex. Pod $\frac{1}{8}$in. long, obovate or broadly pyriform, compressed vertically, broader than deep, turgid; beak very short, subulate, oblique; valves reticulate. Seeds 1 or 2, mottled.—*Carmichaelia compacta*, Petrie in Trans. N.Z.I. xvii. (1884) 272.

SOUTH Island: Otago: local; Clutha Valley, between Lake Wakatipu and Clyde, *Petrie!* Nov., Dec.

Easily recognised by the crowded terete branchlets and lax racemes of pinkish-white flowers, which are very fragrant.

2. **H. curta.** Erect, 1ft.–2ft. high, sparingly branched. Branchlets elongated, slender, subterete, the upper portion compressed, striated. Leaves not seen. Racemes distant, 6–10-flowered. Rhachis silky, elongating after the flowers have fallen; pedicels short, puberulous or silky. Flowers $\frac{1}{6}$in. long. Calyx puberulous or pubescent, campanulate, 2-bracteolate at the base; teeth acute, villous within. Standard broadly obovate, broader than long, shortly retuse, margins revolute; wings with a large callosity at the auricle; keel with a short claw. Ovary pubescent or silky; ovules 4. Pods $\frac{1}{6}$in. long, pendulous, turgid, glabrous when mature, oblong-obovate; beak short, oblique. Seeds 2–4, mottled.—*Carmichaelia curta*, Petrie in Trans. N.Z.I. xxv. (1892) 271.

SOUTH Island: near Duntroon and Kurow, Otago, *Petrie!*

The flowers are larger than in *C. juncea*, and the standard is less reflexed. The callosity at the base of the wings is unique in the genus, and is not found in *Carmichaelia* or other allied New Zealand genera.

3. **H. juncea.** Erect, 1ft.–2ft. high, rarely prostrate. Branchlets strict or curved, almost terete or more or less compressed, grooved. Leaves not seen. Racemes 2–8-flowered. Rhachis and pedicels silky; pedicels very short. Bracts and bracteoles linear, acute, silky, glabrous or puberulous. Flowers minute. Calyx turbinate; teeth subacute, silky. Standard exceeding the keel, broader than long; wings shorter than the keel. Ovary pubescent or glabrous, 2–4-ovulate. Pod 1in.–1½in. long, glabrous or pubescent, broadly oblong, turgid or almost inflated, 1–3-seeded; beak strongly curved upwards, slender, oblique; valves very thin.—*Carmichaelia juncea*, Col. *ex* Hook. f., Fl. N.Z. i. 51; Handbk. 50.

NORTH Island: East Cape, *Sinclair*. Hawke's Bay and Taupo, *Colenso*. Rotorua, *T. K.* (sterile). SOUTH Island: Akaroa, *Raoul*. Canterbury Plains, *Haast*. Interior of Otago, *Petrie*. 1,000ft. to 2,000ft.

This was originally described as a small tree, "8ft.–14ft. high," but I have not seen specimens more than 2ft. high. The beak of the Otago plant usually forms nearly a right angle with the axis of the pod. Unfortunately, I have not had the opportunity of examining flowers of the North Island plant, and, as the original description is very imperfect, it is possible that two species may be confused under *H. juncea*.

4. **H. prona.** A small species, with prostrate stems and branches closely appressed to the ground, 4in.–12in. or more. Branchlets $\frac{1}{20}$in.–$\frac{1}{12}$in. broad, compressed, rarely with few transverse articulations. Leaves unifoliolate or pinnately 3–5-foliolate, silky; terminal leaflets much the longest, narrow-cuneate-oblong, emarginate. Flowers small, in dense 3–7-flowered racemes. Rhachis silky; pedicels short. Calyx campanulate; teeth acute. Standard retuse; wings shorter than the keel. Pods $\frac{1}{8}$in. long, broadly oblong, turgid, with a short upturned beak, which forms a right angle with the pod, 1-seeded.—*Carmichaelia prona*, T. Kirk in Trans. N.Z.I. xxvii. (1894) 350.

SOUTH Island: Lake Lyndon, 2,800ft., *Enys* and *T. K.* Dec., Jan.

Most nearly related to *H. juncea*, from which it is distinguished by the prostrate habit, compressed branches, and numerous leaves.

4. NOTOSPARTIUM, Hook. f.

Calyx turbinate or campanulate; teeth 5, short, obtuse or subacute. Standard obovate-obcordate, not auricled, longer than the keel, shortly reflexed; wings oblong, shorter than the keel, with an incurved auricle at the base; keel hatchet-shaped. Upper stamen free. Ovary sessile or nearly so, linear; style short, curved; ovules 7–10. Pod shortly stipitate, straight or falcate, linear-elongate, shortly beaked, compressed, membranous, torulose, 5–10-jointed, indehiscent. Seeds solitary in the cells, oblong, with a doubly bent and twisted radicle, thickened below. Shrubs, with slender compressed flexuous or pendulous branchlets. Leaves only developed on very young plants, 1-foliolate. Stipules minute. Flowers racemose.

This curous endemic genus, which comprises only two species, has the compressed branchlets of *Carmichaelia*, but differs in habit, and especially in the structure of the pod. In the Handbook the style is said to be ciliated on the upper surface, but I do not find it so.

Flowers red. Pods crowded, straight .. .. 1. *N. Carmichaelia.*
Flowers purple. Pods distant, falcate .. .. 2. *N. torulosum.*

1. **N. Carmichaeliae,** *Hook. f. in Kew. Journ. Bot.* ix. (1857) 176, *t.* 3. A much-branched shrub, 3ft.–9ft. high. Leaves only on young plants, 1-foliolate, varying from orbicular to oblong-emarginate, retuse or entire, mucronate. Branchlets $\frac{1}{20}$in.–$\frac{1}{10}$in. broad, elongated, pendulous, much compressed. Racemes 1in.–2in. long, 8–20-flowered or more. Pedicels very slender. Calyx-teeth ciliate. Free portion of filament short. Seeds 2–5, orbicular, reniform.

SOUTH Island: in ravines and river-valleys. Marlborough: Upper Awatere, *Sinclair.* Medway Creek, *T. K.* Waihopai River, *Monro.* Kaikoura Ranges, *Buchanan!* Nelson: Mount Fyffe, *Spencer!* "Not found in Canterbury," *L. Cockayne!* 800ft. to 1,800ft.

A beautiful plant, which has now become extremely rare. *Pink broom.* Jan.

2. **N. torulosum,** *n. s.* Much branched, 3ft.–8ft. high. Branches in the mature state pendulous, flexuous or trailing in young plants; branchlets very long, $\frac{1}{20}$in.–$\frac{1}{18}$in. broad, slender, almost terete, or much compressed, whip-like, striated. Leaves on seedling plants 1-foliolate, obovate, emarginate, jointed to the petiole. Flowers distant, in strict 3–8-flowered racemes; rhachis filiform, 1in.–2in. long; pedicels very short. Calyx glabrous, campanulate; teeth obtuse. Standard narrower than in the preceding species, reflexed; wings slightly exceeding the keel. Pods $\frac{3}{4}$in.–1in. long, $\frac{1}{16}$in. wide, distant, falcate, compressed, indehiscent, about 10-seeded; beak laterally compressed at the base, subulate above, curved; valves torulose, dilated above each seed. Seeds small, reniform, compressed, mottled.—*Carmichaelia gracilis,* J. B. Armst. in Trans. N.Z.I. xiii. (1880) 336 (flowers only).

SOUTH Island: Nelson and Canterbury: gorge of the Mason River, Amuri, *Haast! Rev. F. H. Spencer! S. D. Barker! L. Cockayne!* Waikari, *S. D. Barker!* Ravines at the base of Mount Peel, *W. E. Barker!*

I am greatly indebted to all the botanists named for assistance in elucidating this species. S. D. Barker was the first to discover the pod and point out its genus. He states that the plant has the "round-headed habit of a weeping-willow." My specimens are few and imperfect. Distinguished from *N. Carmichaeliae* by the glabrous calyx, purple flowers, and the torulose or almost torose falcate distant pods. It is said to have been found "on the site of the City of Christchurch," but there is no evidence to support the statement.

5. CLIANTHUS, Banks and Sol.

Calyx campanulate, 5-toothed. Standard sharply reflexed, acuminate; wings oblong or lanceolate, about half the length of the keel, auricled at the base; keel boat-shaped, produced into a long curved point. Ovary stipitate, many-ovuled; style subulate, incurved, ciliated below the apex. Pod stipitate, oblong-acuminate, terete, falcate, 2-valved, beaked, turgid; seeds many. Herbs. Stems and branches woody below. Leaves alternate, unequally pinnate; leaflets in many pairs. Stipules small. Flowers large, crimson, in pendulous racemes.

A small genus of beautiful flowering-plants, comprising, besides the present species, which is endemic, one or perhaps two from Australia, and one from the Island of Ceram.

1. **C. puniceus,** *Banks and Sol.* ex *Lindl. in Bot. Reg. t.* 1775. A much-branched suffruticose silky pubescent shrub, 2ft.–6ft. high. Leaves unequally pinnate; leaflets sessile, in 8–14 pairs, linear-oblong, obtuse retuse or apiculate. Branches spreading. Racemes many-flowered. Flowers bright-scarlet. Standard ovate, acuminate; wings lanceolate, falcate, acute or sub-acute, less than half the length of the keel; keel falcate, acuminate; style wholly included in the keel. Pod turgid, many-seeded.

Var. **maximus,** (*sp.*), Col. in Trans. N.Z.I. xviii. (1885) 294. Leaflets large, sometimes 1½in. long, often shining. Flowers rather smaller. Standard broadly ovate, acuminate, often with a dark blotch at the base; wings oblong, broad, rounded at apex.—Lindl. in Bot. Reg. t. 1775; A. Cunn. in Trans. Hort. Soc. Ser. II. i. (1835) 521; Precurs. n. 572; Hook. f., Fl. N.Z. i. 49; Handbk. 52; Banks and Sol. MSS. *Donia punicea*, Don., Syst. Gard. ii. 468.

NORTH Island: Great Barrier Island, *T. K.* On one or two islets off the East Coast, and inland, *Bishop Williams!* Near Waimarama, *Nairn.* Collected by Banks and Solander at Motuarohia, Bay of Islands, Anaura, and Tolaga Bay, East Cape, according to their MSS. Also at Mercury Bay, as stated by Cunningham. Collected also at the Thames in 1869, but is certainly extinct in that locality. Formerly cultivated by the Maoris. *Kowainguta. Kaka.* Aug. to Oct.

The shape of the standard and wings differs materially in different plants. Banks and Solander's fine plate represents exactly *C. maximus*, Col., with the wings broadly rounded at the apex; their collection, however, contains specimens with very narrow pointed wings. There are all degrees of difference between the two forms; but either form of wing may be associated with either form of standard. Mr. Colenso's interesting paper should be carefully read, although to Sir Joseph Hooker and myself the differences pointed out appear to pass into each other by insensible gradations.

6. SWAINSONA, Salisb.

Calyx campanulate; teeth 5, nearly equal. Standard nearly orbicular, shortly clawed; wings auricled at the base, falcate, free. Upper stamen free. Ovary sessile or stipitate, many-ovuled; style incurved, slender, bearded on the upper surface; stigma minute. Pod inflated or turgid, membranous or coriaceous, acute, several-seeded. Herbs, with prostrate erect or climbing stems, rarely suffruticose. Leaves unequally pinnate; leaflets usually numerous. Stipules deciduous. Flowers in axillary racemes.

SPECIES, about 30, all restricted to Australia, except one endemic in New Zealand.

1. **S. Novae-Zelandiae,** *Hook. f., Handbk. N.Z. Fl.* 51. Stem forming slender rhizomes below; much branched above; 2in.–5in. high. Branches and leaves clothed with silky pubescence. Leaves 1in.–2in. long; leaflets in 6–8 pairs, narrow-obovate or oblong, rounded or retuse. Peduncles 3–8-flowered.

Pedicels short. Calyx 2-bracteolate at the base; teeth linear, shorter than the tube, ciliate, villous inside. Pod large, inflated, 1in. long, narrowed at both ends; valves coriaceous, thin. Seeds 6–9, small.

SOUTH Island: Nelson: above Fowler's Pass, Amuri, *Travers!* Marlborough: Kaikoura Ranges, &c., *Buchanan!* Canterbury: source of the Kowhai, *Haast!* mountains above Broken River and Lake Coleridge, *Enys!* Otago: Mount St. Bathan's, *Petrie.* 2,000ft. to 4,500ft. Dec., Jan.

A remarkably local plant, but doubtless often overlooked. The flowers are purple.

* ONOBRYCHIS, Tourn.

Calyx-teeth 5, subulate, nearly equal. Leaves shorter than the obliquely-truncate keel. Upper filament free. Pod indehiscent, 1-celled, 1-seeded; lower suture curved, spinous or winged. Perennial herbs or shrubs, with unequally pinnate leaves and axillary spikes or racemes.

* **O. viciaefolia**, *Scop., Fl. Carn.* ed. 2, ii. 76. Stems erect, 1ft.–2ft. high, pubescent. Leaflets ½in.–¾in. long, narrow-oblong or obovate, shortly petiolate, mucronate. Stipules membranous, ovate-acuminate. Racemes on long erect peduncles, compact. Calyx silky. Keel equalling the standard. Pods almost semicircular; valves coarsely reticulated; lower suture toothed or tuberculate.

SOUTH Island: near Lincoln and Springston. Scarcely naturalised. *Sainfoin.* Dec., Jan. West and South Europe, North Asia.

* VICIA, Linn.

Calyx 5-fid or 5-toothed; teeth unequal or nearly equal. Wings adnate to the keel. Stamens diadelphous; tube obliquely truncate. Style inflexed, filiform, glabrous or downy or bearded underneath below the stigma. Ovules few or many. Pods compressed; seeds globose, often arillate. Annual or perennial climbing or spreading herbs. Leaves equally pinnate, usually terminating in a simple or branched tendril. Flowers in axillary racemes.

Stems filiform or very slender. Peduncles long, few-flowered.

Peduncles 1–2-flowered. Pods 3–4-seeded, glabrous .. .. * *V. gemella.*

Peduncles 1–4-flowered. Pods 5–8-seeded, glabrous .. .. * *V. gracilis.*

Peduncles 1–6-flowered. Pods 2-seeded, hairy .. .. * *V. hirsuta.*

Stems stout. Peduncles long, or flowers sessile.

Peduncles many-flowered. Flowers many, drooping .. .. * *V. Cracca.*

Flowers sessile or subsessile. Pods narrow .. * *V. sativa.*

Flowers sessile or subsessile. Pods broad .. .. * *V. Narbonensis.*

* **V. gemella**, *Crantz, Stirp. Austr. ed.* 2, *fasc.* v. 389. Annual. Stems filiform, 2ft.–3ft. long, nearly glabrous. Leaflets in 3–6 pairs, narrow-linear-oblong, mucronate, sometimes truncate. Stipules small, entire or 2-fid. Peduncles capillary, equalling or exceeding the leaves, 1–2-flowered. Calyx-teeth nearly equal, the lower slightly exceeding the upper. Standard emarginate. Pod linear-oblong, 3–4-seeded.

NORTH and SOUTH Islands: commonly naturalised on the borders of fields and in bushy places. *Smooth tare.* Nov. to Jan. Europe.

* **V. gracilis**, *Loiset, Fl. Gall. ed.* i, 460. Similar to the preceding, but the stamens are rather stouter and the leaflets longer, in 3–4 pairs, acute, or truncate and mucronate. Stipules ovate-acuminate, often toothed at the base. Stem, rhachis, leaves, and peduncles usually pubescent. Peduncles exceeding the leaves in fruit, 1–4-flowered. Flowers larger than in the preceding. Calyx-teeth unequal, broad at the base, shorter than the tube. Standard emarginate. Pod obliquely acute. Seeds 5–8-seeded.

SOUTH Island: naturalised. Otago: Taieri Plain, *G. M. Thomson!* Europe.

*V. **hirsuta**, *S. F. Gray, Nat. Arr. Brit. Pl.* ii. 614. Stems, &c., glabrate or pubescent, 1ft. high or more. Leaflets in 6–8 pairs, obtuse or truncate, mucronate. Stipules lobed or toothed at the base. Peduncles 1–6-flowered, usually exceeding the leaves. Calyx-teeth linear-subulate, equalling the tube, nearly equal, silky. Standard entire. Pods sessile, oblong, hairy, 2-seeded.

NORTH and SOUTH Islands: naturalised in cultivated fields, waste places, &c. *Hairy tare.* Nov. to Feb. Europe.

*V. **Cracca**, *L., Sp. Pl.* 735. Perennial, much branched. Stems diffuse, 2ft.–4ft., glabrate or puberulous. Leaves sessile; leaflets numerous, linear-lanceolate, acute or mucronate. Stipules entire. Racemes many-flowered. Flowers drooping, blue or purplish. Calyx-teeth unequal, shorter than the tube. Pod linear-oblong, ¾in.–1in. long. Seeds many.

SOUTH Island: cultivated land by the Opawa, Marlborough, in considerable quantity, but not seen elsewhere. Jan. Europe.

*V. **sativa**, *L., Sp. Pl.* 736. Annual. Stems 1ft.–2ft. high or more, simple, pubescent or hairy, prostrate or ascending. Stipules with long teeth. Leaflets 10–14, oblong-obovate, truncate, mucronate or emarginate. Flowers 1–3, sessile or nearly so. Calyx-teeth linear-subulate, acute, equalling or shorter than the tube, hairy. Pods linear, 1in.–3in. long, hairy or silky.

Only known under cultivation. *Common vetch.*

Var. **segetalis.** Stem rather stout. Leaflets of upper leaves linear-oblong, obovate, obtuse or truncate, mucronate. Flowers 1 or 2, rarely 3, sessile or scarcely pedicellate. Pods 1½in.–2in. long, ascending or spreading, pubescent.

Var. **Bobartii.** Stem slender. Leaflets of the upper leaves linear. Flowers 1, rarely 2. Pods patent.

NORTH and SOUTH Islands: naturalised in cultivated and waste places. Nov. to Jan. Europe, North Africa.

*V. **Narbonensis**, *L., Sp. Pl.* 737, *var.* **serratifolia.** Stems robust, erect or ascending, 12in.–18in. high, pubescent or hairy. Stipules strongly toothed or laciniate. Leaflets in 3–4 pairs, obovate, obtuse, strongly toothed. Flowers 1–3, axillary, sessile. Calyx-teeth ovate-lanceolate, ciliated, shorter than the calyx-tube. Pod broad, 2in.–3in. long, compressed, obliquely narrowed into a stout beak, sparingly ciliated on both margins, 6–8-seeded.

KAIKOURAS Island, Port Fitzroy: in considerable quantity, 1867. Dec. Austria, Hungary.

*LENS, Grenier and Godron.

Calyx deeply 5-lobed; lobes nearly equal, linear, valvate. Corolla and stamens as in *Vicia*. Ovary shortly stipitate or subsessile; ovules 2; style inflexed, scarcely flattened above, minutely hairy on the inner face; stigma minute, capitate. Pod compressed, 1-celled, 2-valved. Seeds 1–2, compressed, lenticular, the funicle expanded into a small aril. Slender herbs, with pinnate leaves terminating in tendrils. Stipules almost free. Flowers small, in few-flowered racemes or solitary.

*L. **esculenta**, *Moench., Meth.* 131. Annual. Stems slender, branched, puberulous. Stipules lanceolate, ciliate; leaflets in about 8 pairs, oblong, subacute. Peduncles naked, equalling or exceeding the leaves, 2-flowered. Pods glabrous, compressed, 2-seeded.

NORTH Island: Auckland, *Cheeseman.* *Lentil.* South Europe, Western Asia, &c.

I have not seen New Zealand specimens.

* LATHYRUS, Linn.

Calyx 5-fid or toothed. Staminal tube transversely truncate. Style flattened or dilated upwards, hairy below the stigma. Upper stamen free. Herbs, with sub-erect or climbing rarely winged stems. Stipules often foliaceous.

Peduncles many-flowered. Pods glabrous .. .. * *L. latifolius.*
Peduncles 2–3-flowered. Pods hirsute or pubescent .. * *L. odoratus.*

* **L. latifolius,** *L., Sp. Pl.* 733. Glabrous. Stems prostrate or climbing, much branched, broadly winged. Stipules broad, acute, adnate to the base of the petiole. Leaflets in 1 pair. Peduncles exceeding the leaves, many-flowered. Pods glabrous, compressed, reticulate.

NORTH Island: Auckland, *Cheeseman.* Wellington: a garden escape, not naturalised. *Everlasting pea.* Dec., Jan. Europe.

* **L. odoratus,** *L., Sp. Pl.* 732. Annual. Stems winged, climbing or prostrate, pubescent or hairy. Stipules adnate to the base of the petiole, linear-acute or acuminate. Leaflets in 1 pair, ovate-oblong, mucronate. Peduncles 5in.–8in. long, 2–3-flowered. Calyx pubescent or hairy; teeth linear, acute or acuminate. Standard exceeding the keel. Pod linear-oblong, hirsute, many-seeded.

NORTH Island: naturalised, Auckland, Taranaki, Wellington; plentiful in one or two localities. *Sweet pea.* Dec., Jan. Italy.

7. CANAVALIA, DC.

Two upper lobes of calyx united into a large obtuse entire or 2-lobed lip, the others into a smaller entire or 3-lobed lower lip. Standard broad, reflexed, with 2 callosities inside above the claw; wings oblong or linear, falcate or twisted, free. Stamens all united, forming a sheath round the ovary, the upper one free at the base only. Pod oblong or linear, broad, 2-valved. Herbs, with trailing or twining stems, large 3-foliolate stipellate leaves, and flowers in axillary racemes. Stipules minute or 0.

SPECIES, about 12, widely distributed in tropical countries.

1. **C. obtusifolia,** *DC., Prod.* ii. 404. Stems trailing or twining, glabrous or pubescent. Leaflets broadly obovate or suborbicular, obtuse or retuse, 2in.–3in. long. Peduncles erect, 6in.–12in. long. Pedicels short; bracteoles orbicular, deciduous. Flowers pink or white. Standard ¾in. in diameter. Pod 5in.–8in. long, 1in. broad, with narrow longitudinal wings. Seeds 2–8.—Benth., Fl. Austr. ii. 256. *C. Baueriana,* Endlich., Prod. Fl. Norf. n. 150.

KERMADEC Islands, *Cheeseman.* Also on the coasts of tropical South America, Africa, and Asia, Polynesia.

* DOLICHOS, Linn.

Calyx-lobes short, the two uppermost united into one broad lobe, entire or emarginate. Standard orbicular, recurved or spreading, with 2 inflexed auricles at the base and 2 callosities inside; wings falcate, adhering to the incurved keel. Upper stamen free. Style thickened upwards. Pod flat, compressed, falcate or nearly straight, 2-valved. Herbs, often suffruticose. Leaves pinnately 3-foliolate; leaflets stipellate. Flowers in axillary racemes or solitary.

* **D. Lablab**, *L.*, *Sp. Pl.* 725. Stems slender, much branched from the base, twining, glabrate or puberulous. Stipules very small. Leaflets stipellate, rhomboid or ovate-acuminate. Peduncles 2–5-flowered, exceeding the leaves. Flowers shortly pedicelled, almost umbellate. Calyx-teeth very short. Pods flattened, falcate, 3–5-seeded; seeds mottled.

NORTH Island: on the sites of deserted homesteads from Auckland to Wellington. Not naturalised. Oct. to Feb. East Indies.

8. SOPHORA, Linn.

Calyx gamosepalous, campanulate, inflated; limb oblique, obscurely 5-toothed. Standard broadly obovate, shortly clawed; wings oblong, shorter than the keel. Stamens 10, perigynous, all free. Ovary stipitate, linear; style slender, glabrous; stigma minute; ovules few or many. Pod stipitate, moniliform, interrupted, terete, 4-winged or -angled, indehiscent or 2-valved. Seeds few or many. Trees, shrubs, or rarely herbs. Leaves exstipulate, unequally pinnate. Flowers in pendulous axillary racemes.

SPECIES, about 25. Distributed through the warmer regions of the earth from Japan to Chili and New Zealand.

1. **S. tetraptera**, *J. Mill.*, *Ic. Pl. t.* i. A shrub or small tree, 30ft.–40ft. high; trunk 6in.–24in. in diameter. Branches in young plants slender, flexuous, and often interlaced; in mature state straight and more or less clothed with fulvous tomentum. Leaves unequally pinnate, 1in.–7in. long; leaflets in from 4–40 pairs, sessile or shortly petiolulate, silky or hairy, orbicular obcordate or linear-oblong. Flowers large, yellow, pendulous. Racemes 2–8-flowered. Peduncles short. Calyx inflated, obscurely and unequally 5-toothed; mouth oblique. Standard broadly obovate, scarcely reflexed, obtuse. Stamens free, exceeding the keel. Ovary stipitate, silky; style long, slender. Pod moniliform, interrupted, the margins produced into 4 narrow longitudinal wings, scarcely dehiscent. Seeds 3–7.—G. Forst., Prod. n. 183; L. f., Supp. 230; Hook. f., Handbk. N.Z. Fl. 53.

Also in South Chili and Juan Fernandez, Easter Island, and Lord Howe Island. *Kowhai.* Aug., Sept.

A variable plant. The leaves and flowers of vars. *grandiflora* and *microphylla* present several points of difference, and the pods of the former are more robust than those of the latter.

Var. **grandiflora.** Leaflets in 10–25 pairs, linear-oblong, entire or emarginate. Flowers large, deeply coloured. Standard one-fourth shorter than the wings, slightly but obviously reflexed. Pods more robust than in var. *microphylla*.—Hook. f., Handbk. 53; T. Kirk, Forest Fl. N.Z. t. 50. *S. tetraptera*, Curt., Bot. Mag. t. 167; Banks and Sol. MSS. *Edwardsia grandiflora*, Salisb., Trans. Linn. Soc. ix. 299; DC., Prod. ii. 97; A. Rich., Fl. N.Z. 344; A. Cunn., Precurs. n. 571; Hook. f., Fl. N.Z. i. 52. NORTH Island: from the East Cape district southwards; chiefly on the eastern side and in the centre of the Island. SOUTH Island: said to occur at Banks Peninsula, *J. B. Armstrong*, and in Otago, *Lindsay*, but I have not seen South Island specimens. Sea-level to 1,500ft.

Var. **microphylla.** Leaflets in 25–40 pairs, orbicular, shortly oblong or obovate. Standard equalling the wings or nearly, not reflexed, narrower than in var. *grandiflora*. Stamens exserted.—Hook. f., Handbk. 53; T. Kirk, Forest Fl. N.Z. t. 51. *S. microphylla*, Ait. Hort. Kew. ed. 1, vii. 42; Curt., Bot. Mag. t. 1442; Banks and Sol. MSS. *Edwardsia grandiflora*, var. *microphylla*, Hook. f., Fl. N.Z. i. 52. *E. microphylla*, Salisb., Trans. Linn. Soc. ix. 299; DC., Prod. ii. 97; A. Rich., Fl. N.Z. 344; A. Cunn., Precurs. n. 570. *E. Macnabiana*, Curt., Bot. Mag. t. 3735. NORTH and SOUTH Islands: Mongonui to Southland. Sea-level to fully 2,000ft.

Var. **prostrata.** Stems prostrate. Leaflets in 2–4 pairs. Flowers small, solitary or in pairs. Standard scarcely shorter than the wings. Stamens exserted. Pod unknown.—T. Kirk, Forest Fl. N.Z. t. 53. *S. prostrata*, Buch. in Trans. N.Z.I. xvi. (1883) 395, t. 36. SOUTH Island: Kaikoura Ranges, *Buchanan!* Mount Torlesse, &c., *N. T. Carrington!* 2,000ft. to 2,500ft.

* ACACIA, Linn.

Flowers 4–5- rarely 3–6-merous. Calyx toothed or lobed, or sepals rarely 0. Petals united or rarely free, often adnate to the stamens. Stamens ∞, usually free, hypogynous or perigynous. Pod flat or almost cylindrical, straight or curved or twisted, coriaceous, 2-valved or indehiscent, rarely septate, many-seeded. Funicle often dilated into a fleshy aril. Trees, shrubs, rarely climbers or herbaceous. Leaves 2-pinnate with almost minute leaflets, often reduced to a flattened phyllodium. Flowers hermaphrodite or polygamous, crowded in globose heads or cylindrical spikes, rarely racemose.

Glabrous, never glaucous. Pod contracted between the seeds .. .. .. * *A. decurrens.*
Glaucous or clothed with silvery hairs. Pod not contracted between the seeds .. * *A. dealbata.*

* **A. decurrens,** *Willd., Sp. Pl.* iv. 1072. A shrub or small tree, glabrous, pubescent or almost tomentose. Branchlets angled. Leaves 2-pinnate; pinnae 7–20 pairs; leaflets 30–40 pairs or more, 3–4 lines long, linear; glands on the rhachis few or many. Flower-heads globose, small, arranged in axillary simple or branched racemes, or the upper paniculate; flowers 20–30 in each head. Calyx 5-lobed, ciliate. Petals 5, spreading. Pod 3in.–4in. long or more, ¼in. broad, contracted between the seeds.

NORTH Island: various places in Hokianga Harbour; naturalised. *Black wattle.* Nov. Australia.

Now largely cultivated for tanning purposes.

* **A. dealbata,** *Link., Enum. Hort. Berol.* 445. A tree, giving off numerous suckers. Twigs and leaves glaucous, more or less clothed with silvery hairs. Leaves 2-pinnate; pinnae in 10–20 pairs or more; pinnules in 30–40 pairs or more, linear, obtuse, with a rounded gland at the base of each pair of pinnae. Heads globose, 20–30-flowered, shortly pedicelled and arranged in axillary or terminal racemes or panicles. Calyx 5-lobed, ciliate. Petals 5, free; tips spreading. Stamens free. Pods 3in.–4in. long, not constricted between the seeds.

NORTH Island: chiefly in the Auckland District. *Silver-wattle.* July, Aug. Australia.

* ALBIZZIA, Durazz.

Calyx 5- or rarely 4-toothed. Corolla tubular, 5- or rarely 4-lobed. Stamens very long, ∞, monadelphous, forming a tube sheathing the ovary. Pod oblong, straight, thin, flat, 2-valved, rarely indehiscent. Seeds orbicular. Leaves 2-pinnate. Flowers hermaphrodite, in globular heads or cylindrical spikes. Shrubs or trees.

* **A. lophantha,** *Benth. in Hook. Lond. Journ. Bot.* iii. 86. A shrub or small tree, with rather distant spreading branches, branchlets deeply grooved, and with the young leaves softly pubescent. Petiole turgid at the base. Leaves 2-pinnate; pinnae in about 10 pairs; leaflets 20–40 pairs, finely pubescent above, glabrous beneath, linear, about 3 lines long, unequally subacute, the nerve just within the upper margin; an oblong gland lies at the base of the lowest pair of pinnae. Flowers in lax axillary racemes, 3in.–4in. long; petals very short. Calyx shortly campanulate, 5-toothed, silky. Corolla tubular, 5-lobed, silky. Stamens united by their filaments. Ovary shortly stipitate, often coloured. Pods pendulous, 2-valved, flat, several-seeded. Seeds oblong.

NORTH Island: naturalised in many places from Auckland to Wellington. *Brush-wattle.* Aug., Sept. West Australia.

ORDER XXIV.—ROSACEAE.

Calyx of 5, 4, or rarely 3 sepals, imbricate or valvate in æstivation, united at the base, inferior or superior. Petals as many as the sepals, rarely 0, free, inserted on the calyx, imbricate in æstivation. Stamens ∞ or rarely few, inserted with the petals. Carpels 1 or more, connate, or adnate to the calyx-tube; styles free or connate; stigma simple, rarely plumose or winged; ovules anatropous; embryo straight. Endosperm 0 or rarely scanty. Fruit various, of 1 or many drupes, achenes, or follicles, rarely a pome, berry, or capsule. Radicle short, straight. Cotyledons plano-convex. Herbs, shrubs, or trees, the stems often armed with prickles or thorns. Flowers hermaphrodite or unisexual.

A large order, containing about 1,250 species, arranged under 80 genera, distributed through nearly every country in the world, from Lapland to Macquarie Island. Most of the New Zealand species are endemic.

* *Ripe carpels not enclosed within the calyx-tube.*

Tribe I. PRUNEAE.

* PRUNUS. Stem woody. Calyx deciduous. Fruit a drupe.

II. RUBEAE.—Calyx persistent, ebracteolate.

1. RUBUS. Stems prickly. Fruit of many small drupes on a dry receptacle.

III. POTENTILLEAE.—Calyx persistent, bracteolate. Carpels 1-ovuled.

2. GEUM. Leaves pinnate. Scape 1- or many-flowered. Styles hooked.

* FRAGARIA. Leaves 3-foliolate. Receptacle pulpy, bearing achenes on its surface.

3. POTENTILLA. Leaves pinnate or digitate. Achenes on a dry receptacle.

** *Ripe carpels enclosed in the calyx-tube.*

IV. POTERIEAE.—Carpels 1-3; ovule 1. Achenes enclosed in the dry calyx-tube, which is constricted at the mouth. Petals 0.

* ALCHEMILLA. Calyx 4-5-lobed. Stamens 1-4.

* POTERIUM. Calyx-lobes 4, petaloid. Flowers capitate.

4. ACAENA. Fruiting-calyx usually armed with barbed spines. Stamens few. Stigmas plumose.

V. ROSEAE.—Carpels many, included in the globose receptacular cavity; ovule 1. Stipules adherent with the petiole.

* ROSA. Shrubs with unequally pinnate leaves and prickly stems.

VI. POMEAE.—Carpels 1-5; ovules 2 in each carpel. Fruit a pome or drupe, inferior. Stipules free.

* CRATAEGUS. Usually thorny shrubs or trees. Fruit resembling a drupe, inferior, with 1-5 bony stones or nuts.

* PRUNUS, Linn.

Calyx deciduous, 5-lobed, imbricate. Petals 5, perigynous. Stamens 15-20 free. Carpel 1; ovules 2, pendulous. Fruit an indehiscent fleshy drupe with a smooth or rugose bony nut. Seed pendulous. Endosperm 0 or scanty. Shrubs or trees, with alternate leaves, and usually with edible fruit.

Flowers on slender peduncles. Endocarp smooth .. .. * *P. cerasus.*
Flowers sessile. Endocarp rugose .. .. * *P. Persica.*

* **P. cerasus**, *L.*, *Sp. Pl.* 474. An erect shrub, 5ft.-9ft. high or more. Leaves alternate, erect, with two glands on the short petiole, obovate-oblong, acuminate, doubly crenate, serrate. Flowers on short peduncles, solitary or in few-flowered umbels. Calyx-tube not constricted. Drupe small, rounded, not glaucous.

NORTH and SOUTH Islands: increasing by suckers, and sometimes forming small groves. *Small cherry.* Sept. Europe, &c.

* **P. Persica,** *Stokes, Bot. Mat. Med.* iii. 100. A shrub or small tree. Leaves convolute in bud, narrow-lanceolate, serrate, acute. Flowers sessile or shortly pedicelled, solitary, appearing before the leaves. Calyx-tube coloured; lobes downy. Fruit a drupe, with a soft velvety membranous epicarp and stony furrowed and rugose endocarp.

NORTH Island: naturalised in many localities in the northern districts; formerly abundant. *Peach.* Aug., Sept. Temperate Asia.

1. RUBUS, Linn.

Calyx 5-lobed, ebracteolate; lobes imbricate in bud, persistent. Petals 5, erect or spreading. Stamens many. Disk lining the calyx-tube. Carpels many, with 2 pendulous ovules in each; style terminal or subterminal. Fruit of many minute 1-seeded drupes, crowded on a dry conical receptacle. Shrubs, rarely herbs, with prickly scrambling stems. Leaves alternate, simple or pinnately or palmately divided into lobes or leaflets. Stipules adnate to the petiole. Flowers in terminal or axillary panicles, rarely solitary, unisexual or hermaphrodite.

SPECIES, about 150, but, according to some authors, over 600; distributed through most parts of the globe. Several species have edible fruits.

A. FLOWERS UNISEXUAL. STEM UNARMED OR WITH FEW PRICKLES.

† *Leaves 3–5-foliolate.*

A lofty climber. Leaflets cordate or truncate at base. Flowers white. Fruit red	1. *R. australis.*
A bush or climber. Leaflets rounded at base. Flowers yellowish. Fruit red	2. *R. cissoides.*
A bush or low climber. Leaflets oval or orbicular-ovate. Fruit amber-coloured	3. *R. schmidelioides.*

†† *Leaves 1-foliolate.*

Prostrate. Leaves dentate. Fruit deep-red, large	4. *R. parvus.*

B. FLOWERS HERMAPHRODITE. STEM WITH STRAIGHT OR LARGE HOOKED PRICKLES.

a. *Leaves pinnately 3–5-foliolate. Prickles straight.*

Leaflets small, white beneath. Fruit red	* *R. Idaeus.*

b. *Leaves digitately 3–5-foliolate. Prickles hooked.*

Leaflets flat or convex, green or white beneath	* *R. fruticosus.*
Leaflets convex, white beneath. Fruit reddish	* *R. rusticanus.*
Leaflets flat, white beneath, ovate or obovate	* *R. leucostachys.*
Leaflets pale-green, roundly obovate or cuspidate. Sepals with long leafy points	* *R. macrophyllus.*
Leaflets deeply laciniate	* *R. laciniatus.*

1. **R. australis,** *G. Forst., Prod. n.* 224. Stems unarmed or with a few scattered prickles, climbing to the tops of the loftiest trees. Branches slender, drooping. Petioles and midribs armed with recurved prickles. Leaves palmately or rarely pinnately 3–5-foliolate; leaflets usually on long petioles, coriaceous, linear, oblong, oblong-lanceolate or broadly ovate-lanceolate, acute or acuminate, sharply serrate, truncate or cordate at base. Panicles unisexual, axillary or terminal, 4in.–24in. long, and leafy at the base. Pedicels short, glandular, pubescent or puberulous, with a linear bracteole at the base of each.

Petals broadly ovate, white. Fruits small, red, austere.—DC., Prod. ii. 556; A. Rich., Fl. N.Z. 340; A. Cunn., Precurs. n. 567. *R. australis*, var. *glaber*, Hook. f., Fl. N.Z. i. 53, t. 14; Handbk. 54.

From the Three Kings Islands to Stewart Island. Ascending to nearly 3,000ft. *Tataramoa. Taraheke. Bush-lawyer.* Sept., Oct.

Varying greatly in the length of the petioles and shape of the leaflets.

2. **R. cissoides,** *A. Cunn., Precurs. n.* 569. Much smaller than *R. australis*, usually forming dense bushes. Stem unarmed. Petioles and midribs with fewer prickles or the latter quite unarmed; petioles varying greatly in length. Leaves usually 3- rarely 4–5-foliolate; leaflets smooth, ovate-acuminate or linear-lanceolate-acuminate, always rounded at the base, sharply unequally serrate or irregularly lobed. Panicles axillary, slender, 2in.–4in. long, often reduced to racemes, glabrous, quite unarmed. Bracteoles longer, acute, pubescent. Sepals broadly ovate, hoary. Petals linear, acute, yellowish. Fruits as in *R. australis*, but more crowded.—*R. australis*, Banks and Sol. MSS. *R. australis*, var. *cissoides*, Hook. f., Fl. N.Z. i. 53; Handbk. 54.

North Cape to Stewart Island, chiefly in lowland districts. Sept., Oct.

Var. **pauperatus.** Leaves wholly reduced to prickly midribs, or sometimes with a reduced leaflet at the apex.

Var. **coloratus.** Young shoots hispid, setose or pubescent. Leaflets rugose, white beneath with appressed tomentum.

These varieties are permanent, but less frequent than the type; they are most plentiful in the South Island.

3. **R. Schmidelioides,** *A. Cunn., Precurs. n.* 568. Usually a rather dense bush or dwarf climber. Young shoots pubescent or setose. Stems and midribs unarmed. Petioles with a few hooked prickles. Leaves 3- rarely 5-foliolate, the terminal petiole much longer than the others; leaflets rounded below or cordate, oval or orbicular-ovate or rarely elliptical-oblong, acute or obtuse, coarsely or rarely acutely toothed, very coriaceous. Panicles axillary, 2in.–6in. long; branchlets short, stout, hispid, setose or pubescent, the upper portion prickly. Sepals broad, silky or downy. Petals broad, rounded. Fruit larger, juicy, amber-coloured, edible.—*R. australis*, var. *Schmidelioides*, Hook. f., Fl. N.Z. i. 58; Handbk. 54.

North Cape to Stewart Island, in lowland districts. Sept. to Dec.

Separated from the preceding by the oval or almost orbicular coriaceous leaves and amber-coloured edible fruit.

4. **R. parvus,** *Buch. in Trans. N.Z.I.* vi. (1873) 243, *t.* 22, *f.* 2 *and* 3. A small prostrate shrub. Stems slender, 12in.–18in. long, with few prickles; sometimes rooting at the nodes or subterranean; bark red. Leaves 1-foliolate, 1in.–3in. long, bronze-coloured, coriaceous, linear-oblong-lanceolate, acute, slightly cordate at the base; margins acutely dentate. Petiole very short and, like the midrib, with a few prickles or unarmed. Stipules linear, entire or

laciniate. Flowers unisexual, solitary or in few-flowered terminal or axillary panicles. Sepals acuminate, ultimately reflexed. Fruit ½in.–1in. long, red, edible.

SOUTH Island: source of the Heaphy River; valley of the Teremakau, near Jackson's; valley of the Grey; mountains above Lyell Creek; valley of the Buller. 200ft. to 4,000ft.

This curious species may be an arrested state of *R. australis*. The ripe fruits resemble large raspberries.

* **R. Idaeus,** *L.*, *Sp. Pl.* 492. Stems 2ft.–5ft. high, slender, erect or suberect, giving off suckers; prickles straight. Leaves pinnately 3–5-foliolate; leaflets ovate-acuminate, sharply serrated, white beneath. Prickles of flowering-shoots deflexed. Flowers corymbose, axillary or terminal, drooping, white. Ripe fruits separating from the receptacle, red.

NORTH and SOUTH Islands: sparingly naturalised. *Raspberry.* Oct., Nov. Europe, North Africa.

* **R. fruticosus,** *L.*, *Sp. Pl.* 493. Stem arcuate, prickly, usually furrowed. Leaves palmately 3–5-foliolate. Flowers in terminal much-branched and often elongated leafy panicles. Fruit large, black or purplish.

Blackberry. Bramble. Europe.

An extremely variable plant, the numerous forms of which are treated as species by many botanists, while others regard them merely as varieties. The three following are widely naturalised.

* **R. rusticanus,** *Weihe.* Stem prostrate or arched, angular, downy; prickles deflexed from a large compressed base. Leaves small, 5-foliolate, convex above; terminal leaflet obovate-acuminate; white with closely appressed tomentum beneath. Panicle long and very narrow, leafy at the base, prickly. Calyx white with densely appressed tomentum. Corolla pink. Fruit reddish, scarcely edible.

NORTH and SOUTH Islands: naturalised in many localities. Dec., Jan. Europe.

* **R. leucostachys,** *Sm.* Forming dense bushes, 2ft.–4ft. high. Stem arched or nearly prostrate, angular; hairs spreading, rarely glandular; prickles numerous, slender, patent. Leaflets quinate, hairy, flat, shining, white with felted hairs beneath, finely toothed; terminal leaflet ovate or obovate. Panicle elongated, narrow; branches short, few-flowered, hairy, setose; prickles slender, deflexed. Calyx hairy, setose. Petals white or pinkish.

NORTH Island: Auckland, Wellington. SOUTH Island: Westland. Naturalised; probably common. Dec., Jan.

* **R. macrophyllus,** *Weihe.* Forming dense bushes, 6ft.–9ft. high. Stem angular, arched or nearly prostrate, more or less pilose; prickles short, slender, conical-compressed, hooked, base large, dilated. Leaflets 5-nate, irregularly serrate or dentate, pilose above, pale-green, soft and velvety beneath; terminal leaflet roundly obovate or cuspidate or acuminate, cordate. Panicle hairy, felted, setose, corymbose; lower branches axillary and ascending; prickles declining. Sepals ovate, alternate, hairy, loosely reflexed in fruit. Petals white, large.

NORTH and SOUTH Islands: naturalised, Wellington, Nelson, Westland.

Var. **Schlechtendahlii** (*sp.*). Weihe and Nees Prickles short, springing from a large base. Leaflets quinate, doubly dentate, hairy on the veins only; terminal leaflet long, obovate, acuminate, cuneate or subcordate at base. Prickles of the panicle strong. Sepals with a long point. Corolla white. SOUTH Island: Brightwater. Identified by the Rev. W. Moyle Rogers.

* **R. laciniatus,** *Willd.*, *Hort. Berol. t.* 82. A low bush, 2ft.–3ft. high, with stout arcuate or subprostrate grooved angular stems. Prickles deflexed from an elongated base, rather stout, thin. Leaflets quinate, with long slender petioles,

glabrous above, hairy beneath, ternately divided, the divisions cuneate at the base, deeply and irregularly cut into narrow toothed acute lobes. Panicle narrow, with short hooked prickles on the rhachis and branches; branchlets pubescent, 1–2-flowered. Sepals with long leafy points, often pinnate, downy or pubescent; bristles few, deflexed. Petals pink. Fruit purplish, black, large.

NORTH Island: naturalised in several localities on the Rimutaka Range and elsewhere. Often cultivated. *Canadian blackberry*. Dec., Jan.

2. GEUM, Linn.

Calyx-tube short, flat or turbinate; lobes 5, imbricate, the lobes alternating with as many accessory bracteoles. Petals 5. Stamens usually indefinite. Carpels numerous, crowded on a short receptacle; ovule solitary, erect; style terminal, elongating after flowering, straight, hooked, or geniculate. Achenes hairy or villous, rarely naked. Perennial herbs, glabrate or pilose. Rootstock stout. Radical leaves rosulate, pinnate or pinnatisect, toothed or incised, the terminal leaflet usually larger than the others. Flowers in loose corymbiferous panicles or terminal and solitary.

SPECIES, about 35, distributed through the cooler temperate regions of the globe. Except the first and second, all the New Zealand species are endemic. The name is of unknown derivation.

* *Achenes villous.*

Stems leafy. Petals yellow. Styles spreading .. .. .. 1. *G. urbanum*, v. *strictum*.

† *Cauline leaves reduced to bracts. Petals white except in 3.*

Styles much longer than the pilose ovaries .. .. 2. *G. parviflorum*.

Terminal leaflet rounded-reniform. "Flowers small, yellow" .. 3. *G. alpinum*.

Styles much shorter than the silky ovaries .. .. 4. *G. sericeum*.

Terminal leaflet ovate-reniform. Scapes 1-flowered. Flowers large .. 5. *G. uniflorum*.

** *Achenes glabrous. Petals white.*

Terminal leaflet ovate. Flowers in cymose panicles.. .. .. 6. *G. leiospermum*.

Leaves obovate in outline. Scapes simple, 1-flowered .. .. 7. *G. pusillum*.

1. **G. urbanum**, *L., var.* **strictum**. Erect, 1½ft.–3ft. high, leafy, the entire plant softly pubescent, silky or villous. Radical leaves 3in.–7in. long, on rather long petioles, pinnate; segments 3–5, very unequal, large with smaller intermixed, variable, ovate or obovate, cuneate below, toothed, lobed or pinnatifid; cauline leaves smaller, with few narrow and more acute segments. Stipules leafy, much divided. Panicles spreading. Peduncles few, strict. Flowers ½in.–1in. in diameter, yellow. Calyx-segments triangular-ovate, acute or acuminate. Petals usually larger than the sepals. Carpels villous, forming a subglobose head in fruit; styles deflexed, hooked at the tip. At first the style is abruptly bent below the middle, but the lower part elongates and becomes nearly glabrous, while the upper usually falls away.—Handbk. 55; Benth., Fl. Austr. ii. 428. *G. Magellanicum*, Comm. ex. Pers. Syn. ii. 57; Hook. f., Fl. N.Z. i. 55.

NORTH and SOUTH Islands: from the Hunua (Auckland) to Southland. Nov. to Jan.

The New Zealand plant differs from the typical form chiefly in its more robust habit, copious pubescence, and larger flowers. It is found also in North and South America, temperate Asia, and the East Indies.

2. **G. parviflorum**, *Sm. in Rees. Cycl.* v. *n.* 12. Erect or suberect, 6in.–18in. high, softly pubescent, silky or villous in all its parts. Radical leaves membranous, 3in.–5in. long; terminal leaflet 1in.–2in. broad, rounded-reniform, obscurely 3–5 lobed, hairy on both surfaces, toothed or crenate; lower leaflets 6–8, solitary or paired, very small, lobed or toothed. Cauline leaves few, small, toothed. Panicle lax, open. Peduncles slender. Flowers ½in. in diameter, white. Calyx-lobes ovate, subacute or obtuse; bractlets narrow-linear. Receptacle very silky. Achenes stipitate, clavate, villous; style slender, spreading, longer than the ovary, villous at the base, hooked at the tip.—DC., Prod. ii. 553; Hook. f., Fl. Antarc. ii. 263; Fl. N.Z. i. 56; Handbk. 55.

NORTH and SOUTH Islands: in mountain districts, East Cape and Ruahine Range to Southland. 1,500ft. to 4,000ft. Dec., Jan. Also in Chili and Fuegia.

3. **G. alpinum**, *Buch. in Trans. N.Z.I.* xix. (1886) 216 *(not of Mill.)*. Rootstock stout. Leaves all radical on rather stout pubescent petioles, pinnate or consisting of a terminal rounded reniform lobe 1in. or more in diameter, pilose beneath, obscurely 3–5-lobed; lobes crenate-toothed; lower leaflets when present minute, restricted to 1 or 2 solitary or paired segments, toothed. "Flowers minute, yellow, on numerous branches towards the end of the stems," ⅕in. in diameter. Carpel unknown.

SOUTH Island: Otago, *Buchanan!*

A doubtful plant, apparently founded on imperfect specimens of *G. uniflorum*. Although Buchanan's description of the flowers differs widely from that of any other New Zealand species, it is to be feared that some error has crept in. I have not seen flowering specimens.

4. **G. sericeum**, *n. s.* Pubescent, silky or villous in all its parts. Leaves all radical, ¾in.–1in. long, including the petiole; terminal segment orbicular-cordate or reniform, minutely lobed or crenate-toothed, pubescent and rugose beneath, silky above; lower leaflets minute or 0. Scape strict, downy, 2in.–4in. high, with 1–3 toothed bracts. Flowers few, small, white, racemose or solitary and terminal. Calyx-tube open, silky; segments narrow, ovate, subacute; bractlets short, ovate. Petals slightly exceeding the calyx, retuse. Receptacle glabrous. Achenes stipitate, obliquely ovate, villous, compressed; style much shorter than the ovary, hooked at the tip. Heads not spreading. *Sieversia albiflora*, Hook. f., Fl. Antarc. i. 9, t. vii.

AUCKLAND Islands, *Hook. f.* Highest parts of Adam's Island, *T. K.* 1,000ft. to 2,000ft.

Most nearly related to *G. parviflorum*, from which it is separated by the short ovate bractlets, compressed oblique achenes, with the very short styles, silky nearly to the apex. The leaves also are silky above and rugose beneath, and the heads are not spreading in fruit.

5. **G. uniflorum**, *Buch. in Trans. N.Z.I.* ii. (1869) 88. Leaves all radical, 1in.–3in. long, membranous, pinnate; terminal leaflet ovate- or rounded-reniform, obscurely lobed, crenate-toothed, strongly ciliated, with few strigose hairs on the upper surface; lower segments very small, 1–4, rarely 0, toothed and ciliated. Scapes 4in.–6in. high, very slender, downy; bracts 1 or 2, small, entire or toothed. Flowers solitary, ½in.–¾in. in diameter, white. Calyx-tube

broad; segments linear-oblong; bractlets very short. Petals large, broadly rounded. Achenes silky; style much longer than the silky ovary, glabrous, minutely hooked at the apex.

SOUTH Island: local. Nelson: Discovery Peaks, *H. H. Travers!* Canterbury: mountains above Arthur's Pass, *Cheeseman!* 3,000ft. to 4,500ft.

A very distinct species, easily distinguished by the slender 1-flowered scapes and large white flowers with narrow-linear calyx-lobes.

6. **G. leiospermum,** *Petrie in Trans. N.Z.I.* xxvi. (1893) 267. Pubescent, silky or villous. Scapes 3in.–6in. long, erect or suberect, slender, ultimately strict, pubescent or downy, with longer hairs intermixed. Leaves all radical, 1in.–2in. long, pinnate, rugose and silky below, glabrate or silky above; terminal leaflet ovate, rounded, unequally sharply toothed, ciliate; segments 5–8, much smaller. Bracts toothed or laciniate. Flowers small, forming a few-flowered lax cymose panicle. Peduncles elongating in fruit, strict. Calyx with a few long scattered hairs, turbinate; segments deltoid; bractlets minute. Petals small, white. Stamens 10 or more. Fruiting receptacle shortly elongated, silky. Achenes perfectly glabrous, or rarely with 1 or 2 hairs on the back, slightly turgid, narrow-oblong, $\frac{1}{14}$in.–$\frac{1}{12}$in. long; style very short, recurved, glabrous.

SOUTH Island: Canterbury: Broken River Basin, *Enys!* Otago: Mount Cardrona, Dunstan Mountains, St. Bathan's, Upper Waipori, *Petrie*. STEWART Island, *Thomson!*

A singular species, distinguished from all others except *G. pusillum* by the glabrous achenes. Mr. Petrie describes the achenes as "slightly compressed," but those on my specimens are only compressed in the young state, the mature specimens being slightly turgid, resembling those of some *Ranunculi*.

7. **G. pusillum,** *Petrie in Trans. N.Z.I.* xxviii. (1895) 538. Leaves all radical, pinnate, obovate in outline, 1in. long or less, more or less strigose above, almost glabrous beneath; terminal leaflet $\frac{1}{4}$in. broad, rounded-ovate, crenate-toothed; segments 3–5, minutely toothed. Scapes yellowish, downy, 1in.–2in. high, naked or with 1 or 2 minute bracts. Flowers solitary, white. Calyx-lobes ovate or deltoid-ovate; bractlets minute. Petals small, 5–6. Fruiting receptacle shortly elongated, hairy. Achenes perfectly glabrous, obliquely oblong; style minute, shortly recurved.

SOUTH Island: Otago: Old-man Range, 5,000ft., *Petrie!*

Distinguished from the preceding by the simple 1-flowered scape and minute styles, and from all others by the very small flowers and glabrous achenes. My specimens, for which I am indebted to Mr. Petrie, are imperfect.

* FRAGARIA, Linn.

Calyx persistent, 5-lobed, with 5 bractlets at its base, valvate in bud. Petals 5. Stamens numerous. Carpels numerous on a convex receptacle; style lateral; ovule 1. Fruit of numerous achenes partially imbedded in a large pulpy receptacle. Perennial herbs, usually with 3-foliolate stipulate leaves, erect simple or branched scapes, and hermaphrodite or polygamous flowers and creeping stolons.

* **F. vesca,** *L., Sp. Pl.* 494. Leaves all radical, on rather long petioles; leaflets sessile, coarsely toothed. Stolons with a rosette of leaves at the apex. Scapes few-flowered. Hairs on the peduncles spreading on the pedicels, adpressed

upwards. Flowers hermaphrodite, cymose, with a foliaceous bract at the base of the cyme and ovate bracteoles at the base of the pedicels. Calyx-lobes spreading. Receptacle of the fruit globose or ovoid, pulpy, thickly dotted with achenes.

NORTH and SOUTH Islands: a garden escape, not fully naturalised. *Wild strawberry.* Oct. to Dec. Europe, North Africa, West Asia.

* **F. elatior,** *Ehrh., Beitr.* vii. 23. Larger than the preceding and more hairy. Leaflets petiolulate. Flowers partly diœcious. Pedicels with spreading hairs.

NORTH Island: Auckland, *Cheeseman.* A garden escape. I have not seen this in a wild state. *Hautbois strawberry.* Nov. to Jan.

3. POTENTILLA, Linn.

Calyx persistent; lobes 5 or 4; bractlets as many, valvate in bud. Petals 5 or 4. Stamens usually numerous. Carpels many; styles lateral or terminal; ovule 1, pendulous. Achenes many, forming a head on the dry glabrate pubescent or hairy receptacle. Perennial herbs or rarely shrubs, with compound leaves, stipules adnate to the petiole, and solitary or cymose flowers.

SPECIES, about 160. Chiefly distributed through the temperate and arctic zones of the Northern Hemisphere; extremely rare in the Southern.

Leaves all radical, pinnate .. .. 1. *P. Anserina.*
Leaves palmately 3–5-foliolate .. * *P. reptans.*

1. **P. Anserina,** *L., Sp. Pl.* 495. Stemless, silvery with white tomentum, rootstock giving off numerous jointed runners. Leaves 3in.–6in. long, unequally pinnate; leaflets in 5–20 pairs, oblong, obovate or rounded, with very small ones intermixed, sharply toothed or incised, silky, tomentose beneath or on both surfaces. Scapes erect, 1-flowered, equalling the leaves. Flowers ½in.–1in. in diameter, yellow. Receptacle villous. Achenes silky or rarely glabrous.—Hook. f., Fl. N.Z. i. 54; Handbk. 54.

Var. **anserinoides,** Raoul, (*sp.*), Choix 28. Leaflets small, ¼in.–½in. long, rounded, sessile or petioled; intermediate leaflets minute.

NORTH and SOUTH Islands: Auckland Isthmus to Southland; ascends to 3,000ft. CHATHAM Islands. *Silver-weed.* Dec., Jan.

Also in the arctic regions, and in most temperate countries in both hemispheres. Var. *anserinoides* endemic in New Zealand.

* **P. reptans,** *L., Sp. Pl.* 499. Stems pubescent, filiform, creeping and rooting at the nodes. Leaves on long slender petioles, digitately 3–5-foliolate, solitary or in pairs; leaflets obovate, bluntly serrate. Stipules almost free. Peduncles slender, equalling or exceeding the leaves. Flowers yellow, ¾in. in diameter. Achenes granulate.

NORTH Island: Waikato, Auckland; Wellington: a few plants only. SOUTH Island: Akaroa, naturalised. *Creeping cinquefoil.* Nov. to Jan. Europe, North and West Asia.

* ALCHEMILLA, Linn.

Calyx persistent, urceolate, 4–5-lobed, valvate in bud; bracteoles 4–5. Petals 0. Stamens 1–4, perigynous. Disk lining the calyx-tube, thickened at the mouth. Carpels 1–5, enclosed in the calyx-tube; style basal or ventral. Achenes 1–4. Seeds ascending. Low herbs, with palmately-lobed or -divided leaves and greenish cymose or corymbose flowers.

* **A. arvensis**, *Scop.*, *Fl. Carn. ed.* 2, i. 115. Annual, 1in.–5in. high, pubescent or hairy. Leaves fan-shaped, deeply divided into 3 cuneate segments, which are cleft into obtuse linear lobes. Flowers in sessile clusters opposite the axils, almost hidden by the toothed or incised stipules. Calyx 4-toothed. Achenes 2–3.

NORTH and SOUTH Islands: naturalised from Auckland to Otago, but often local. When growing on loose sand it forms a compact turf less than 1in. in height. *Parsley piert.* Oct. to Feb. Europe, North Asia.

4. ACAENA, Linn.

Calyx-tube urceolate, campanulate or obconic, contracted at the mouth, rounded, compressed or 4-angled; lobes usually 4 or 5, rarely 3–7, valvate, small. Petals 0. Stamens 1–10, rarely 30–40, opposite the sepals. Carpels 1, rarely 2, enclosed in the calyx-tube; ovule 1, pendulous; style subterminal, very short; stigma dilated longitudinally, fringed or plumose, rarely penicillate. Achene solitary, enclosed in the dry hardened closed calyx-tube, which is usually armed with bristles. Pericarp bony or membranous. Herbaceous or suffruticose plants, with pinnate radical or alternate leaves and toothed leaflets. Stipules adnate to the petiole. Flowers hermaphrodite or unisexual, arranged in interrupted spikes or dense heads, with a linear ciliated or laciniate bractlet at the base of each flower.

SPECIES, about 35, distributed throughout the temperate and cooler regions of the Southern Hemisphere. All the New Zealand species are endemic except *A. Sanguisorbae* and *A. ovina*, the latter being an introduction from Australia, now thoroughly naturalised. The capitula usually have one or more scarious laciniated fugacious bracts at the base.

A. CALYX-TUBE WITH NUMEROUS SPINES OR BRISTLES.

Flowers in an erect interrupted spike	* *A. ovina.*
Flowers chiefly in a compact head	* var. *ambigua.*

B. FRUITING-CALYX WITH A SPINE AT EACH ANGLE, RARELY SPINELESS.

* *Calyx-tube longer than broad. Achene linear.*

Leaves glabrate or silky. Achene acute with a truncate base ..	1. *A. Sanguisorbae.*
Leaves glabrous. Achene narrowed at both ends, bony	2. *A. adscendens.*
Leaves glabrate or silky. Heads large. Spines long, reddish-purple	3. *A. Novae-Zelandiae.*

** *Fruiting-calyx turbinate, shorter than broad.*

Heads pedunculate or sessile. Spines bright-red, rarely 0. Achene broadly ovoid, obscurely angled	4. *A. microphylla.*
Heads sessile. Achene broadly turbinate, 4-angled ..	5. *A. Buchanani.*
C. CALYX-TUBE MUCH COMPRESSED. SPINES 0	6. *A. glabra.*

* **A. ovina**, *A. Cunn. in Field, New South Wales*, 358. Erect or ascending, suffruticose, 1ft.–2ft. high, leafy, simple or branched at the base, pubescent or silky or nearly glabrous. Leaves 1in.–2in. long; leaflets in 5–8 pairs, ovate or oblong, sessile, crenate-toothed or almost pinnatifid, glabrate or silky beneath. Flowers polygamous, sessile or shortly pedicelled, forming an interrupted spike, often crowded near the apex. Calyx-tube downy, clothed with short bristles varying in length, barbed at the tips; lobes 4 or 5, rarely 3, 6, or 7. Stamens 2, 4–5, rarely 8–10. Ovules 1, rarely 2; style fimbriate. Fruiting-calyx ovoid or rounded, glabrous or silky. Achenes linear-ovate, coriaceous.

Var. ambigua. Flowers in a compact terminal globose head, with or without one or more solitary flowers on the peduncle. Calyx-tube pilose, the 4 upper spines longer and stouter than the others. Calyx-segments 4 or 3, scarcely coherent. Stamens 2. Stigma dilated transversely. Achene slightly compressed, linear.

NORTH and SOUTH Islands: abundantly naturalised in many localities. Dec., Feb. Australia. Var. *ambigua*, near Wellington. Mr. Buchanan considers this variety to be a hybrid between *A. ovina* and *A. Sanguisorbae*.—See Trans. N.Z.I. iii. (1870) 208.

1. **A. Sanguisorbae,** *Vahl., Enum.* i. 294. Stems prostrate, woody at the base, much branched; tips ascending, leafy. Leaves glabrate, pubescent, silky or pilose; leaflets ¼in.–¾in. long or more, orbicular-oblong or obovate, glabrous above, the upper pairs rather large, margins coarsely serrate. Flowers in globose heads on slender terminal peduncles 3in.–6in. long or more, naked or 1–2-leaved at the base. Calyx-lobes 4, united at the base, persistent. Stamens 2, free. Fruiting-calyx 4-angled, with a long barbed bristle at each angle. Achene linear-oblong, acute, truncate below, bony.—DC., Prod. ii. 592; A. Cunn., Precurs. n. 566; Hook. f., Fl. N.Z. i. 54; Handbk. 56; Benth., Fl. Austr. ii. 434. *A. diandrum*, Forst., Prod. n. 52. *Ancistrum decumbens*, Banks and Sol. MSS.

KERMADEC Islands to STEWART Island; CHATHAM Islands, AUCKLAND, CAMPBELL, ANTIPODES, and MACQUARIE Islands. Sea-level to 3,100ft. Also in Australia and Tristan d'Acunha.

Var. **pilosa.** Leaves with coarser teeth, white with appressed silky hairs. Alpine districts in the SOUTH Island, AUCKLAND, CAMPBELL, ANTIPODES, and MACQUARIE Islands.

2. **A. adscendens,** *Vahl., Enum.* i. 297. Suberect or almost prostrate. Branches ascending, usually with few hairs on the stem and leaves, rarely glabrous. Leaves 2in.–5in. long or more; leaflets 4–6 pairs, ovate or suborbicular, rounded or cuneate at the base, membranous, often glaucous, deeply crenate-toothed, each tooth tipped with a short pencil of hairs. Peduncles 4in.–8in. long, usually pubescent, rarely with 1 or more solitary flowers below the head. Bractlets linear, ciliated. Calyx-tube obconic, 4-angled, pilose; segments united for nearly half their length. Stamens 2. Stigma fimbriate. Fruiting-calyx narrow, obconic, glabrate; bristles 4, short, stout, spreading, barbed. Achene spindle-shaped, the calycine midribs and bristles sometimes persistent; pericarp bony.—DC., Prod. ii. 593; Hook. f., Fl. Antarc. i. 10, ii. 268, t. 96; Fl. N.Z. i. 54; Handbk. 56. *Ancistrum humile*, Pers., Ench. i. 30.

SOUTH Island: Marlborough to Southland. Not infrequent in mountain districts, 2,800ft. to 5,000ft. Jan. MACQUARIE Island: near sea-level. Also in Chili, Fuegia, and the Falkland Islands.

Nearly related to *A. Sanguisorbae*, but distinguished by the rounded glaucous leaflets, long strict peduncles, short bristles, and spindle-shaped achene.

3. **A. Novae-Zelandiae,** *T. Kirk in Trans. N.Z.I.* iii. (1871) 177. Suffruticose, branched at the base. Stems ascending or erect, leafy, glabrous or with scattered silky hairs. Leaves 2in.–3in. long, pinnate; leaflets sessile or shortly petioled, elliptical, rounded at both ends or slightly obovate, serrate or crenate-serrate. Peduncles terminal, 3in.–6in. long. Calyx-tube pilose, linear, obconic, 4-angled; segments free for half their length. Stamens 2 or 3. Stigma fimbriate. Fruiting-calyx silky, red, slightly winged; bristles 4, long, reddish-purple, barbed. Achene coriaceous, linear-oblong, narrowed at both ends.—*A. macrantha*, Colenso in Trans. N.Z.I. xxiii. (1890) 382!

Var. pallida. More robust and less silky than the type, forming large masses, with pale-green leaves and stout erect green peduncles. Heads 1½in. in diameter, including the spines, which are usually green. Calyx-tube narrow-linear, much longer than in the type, greenish.

NORTH and SOUTH Islands: Auckland to Southland. Nov. to Jan. *Red piripiri.* Var. *pallida*, Port Nicholson.

This species is distinguished from *A. Sanguisorbae* by the large heads, the longer spines, and the achene narrowed at both ends. Var. *pallida* differs widely from the type in appearance, but cannot be specifically distinguished. It is only found on blown sand.

4. **A. microphylla,** *Hook. f., Fl. N.Z.* i. 55. Suffruticose at base. Branches slender, short, and tufted or spreading, 3in.–18in. long. Leaves glabrous or sparingly silky, ¾in.–1in. long, narrow-obovate; leaflets 3–6 pairs, almost orbicular, often cuneate at base and truncate above, with 3 subacute or crenate teeth on each side, and a smaller apical tooth. Peduncles slender, 1in.–3in. long, or the heads sessile. Calyx-tube broadly turbinate, nearly glabrous or silky at the angles; segments free nearly to the base. Stamens 2. Fruiting-calyx broader than high, 4-angled; bristles 4, spreading, barbless, bright-red. Achene 1, bony, broadly ovoid, obscurely angled.

NORTH and SOUTH Islands: common in mountain districts from the East Cape to Southland. Sea-level to 3,500ft. Nov. to Jan.

Var. depressa. Branches rather stout or very slender. Leaves smaller than in the type. Heads sessile or shortly peduncled, few-flowered. Achenes with ridges rather more prominent than in the type.—*A. depressa*, T. Kirk in Trans. N.Z.I. ix. (1877) 548. NORTH Island: Poverty Bay. SOUTH Island: Otago &c., Southland.

Var. inermis. More slender than the type. Leaves 1in.–4in. long; leaflets larger, often glaucous. Peduncles very slender. Fruiting-calyx without bristles. Achenes as in the type.—*A. inermis*, Hook. f., Fl. N.Z. i. 54. SOUTH Island: in mountain districts, Nelson to Southland.

From the total absence of spines this plant appears at first sight very distinct from the type, but there is no character by which it can be separated, as armed and unarmed fruits may be found on the same plant or in the same head.

5. **A. Buchanani,** *Hook. f., Handbk.* 57. Suffruticose, 1in.–2in. high. Branches short, closely appressed to the ground. Leaves silky or densely villous, ¾in.–1in. long; leaflets in 3–5 pairs; teeth minute. Heads sessile, 3–10-flowered. Bractlets ciliated. Calyx-tube broadly turbinate, 4-angled; segments cohering at the very base. Stamens 2. Fruiting-calyx pilose, with 4 prominent ridges and 4 bristles, barbed when young. Achene bony, broadly turbinate, 4-angled.

SOUTH Island: Otago: Lake district and upper part of the Clutha Valley, *Buchanan! Petrie!*

Best distinguished by the small leaves, sessile heads, and broadly turbinate bony achene.

6. **A. glabra,** *Buch. in Trans. N.Z.I.* iv. (1871) 226, *t.* 14. Suffruticose, rather stout, glabrous. Branches erect or ascending. Leaves ¾in.–1in. long; leaflets in 3 or 4 pairs, obovate or deltoid, cuneate below, with about 3 deep crenatures on each side. Peduncles 3in.–5in. long. Heads often unisexual or polygamous. Flowers shortly pedicellate; bractlets ciliate or laciniate. Calyx-tube much compressed, the lateral angles produced upwards into wing-like processes, the anterior and posterior angles short and depressed; segments free or coherent at the very base only. Stamens 25–40, all perfect, in the male

flowers, but only 1 or 2, and usually imperfect, in the female flowers, with fimbriate stigmas. Some imperfect female flowers have 2 penicillate stigmas. Fruiting-calyx red, unarmed. Achene subulate, coriaceous.

SOUTH Island: Marlborough: Mount Mouatt and other mountains in the Awatere, *T. K.* Nelson: Wairau Gorge, *Rough! Travers!* Fowler's Pass and Mount Captain, Amuri, *T. K.* Otago: Mount Ida, *Petrie.* Mountains above Lake Harris, *T. K.* 3,000ft. to 4,500ft. Jan.

The perfectly glabrous habit and compressed spineless calyx distinguish this species from all others, while the numerous stamens of the perfect male flowers and the two penicillate stigmas are exactly those of *Poterium*, and are, I believe, unique in *Acaena*, to which the plant must clearly be referred on account of the fimbriate stigmas of the perfect female flowers.

*POTERIUM, Linn.

Calyx persistent, turbinate, constricted at the mouth; lobes 4, broad, spreading, imbricate in bud. Petals 0. Stamens 4–20 or more, perigynous. Disk lining the calyx-tube and thickened at its mouth. Carpels 1–3; style terminal; stigma penicillate; ovule 1. Achenes 1, rarely more, enclosed in the hardened tetragonous calyx-tube. Perennial herbs, with unequally pinnate leaves, stipules adnate to the petiole, and terminal heads or spikes of diœcious or polygamous flowers.

Fruiting-calyx with netted veins .. .. .. * *P. Sanguisorba.*
Fruiting-calyx pitted and muricate .. .. * *P. polygamum.*

* **P. Sanguisorba,** *L., Sp. Pl.* 994. Stems erect, 6in.–18in. high. Radical leaves 3in.–6in. long; leaflets shortly stalked, ovate, coarsely toothed. Heads with the lower flowers male or hermaphrodite, upper female; bracteoles ciliated. Stamens 12–20, with pendulous capillary exserted filaments. Fruiting-calyx hard, 4-angled, with netted veins and 4 thin wings.—*Acaena Huttonii*, R. Br., ter. in Trans. N.Z.I. xvi. (1883) 382.

NORTH and SOUTH Islands: observed in Auckland, Hawke's Bay, Wellington, Nelson, Canterbury, and Otago, but scarcely naturalised. *Burnet.* Dec., Jan. Europe, &c.

* **P. polygamum,** *Waldst. and Kit., Pl. Rar. Hung.* ii. 217, *t.* 198. Usually more robust than the preceding species, the stems more deeply furrowed, the leaves longer and more deeply toothed, heads much larger, and fruiting-calyx with 4 denticulate or entire wings; the tube deeply pitted or reticulated; ridges muricated.

SOUTH Island: near Lake Ellesmere and other places on the Canterbury Plains. Scarcely naturalised. Central and South Europe.

* ROSA, Linn.

Calyx urceolate, persistent, contracted at the mouth, at length fleshy; lobes 5, imbricate, spreading in flower, leafy. Petals 5. Stamens numerous, perigynous. Disk lining the calyx-tube, thickened at the mouth. Styles subterminal. Ovaries hairy, becoming bony achenes enclosed in the fleshy calyx-tube. Shrubs, erect or climbing, with prickly stems, pinnate leaves with toothed leaflets and adnate stipules, and terminal solitary or corymbose flowers.

Prickles hooked, numerous, intermixed with bristles .. .. * *R. rubiginosa.*
Prickles many, hooked. Setae 0 .. .. * *R. canina.*
Prickles few. Leaves soft, rugulose .. .. * *R. multiflora.*

* **R. rubiginosa,** *L., Mant.* ii. 564. An erect shrub, 4ft.–6ft. high. Stipules, leaves, and sepals copiously glandular, pubescent, fragrant. Prickles on stem and branchlets slender, from a linear dilated base, hooked. Leaflets at length glabrous above, downy beneath, doubly serrate. Flowers 1–4. Peduncles setose. Sepals elongate, pinnate, glandular. Fruit globose or ovoid, crowned with a disk.

NORTH and SOUTH Islands: abundantly naturalised, and destructive to pasturage. *Sweet-briar.* Oct. to Dec. Europe, &c.

* **R. canina,** *L., Sp. Pl.* 492, var. **sarmentacea.** An erect bush, with arching branches 6ft.–8ft. high, eglandular. Branches with stout hooked prickles from a dilated base. Midrib and stipules with a few glandular hairs or glabrous. Leaflets glabrous on both surfaces, flat, acute or acuminate, sharply toothed. Flowers 1–4. Peduncles glabrous. Sepals pinnate, reflexed, deciduous, pubescent or downy. Fruit subglobose or ovoid.

NORTH and SOUTH Islands: naturalised, but more local than *R. rubiginosa.* Nov. to Jan. *Dog-rose.* Europe, North Africa, &c.

* **R. multiflora,** *Thunb., Fl. Jap.* 214. A much-branched shrub, with long spreading shoots. Bark downy; prickles few; stipules pectinate. Leaflets in 2 or 3 pairs, ovate-lanceolate, serrate, soft, rugose. Flowers in terminal corymbs. Peduncles and calyx downy or pubescent. Petals numerous, crowded.

NORTH Island: chiefly on the site of deserted homesteads, &c., in the Auckland and Wellington districts. Often planted for hedges. Not naturalised. China.

* CRATAEGUS, Linn.

Calyx-tube urceolate; segments 5, acute. Petals 5, perigynous. Carpels 1–5, adnate to the calyx-tube; styles 1–5; ovules 2 in each cell. Fruit globose or ovoid, with 1–5 bony 1-seeded stones or with a bony 5-celled stone. Small trees or shrubs, often with spiny branches, deciduous stipules, lobed entire or pinnatifid leaves, and terminal cymose or subcorymbose flowers.

* **C. Oxyacantha,** *L., Sp. Pl.* 477. An erect, spinose, much-branched shrub or small tree. Leaves on short petioles, obovate, cuneate at the base, 3–4-lobed; lobes incised or toothed. Cymes many-flowered, often corymbose. Pedicels strict, puberulous or glabrous. Calyx-lobes acute or acuminate. Carpels 1–3. Fruit subglobose or ovoid, crimson.

NORTH Island: occasionally met with in thickets and on the margins of forests, &c. *Hawthorn.* Oct., Nov. Europe, &c.

ORDER XXV.—SAXIFRAGEAE.

Calyx inferior or superior; lobes 4 or 5, valvate or imbricate. Petals 4 or 5, valvate or imbricate, often small or 0. Stamens 4 or 5, or twice as many as the petals, rarely more or fewer, inserted on or outside a perigynous or epigynous disk, rarely hypogynous. Ovary 2–5-celled, or with 2–5 parietal placentas, rarely apocarpous; styles as many as cells, free or united; stigmas capitate; ovules several in each cell. Fruit usually a capsule or rarely subbaccate and indehiscent. Seeds usually small, with copious endosperm; embryo terete. Herbs, shrubs, or trees. Leaves stipulate or exstipulate, alternate or opposite, simple or compound. Flowers usually regular and hermaphrodite.

GENERA, 75. SPECIES, about 680. Represented in nearly all countries. All the New Zealand species are evergreen shrubs or trees. Two New Zealand genera are monotypic and endemic; two extend to Australia, and one is represented in most warm countries.

Tribe I. ESCALLONIEAE.—Leaves alternate, simple, exstipulate. Stamens as many as calyx-lobes. Style more or less united.

1. QUINTINIA. Flowers racemose. Petals imbricate. Ovary inferior.
2. IXERBA. Flowers panicled. Petals large, imbricate. Ovary superior.
3. CARPODETUS. Flowers panicled. Petals valvate. Ovary inferior.

Tribe II. CUNONIEAE.—Leaves opposite, stipulate. Stamens twice as many as calyx-lobes. Styles free.

4. ACKAMA. Leaves unequally pinnate. Flowers panicled. Calyx valvate. Stamens 10.

5. WEINMANNIA. Leaves unequally pinnate. Flowers racemose. Calyx imbricate. Stamens 10.

* RIBESIEAE.—Leaves alternate, simple. Stipules adnate to the petiole or 0. Fruit a berry.

* RIBES. Calyx superior. Ovary 1-celled.

1. QUINTINIA, DC.

Calyx-tube obconic, superior; teeth 5, persistent. Petals 5, deciduous, imbricate. Stamens 5. Ovary inferior, 3–5-celled, the apex broadly conical, narrowed into a conical furrowed persistent style; stigma capitate, lobed; ovules numerous. Fruit a 1-celled capsule, opening at the apex in valves, which separate up to the stigma. Seeds numerous, with a loose winged testa. Shrubs or trees, with alternate exstipulate leaves and numerous axillary or terminal racemes.

The genus contains 2 species endemic in New Zealand, and 2 or 3 others in Australia. Named in honour of *La Quintinie*, a French botanist.

Leaves oblong, 3in.–6in. long, remotely serrate. Racemes often crowded .. 1. *Q. serrata.*
Leaves 2in. long, very obtuse, obscurely sinuate-serrate .. .. .. var. *elliptica.*
Leaves broadly obovate, acute. Petals broad .. .. 2. *Q. acutifolia.*

1. **Q. serrata,** *A. Cunn., Precurs. n.* 315. A shrub or small tree, 12ft.–20ft. high. Branchlets, leaves, and racemes clothed with small scurfy scales, excessively viscid when young. Leaves petioled, very coriaceous, 3in.–6in. long, ¾in.–1in. broad, linear-lanceolate or oblong, subacute, margins remotely irregularly obtusely serrate, yellow-brown when dry. Racemes 3in.–4in. long, crowded in the axils of the upper leaves, many-flowered. Flowers lilac-coloured. Capsule ⅕in. long, woody.—Hook., Ic. Pl. t. 558; Hook. f., Fl. N.Z. i. 78; Handbk. N.Z. Fl. 58; A. Gray, Bot. U.S. Expl. Exped. i. 666; T. Kirk, Forest Fl. N.Z. t. 125.

NORTH Island: Auckland, Hawke's Bay, and Taranaki. Sea-level to 3,000ft. *Tawheuwheu. Kumarahou. New Zealand lilac.* Oct., Nov.

Var. **elliptica.** "Very similar to *Q. serrata*, but smaller. Leaves 1½in.–2in. long, on rather longer petioles, very obtuse, broader and very obscurely sinuate-serrate." Flowers unknown.—*Q. elliptica*, Hook f., Fl. N.Z. i. 78; Handbk. 59. NORTH Island: east coast, *Colenso, ex* Handbk. Sir Joseph Hooker states that Mr. Colenso's "specimens are only in bud and fruit, neither of which shows any difference from *Q. serrata*." I have leaves from the Urewera Country which may belong to this form, but have not seen flowers.

2. **Q. acutifolia,** *n. s.* A shrub or tree, often 20ft.–40ft. high; trunk 1ft.–2ft. in diameter. Branchlets, leaves, and racemes clothed with small scurfy scales, viscid when young. Leaves petioled, broadly oblong or obovate-lanceolate, always narrowed below, 3in.–5in. long, 1in.–1¾in. broad, acute or shortly acuminate, submembranous or coriaceous, obscurely sinuate-serrate. Racemes solitary, axillary, much shorter than the leaves. Petals rounded-

ovate. Filaments shorter than the anthers. Capsule rather larger than in *Q. serrata*.—*Q. serrata β*, T. Kirk, Forest Fl. N.Z. t. 125, f. 6 and 7 (not of Hook. f.).

SOUTH Island: West Coast, from Nelson to Hokitika, *T. K.* Dec., Jan.

This must be regarded as a somewhat critical species, and is advanced with considerable hesitation. It differs from *Q. serrata* in the large dimensions, broader and less coriaceous acute leaves, shorter racemes, and especially in the broader petals and very short filaments.—*Q. serrata β*, Hook. f., Handbk. 59.

Var. **lanceolata**. Leaves broadly lanceolate or ovate-lanceolate, narrowed at both ends, acute. East Cape, *Bishop Williams!* Nelson, *Cheeseman!*

2. IXERBA, A. Cunn.

Calyx-tube adherent with the base of the ovary; lobes 5, imbricate. Disk 5-lobed. Petals 5, large, inserted beneath the disk, clawed, imbricate. Stamens 5. Ovary conical, 5-lobed, 5-celled, narrowed into a twisted beak-like furrowed style; stigma acute; ovules 2 in each cell. Capsule thick, coriaceous, shortly ovoid, 5-celled, dehiscing loculicidally through the style; valves cohering below, 2-fid above; cells 1–2-seeded. Seeds large, oblong, compressed, shining; funicle thick; endosperm scanty. A shrub or tree, with alternate opposite or whorled exstipulate leaves and cymose or panicled flowers.

The only species; endemic. Etym. An anagram of *Brexia*.

1. **I. brexioides**, *A. Cunn., Precurs. n.* 580. A large shrub or tree, sometimes 50ft.–70ft. in height, with trunk 1ft.–2ft. in diameter. Leaves 3in.–7in. long, ½in.–1in. broad, coriaceous, linear or linear-lanceolate, acute or subacute, distantly obtusely glandular-serrate, glabrous. Flowers large, in terminal cymes or panicles. Pedicels articulated, downy. Calyx-lobes broadly ovate, downy or pubescent.—Hook., Ic. Pl. t. 577–578; Hook. f., Fl. N. Z. i. 82; Handbk. 59; T. Kirk, Forest Fl. N.Z. t. 48.

NORTH Island: from Ahipara and Whangaroa North to the Urewera Country and northern portion of Hawke's Bay. Sea-level to fully 3,000ft. *Tawari*. The flowers are called *Whakou*. Nov. to Jan.

Perhaps the most beautiful tree in the flora.

3. CARPODETUS, Forst.

Calyx-tube adnate with the ovary; lobes 5 or 6, small. Petals 5–6, spreading, inserted beneath the epigynous disk. Stamens 5 or 6, inserted with the petals; filaments short. Ovary inferior or half-inferior, 3–5-celled, with numerous ovules in each cell. Fruit globose, almost fleshy, indehiscent, girt round the middle by the cicatrix of the calyx-limb, 3–5-celled, many-seeded. Seeds small, pendulous; embryo minute; endosperm fleshy. A shrub or small tree, with alternate exstipulate leaves. Flowers in axillary cymes.

The only species; endemic. Etym. From the Greek, in allusion to the fruit being girt by the calyx-limb.

1. **C. serratus**, *Forst., Char. Gen.* 34, *t.* 17a. A shrub with spreading branches, rarely a small tree with slender stem, 15ft.–30ft. high; trunk 5in.–6in. in diameter. Branchlets, petioles, leaves, and pedicels pilose or pubescent.

Leaves petioled, ovate-oblong, acute or obtuse, sharply serrate, ¾in.–1½in. long. Flowers small, white, in broad axillary cymes or panicles, shorter than the leaves. Calyx-lobes pubescent. Fruit the size of a small pea, black, shining when fully ripe.—G. Forst., Prod. n. 11; A. Rich., Fl. N.Z. 366; A. Cunn., Precurs. n. 575; Hook., Ic. Pl. t. 564; Hook. f., Fl. N.Z. i. 78; Handbk. 59; T. Kirk, Forest Fl. N.Z. t. 47.

North Cape to Stewart Island. Sea-level to 3,000ft. *Putaputaweta. Punaweta.* Nov. to Jan.

4. ACKAMA, A. Cunn.

Flowers unisexual. Calyx-tube 5-lobed; lobes valvate. Disk crenate. Petals 5, inserted beneath the disk. Stamens 10, inserted with the petals; filaments subulate, unequal. Ovary superior, hairy, 2-celled; styles 2, persistent; ovules on parietal placentas, numerous. Capsule small, coriaceous, turgid, 2-celled, septicidally dehiscent. Seeds ovoid, apiculate, pilose; embryo cylindrical; endosperm fleshy. Diœcious or polygamous trees, with unequally pinnate stipulate leaves and cymose or panicled flowers.

Besides the present species, which is endemic, another is endemic in Australia. ETYM. An anagram of the Maori name.

1. **A. rosaefolia,** *A. Cunn., Precurs. n.* 520. A large shrub or tree, 20ft.–40ft. high; trunk sometimes 2ft. in diameter. Branchlets, midribs, leaves, and pedicels more or less clothed with brown pubescence. Leaves opposite, 3in.–10in. long; leaflets 3–10 pairs, oblong, decreasing in size downwards, 1in.–3in. long, acute, margins sharply toothed. Stipules interpetiolar, ovate, deciduous, toothed. Flowers unisexual, minute, sessile on the slender branches of axillary panicles, which are longer or shorter than the leaves. Ovary pilose or silky.—Hook. f., Fl. N.Z. i. 79; Handbk. 60; T. Kirk, Forest Fl. N.Z. t. 63. *Weinmannia rosaefolia*, A. Gray, Bot. U.S. Expl. Exped. 671, t. 84.

NORTH Island: from Ahipara and Mongonui to Hokianga and Whangarei, but often local. *Makamaka.* Sept. to Nov.

5. WEINMANNIA, Linn.

Calyx inferior, deeply 4–5-lobed, imbricate. Petals 4 or 5, inserted beneath the lobed disk. Stamens 8–10, inserted with the petals. Ovary free, 2- rarely 3-celled; ovules several in each cell, pendulous; styles filiform, distinct; stigma terminal or decurrent. Capsule coriaceous, 2-celled, oblong or ovoid, septicidally dehiscent. Seeds oblong, reniform or nearly globular, pilose. Embryo small, terete. Endosperm fleshy. Shrubs or trees, with opposite stipulate 1–3-foliolate or pinnate leaves and terminal or axillary racemose hermaphrodite flowers.

The genus comprises about 50 species, which are distributed through the Malayan Archipelago, Polynesia, tropical and temperate South America, South Africa, with a doubtful species in Australia and two in New Zealand.

Named in honour of *J. J. Weinmann*, a noted German writer.

Mature leaves pinnate or 3-foliolate. Capsules glabrous .. .. .. 1. *W. sylvicola.*
Mature leaves unifoliolate. Capsules pubescent .. .. 2. *W. racemosa.*

1. **W. sylvicola,** *Sol.* ex *A. Cunn., Precurs. n.* 518. A shrub or tree, sometimes 70ft. high, with trunk 1ft.–3ft. in diameter, but usually much smaller. Branchlets, petioles, and midribs beneath more or less pubescent. Leaves opposite, unequally pinnate or 3-foliolate, rarely 1-foliolate; leaflets in 1–9 pairs, linear-lanceolate, lanceolate, or ovate, acute or acuminate, coarsely toothed. Stipules free, leafy, toothed. Flowers $\frac{1}{12}$in. in diameter, in terminal racemes 2in.–6in. long or in racemose panicles. Ovary 2- rarely 3-celled, glabrous; styles 2 or 3. Capsule usually glabrous, $\frac{1}{8}$in.–$\frac{1}{6}$in. long. Seeds with a tuft of hairs at each extremity.—Hook. f., Fl. N.Z. i. 79; Handbk. 60; A. Gray, Bot. U.S. Expl. Exped. 671; T. Kirk, Forest Fl. N.Z. t. 72. *W. betulina* and *W. fuchsioides*, A. Cunn., Precurs. n. 516, 517.

NORTH Island: North Cape to Rotorua, East Cape, &c., but the exact limit unknown. Apparently not found in Hawke's Bay. Sea-level to 3,000ft. *Tawhero*. Jan., Feb.

Leaves with 5 or more pairs of leaflets are characteristic of young plants, and are usually membranous. The bark contains a high percentage of tannin.

2. **W. racemosa,** *Linn. f., Supp.* 227. A shrub or large tree, often from 70ft.–90ft. high; trunk 1ft.–3ft. in diameter. Branchlets, midribs, and leaves glabrous, except on very young plants, which are pubescent or hirsute and diaphanous, 1–3- or 5-foliolate, sharply toothed; on mature plants very coriaceous, 1in.–3in. long, 1-foliolate, coarsely and obtusely serrate, ovate, oblong-ovate or oblong-lanceolate, obtuse or subacute, punctate beneath. Racemes 1in.–4in. long, axillary or terminal, often panicled, pubescent; pedicels stout. Ovary pubescent; styles 2 or 3, often united to the apex. Capsule 2–3-celled, hirsute. Seeds hairy.—Forst., Prod. n. 173; DC., Prod. iv. 8; A. Rich., Fl. N.Z. 321; Hook. f., Fl. N.Z. i. 80; Handbk. 61; T. Kirk, Forest Fl. N.Z. t. 73. *W. spatiosa*, Banks and Sol. MSS. *Leiosperma racemosa*, Don in Edinb. New Phil. Journ. (1830) 8; A. Cunn., Precurs. n. 519.

NORTH and SOUTH Islands: from the Hauraki Gulf and Middle Waikato to Stewart Island. Ascends to 3,000ft. *Kamahi*. Jan.

Slender flexuous shoots produced from the trunks of old trees occasionally develop 3-foliolate leaves.

*RIBES, Linn.

Calyx-tube adnate to the ovary, 4–5-lobed, imbricate or valvate in bud. Petals 4–5. Stamens 4–5, perigynous. Ovary 1-celled, inferior; styles 2; ovules on parietal placentas. Berry globose or oblong, crowned with the persistent calyx, pulpy. Seeds horizontal, few or many. Endosperm adhering to the testa. Shrubs, with entire or lobed leaves and axillary solitary or racemose flowers.

* **R. grossularia,** *L., Sp. Pl.* 201. A spreading shrub, 1ft.–2ft. high, with 1–3 straight spines below each axil. Leaves fascicled at the tips of abortive branches, rounded, 3–5-lobed, pubescent beneath. Flowers pendulous. Peduncles 2-bracteolate, pubescent. Calyx-lobes reflexed. Petals minute, erect. Berry glabrous or glandular, hairy, many-seeded.

Not unfrequent on the margins of woods in many localities. Europe, &c.

Order XXVI.—CRASSULACEAE.

Calyx 3–20-toothed or divided, inferior. Petals as many as the sepals or twice as many, perigynous. Stamens as many as the sepals or some multiple of them. Carpels as many as the sepals, free, frequently with a small flat scale at the base of each; ovules several; styles very short, or the stigmas sessile. Fruit a follicle, opening down the ventral suture. Seeds few or many. Endosperm thin and fleshy. Embryo straight. Annual or perennial succulent herbs, often minute, rarely shrubs. Leaves succulent, exstipulate. Flowers usually terminal, paniculate or cymose, rarely in axillary clusters.

Genera, about 15. Species, about 420. Widely distributed.

1. TILLAEA, Linn.

Sepals, petals, stamens, and carpels 3 or 4, rarely 5, all free except the sepals, which are usually coherent at the base. Carpels usually with a minute scale at the base of each. Fruit of 3 or more minute follicles, 1- or several-seeded. Flowers minute, axillary, solitary or clustered, rarely in terminal leafy cymes or panicles. Small annual rarely perennial herbs, with opposite more or less succulent leaves. Often minute.

Species, about 25: widely distributed. Two of the New Zealand species extend to Australia, another to South Chili and the Falkland Islands; the others are endemic, except the naturalised *T. trichotoma*. Some of the species vary greatly when growing in water.

Named in honour of *Michael Angelo Tilli*, an Italian botanist.

* *A scale at the back of each carpel.*

Stem trichotomously branched .. .. * *T. trichotoma.*

† *Flowers large.*

Stems rather stout, red. Flowers $\frac{1}{6}$in.–$\frac{1}{5}$in. in diameter .. .. 1. *T. moschata.*
Stems brown. Flowers $\frac{1}{12}$in. in diameter .. .. 2. *T. Helmsii.*

†† *Flowers minute (except in 7). Stems filiform or capillary, matted.*

Leaves close-set or imbricate.. .. .. 3. *T. Sinclairii.*
Leaves spreading, distant. Style reflexed .. .. 4. *T. Novae-Zelandiae.*
Leaves spreading, obtuse. Peduncles thickened upwards .. .. 5. *T. pusilla.*
Leaves acute. Sepals exceeding the petals .. .. .. 6. *T. acutifolia.*
Leaves ovate-subulate. Carpels 8-seeded .. .. .. 7. *T. multicaulis.*

** *Scales 0.*

Erect. Flowers crowded in the axils of the leaves, solitary .. .. 8. *T. Sieberiana.*
Prostrate, minute, matted, green .. .. 9. *T. debilis.*
Suberect or prostrate, red .. .. 10. *T. diffusa.*
Flowers on capillary peduncles .. .. 11. *T. purpurata.*

* **T. trichotoma,** *Walp. Rep.* ii. 251. Stems trichotomously branched from the base, tufted, about 1in. high. Leaves linear-oblong, $\frac{1}{20}$in.–$\frac{1}{16}$in. long, flat above, convex below. Flowers in 3–5-flowered terminal cymes. Sepals 4, coherent at the base. Petals 4, ovate, slightly shorter than the sepals. Stamens 4. Scales 4. Carpels 4, with a short oblique style. Seeds 6–7; testa minutely reticulate.

North Island: naturalised. Auckland, *Cheeseman!* Whanganui, *Andrew!* Sept., Oct. South Africa.

1. **T. moschata,** *DC., Prod.* iii. 382. Stems red, 3in.–7in. long, rather succulent, creeping or ascending, emitting roots from the axils. Leaves about ⅛in. long, entire, oblong-spathulate or narrowly obovate, obtuse. Flowers ⅛in.–⅕in. in diameter, on short peduncles. Calyx deeply 4-lobed; lobes obtuse, shorter than the rounded petals. Stamens 4; filaments dilated below. Scales 4, truncate. Carpels 4, turgid, obtuse, many-seeded; styles very short, recurved.—Hook., Ic. Pl. t. 535 (but scales not shown); Hook. f., Fl. N.Z. i. 76; Handbk. 75; Phill., Cat. Pl. Vasc. Chil. 91. *T. pulchella*, Banks and Sol. MSS. *Bulliarda moschata*, D'Urv. in Mem. Soc. Linn. Par. iv. 618; Hook. f., Fl. Antarc. i. 13. *Crassula moschata*, Forst., Comm. Gotting. ix. 26.

Northern shore of Cook Strait, from Cape Terawhiti to Cape Palliser. Southern shore: Queen Charlotte Sound, *Banks* and *Solander!* Banks Peninsula (?); east, south, and west coasts of Otago; Chatham Islands; Stewart Island; the Snares; Auckland, Campbell, Macquarie, and Antipodes Islands. Also in South Chili, Fuegia, Falkland Islands, Kerguelen's Land, and Marion Island. Nov. to Jan.

Much the largest New Zealand species. The scale is narrowed into a short stipes.

2. **T. Helmsii,** *n. s.* Stems 2in.–4in. long, prostrate or ascending. Leaves distant, connate at the base, ⅛in.–¼in. long, linear, uppermost acute, lower obtuse. Flowers axillary, solitary. Peduncles very short. Calyx deeply 4-lobed; lobes ovate, acute. Petals narrow, ovate-oblong, obtuse. Scales 4. Carpels 4, turgid; style oblique, scarcely recurved. Seeds 4 (?).

SOUTH Island: West Coast: Karamea, *Rev. F. H. Spencer!* Near Greymouth, *R. Helms!*

This species is closely related to the Australian *T. recurva*, Hook. f., but differs in the larger size, longer leaves and pedicels, more especially in the longer scales and in the acuminate sepals and petals. I have only scraps of this species. Better specimens must be obtained before a satisfactory diagnosis can be drawn.

3. **T. Sinclairii,** *Hook. f., Handbk.* 62. A matted prostrate or erect slender species, with delicate stems, rarely exceeding 1in. in height. Leaves $\frac{1}{14}$in.–$\frac{1}{12}$in. long, oblong, subacute or acute. Flowers minute, on solitary axillary peduncles, about $\frac{1}{20}$in.–$\frac{1}{15}$in. in diameter, white. Calyx deeply divided; segments 4, ovate-oblong, obtuse. Petals 4, exceeding the sepals, obtuse. Scales 4. Carpels 4, turgid; style short, oblique. Seeds 4, rarely 8.

SOUTH Island: in damp situations. Nelson: Lake Rotoiti. Canterbury: Lake Lyndon, Broken River basin, Lake Pearson, &c. Otago: more frequent, especially in the southern and central districts; Lake Hayes, &c. 100ft. to 3,000ft. Dec., Jan.

When growing in water the stems are greatly elongated, and the leaves much larger and obtuse. The upper leaves are invariably acute, the lower often obtuse.

4. **T. Novae-Zelandiae,** *Petrie in Trans. N.Z.I.* xxv. (1892) 270. Stems very slender, matted, prostrate, 1in. long. Leaves shortly connate, fleshy, linear-ovate, acute or subacute. Flowers solitary, axillary. Peduncles shorter or longer than the leaves. Calyx 4-lobed; lobes very short, obtuse. Petals exceeding the sepals, ovate, acute. Scales 4, minute and very thin. Carpels 4, very obtuse, oblique, turgid, 4-seeded; style almost lateral, reflexed.

SOUTH Island: Waipahi; Te Anau, *Petrie.*

Very close indeed to *T. Sinclairii*, of which it will probably prove an aquatic variety.

Var. **obtusa.** Very similar to the type, but the stems are 1in.–2in. long, leaves more acute, and petals larger, rounded at the apex. Carpels 4-seeded. Otago: Lake Waihola, *Petrie!*

5. **T. pusilla,** *n. s.* Stems extremely slender, delicate, matted, prostrate or ascending, 1in. long or more, pale-green. Leaves in rather distant pairs, connate at the base, $\frac{1}{16}$in.–$\frac{1}{12}$in. long, linear, obtuse. Flowers minute, solitary, axillary. Peduncles usually longer than the leaves, thickened upwards. Calyx deeply 4-lobed; lobes ovate-oblong, shorter than the oblong petals. Stamens 4. Scales 4, oblong-spathulate. Carpels 4, turgid, as long as the sepals; style recurved. Seeds 2 or 4.

NORTH Island: banks of streams, &c. Kawakawa, Bay of Islands, *T. K.* Auckland, *Cheeseman!* Dec.

Var. **brevia.** Peduncles usually shorter than the leaves, not thickened upwards. Carpels more obtuse. NORTH Island: Wairoa Falls, Hunua, *T. K.* Near *T. Sinclairii*, but even more delicate, the carpels less obtuse and the style oblique.

6. **T. acutifolia,** *n. s.* Stems 1in.–2in. long, matted, intricate, almost capillary, green. Leaves in rather distant pairs, spreading, shortly connate at the base, linear, acute or apiculate, $\frac{1}{16}$in. long. Flowers axillary, solitary, sessile or on peduncles shorter than the leaves, minute. Calyx deeply divided; segments 4, linear-lanceolate-acuminate. Petals linear-ovate, acute, shorter than the sepals. Scales 4. Carpels 4, ovate, turgid; styles recurved. Seeds ?

NORTH Island: Hurunuiorangi (flowers not seen). SOUTH Island: Winton Forest, Southland, *T. K.* Dec.

Very close to *T. debilis*, from which it is distinguished by the acute leaves, acuminate sepals, and the presence of scales. From *T. pusilla* it differs in the narrow sepals, which are longer than the petals.

7. **T. multicaulis,** *Petrie in Trans. N.Z.I.* xxv. (1892) 270. Stems slender, reddish-purple, branched from the base, $\frac{3}{4}$in.–2in. long, ascending. Leaves rather distant below, close-set and often imbricating above, ovate-subulate, connate at the base, fleshy, concave above, keeled below. Flowers on short pedicels. Calyx turbinate, divided fully half its length; segments broadly subulate, acute. Petals 4, broadly rounded, exceeding the sepals, white. Stamens 4. Scales 4. Carpels 4, obliquely ovate; style slender. Seeds 8.

SOUTH Island: Canterbury: Mount Torlesse and Broken River basin, *Enys* and *T. K.* (1876). Otago: Maniototo and Manuherikia Plains, *Petrie!* 1,200ft. to 3,000ft. Dec., Jan.

The flowers are larger than those of any other New Zealand species except *T. moschata.*

8. **T. Sieberiana,** *Schult., Mant.* iii. 345. Stems erect, branched from the base or simple, 1in.–5in. high, usually reddish-brown. Leaves about $\frac{1}{8}$in. long, concave above, ovate-lanceolate, connate at the base, subacute, fleshy. Flowers densely crowded in the axils of the leaves, sessile or on very short pedicels. Sepals 4, free nearly to the base, ovate, acute. Petals very narrow, linear, shorter than the carpels, acute. Scales 0. Carpels linear-oblong, about as long as the petals; style short, slender. Seeds 2, rarely 1.—*T. verticillaris*, DC., Prod. iii. 382; A. Cunn., Precurs. n. 521; Hook. f., Fl. N.Z. i. 75; Handbk. 62; Benth., Fl. Austr. ii. 451. *T. muscosa*, G. Forst., Prod. n. 61.

Three Kings Islands and North Cape to Southland: chiefly in lowland districts. Oct. to Jan. Also in Australia and Tasmania.

9. **T. debilis,** *Col.* ex *Hook. f., Fl. N.Z.* i. 75. A very small delicate species. Stems intricate, filiform or capillary, prostrate, 2in.–3in. long. Leaves

in scattered pairs, minute, $\frac{1}{16}$in.–$\frac{1}{12}$in. long, ovate-oblong or linear-oblong. Flowers minute, 1 or 2 in the axils of the leaves, sessile or on slender peduncles. Sepals 4, oblong, subacute. Petals ovate-acuminate, shorter than the sepals. Scales 0. Carpels ovate-lanceolate, 1–2-seeded.

NORTH Island: East Coast, *Colenso*.

Not having seen any plant which could be satisfactorily identified with this, I have copied the original description.

10. **T. diffusa,** *T. Kirk in Trans. N.Z.I.* xxiv. (1891) 424 (not of Willd.). A very slender matted species. Stems reddish, 1in.–3in. high, erect or prostrate. Leaves fleshy, $\frac{1}{12}$in.–$\frac{1}{8}$in. long, concave above, convex beneath, linear-oblong, minutely apiculate, connate at the base. Flowers on short axillary peduncles, solitary, tetramerous. Calyx-segments broadly ovate, obtuse. Petals equalling the sepals. Scales 0. Carpels ovate, 2–4-seeded, enclosed in the persistent perianth.

NORTH Island: Miramar, Port Nicholson; in places where water has stagnated during the winter. STEWART Island, *T. K.*

Nearly related to *T. debilis*, from which it is distinguished by its larger size, broader leaves, broad obtuse sepals, and 2–4-seeded carpels.

11. **T. purpurata,** *Hook. f. in Lond. Journ. Bot.* vi. (1847) 472. A slender tufted fugacious plant, about 1in.–2in. high, sparingly branched, erect or suberect. Leaves linear-acuminate, concave above, $\frac{1}{10}$in.–$\frac{1}{8}$in. long. Flowers on spreading capillary peduncles, exceeding the leaves and elongating in fruit, minute. Calyx-segments 4, short, obtuse. Petals 4, exceeding the sepals, but narrow. Scales 0. Carpels obtuse, as long as or shorter than the sepals. Seeds several.—Hook. f., Fl. N.Z. i. 75; Handbk. 62; F. Muell., Pl. Vict. ii. t. 19; Benth., Fl. Austr. ii. 451.

NORTH Island: Cape Palliser, *Colenso*. SOUTH Island: Lake Wanaka, *Petrie!* Sea-level to 1,000ft. Nov.

Probably not infrequent, but speedily disappears under the influence of sunshine. Common in various parts of Australia.

ORDER XXVII.—DROSERACEAE.

Calyx inferior; lobes 4 or 5, rarely 8, imbricate, persistent. Petals 4–5, hypogynous or rarely perigynous, convolute. Stamens 4 or 5, rarely more; anthers opening outwards. Ovary superior, ovoid or globose, 1-celled with 2–5 parietal placentas, or 2–3-celled; styles 3–5, simple or clavate or divided; ovules numerous. Capsule 1-celled, loculicidally 3–5-valved. Seeds numerous, minute; testa lax, sometimes produced into a wing; endosperm fleshy; embryo cylindrical. Herbs, with radical or cauline leaves, circinate in vernation and clothed with glandular hairs which secrete a viscid fluid capable of abstracting the albuminoid matter from insects. Flowers scapigerous, racemose, or solitary and terminal.

GENERA, 6. SPECIES, about 125. Widely distributed. Some genera have connate petals, epipetalous stamens, and a shrubby habit.

1. DROSERA, Linn.

Calyx-segments, stamens, and petals 4, 5, or rarely 8. Ovary 1-celled, with 2, 3, or 5 parietal placentas. Styles as many as placentas, simple or branched. Capsule with as many valves as placentas. Herbs, with glandular leaves and scapigerous or racemose flowers. Rootstock fibrous or tuberous. Leaves circinate in vernation.

SPECIES, upwards of 100, distributed over the area of the order. Abundant in Australia, but rare in Polynesia. One New Zealand species is endemic; the others extend to Australia.

Scape 1-flowered. Flowers white.

Leaves spathulate. Calyx campanulate .. .. 1. *D. stenopetala.*
Leaves linear. Calyx divided nearly to the base .. .. .. 2. *D. Arcturi.*
Minute. Leaves orbicular. Calyx 4-lobed .. .. 3. *D. pygmaea.*

Scape many-flowered. Flowers white.

Leaves spathulate. Styles 3, 2-partite .. .. 4. *D. spathulata.*
Leaves divided into narrow-linear acute segments .. .. .. 5. *D. binata.*

Scape leafy, many-flowered.

Flowers purple .. .. 6. *D. auriculata.*

1. **D. stenopetala,** *Hook. f., Fl. N.Z.* i. 19, *t.* 9. Stemless. Early leaves oblong, ½in.–¾in. long, recurved and fringed with short glands; mature 1in.–5in. long; petiole slender, glabrous; blade spathulate, the upper surface thickly clothed with glandular hairs, which are shorter than those at the margins. Scape 1in.–6in. long, glabrous, 1-flowered. Calyx campanulate, 5-lobed; lobes short. Petals long, narrow-linear-spathulate. Styles 3, deeply 3-fid.—Handbk. 63.

NORTH Island: Ruahine Range, *W. F. Howlett!* SOUTH Island: Nelson: Mount Arthur Plateau, *Cheeseman.* Mount Rochfort, *Gaze!* Westland: Kelly's Hill, &c., *Petrie!* Canterbury: Arthur's Pass, *T. K.* Otago: Milford Sound, *Hector.* Port Preservation, *Lyall.* Longwood Range, *T. K.* STEWART Island: common: *T. K.* AUCKLAND Islands, *T. K.* Descends to 150ft. on the Auckland Islands. Ascends to 4,500ft. Dec., Jan.

The flat oblong leaves of the early state form a conical rosette. The petals are persistent.

2. **D. Arcturi,** *Hook. in Journ. Bot.* i. (1834) 247; *Ic. Pl. t.* 56. Stock usually tufted, very short or 1in.–2in. long. Leaves 1in.–4in. long, linear-ligulate, obtuse, the early leaves short and quite glabrous, sometimes narrow-linear-ovate, the mature leaves glandular for half their length, narrowed below and glabrous. Scape exceeding the leaves, 1- or rarely 2-flowered. Calyx divided nearly to the base; segments 4, linear. Petals obovate, white, scarcely exceeding the calyx. Stamens 4. Styles 3, rather stout, short; stigma broad.—Hook. f., Fl. N.Z. i. 20; Handbk. 63; Benth., Fl. Austr. ii. 456.

NORTH Island: Ruahine Range, *Colenso.* SOUTH Island: Nelson to Otago, in mountain districts. STEWART Island. Descends nearly to sea-level on Stewart Island; chiefly from 2,500ft. to 4,000ft. elsewhere. Dec., Jan. Also in Australia and Tasmania.

Var. **polyneuron.** Leaves of two forms, some 4 lines wide and many-nerved, the others 1½ lines, erect. Scape longer or shorter than the leaves, with a linear basal leaf ¾in.–1in. long. Stamens 5; filaments flat. Styles 4.—*D. polyneuron,* Col. in Trans. N.Z.I. xxii. (1889) 460. NORTH Island: base of Mount Tongariro, Taupo, *H. Hill.*

3. **D. pygmaea,** *DC., Prod.* i. 317. A minute species, forming depressed rosettes less than ½in. in diameter. Petiole slender; blade orbicular, subpeltate, glandular. Stipules scarious, forming a silvery cone in the centre of the rosette. Scapes 1–4, capillary, ½in.–¾in. high, 1-flowered. Flowers minute. Calyx 4-lobed. Petals scarcely exceeding the sepals. Styles 4, short, clavate. Capsule 4-valved; seeds few.

NORTH Island: Cape Maria van Diemen, *Colenso*. Te Paua, Parengarenga, *Cheeseman!* SOUTH Island: Bluff Hill, *T. K.* Dec., Jan.

A charming little gem. Also found in Australia.

4. **D. spathulata,** *Labill, Nov. Holl. Pl.* i. 79, *t.* 106, *f.* 1. Stemless. Leaves rosulate, crowded, obovate or spathulate, narrowed into a long or short often broad petiole, densely glandular. Stipules scarious, laciniate at the tip or fimbriate. Scape slender, 1in.–6in. long. Flowers 1–6, shortly pedicelled, forming a 1-sided or spiral raceme, or often solitary. Calyx deeply 5-lobed; lobes linear-oblong. Petals 5, equalling the calyx. Stamens 5. Styles 3, 2-partite.—DC., Prod. i. 318; Hook. f., Fl. N.Z. i. 20; Handbk. 63; Benth., Fl. Austr. ii. 459. *D. propinqua*, R. Cunn. in Precurs. n. 620. *D. minutula*, Col. in Trans. N.Z.I. xxi. (1880) 81.

Mangonui and Bay of Islands to Stewart Island, but very local in many districts. Sea-level to 4,500ft. Dec., Jan. Also in Australia.

Var. **triflora.** Petioles numerous, short, broad; blade orbicular. Scape 1 or 2, stout, ¼in.–¾in. long, 2–3-flowered.—*D. triflora*, Col. in Trans. N.Z.I. xxii. (1889) 461. Base of Tongariro, *H. Hill!*

5. **D. binata,** *Labill, Nov. Holl. Pl.* i. 79, *t.* 105, *f.* 1. Stemless. Root of few or many fleshy fibres. Leaves all radical or naked; petioles 2in.–4in. long; the blade divided into 2 linear acute lobes, 1in.–3in. long, simple or again divided, and clothed above with glandular hairs, glabrous beneath. Stipules toothed or laciniate at the apex. Scapes 6in.–18in. high, branched at the apex. Flowers on slender pedicels, forming a lax few- or many-flowered cyme. Sepals 5 or 4, acute, glabrous. Petals 5 or 4, obovate, more than twice as long as the sepals. Stamens 5 or 4. Styles usually 3, penicillate. Capsule globose.—DC., Prod. i. 319; Bot. Mag. t. 3082; Hook. f., Fl. N.Z. i. 20; Handbk. 64; Benth., Fl. Austr. ii. 461. *D. intermedia*, R. Cunn. *ex* Precurs. n. 621. *D. dichotoma*, Banks and Sol. MSS.

From the North Cape to the Bluff; Stewart Island. Sea-level to nearly 2,000ft. *Fly-catcher.* Nov. to Feb. Rarely the leaves are multi-partite.

Var. **flagellifera.** Smaller than the typical form. Leaf-segments very narrow, simple or forked. Calyx-lobes divided nearly to the base, oblong-truncate, unequally laciniate at the tips. Styles 5 or more, much branched and forked.—*D. flagellifera*, Col. in Trans. N.Z.I. xxiii. (1890) 384. Tahoraite, Waipawa. Also in Australia.

6. **D. auriculata,** *J. Backh.* ex *Planch. in Ann. Sc. Nat. Ser.* 3, ix. (1848) 295. Stock 1in.–2in. long, slender, springing from a spherical tuber. Stems 1 or more, very slender, erect, simple or sparingly branched, glabrous, leafy. Radical leaves rosulate; petioles flattened, short; blade with long glands near the margins, orbicular, subreniform, peltate. Cauline leaves on longer capillary or filiform petioles, lunate, or truncate on the upper side, the two angles

produced into glandular appendages. Flowers purple, 4–8 or more; pedicels slender. Sepals 5, entire or minutely toothed at the apex. Petals 5, obovate or obcordate, fully twice as long as the sepals. Styles 3, divided from below the middle into many short dichotomous lobes. Capsule globose. Seeds numerous, linear; testa loose, produced into a short wing at each end.—Hook. f., Fl. N.Z. i. 21; Handbk. 64; Benth., Fl. Austr. ii. 465. *D. peltata*, Banks and Sol. MSS.

Three Kings Islands southward to Banks Peninsula. Sea-level to nearly 1,800ft. Nov. to Jan. Also in Australia.

The only New Zealand species with purple flowers. The stem is sometimes nearly 5ft. in length, climbing by means of the glandular leaves and slender petioles, which are often 1½in. long.

D. circinervia, Col. in Trans. N.Z.I. xxvi. (1893) 314, is a form occasionally met with, in which the lower and middle peduncles become elongated, forming a lax few-flowered corymb, which gradually assumes a racemose character with the lengthening of the axis.

Order XXVIII.—HALORAGEAE.

Calyx-tube adnate with the ovary; limb 2–4-lobed or toothed or 0. Petals as many as the sepals or 0, valvate, induplicate or slightly imbricate. Stamens 2–4, rarely 1–3, epigynous. Ovary inferior, 2–3- rarely 4-celled, with 1 ovule in each cell, or 1-celled with 4 ovules; ovules pendulous; styles as many as ovules or 0; stigmas plumose or papillose, sometimes sessile. Fruit inferior, small, indehiscent, with 1–4 cells, or separating into 2–4 indehiscent 1-seeded carpels, or a small drupe. Seeds pendulous; testa membranous; endosperm fleshy; embryo minute, cylindrical or ovoid; radicle superior. Herbs, often aquatic or uliginal, with opposite alternate or whorled exstipulate leaves. Flowers hermaphrodite or unisexual, usually axillary, rarely in terminal spikes, racemes, or panicles.

A small order, containing fewer than 100 species, arranged under 9 genera widely scattered over the earth. The precise affinities of *Gunnera* and *Callitriche* have not been clearly ascertained: the latter especially must be considered a doubtful member of the order.

1. Haloragis. Calyx 4-lobed. Petals induplicate. Fruit 4–8-angled. Terrestrial.
2. Myriophyllum. Calyx-lobes inconspicuous. Petals imbricate. Fruit separating into 2 or 4 carpels. Aquatic.
3. Gunnera. Flowers in terminal spikes or panicles. Fruit a drupe. Uliginal or terrestrial.
4. Callitriche. Perianth 0. Stamen 1. Styles 2. Fruit 4-lobed. Uliginal or aquatic.

1. HALORAGIS, Forst.

Calyx-tube 4–8-angled or winged; lobes 4, rarely 3, short, erect. Petals small, as many as calyx-lobes, deciduous, induplicate, usually 0 in female flowers. Stamens 4 or 8, on short filaments, usually 0 in female flowers. Ovary 2–4- rarely 5-celled, with 1 ovule in each cell; styles short; stigmas simple or plumose. Fruit a small 2–4-celled nut, sometimes drupaceous; the adherent calyx-tube 4–8-ribbed or winged, smooth or muricate. Herbs or suffruticose plants, with alternate or opposite entire toothed or lobed leaves and minute hermaphrodite or unisexual racemose flowers. The fruits are sometimes 1-celled by absorption.

SPECIES, about 50, of which 32 are endemic in Australia; 4 others are common to Australia and New Zealand; the remainder are distributed through Eastern Asia, South Africa, and temperate South America.

Leaves sharply serrate, 1in.–1½in. long. Flowers drooping, crowded .. 1. *H. alata*.
Floral leaves alternate. Fruits solitary, erect, rugose .. .. 2. *H. tetragyna*.
Floral leaves opposite. Fruits solitary, erect, smooth.. .. .. 3. *H. depressa*.
Flowers paniculate. Fruits terminal, smooth .. .. .. 4. *H. spicata*.
Floral leaves 0. Fruits drooping, smooth .. .. 5. *H. micrantha*.

1. **H. alata,** *Jacq., Misc.* ii. 332. Herbaceous, 1ft.–3ft. high, erect or suberect. Stems tetragonous, apparently glabrous but minutely scabrid. Leaves opposite, very shortly petioled, ovate-lanceolate or nearly oblong, ½in.–1½in. long, sharply serrate. Flowers minute, drooping, green, in terminal leafy racemes, solitary or whorled. Pedicels short, curved. Calyx 4-angled; lobes small, broad. Petals exceeding the calyx-lobes. Stamens 8. Stigmas 4. Nut ovoid; wings 4, narrow; interspaces smooth or rugose.—G. Forst., Prod. n. 180; Hook. f., Fl. N.Z. i. 62; Handbk. 65; Benth., Fl. Austr. ii. 479. *Cercodia erecta,* Ait. Hort. Kew. ed. 1, ii. 57; A. Rich., Fl. N.Z. 324; A. Cunn., Precurs. n. 526; Banks and Sol. MSS. *C. alternifolia,* A. Cunn., Precurs. n. 527.

From the KERMADEC Islands to Otago; STEWART Island. Sea-level to 2,300ft. Nov. to Jan. Also in Australia and Juan Fernandez.

2. **H. tetragyna,** *Hook. f., Fl. N.Z.* i. 63. Stems 6in.–15in. high. Branches decumbent at base, ascending; branchlets numerous, very slender, angular, scabrid. Leaves opposite, coriaceous, scabrid with appressed hairs, linear-lanceolate or ovate-lanceolate, sharply serrate. Flowers minute, solitary, sessile or nearly so in the axils of leafy bracts forming lax slender terminal leafy spikes, which are sometimes paniculate. Stamens 8. Styles 4; stigmas plumose. Fruits shortly ovoid, 4–8-angled, rugose.—Handbk. 65; Benth., Fl. Austr. ii. 484. *Goniocarpus tetragyna,* Labill, Pl. Nov. Holl. i. 39, t. 53. *G. tetragynus,* DC., Prod. iii. 56; A. Cunn., Precurs. n. 529.

NORTH Island: local. Auckland: on dry hills, Bay of Islands, Whangaroa North, Mount Carmel, Kaitaia, &c.

Var. **diffusa,** Handbk. 65. Stems spreading, slender, prostrate. Leaves ovate or ovate-lanceolate, with fewer teeth. Three Kings Islands, *Cheeseman,* and the North Cape to Stewart Island.

Var. **incana.** Leaves oblong or oblong-ovate, pubescent or villous.—*Cercodia incana,* A. Cunn., Precurs. n. 528. Bay of Islands. Also in Australia. The alternate floral leaves readily distinguish this from *H. depressa.*

3. **H. depressa,** *Walp. Rep.* ii. 99. A very slender wiry plant. Stems 1in.–2in. long, suberect or prostrate. Rhizomes extensively creeping. Leaves all opposite, on very short petioles or sessile, broadly ovate or almost cordate, rounded at the apex or subacute or shortly mucronate, with 1–3 teeth on each side, coriaceous, slightly scabrid on one or both surfaces. Floral leaves similar but smaller. Flowers sessile, solitary, axillary, forming short interrupted terminal spikes. Nut ⅟₁₆in. long, obtusely 4–8-angled, nerved, the interspaces

smooth and shining.—Hook. f., Fl. N.Z. i. 63; Handbk. 65; Benth., Fl. Austr. ii. 485. *H. bibracteolata,* Col. in Trans. N.Z.I. xxii. (1889) 462. *H. montana,* Hook. f. in Hook. Journ. Bot. vi. 475. *Goniocarpus depressus,* A. Cunn., Precurs. n. 531.

Var. **aggregata.** Leaves oblong-ovate. Flowers in small terminal umbels.—*H. aggregata,* Buch. in Trans. N.Z.I. iv. (1871) t. 13.

Var. **serpyllifolius.** Stems 1in.–4in. long, often forming a close sward. Leaves 1–3 lines long, lanceolate or ovate-lanceolate, narrowed below.—*H. serpyllifolia* and *H. vernicosa,* Walp., Rep. ii. 90. *Goniocarpus serpyllifolius* and *G. vernicosus,* Hook. f. in Hook. Ic. Pl. t. 290 and 311. *H. uniflora,* T. Kirk in Trans. N.Z.I. ix. (1876) 540, is a form with solitary terminal flowers, which sometimes prevails over wide areas.

NORTH and SOUTH Islands: Three Kings Islands and the North Cape to STEWART Island. Ascends to 3,500ft.

Best distinguished from *H. tetragyna* by the opposite floral leaves and the smooth interspaces of the fruit.

4. **H. spicata,** *Petrie in Trans. N.Z.I.* xix. (1885) 325. Stems sub-erect or ascending, 4in.–10in. high, pubescent or silky. Leaves opposite, very shortly petioled, subcoriaceous, ovate or oblong-lanceolate, acute or subacute, serrate, pubescent. Flowers in terminal slender-branched panicles, sessile in the axils of minute alternate or opposite bracts, the terminal flower or sometimes the two highest flowers female, the lower bracts flowerless or with a minute male flower. Anthers 4; filaments very short. Fruit about $\frac{1}{10}$in. long, 4-angled; interspaces smooth.

SOUTH Island: terraces, north end of Lake Hawea, 1,150ft., *Petrie.*

Although the paniculate inflorescence differs widely from that of any other New Zealand species, I have no doubt that this plant is an aberrant form of *H. depressa,* with which its leaves and fruits exactly agree. All the specimens dissected by me have 1-celled *seedless* fruits—a peculiarity very common in single-flowered specimens of *H. depressa.* Its appearance, however, is so very singular that for the present it seems worth while to allow it specific rank.

5. **H. micrantha,** *R. Br.* ex *Sieb. and Zucc. Fl. Jap. Nat.* i. 25. A small tufted species, with filiform stems and branches 1in.–6in. long, glabrous or rarely scaberulous, diffuse or ascending. Leaves all opposite, very shortly petioled or sessile, orbicular-cordate or ovate, rarely more than $\frac{1}{4}$in. in diameter, crenate or obscurely crenate. Flowers few, in short terminal leafless racemes. Pedicels very short. Petals 4. Styles and ovules 4. Fruit pendulous, 8-ribbed; interspaces smooth.—Hook. f., Fl. Tasm. i. 121; Handbk. 66; Benth., Fl. Austr. ii. 482. *H. tenella,* Brong. in Duper. Voy. Coq. Bot. t. 68, f. B; Hook. f., Fl. N.Z. i. 63. *H. minima,* Col. in Trans. N.Z.I. xviii. (1885) 250! *Goniocarpus citriodorus,* A. Cunn., Precurs. n. 530.

NORTH and SOUTH Islands: from the North Cape to the Bluff; STEWART Island. Sea-level to 2,300ft. Nov. to Jan. Also in Australia, eastern Himalaya, Bengal, and Japan.

2. MYRIOPHYLLUM, Linn.

Flowers unisexual or hermaphrodite. Calyx-tube very short or 0 in the male flower, 4-toothed or lobed; petals 4, imbricate; stamens 4–8, filaments short. Female: calyx-tube ovoid, lobes minute or 0; petals 0; ovary 2- or 4-celled; ovules 1 in each cell, pendulous; styles 2 or 4, often plumose. Fruit 2–4-celled, furrowed between the carpels, ultimately dividing into 2–4 hard

1-seeded nuts. Embryo terete; radicle superior. Aquatic or marsh plants, with terete sparingly-branched stems, whorled or rarely opposite or alternate leaves, which when submerged are divided into capillary segments, the upper toothed or pinnatifid or entire. Flowers mostly unisexual, axillary, minute, the male usually in the upper axils.

A widely-distributed genus, comprising about 20 species. One of the New Zealand species is endemic, the others extend to Australia.

ETYM. From the Greek, signifying *a thousand* and *a leaf*, in reference to the highly-dissected leaves of some species.

* *Flowering-stems submerged or aerial.*

Upper leaves whorled or opposite, linear, serrate or pinnatifid .. .. 1. *M. intermedium.*
Upper leaves pinnatifid. Stems slender .. .. 2. *M. verrucosum.*
Upper leaves whorled or opposite, broad, entire .. .. 3. *M. elatinoides.*

** *Flowering-stems always aerial.*

Erect, 9in.–18in. high. Leaves 5, in a whorl .. .. 4. *M. robustum.*
Erect or matted, 2in.–4in. high. Leaves opposite .. .. 5. *M. pedunculatum.*

1. **M. intermedium,** *DC., Prod.* iii. 69. Stems 6in.–30in. long, according to the depth of water. Leaves in whorls of 3–8, usually 4–6, those submerged pectinately divided into capillary lobes, the aerial floral leaves narrow-linear, entire, serrate, or pinnatifid. Male flowers very shortly pedicelled; calyx-lobes 4; petals white; stamens 8. Female: calyx-teeth and petals 0; carpels 4, smooth.—*M. variaefolium*, Hook. f. in Hook. Ic. Pl. t. 289; Fl. N.Z. i. 65; Handbk. 66; A. Cunn., Precurs. n. 532.

NORTH and SOUTH Islands: Mongonui to STEWART Island: common in still waters from sea-level to 3,000ft. Dec., Jan. Also in Australia.

Owing to the evaporation of the shallow winter ponds in which it often grows, this species is sometimes terrestrial, when the uppermost leaves are reduced to a few opposite pairs, linear and quite entire.

2. **M. verrucosum,** *Lindl. in Mitch. Trop. Austr.* 384. Stems slender, 12in.–18in. long. Leaves 3 or 4 in a whorl, the submerged pectinately divided into short capillary segments; the aerial floral leaves sessile, oblong or lanceolate, pinnatifid; lobes short. "Calyx-lobes short, but perceptible in both sexes, very deciduous in the females. Petals in the male under 1 line long. Stamens 8. Females without petals. Styles 4, very short. Carpels 4, rarely above ½ line long, obtuse on the back, more or less tuberculate."—Benth., Fl. Austr. ii. 488.

NORTH Island: in ponds near Tauranga, *T. K.* April, 1865.

The identification must be considered open to doubt, as the flowers have not been seen, and the fruiting specimens collected by me are very imperfect. Dr. Berggren, who saw the plant growing, agrees with me in the identification.

3. **M. elatinoides,** *Gaudich. in Ann. Sc. Nat. Ser. I.* v. (1825) 105. Stems 6in.–30in. long. Leaves 4 in a whorl, rarely more, the submerged pectinately pinnatifid; segments capillary. Flowers in a leafy spike or panicle, 2in.–6in. long. Floral leaves in whorls of 4 or 3, those of the male flowers opposite, broadly lanceolate or ovate-lanceolate, entire or toothed. Male:

calyx-lobes very minute; petals 4; stamens 8. Female: calyx-lobes and petals 0. Fruits smooth, shorter than the floral leaves.—DC., Prod. iii. 68; Hook. f., Fl. N.Z. i. 63; Handbk. 66; Benth., Fl. Austr. ii. 487.

NORTH and SOUTH Islands: from the Waitemata (Auckland) to STEWART Island. In the South Island it ascends to upwards of 3,000ft. *Water-milfoil.* Nov. to. Jan. Also found in Australia and temperate South America.

4. **M. robustum,** *Hook. f., Handbk.* 67. Stems usually erect, stout, 9in.–20in. high or more. Leaves 5 in a whorl, rarely more, 1in.–2in. long, pectinately pinnatifid; segments acute. (Submerged leaves with longer capillary segments are rarely seen.) Flowers 1, rarely 2, with 2 jagged or laciniate bracts at the base of each. Calyx-lobes 4, deltoid, jagged. Male: petals 4; stamens 8, rarely 4–6. Female: styles usually 4; stigmas plumose; carpels 4, rounded at the back and compressed laterally, smooth or minutely tuberculate. —*M. variaefolium, β,* Hook. f., Fl. N.Z. i. 64.

NORTH Island: in swamps; local. Bay of Islands to middle Waikato, &c., *T. K.* Te Aroha Mountain, *Adams.* East Coast, *Colenso.* Wellington: Mungaroa Swamp, *T. K.* SOUTH Island: Marlborough: Awatere, *T. K.* Nelson: Moutere, *Cheeseman.* Hokitika, *Tipler!* Okarito, *A. Hamilton!* Sea-level to 900ft. Dec. to Feb.

A handsome plant, resembling a miniature pine. The fruits are crowded, presenting a whorled appearance, rarely consisting of 5 or even 6 carpels. Sometimes 1 or 2 male flowers are found amongst the female, and more rarely 1 or 2 female amongst the lowest males. In the late autumn the old flowering-stems are submerged by floods, and towards spring develop erect stems from the nodes.

5. **M. pedunculatum,** *Hook. f., Lond. Journ. Bot.* vi. (1847) 474. Stems tufted, 2in.–4in. high, simple or branched, erect or matted. Leaves very small, opposite, narrow-linear or linear-oblong, entire, $\frac{1}{8}$in.–$\frac{1}{4}$in. long. Flowers axillary. Male usually on short peduncles, with 2 linear bracts at the base; calyx-lobes 4, linear; petals 4; stamens 8. Female sessile; bracts 2, minute, acute; calyx-lobes 4, minute; petals 0; styles 4; carpels small, rugose or minutely tuberculate.—Hook. f., Fl. Tasm. i. 123, t. 23, f. B; Handbk. 67; Benth., Fl. Austr. ii. 489.

NORTH and SOUTH Islands: from the Waitemata (Auckland) to STEWART Island, but often local. Sea-level to 2,800ft. Dec., Jan. Also in Australia. Often diœcious.

3. GUNNERA, Linn.

Flowers unisexual, rarely hermaphrodite. Male: calyx usually reduced to 2 or 3 minute teeth; petals 2 or 3 or 0; stamens 2 or 3; anthers large. Female: calyx-tube terete or ovoid, lobes 2 or 3, small; petals 2 or 3 or 0; ovary 1-celled; ovule solitary, pendulous; styles 2 rarely 4, linear, stigmatic from the base. Fruit a small fleshy drupe, often minute; seed adherent to the pericarp; endosperm fleshy; embryo minute; radicle superior. Stemless herbs, with creeping rhizomes often forming matted patches. Leaves petioled. Flowers small, in simple or branched spikes or panicles.

Usually inhabiting swamps and watery places, often forming dense masses amongst sphagnum. *G. arenaria* exhibits a curious dimorphism in the flowers and fruits. *G. mixta* is probably dimorphic also. In most species the fruits require from four to six months for maturation.

Species, about 25, distributed through the cooler regions of the Southern Hemisphere, from South Africa to Chili and the Falkland Islands, Juan Fernandez, Java, Sandwich Islands, and Tasmania. All the New Zealand species are endemic. Some exotic species have leaves from 3ft. to 6ft. in diameter.

Named in honour of *Bishop Gunner*, a famous Swedish botanist.

* *Scapes bisexual. Female flowers at the base.*

Leaves orbicular, lobed. Scapes simple or branched .. .. 1. *G. monoica.*
Leaves entire. Scapes simple, slender, exceeding the leaves .. .. .. 2. *G. mixta.*
Scapes simple or branches very short, hidden amongst the leaves .. .. 3. *G. microcarpa.*

** *Scapes unisexual.*

Leaves ovate, ovate-cordate, or oblong. Scapes shorter than the leaves. Drupes broadly obconic, red .. .. 4. *G. prorepens.*
Scapes red, usually exceeding the leaves. Drupes obconic, red or yellow .. 5. *G. flavida.*
Leaves ovate-lanceolate or ovate, acute, dentate .. .. 6. *G. dentata.*
Leaves orbicular, cordate, acutely toothed. Scapes shorter than the leaves .. 7. *G. densiflora.*
Leaves with long sheathing petioles, ovate-cuneate. Flowers often dimorphic .. 8. *G. arenaria.*
Leaves broadly deltoid, cuneate, coriaceous. Scapes shorter than the leaves .. 9. *G. Hamiltonii*

1. **G. monoica,** *Raoul in Ann. Sc. Nat. Ser. III.* ii. (1844) 117. A slender tufted plant, with creeping rhizomes, glabrous or almost strigose. Leaves reniform or orbicular, with few short hairs on both surfaces, ⅓in.–⅔in. in diameter, 3–5-lobed or crenate or crenate-dentate only; petioles slender, 1in.–2in. long. Panicle 1in.–5in. long, very slender, male flowers above, female with few short often crowded branches. Male: perianth of 2 minute narrow segments; stamens 2. Female: segments 2, acute; styles elongated, capillary. Fruits red or white, minute, spherical, forming a compact spike or panicle hidden amongst the leaves.—Raoul, Choix de Pl. de la Nouv. Zélande xv. t. 8; Hook. f., Fl. N.Z. i. 65; Handbk. 67; DC., Prod. xvi. ii. 599. *G. prorepens,* Hook. f., Fl. Antarc. ii. 274 (in note, not of Fl. N.Z. i. 66).

NORTH and SOUTH Islands: Mangonui to STEWART Island: in cool, moist situations, but often rare and local. Sea-level to 3,000ft. Oct., Nov.

Var. **strigosa.** Whole plant more hairy, especially the rhizomes and petioles, which are often strigose. Leaves sometimes crowded at the nodes.—Sp., Col. in Trans. N.Z.I. xv. (1882) 322. NORTH Island: in rather dry places. Hawke's Bay, *Colenso!* Mungaroa, Wellington, *T. K.*

Var. **ramulosa.** Stems rather stout, clothed below with the bases of old petioles. Panicle much branched, upper branches 1in.–1½in. long. Flowers densely crowded. Drupes not seen. SOUTH Island: Broken River, *Enys!* 2,800ft.

Var. **albocarpa.** Rhizomes robust, sometimes thicker than a goose-quill. Leaves larger than in the type. Panicle sometimes 6in. long or more, lax or compact, much branched; branches sometimes 3in. long. Drupes minute, spherical, milk-white, tipped with the black perianth-segments. SOUTH Island: Southland; Stewart Island. A remarkable form of this plant collected in Southland has erect or suberect branches, 12in.–18in. long, with from 4 to 6 shortly-petioled fleshy leaves crowded at each node.

2. **G. mixta,** *T. Kirk in Trans. N.Z.I.* xxvii. (1894) 344. Rhizomes slender. Leaves 1½in.–2½in. long, with weak scattered hairs on petiole and blade; blade ovate or slightly cordate, rounded at the apex, crenate but not lobed. Scape very slender, unbranched, exceeding the leaves. Upper flowers lax, sessile or shortly pedicelled. Staminate, pistillate, and hermaphrodite flowers intermixed; staminate mostly pedicelled. Perianth-segments 2, narrow-

linear-oblong, obtuse; female perianth-segments 4, ovate, unequal, sometimes with 2 liner-oblong bracts at the base. Hermaphrodite flowers with 2 ovate segments. Filaments short; anthers apparently abortive. Fruit not seen.—*G. ovata,* Petrie in Trans. N.Z.I. xxv. (1892) 274! (in part).

SOUTH Island: Otago, *Buchanan! A. Hamilton! Petrie!*

A curious plant, resembling a small state of *G. prorepens*, but distinguished by the simple lax subracemose scape. The flowers may be either male or female primarily, but in either case one or two of the complementary form and one or two hermaphrodite will be found on each rhachis. More copious material is required for a satisfactory diagnosis.

3. **G. microcarpa,** *T. Kirk in Trans. N.Z.I.* xxvii. (1894) 348. Rhizomes slender, tufted. Leaves 2in.–4in. long, slender, hairy or strigose; blade about 1in. long, broadly ovate or ovate-cordate, with scattered hairs on both surfaces, crenate or crenate-lobed. Flowering-spike or raceme lax, interrupted, monœcious. Male: sessile or shortly pedicelled, with two linear concave fringed obtuse bracts at the base; perianth-lobes 2; stamens 2, equalling the bracts; filaments broad at the base; anthers compressed, broadly oblong; connective apiculate. Female: flowers few, trichotomous; perianth 2-lobed. Fruiting-scapes 1in.–1½in. long, weak, almost filiform, hidden amongst the leaves. Bracts linear, ciliate at the apex. Styles filiform. Drupes sessile, erect, about the size of mustard seeds, yellow or red.

SOUTH Island: Southland, *T. Waugh!* Dec.

My flowering specimens are few and imperfect, while their identity with the fruiting specimens is not absolutely proved. The fruiting-scapes are sometimes slightly branched at the base. The plant differs from *G. prorepens* in the weak flexuous scapes, minute drupes, and filiform styles.

4. **G. prorepens,** *Hook. f., Fl. N.Z.* i. 66. Rhizomes rather stout. Leaves 3in.–5in. long, ovate or ovate-oblong or rarely elliptic, more or less clothed with scattered hairs; petioles strigose or hairy. Scapes much shorter than the petioles, wholly hidden amongst the leaves, strigose, stout, unisexual. Male flowers not seen. Female: perianth-segments 4, linear-obovate, obtuse, concave, cucullate; styles short, stout, slightly compressed. Drupes sessile or very shortly pedicelled, forming a compact spike less than 1in. in length, broadly obconic, with a shallow groove at the apex, red.—Hook. f., Handbk. 68; DC., Prod. xvi. ii. 599.

NORTH and SOUTH Islands; STEWART Island: from Taupo southward.

The short stout spike and broadly obconic drupes distinguish this species from all others, although several exhibit similar leaves.

5. **G. flavida,** *Col. in Trans. N.Z.I.* xviii. (1885) 261. Densely tufted. Rhizomes rather stout. Leaves 1in.–4in. long; blades ovate or slightly ovate-cordate, rarely elliptic, rounded at the apex; margins entire or slightly sinuate, or rarely crenate at base, glabrous or with few weak hairs on petiole; petioles and peduncles red. Scapes unisexual, 1in.–4in. long, usually exceeding the leaves when in fruit. Male flowers not seen. Female: perianth-segments 4, the two innermost ovate or oblong, ciliate at the apex; outer larger, obovate, cucullate, deciduous. Drupes obconic, sessile or shortly pedicellate, crimson

or lemon-coloured, forming a lax spike, sometimes 2in. long, overtopping the leaves.—*G. ovata*, Petrie in Trans. N.Z.I. xxv. (1892) 274 (in part—fruits only). *G. elongata*, T. Kirk, MSS.

NORTH Island: (collected by *H. Hill*) *Colenso!* Probably not uncommon in the Taupo district. SOUTH Island: Otago and Southland. STEWART Island, *T. K.*

Frequent amongst sphagnum. The fruits are often glaucous. Easily recognised by the red petioles and peduncles, and especially by the obconic fruits, which are elevated above the leaves. I am indebted to the Rev. W. Colenso for one of his type specimens.

6. **G. dentata,** *T. Kirk in Trans. N.Z.I.* xxvii. (1894) 346. Rhizomes rather stout, tufted. Leaves 1in.–2in. long or more; petioles hairy or strigose at base; blade ½in. long or less, membranous, ovate-lanceolate or ovate, acute, with scattered hairs on both surfaces or nearly glabrous; margins dentate. Scapes unisexual. Male: very slender, slightly exceeding the leaves; flowers sessile; perianth-segments linear-oblong, cucullate; anthers broad, obtuse. Female: about ½in. long, hidden amongst the leaves, silky at the base, spike ¼in. long; perianth-segments 2, broadly oblong, obtuse; style very long, stout, much compressed at the base; scape either elongating in fruit, 1in.–2in. high, lax, or sessile, short and compact. Drupes sessile or on stout pedicels, clavate, patent or pendulous.—*G. prorepens, β,* Hook. f., Handbk. 68.

NORTH Island: *Colenso* (in Handbk.). Taupo, *Petrie!* SOUTH Island: Nelson to Southland, but often local. 1,000ft. to 3,000ft.

7. **G. densiflora,** *Hook. f., Handbk.* 68. Rhizome rather stout, tufted. Leaves 1in.–1½in. long; petiole as long as the blade, strict, strigose, villous or glabrescent; blade ½in.–1in. in diameter, orbicular or orbicular-ovate, cordate, doubly serrate; teeth acute. Scapes unisexual. Male not seen. Female: ¼in.–⅜in. long; peduncle short, stout, villous; flowers sessile or subsessile, crowded; ovary narrow-oblong; perianth-lobes subulate, acute; bracts subulate; styles 2, spreading. Fruiting-scape shorter than the leaves. Drupes linear-oblong, $\frac{1}{10}$in. long, pendulous, purple.

SOUTH Island: Acheron and Clarence Rivers, 4,000ft., *Travers* in Handbk. Craigieburn Mountains, Canterbury, 3,500ft., *Cockayne!*

The description must remain imperfect until more copious material can be obtained; but the species is easily distinguished by the acutely-toothed leaves and pendulous purple fruits.

8. **G. arenaria,** *Cheesm.* ex *T. Kirk in Trans. N.Z.I.* xxvii. 348. Rhizomes slender or stout, tufted. Whole plant rather fleshy, glabrous or glabrate, with thin flattened hairs on the petiole. Leaves 1in.–2in. long; blade ovate or ovate-cuneate or almost cordate, crenate or crenate-dentate; petioles sheathing at the base. Scapes unisexual, dimorphic. Male: in the early spring slender, exceeding the leaves, 1in.–1½in. long; anthers 2, nearly orbicular, sessile, with one or two narrow-linear fimbriate bracts at the base of each flower. Female: very short, hidden amongst the leaves; spike ¼in.–⅜in. long, dense; perianth 2-lobed, lobes minute, acute; styles rather stout, spreading, attenuated to the apex. Drupes hidden amongst the leaves, crowded, sessile or on very short

pedicels, forming compact spikes or heads ½in. long or more, ovoid, slightly narrowed below. Nut ovoid, acute. In the summer both male and female scapes are 1in.–3in. high, stout. Male flowers on stout pedicels or sessile, with 2 linear-spathulate cucullate bracts; anthers slightly mucronate. Female not seen. Fruiting-scapes forming erect racemes. Drupes clavate, pendulous, fleshy; nut linear, narrowed at both ends.—*G. densiflora*, T. Kirk in Trans. N.Z.I. *l.c.* (not of Hook. f.).

NORTH and SOUTH Islands: on sandy beaches. Cape Maria van Diemen and mouth of the Waitakerei River, Auckland, *Cheeseman!* Cape Farewell, Nelson, *T. K.* New Brighton, *Cockayne!* Seventy-mile Beach, Canterbury, *Buchanan!*

The remarkable dimorphism exhibited by this species in the form of the flowers, the drupes, and even the nuts, has only been observed in three of the habitats recorded here. The rhizomes of the Waitakerei specimens are mostly clothed with the ragged bases of the petioles, and the plant is much larger; but the flowers have not been observed, although there seems no doubt as to its identity. The short flattened hairs on the upper part of the petiole resemble those found on some Cruciferae.

9. **G. Hamiltonii**, *T. Kirk in Trans. N.Z.I.* xxvii. 347. Rhizome as thick as a goose-quill, tufted. Leaves forming dense rosettes 2in.–4in. in diameter, coriaceous, brown, glabrous or the petiole glabrescent; blade usually exceeding the winged petiole, ¾in.–1in. long, cuneate-ovate or almost deltoid, acute, gradually narrowed into the short broad winged petiole; margins minutely toothed, strongly nerved beneath. Scapes unisexual. Male: flowers sessile on stout scapes, not crowded. Female: spikes 1in.–1¼in. long, lower half naked, upper densely crowded; bracts large, broadly ovate, laciniate; perianth-segments 4; stigma stout. "Drupes almost sunk in the fleshy peduncle."

SOUTH Island: hills near the mouth of the Oreti River, *W. S. Hamilton!* STEWART Island: Mason Bay, *W. Traill!*

"The extremely coriaceous strongly-ribbed leaves support the foot, and, spreading from a hollow centre, give the ground a bird-nested appearance."—*W. S. H.* Better specimens must be obtained before a good diagnosis of this unique plant can be drawn.

4. CALLITRICHE, Linn.

Flowers minute, monœcious. Perianth 0 or represented by a pair of minute bracts. Male flowers of a single stamen, with a 4-celled anther opening by lateral slits, which by confluence assume a crescentic form, and the anther becomes 1-celled. Female sessile or shortly stalked; ovary 4-celled; styles 2, filiform, stigmatic for their entire length; ovules 1 in each cell. Fruit compressed, 4-lobed, 4-celled, notched at the apex and separating into 4 1-celled 1-seeded carpels. Embryo small, terete. Endosperm fleshy or oily. Slender delicate herbs, with opposite entire linear-spathulate obovate or rhomboid leaves and axillary solitary or geminate flowers.

In stagnant waters or on damp soil.

By some botanists all the forms are united under a single species, which is found in nearly all parts of the world. Others recognise thirty or forty species. As the three New Zealand species are constant in the characters indicated, and can be easily recognised, it seems preferable to consider them distinct.

Leaves linear-spathulate, retuse. Fruits rounded or keeled .. .. 1. *C. verna.*

Leaves obovate-spathulate, fleshy. Fruits keeled .. .. .. 2. *C. antarctica.*

Leaves rhomboid or rhomboid-spathulate. Fruits broadly winged .. 3. *C. Muelleri.*

1. **C. verna,** *L., Fl. Suec. ed.* 2, ii. *n.* 3. Stems long, slender. Leaves linear or oblong-spathulate or linear-obovate, minutely retuse. Fruits convex, slightly keeled at the back.—Hook. f., Fl. N.Z. i. 64; Handbk. 68.

NORTH and SOUTH Islands: floating in still water. Mongonui to Southland. Oct. to Feb.

2. **C. antarctica,** *Engelm.* ex *Hegelm. in Verh. Bot. Ver. Brandent.* ix. (1867) 20. Stems 2in.–5in. long, succulent, densely matted, rooting at the nodes. Leaves fleshy, linear-obovate-spathulate, rounded at the apex. Fruits obcordate, convex on the sides, scarcely keeled at the back.—*C. verna β terrestris,* Hook. f., Fl. Antarc. i. 11.

The SNARES, AUCKLAND and CAMPBELL Islands, ANTIPODES Island, MACQUARIE Island. Also in Kerguelen's Land, &c., Falkland Islands, and South Georgia. On damp soil. Dec., Jan.

3. **C. Muelleri,** *Sond. in Linn.* xxviii. (1856) 229. Stems filiform or capillary, densely matted. Leaves mostly rhomboid or rhomboid-spathulate, usually with a minute tooth on each side, sharply narrowed into a distinct petiole. Female flowers usually with an abortive stamen at the base. Fruits orbicular-obcordate, with a deep notch and 4 membranous wings.—*C. verna, β,* Hook. f., Fl. N.Z. i. 64. *C. microphylla,* Col. in Trans. N.Z.I. xx. (1887) 190. *C. tenella,* Banks and Sol. MSS.

Var. **obtusangula.** Leaves obovate, narrowed into a petiole. Fruits larger and more orbicular, with a very narrow wing.—Sp., Le Gall. *ex* Hegelm. Monog. Call. 54.

NORTH and SOUTH Islands: North Cape to STEWART Island; CHATHAM Islands. On damp ground. Var. *obtusangula* most frequent in the South.

It is worthy of remark that *Callitriche* is not mentioned by A. Richard, Cunningham, or Raoul.

Order XXIX.—MYRTACEAE.

Calyx-tube adnate with the ovary at least to the insertion of the petals, sometimes produced beyond it; limb 4–5-cleft; lobes usually imbricate in bud. Petals 4–5, imbricate in bud, equal or nearly equal when expanded, sometimes coherent, forming an operculum, rarely 0. Stamens ∞, usually numerous, rarely few, inserted on a disk lining the calyx-tube and forming a thickened margin at its mouth; filaments incurved in bud; anthers small, didymous. Ovary inferior, often wholly enclosed in the calyx-tube, 4–5-celled; style simple; stigma capitate, rarely lobed; ovules numerous or rarely few, on parietal placentas. Fruit inferior, 1 or more celled, often crowned by the persistent calyx-limb, capsular and dehiscing loculicidally at the apex, or indehiscent, 1-seeded, dry or fleshy. Perfect seeds few. Endosperm usually 0. Embryo straight or curved. Shrubs or trees, sometimes scandent, with opposite or alternate exstipulate entire leaves, which are dotted with pellucid oil-glands. Flowers hermaphrodite, solitary or in terminal racemes, cymes, or panicles, often of great beauty.

Genera, 80. Species, about 1,850. Plentiful in the tropics; rare in the north temperate zone, but more frequent in the south, extending to Cape Horn and Campbell Island.

Tribe I. LEPTOSPERMEAE.—Ovary inferior or half-inferior, 2-5-celled, rarely more. Capsule dehiscing at the apex by as many valves as there are cells. Seeds usually numerous, often imperfect.

1. Leptospermum. Leaves alternate. Flowers solitary or crowded.

* Eucalyptus. Leaves alternate. Calyx truncate, entire. Petals forming a convex operculum.

2. Metrosideros. Leaves opposite. Flowers in terminal cymes or umbels.

II. MYRTEAE.—Ovary 2-celled, with 2 parietal placentas, or rarely 1-celled. Fruit a berry or drupe, indehiscent.

3. Myrtus. Flowers mostly solitary. Berry few- or many-seeded. Embryo long and narrow.

4. Eugenia. Flowers cymose. Berry with 1 or more large angular seeds.

1. LEPTOSPERMUM, Forst.

Calyx-tube adnate to the base of the ovary, turbinate; lobes 5, persistent or deciduous. Petals 5, rounded, exceeding the calyx-lobes, imbricate. Stamens numerous, free, inserted on the margin of the disk, not exceeding the petals. Ovary enclosed in the calyx-tube, inferior; cells 5, rarely more or fewer; style straight; ovules very numerous. Capsule woody or coriaceous, each cell dehiscing by a valve on the upper surface. Seeds numerous, linear, pendulous, mostly sterile; testa membranous. Shrubs or trees, with small alternate entire leaves and hermaphrodite or polygamous axillary flowers.

Species, about 28, of which 20 are restricted to Australia; 3 are found in New Zealand, the others in New Caledonia and the Indian Archipelago.

Etym. From the Greek, in reference to the slender seeds.

Calyx broadly turbinate; lobes rounded, deciduous. Ovary half exserted .. 1. *L. scoparium.*

Calyx broadly turbinate or campanulate. Ovary wholly included in the calyx-tube. Teeth persistent .. .. 2. *L. ericoides.*

Leaves, pedicels, and calyx-tube white with silky hairs. Ovary deeply sunk in the narrow silky calyx-tube .. .. 3. *L. Sinclairii.*

1. **L. scoparium,** *Forst., Char. Gen.* 48, *t.* 36. Varying from a rigid or slender dwarf shrub to a tree 30ft. high, but usually under 20ft. high. Leaves alternate or scattered, ⅛in.–½in. long, sessile, linear, linear-lanceolate ovate-lanceolate or ovate, erect, spreading or rarely recurved, rigid, acute or pungent, concave, silky when young. Flowers sessile, solitary, axillary or terminal, ¼in.–½in. in diameter. Calyx-tube broadly turbinate; lobes orbicular, deciduous. Petals orbicular, shortly clawed. Capsule woody, girt round the middle with the calyx-limb, 5-valved at the apex.—DC., Prod. iii. 227; A. Rich., Fl. N.Z. 337; A. Cunn., Precurs. n. 553; Hook. f., Fl. N.Z. i. 70; Handbk. 69; Benth., Fl. Austr. iii. 105; T. Kirk, Forest Fl. N.Z. t. 117. *Philadelphus parvifolius,* Banks and Sol. MSS.

NORTH and SOUTH Islands: from the Three Kings Islands and North Cape to STEWART Island. Sea-level to upwards of 3,000ft. CHATHAM Islands, *Cox!* *Manuka. Tea-tree.* Nov. to April. Also in Australia.

Var. **linifolium.** Leaves linear-lanceolate.—T. Kirk, Forest Fl. N.Z. t. 117.

Var. **myrtifolium.** Leaves ovate, spreading or recurved.

Var. **prostratum**. Stems prostrate, ascending at the tips. Leaves ovate or nearly orbicular, close-set, spreading or recurved. On mountains.

Var. **parvum**. 1ft.–2ft. high. Branchlets strict. Bud-scales silky, villous. Leaves ovate, spreading, pungent, very coriaceous. Flowers small, ⅙in. in diameter. Calyx-tube glabrous. Near Wellington.

Flowers white or rarely red. The leaves are occasionally used as a substitute for tea.

2. **L. ericoides**, *A. Rich., Fl. N.Z.* 338. A shrub or tree, 20ft.–60ft. high, with trunk 1ft.–3ft. in diameter. Branchlets very slender. Leaves crowded or fascicled, narrow-linear or linear-lanceolate, acute, glabrous, glabrate or silky. Flowers ⅛in.–¼in. in diameter, axillary, solitary or fascicled, often crowded. Pedicels glabrous, puberulous or silky. Calyx-tube broadly turbinate in flower, glabrous or puberulous; teeth 5, acute, persistent. Petals orbicular, scarcely clawed. Ovary slightly sunk within the calyx-tube, 5-celled; style stout. Fruiting calyx-tube campanulate. Capsule coriaceous or woody, wholly included within the calyx-tube.—A. Cunn., Precurs. n. 554; Hook. f., Fl. N.Z. i. 70; Handbk. 70; T. Kirk, Forest Fl. N.Z. t. 69. *Philadelphus parviflorus*, var. *aromaticus*, Banks and Sol. MS.

NORTH and SOUTH Islands: from the Three Kings Islands and the North Cape to Southland. Ascends to nearly 3,000ft. *Manuka-rauriki*. *Kanuka*. *Maru*. *Tea-tree*. Nov. to Jan.

Var. **lineatum**. Very slender, 2ft.–6ft. high, rarely prostrate. Leaves very narrow linear, $\frac{1}{30}$in.–$\frac{1}{20}$in. broad, acute; margins faintly recurved, glabrate or silky; midrib and dotted glands sunk. Flowers smaller than in the type, and petals more crumpled.—T. Kirk, Forest Fl. N.Z. t. 69, f. 2. NORTH Island: Mangonui to the Waitemata.

3. **L. Sinclairii**, *n. s.* Stems prostrate or suberect, 1ft.–3ft. high, spreading; young shoots and leaves white with loosely appressed silky hairs. Leaves linear-lanceolate or oblong-lanceolate, flat or concave. Flowers larger and pedicels longer than in *L. ericoides*, crowded, fasciculate or umbellate. Pedicels and calyx-tube silky, villous; calyx-tube narrow-turbinate; lobes oblong, subacute or rounded. Petals obovate. Ovary deeply sunk within the calyx-tube; style slender. Fruiting calyx-tube campanulate, silky.

NORTH Island: Great Barrier Island, *Hutton* and *Kirk*. Sea-level to 1,800ft. Nov. to Jan. Originally discovered by Dr. Sinclair.

Nearly related to *L. ericoides*, but distinguished at sight by the white silky leaves, larger flowers, and longer pedicels. The ovary is sunk fully one-third below the narrow calyx-tube, while the sepals and petals are narrower, and the style is extremely slender. The flowers are deliciously fragrant.

*EUCALYPTUS, L'Hérit.

Calyx-tube broadly turbinate or campanulate, adnate to the ovary at the base or rarely to the apex, truncate, entire or minutely 4-toothed, the orifice closed by a conical or convex operculum (formed by the connate and concreted petals), which is forced off by the growth of the stamens. Stamens numerous, in several series, free or shortly cohering at the base; anthers small, versatile. Ovary inferior, 3–6-celled; ovules many in each cell. Fruit a coriaceous or woody capsule, truncate, adnate to the enlarged calyx-tube, 3–6-celled, dehiscing loculicidally at the apex. Seeds numerous, partly sterile, often of 2 forms, ovoid or subglobose, or linear, angular. Shrubs or trees of large dimensions. Leaves in the young state often opposite and horizontal; in the mature state usually alternate, pendulous and rigid. Flowers pedunculate or sessile, solitary or umbellate or racemose.

***E. Globulus,** *Labill, Voy.* i. 153, *t.* 13. A lofty tree. Stem and branches of the young tree 4-angled, with opposite, sessile, ovate-truncate or cordate glaucous leaves. Leaves of the mature state pendulous, alternate, coriaceous, ovate-lanceolate, acuminate, falcate. Flowers very large, solitary or geminate, sessile or very shortly peduncled. Calyx-tube broadly turbinate, thick, rugose or warted, 4-toothed. Operculum rugose, thick, shorter than the tube. Stamens all perfect. Ovary equalling the calyx, convex. Fruit ¾in.–1in. in diameter, 5-celled, the margin rising above the calyx; valves not protruding.

NORTH Island: naturalised on the Te Karaka Flats and in other localities. *Blue-gum.* Australia.

2. METROSIDEROS, Banks.

Calyx-tube adnate with the base of the ovary, globose, campanulate or turbinate, sometimes forming a tubular limb; lobes 5, imbricate. Petals 5, spreading. Stamens very numerous, exceeding the petals; filaments filiform, often unequal. Ovary included in the calyx-tube or half extruded, 3-celled; style slender; ovules numerous in each cell, radiating outwards from axile placentas. Capsule coriaceous, closely invested by the calyx-tube, inferior or half-superior, 3-celled. Seeds very numerous, linear, rarely flat; testa membranous. Embryo straight. Trees or shrubs, often scandent or trailing. Leaves opposite, sometimes distichous, coriaceous. Flowers in terminal racemes, cymes, or umbels, white, red, or crimson, rarely yellow.

SPECIES, about 30, of which 11 are found in New Zealand, all except one being endemic; the others are distributed through the Indian Archipelago, Oceania, South Africa, Chili, New Caledonia, and Australia. The scandent species are restricted to New Zealand, and, on the fall of the supporting tree, usually form large erect bushes, all trace of the climbing habit being lost.

ETYM. From the Greek, in reference to the strength and durability of the timber.

* *Capsule coriaceous or woody, wholly enclosed in the calyx-tube, dehiscing irregularly or by 3 apical valves.*

An erect bush or large climber. Calyx glabrous. Capsule woody .. .. 1. *M. florida.*
A shrub or large tree, with acuminate leaves. Calyx silky .. .. .. 2. *M. lucida.*
A shrub or small tree. Leaves elliptical, ovate, acute .. .. .. 3. *M. Parkinsonii.*

** *Capsule scarcely coriaceous, dehiscing to the base, wholly enclosed in the calyx-tube, which is narrowed above.*

Leaves decussate, acuminate. Flowers white, terminal .. .. .. 4. *M. albiflora.*
Leaves decussate, obtuse. Flowers scarlet .. .. 5. *M. diffusa.*
Leaves distichous, subacute. Capsule glabrous .. .. 6. *M. hypericifolia.*
Leaves distichous, acuminate. Capsules terminal, pubescent .. .. 7. *M. Colensoi.*

*** *Capsule protruded beyond the calyx-tube, 3-valved.*

a. Trees, often of large dimensions.

Leaves decussate, glabrous, obtuse, 1in.–1½in. long. Capsule glabrous .. 8. *M. robusta.*
Leaves decussate, white beneath with appressed tomentum, 2in.–4in. long. Capsule tomentose .. .. 9. *M. tomentosa.*
Leaves decussate, obovate, white with appressed tomentum beneath, ¾in.–1½in. long .. .. 10. *M. villosa.*

b. An erect bush or lofty climber.

Leaves ½in.–¾in. long. Flowers white .. .. 11. *M. scandens.*

1. **M. florida,** *Sm. in Trans. Linn. Soc.* iii. (1797) 269. A large bush, 2ft.–5ft. high, or a lofty climber with a stout main stem. Leaves 1½in.–3in. long, shortly petioled, coriaceous, lanceolate-oblong or elliptic-oblong, obtuse, glabrous; midrib stout. Flowers in terminal branched cymes; pedicels short, articulated with the peduncles; orange-red to crimson. Calyx obconic or turbinate, gland-dotted, exceeding the ovary. Petals 5, exceeding the calyx-lobes, yellowish. Stamens scarlet, very numerous. Ovary adnate with the base of the calyx-tube. Fruit a woody capsule, adnate with and half the length of the 5-ribbed campanulate calyx-tube. Seeds numerous.—A. Rich., Fl. N.Z. 333; A. Cunn., Precurs. n. 559; Hook. f., Fl. N.Z. i. 67, t. 15; Handbk. 70; T. Kirk, Forest Fl. N.Z. t. 127. *M. speciosa,* Col. in Trans. N.Z.I. xxii. (1889) 463. *M. fulgida,* Banks and Sol. MSS. *Melaleuca florida,* G. Forst., Prod. n. 214. *Leptospermum scandens,* Forst., Char. Gen. 72.

Var. **aurata.** Calyx-tube long and narrow. Petals yellow. Stamens longer than usual, and more slender, at first yellow, but changing to reddish-yellow.—*M. aurata,* Col. in Trans. N.Z.I. xxiii. 385.

NORTH Island: Three Kings Islands and the North Cape to Cook Strait. SOUTH Island: Marlborough and Nelson. Sea-level to 3,000ft. Var. *aurata* has been observed on the Mokau River, Taranaki; Wellington; and Collingwood, Nelson; but only a single specimen in each locality. *Aka. Aka-tawhiwhi. Pua-tawhiwhi.* Nov. to April.

J. W. Hall informs me that the seeds require a year or more for their full development.

2. **M. lucida,** *A. Rich., Fl. N.Z.* 333. A shrub or tree, 40ft.–60ft. high, with trunk 2ft.–6ft. in diameter. Leaves silky in the young state, glabrous when mature, shortly petioled, varying from narrow-lanceolate-acuminate to ovate-lanceolate, very coriaceous, shining, the lower surface dotted with oil-glands. Flowers crimson, in short broad terminal cymes. Peduncles and pedicels short, silky. Calyx-tube shortly obconic, silky; lobes 5, oblong, persistent. Petals exceeding the sepals. Ovary sunk in the base of the calyx-tube, 3-celled; style shortly exceeding the stamens. Fruiting-calyx campanulate, longer than the 3-valved woody capsule.—A. Cunn., Precurs. n. 561; Hook. f., Fl. Antarc. i. 12; Fl. N.Z. i. 67; Handbk. 71; T. Kirk, Forest Fl. N.Z. t. 58; Menzies MS. in Herb., Hook. *M. umbellata,* Cav., Ic. iv. t. 337. *Agalmanthus umbellatus,* Homb. et Jacq., Voy. au Pole Sud., Bot. t. i. Dicot. *Melaleuca lucida,* G. Forst., Prod. n. 216 (not of Linn.).

NORTH Island: in hilly or mountain districts. From Whangarei (*Cheeseman*) to Cook Strait, but very local in many districts; Great Barrier Island. SOUTH Island: plentiful in alpine districts. STEWART Island. AUCKLAND Islands: abundant. CAMPBELL Island: very rare. Sea-level to 4,000ft. *Southern rata.* Dec., Jan.

A single specimen with golden-yellow flowers occurs in Arthur's Pass.

3. **M. Parkinsonii,** *Buch. in Trans. N.Z.I.* xv. (1882) 339, *t.* 28, *f.* 1. A prostrate straggling shrub or small tree, 20ft.–25ft. high, with trunk 2in.–6in. in diameter. Leaves opposite, shortly petioled, 1in.–2½in. long, ½in.–1in. broad, elliptical-ovate, acute, glabrous. Flowers in much-branched paniculate cymes, developed on the old wood, rarely axillary. Peduncles and pedicels slender, glabrous. Calyx-tube turbinate; lobes deltoid, subacute. Petals shortly ob-

long, rounded, exceeding the sepals. Ovary adnate with the base of the calyx-tube, 3-celled. Fruiting calyx-tube campanulate, 5-ribbed. Capsule woody, adnate with the calyx-tube for nearly half its length, 3-celled, centre much depressed.

SOUTH Island: Nelson: Whakamarina Ranges, Collingwood; Anatori Ranges, Heaphy River, *J. Dall! W. S. Hayward!* Dec., Jan.

Originally discovered by Mr. Hayward. I am indebted to Mr. Dall for much information respecting this beautiful plant.

4. **M. albiflora,** *Soland.* ex *Gaertn. Fruct.* i. 172, *t.* 34, *f.* 11. A scandent species, glabrous in all its parts. Leaves 1in.–3in. long, opposite, spreading; petiole short; blade elliptic-lanceolate, narrowed at both ends, acute or acuminate, coriaceous, shining. Flowers white, in terminal paniculate cymes or thyrsi. Calyx tubular or campanulate; lobes shortly oblong, obtuse, persistent. Petals exceeding the sepals, orbicular. Stamens exceedingly slender, delicate. Ovary adnate to the base of the calyx-tube. Capsule dehiscing to the base, ⅓in. long, turgid, urceolate, 3- rarely 4-celled and ribbed, crowned by the tubular calyx-limb; lobes reflexed, persistent.—Hook., Ic. Pl. t. 569; Hook. f., Fl. N.Z. i. 67; Handbk. 71. *M. diffusa*, A. Cunn., Precurs. n. 560 (not of Sm.).

NORTH Island: in forests, Mangonui and Hokianga to the East Cape district; Great Barrier Island. Dec., Jan.

The branchlets are very slender, and often pendulous or trailing.

5. **M. diffusa,** *Sm. in Trans. Linn. Soc.* iii. (1797) 268. A strong climber, attaining the tops of the highest trees. Leaves opposite, very shortly petioled, ½in.–1in. long, lanceolate or elliptical-oblong, obtuse. Cymes terminal, rarely axillary, often paniculate and leafy. Pedicels and peduncles glabrate, pubescent or setose; pedicels very short. Flowers numerous, scarlet. Calyx-tube narrow-oblong; limb broadly cup-shaped; lobes orbicular, shorter than the rounded petals. Ovary adnate to the base of the calyx-tube. Capsule thin, coriaceous, turgid, 6-ribbed, crowned with the short calyx-limb, 3-celled, dehiscing loculicidally to the base.—Hook. f., Fl. N.Z. i. 67; Handbk. 71.

NORTH Island: in forests from Ahipara to Te Aroha and the East Cape district, but often rare and local. Tararua Range, *W. B. D. Mantell!* Sept., Oct.

One of the most beautiful plants in the colony. The lower branches of the panicle diverge at an angle of 80°. J. W. Hall states that the seeds of this species mature in four or five months.

6. **M. hypericifolia,** *A. Cunn., Precurs. n.* 562. A rambling climber. Leaves glabrous, distichous, sessile or subsessile, scarcely coriaceous, oblong-ovate or ovate-lanceolate, acute obtuse or apiculate. Flowers in lateral cymes or racemes, small, 5–8-flowered. Pedicels very slender, glabrous. Calyx-tube glabrous or pubescent, obconic or pyriform, constricted below the apex, and abruptly dilated into a short cup-shaped limb; lobes ovate-deltoid. Petals small, orbicular, exceeding the sepals, white or pink. Filaments white or pink. Ovary adnate with the base of the calyx-tube. Capsule less than in. long,

3-lobed, turgid, crowned with the short funnel-shaped calyx-limb, 3-valved to the base.—Hook. f., Fl. N.Z. i. 67, t. 16; Handbk. 71. *M. subsimilis*, Col. in Trans. N.Z.I. xii. (1880) 361. *M. myrtifolia*, Banks and Sol. MSS.

From the North Cape to STEWART Island. Common in damp forests. Nov. to Jan.

The Banksian specimens have narrower and more acuminate leaves than any others seen by me, while the calyx-limb is larger and the segments more acute. In northern specimens the leaves are sometimes broadly ovate and slightly cordate. The flowers are produced on the old wood.

7. **M. Colensoi,** *Hook. f., Fl. N.Z.* i. 68. A very slender climbing shrub. Branchlets and young leaves tomentose or pubescent, glabrous when mature. Leaves distichous, often imbricating, sessile or subsessile, $\frac{1}{3}$in.–$\frac{3}{4}$in. long, ovate or ovate-lanceolate, acuminate, scarcely coriaceous. Flowers small, in terminal or rarely lateral trichotomous cymes, rarely exceeding 1in. in length. Peduncles and pedicels pubescent or silky. Calyx narrow, funnel-shaped, 2–3 times longer than the ovary, silky; lobes subulate, acute, as long as the minute orbicular sepals. Ovary adnate with the base of the calyx-tube. Capsule globose, pubescent, crowned with the funnel-shaped calyx-limb, 3-lobed, dehiscing loculicidally to the base.—Handbk. 72.

NORTH Island: from the Bay of Islands to Cook Strait, but often local. SOUTH Island: Marlborough and Nelson, probably extending to Westland. Banks Peninsula, *J. B. Armstrong*. This habitat requires confirmation.

Var. **pendens.** Often clothing straight trunks to the height of 50ft. or 60ft. with slender almost filiform pendulous branches. Leaves broadly ovate-lanceolate, acute or acuminate. Flowers white or with a faint pink tinge.—*M. pendens*, Col. in Trans. N.Z.I. xii. (1879) 360.—Whangape, Waikato, *T. K.*, and head of the Manawatu River. Jan., Feb. Mr. Colenso's specimens of this variety are the finest I have seen.

8. **M. robusta,** *A. Cunn., Precurs. n.* 557. A much-branched shrub, or gigantic tree, 50ft.–100ft. high; trunk 2ft. to 10ft. in diameter. Branchlets short, angular, puberulous. Leaves decussate, $1\frac{1}{2}$in. long, elliptic-lanceolate or oblong, obtuse, coriaceous; petioles short, glabrate or pubescent. Cymes broad, terminal, many-flowered. Peduncles and pedicels short, stout, pubescent. Calyx-tube obconic, short; lobes 5, deltoid, very short. Petals orbicular, exceeding the sepals. Ovary adnate with the lower part of the calyx-tube, shorter than the tube until the flower has withered. Capsule small, $\frac{1}{4}$in.–$\frac{1}{3}$in. long, half-superior, 3-lobed, the free portion equalling the calyx-tube, loculicidally 3-valved to the base. Seeds erect.—Hook. f., Fl. N.Z. i. 68, t. 17; Handbk. 72; T. Kirk, Forest Fl. N.Z. t. 128. *M. florida*, Hook., Bot. Mag. t. 1471 (not of Smith).

NORTH Island: from the Three Kings Islands and the North Cape to Cook Strait. SOUTH Island: Marlborough, Nelson, and Westland. Sea-level to nearly 3,000ft. *Rata*. Dec., Jan.

Var. **intermedia.** Twigs much stouter. Leaves excessively coriaceous, ovate-lanceolate. Cymes small. Pedicels and peduncles short and stout: Rangitoto Island, *T. K.*

Var. **retusa.** Leaves less than 1in. long, coriaceous, broadly elliptical, ovate, rounded at both ends, retuse. Cymes very broad. Petals narrow. Lowry Bay, Port Nicholson, *T. K.*

The rata is commonly epiphytic in its early stage, and gives off aerial roots, which ultimately form immense trunks from 2ft.–20ft. or more in diameter, and at length destroy the supporting tree. A full description of this peculiarity is given in the "Forest Flora of New Zealand."

9. **M. tomentosa,** *A. Rich., Fl. N.Z.* 336, *t.* 37. A large shrub or tree, 30ft.–70ft. high, with massive spreading arms and stout tomentose branchlets. Leaves decussate, 1in.–4in. long, shortly petioled, narrow-lanceolate, oblong or broadly oblong, usually narrowed to the apex, rounded at the base, the margins flat or recurved, clothed with white tomentum beneath or rarely glabrous. Flowers in large broad terminal cymes, often trichotomous. Peduncles and pedicels stout, and with the calyx densely tomentose. Calyx-tube broadly obconic; lobes short, deltoid, persistent. Petals narrow, exceeding the sepals, pubescent on the back. Base of the ovary adnate with the calyx-tube, 3-celled; style stout, exceeding the stamens. Capsule half-superior, woody, tomentose, 3-lobed, and 3-valved to the base, one-half sunk in the turbinate calyx-tube. Seeds erect.—A. Cunn., Precurs. n. 558; Hook. f., Fl. N.Z. i. 68; Handbk. 72; T. Kirk, Forest Fl. N.Z. t. 118. *M. excelsa,* Banks and Sol. MSS.

NORTH Island: chiefly on maritime cliffs and headlands, from the Three Kings Islands and the North Cape to Poverty Bay, Lake Taupo, and the Mimi River, Taranaki; Waikaremoana, East Coast, *Colenso.* On Tarawera Mountain prior to the eruption. 2,300ft. SOUTH Island: Queen Charlotte Sound, *Banks* and *Solander!* *Pohutukawa. Christmas flower.* Dec., Jan.

A magnificent tree. When growing in forests the leaves are nearly glabrous and the cymes very small. It is to be feared that some mistake has been made as to its occurrence in Queen Charlotte Sound.

10. **M. villosa,** *Sm. in Trans. Linn. Soc.* iii. (1797) 268. A tree, 40ft.–60ft. high, extremely variable in stature and in the shape of the leaves. Branchlets rather slender, downy or tomentose. Leaves decussate, ¾in.–1½in. long, shortly petioled, coriaceous, broadly ovate or obovate, rounded at the tip, white with appressed tomentum beneath; margins recurved. Flowers in small axillary or terminal cymes, the central flowers mostly sessile. Peduncles and pedicels downy or woolly. Calyx-tube short, broadly obconic, woolly; lobes minute, acute, persistent, black at the tip. Petals pubescent on the outer surface, exceeding the sepals. Filaments very slender. Ovary adnate with the base of the calyx-tube, 3-celled. Capsule woolly, 3-lobed, 3-valved, sessile in the broad calyx-tube, which closely invests its lower half.—*M. polymorpha,* Gaud. in Frey. Voy. Bot. 482, t. 85; Hook. f., Handbk. 73; T. Kirk, Forest Fl. N.Z. t. 119.

SUNDAY Island, Kermadec Group, *McGillivray, Cheeseman!* Sea-level to the tops of the hills, 1,700ft., according to Mr. Cheeseman, to whom I am indebted for specimens. Also in Lord Howe's Island, New Caledonia, Fiji, and other Polynesian Islands to the Sandwich Islands. Sept., Oct.

11. **M. scandens,** *Soland.* ex *Gaertn. Fruct.* i. 172, *t.* 34, *f.* 10. A climber, clothing the trunks of the loftiest trees, and giving off spreading branches above. Branchlets tomentose or hispid. Leaves distichous, subsessile, about ⅓in. long, orbicular orbicular-ovate or oblong-ovate, obtuse or rarely acute, very coriaceous, shining, covered with large dotted glands below; petiole and midrib pubescent. Flowers in 3-flowered pedunculate cymes, crowded in the upper axils, forming a more or less elongated terminal leafy panicle. Peduncles and pedicels pubescent; pedicels very short or 0. Calyx broadly turbinate; lobes

very short and broad, obtuse, persistent. Petals short, orbicular, white. Ovary adnate with the base of the calyx-tube, included until after flowering. Capsule globose, ⅙in. long, half-superior, the protruded portion loculicidally 3-valved.—Hook. f., Fl. N.Z. i. 69; Handbk. 73. *M. perforata*, A. Rich., Fl. N.Z. 334. *M. buxifolia*, A. Cunn., Precurs. n. 555. *M. vesiculata*, Col. in Trans. N.Z.I. xvi. (1884) 327. *M. tenuifolia*, Col. *l.c.* xxiv. (1891) 387. *Leptospermum perforatum*, Forst., Char. Gen. 36, n. 4. *Melaleuca perforata*, G. Forst., Prod. n. 212.

NORTH Island: Three Kings Islands and North Cape to Cook Strait. SOUTH Island: Marlborough and Nelson. Banks Peninsula, *J. B. Armstrong*. Lord Auckland Group, *Bolton*, in Handbk. (The two last habitats require confirmation. I suspect some mistake has occurred.) *Aka*. Feb., Mar.

In the young state the plant is often creeping, with slender glabrous stems and glabrous membranous leaves, when it is *M. tenuifolia*, Col. In others the twigs and young leaves are almost hispid, both leaves and calyx-tubes covered with unusually large black glands, when it is *M. vesiculata*, Col.; but both forms are connected by intermediates

3. MYRTUS, Linn.

Calyx-tube subglobose or turbinate; lobes 4–5, imbricate, usually persistent. Petals 4 or 5, spreading. Stamens many, in several series, exceeding the petals. Ovary wholly immersed in the calyx-tube, imperfectly 2–4-celled; ovules several in each cell. Fruit a globose or ovoid berry, crowned with the calyx-limb. Seeds few or many, rounded or reniform, angular; testa bony; embryo terete, curved or annular. Shrubs or small trees, with opposite leaves dotted with pellucid glands. Flowers axillary, solitary, rarely cymose. Peduncles slender.

SPECIES, about 115; most abundant in South America, one extending to Cape Horn, others to Mexico and the West India Islands. *M. communis* is common in South Europe and West Asia. About a dozen are found in Australia and four in New Zealand, all endemic.

Leaves 1in.–2in. long, tumid between the veins. Calyx-lobes acute .. .. 1. *M. bullata*.
Leaves ¾in.–1in. long, flat or slightly tumid. Calyx-lobes rounded .. .. 2. *M. Ralphii*.
Leaves obcordate, ⅛in.–⅓in. long. Calyx 4-lobed .. .. 3. *M. obcordata*.
Leaves obovate or oblong-ovate, ⅛in.–⅓in. long. Calyx 5-lobed .. .. 4. *M. pedunculata*.

1. **M. bullata,** *Soland.* ex *A. Cunn., Precurs. n.* 565. A shrub or rarely a small tree, 20ft.–30ft. high. Branchlets, young leaves, and petioles tomentose. Leaves shortly petioled, reddish-brown, ¾in.–2in. long, broadly ovate or orbicular-ovate, obtuse or acute, subcoriaceous, tumid between the veins. Flowers axillary, solitary. Peduncles longer or shorter than the leaves. Calyx-tube with 2 minute bracteoles at the base; lobes 4, acute. Petals orbicular, white. Berry ovoid, red, 2-celled, crowned with the calyx-limb. Seeds 2-seriate in each cell, reniform; testa bony.—Hook., Ic. Pl. t. 557; Hook. f., Fl. N.Z. i. 70; Handbk. 74; T. Kirk, Forest Fl. N.Z. t. 131.

NORTH Island: North Cape to Cook Strait, frequent. SOUTH Island: Marlborough and Nelson, rare. Sea-level to 1,800ft. *Ramarama*. Dec., Jan.

The petals and sepals are dotted with minute warts.

2. **M. Ralphii,** *Hook. f., Fl. N.Z.* ii. 329. A much-branched shrub, or rarely a small tree, 20ft.–25ft. high. Leaves opposite, mostly orbicular-ovate, flat or slightly tumid between the veins, acute or obtuse. Peduncles slender, longer or shorter than the leaves. Calyx superior, 4-lobed; lobes rounded. Berry 2-celled. Seeds 2-seriate.—Handbk. 74; T. Kirk, Forest Fl. N.Z. t. 94.

NORTH Island: from the South Head of the Manukau Harbour to Cook Strait, but often local. SOUTH Island: Marlborough and Nelson, local. Sea-level to 1,200ft. Dec., Jan.

Similar to *M. bullata*, but the leaves are usually green, smaller; the calyx-lobes are rounded, and the seeds 1-seriate in each cell.

3. **M. obcordata,** *Hook. f., Fl. N.Z.* i. 71. A small shrub or tree, 5ft.–15ft. high. Branchlets pubescent. Leaves opposite or fascicled, less than ½in. long, narrowed into a very short petiole, slightly coriaceous, orbicular or obovate, obcordate, glabrous. Peduncles solitary, equalling or exceeding the leaves. Flowers small, ¼in.–⅓in. long. Calyx superior, 4-lobed; lobes acute. Ovary 2-celled. Berry globose or ovoid, red or violet, 2-celled. Seeds 1 or 2 in each cell; testa bony.—Handbk. 74; T. Kirk, Forest Fl. N.Z. t. 70. *Eugenia obcordata*, Raoul in Ann. Sc. Nat. Ser. III. ii. (1844) 122.

NORTH and SOUTH Islands: head of the Hauraki Gulf to Southland. Often local. (STEWART Island?) Sea-level to 1,500ft. Dec., Jan.

4. **M. pedunculata,** *Hook. f. in Hook. Ic. Pl. t.* 629. A compact or straggling shrub, sometimes 20ft. high. Branchlets glabrous. Leaves opposite, obovate or ovate-oblong, usually rounded at the apex. Peduncle glabrous, longer or shorter than the leaves. Calyx glabrous, 5-lobed. Ovary 2-celled. Berry small, yellow or red, 2-celled, with 2 or 4 bony seeds.—Handbk. 74; T. Kirk, Forest Fl. N.Z. t. 112.

From the Bay of Islands and Hokianga southwards to STEWART Island, but often local. Sea-level to 2,300ft. *Rohutu*. Dec., Jan.

This species closely resembles small-leaved states of *M. obcordata*, but is easily distinguished by the obtuse or rarely subacute leaves, glabrous twigs, and especially by the 5-lobed calyx.

4. EUGENIA, Linn.

Calyx-tube turbinate, not exceeding the ovary; lobes 4 or 5. Petals 4–5, often falling before expansion. Stamens many, in several series. Ovary 2–3-celled. Fruit a berry. Seeds solitary and globose, or several and angular; embryo thick, fleshy; cotyledons usually inseparable. Shrubs or trees, with opposite leaves. Flowers axillary, solitary, or in axillary or terminal cymes or corymbs.

A large genus, comprising 600 species, found in all tropical and subtropical regions. The only New Zealand species is endemic.

1. **E. Maire,** *A. Cunn., Precurs. n.* 564. A small conical tree, 20ft.–50ft. high, glabrous in all its parts; bark white. Branchlets very slender, 4-angled. Leaves opposite, 1in.–2in. long, membranous, oblong-lanceolate or lanceolate, acuminate, narrowed into the slender petiole. Flowers white, sometimes unisexual, in axillary or terminal many-flowered corymbs. Pedicels slender,

thickened upwards, jointed to the rhachis. Calyx with 5 minute deciduous teeth. Petals 5, often falling before expansion. Ovary wholly immersed in the calyx-tube, 2-celled, many-ovuled. Berry red, unequally ovoid, crowned with the thickened rim of the calyx, 1-celled; seed solitary, large; testa coriaceous or almost bony.—Fl. N.Z. i. 71; Handbk. 74; T. Kirk, Forest Fl. N.Z. t. 122.

NORTH Island: in swampy places, Mangonui to Cook Strait, but often local. SOUTH Island: Marlborough and Nelson, rare, *J. Rutland!* Ascends to 1,300ft. *Maire-tawhake.* June, July.

Order—* LYTHRARIEAE.

Calyx-tube inferior, 4–5-toothed, persistent, valvate in bud. Petals 4–5, inserted at the mouth of the tube, rarely 0. Stamens 4–14, inserted below the petals, equal or unequal; anthers versatile. Ovary superior, enclosed in the calyx-tube, 2 or more celled, or 1-celled by absorption of the placentas; style simple. Seeds anatropous, without endosperm. Herbs, shrubs, or trees, usually with opposite entire leaves and hermaphrodite axillary or whorled regular or irregular flowers, which are sometimes di- or tri-morphic.

* PEPLIS, Linn.

Calyx campanulate; teeth 12, 6 subulate, alternating with 6 broader erect teeth. Petals 6, fugacious, or 0. Stamens 6–12, perigynous. Ovary 2-celled; style minute; stigma capitate; ovules numerous; placentas attached to the septum. Capsule 2-celled, 2-valved, many-seeded. Prostrate or suberect annuals. Leaves alternate or opposite, exstipulate. Flowers minute.

* **P. Portula,** *L., Sp. Pl.* 332. Stems brittle, 1in.–6in. long, prostrate or ascending, creeping, angular. Leaves opposite, narrow-obovate, shortly stalked. Flowers axillary, solitary, sessile. Calyx-tube very short, 12-ribbed. Capsule globose, exceeding the calyx-tube.

SOUTH Island: in damp places. Tuapeka, Tapanui, &c., Otago, *Petrie! Water-purslane.* Dec. to Feb. Europe, North Africa.

* LYTHRUM, Linn.

Calyx tubular, striate; teeth 8–12, 4–6 subulate, alternating with broader erect teeth. Petals 4–6, minute, wrinkled in bud, rarely 0. Stamens as many or twice as many as the petals, 1–2-seriate. Capsule 2-celled, 2-valved. Perennial or annual herbs, with entire leaves and solitary or cymose flowers, which are often trimorphic.

Flowers minute. Calyx-tube strongly ribbed in fruit .. .. .. 1. *L. Hyssopifolia.*
Flowers $\frac{1}{4}$in.–$\frac{1}{3}$in. in diameter, showy. Calyx-tube faintly ribbed in fruit .. 2. *L. Graefferi.*

* **L. Hyssopifolia,** *L., Sp. Pl.* 447. Annual, 6in.–18in. high, branched, suberect or procumbent. Leaves sessile, alternate, narrow-linear or oblong-lanceolate, obtuse, cuneate at the base. Flowers minute, axillary, solitary. Calyx-tube with 2 minute subulate bracteoles. Petals small, pale-purple. Stamens 6, included. Calyx-tube strongly ribbed in fruit.

NORTH and SOUTH Islands: Bay of Islands to Otago: copiously naturalised in ditches and watery places. *Small loose-strife.* Dec. to Feb. Europe.

* **L. Graefferi,** *Tenore, Prod. Fl. Nap. p.* lxviii. Annual. Stem 4in.–6in. high, branched, very slender. Leaves alternate, narrow-linear-lanceolate or oblong, acute. Flowers solitary, $\frac{1}{4}$in.–$\frac{1}{6}$in. in diameter, on short pedicels; bracteoles minute. Petals 5–6, obovate, purple, apiculate. Calyx-tube faintly ribbed in fruit.

NORTH Island: naturalised in the Auckland District; local, but increasing. Matakana, Lower Waikato, &c. SOUTH Island: near Greymouth, *R. Helms!*

ORDER XXX.—ONAGRARIEAE.

Calyx-tube often elongated, adnate with the ovary, and rarely produced beyond it; lobes 2–4, rarely more, valvate. Petals 4, free, inserted at the mouth of the calyx-tube, convolute in the bud, rarely 0. Stamens 4 or 8, inserted with the petals. Ovary inferior, 2–4-celled, rarely 1-celled; style straight; stigma 2–4-lobed or capitate or clavate; ovules numerous in the angles of the cells, anatropous. Seeds small, without endosperm. Embryo straight. Herbs, rarely shrubs or trees. Leaves opposite or alternate, exstipulate. Flowers hermaphrodite, symmetrical, axillary, solitary or geminate, racemose or spicate, sometimes trimorphic.

GENERA, about 25. SPECIES, about 360. Most plentiful in cold and temperate countries, but also between the tropics.

1. EPILOBIUM. Herbs. Fruit a capsule. Seeds with a coma.

* OENOTHERA. Herbs. Fruit a capsule. Seeds naked.

2. FUCHSIA. Shrubs or trees. Fruit a berry.

1. EPILOBIUM, Linn.

Calyx-tube adnate with the tetragonous ovary, and rarely produced above it; lobes 4, deciduous. Petals 4, spreading, perigynous. Stamens 8, inserted with the petals. Ovary inferior, 4-celled; style short; stigma obliquely clavate, capitate or 4-lobed. Capsule linear, elongated, 4-celled, 4-valved, many-seeded, dehiscing loculicidally from the apex downwards. Seeds with a tuft of white hairs at the apex. Herbs, rarely suffruticose, erect, prostrate, or creeping. Leaves opposite or alternate, exstipulate. Flowers sometimes drooping in bud, afterwards erect, axillary or forming terminal spikes or racemes. Peduncles often elongating in fruit.

A large genus, common in all temperate countries, but rare in the tropics. By some authors the species are estimated at about 50; by others at upwards of 160. They are extremely variable in all countries, and hybrid forms are not unfrequent. For the New Zealand species I have availed myself largely of the beautiful and elaborate monograph of Professor Haussknecht, and gladly acknowledge the assistance he has kindly rendered in the elaboration of several doubtful forms.

About five of the New Zealand species extend to Australia, and another to the lofty mountains of New Guinea; the others are endemic.

It is desirable that the student should examine good series of specimens before coming to a determination, and that he should devote special attention to the vernal and autumnal states of each species.

The following arrangement of the species is adapted from Haussknecht:—

I. SIMILES. Stems erect, herbaceous or slightly woody at the base. Flowers numerous.

* *Leaves sessile.*

† Seeds smooth.

Leaves ovate, glabrous. Flowers large .. .. 1. *E. chionanthum.*

†† Seeds papillose, tomentose or pubescent.

Leaves distant, broad, toothed .. .. 2. *E. junceum.*

Leaves narrow, toothed .. .. var. *campestre.*

Almost glabrous. Flowers large, white .. .. .. 3. *E. pallidiflorum.*

Glabrous. Stems red. Flowers small, red .. .. .. 4. *E. Billardierianum.*

** *Leaves petioled.*

Erect, with slender branches. Leaves ovate, mostly alternate .. 5. *E. pubens.*

Erect, rigid, simple or sparingly branched. Leaves distant, mostly opposite .. .. 6. *E. perplexum.*

II. MICROPHYLLAE. Herbaceous. Stems slender, creeping, short, ascending at the tips. Flowers few, terminal. Leaves small, suborbicular.

† *Peduncles very short.*

Densely foliaceous. Petioles short, broad .. .. .. 7. *E. confertifolium.*

Branches slender. Leaves distant, $\frac{1}{3}$in. long .. .. 8. *E. Tasmanicum.*

†† *Peduncles elongated.*

Leaves narrow-linear. Peduncles usually solitary, terminal .. .. 9. *E. tenuipes.*

Leaves linear-oblong. Peduncles very slender .. .. .. 10. *E. Hectori.*

Leaves suborbicular or orbicular-ovate .. .. 11. *E. alsinoides.*

Leaves ovate; petioles very short .. .. 12. *E. chloraefolium.*

Leaves opposite, distant, orbicular, membranous .. .. .. 13. *E. rotundifolium.*

Leaves opposite, oblong-ovate, very membranous .. .. .. 14. *E. insulare.*

III. SPARSIFLORAE. Herbaceous, slender, creeping. Leaves small, orbicular or rounded. Flowers few, solitary, in the axils of intermediate leaves. Peduncles elongated.

Leaves orbicular, uniform, sharply toothed .. .. 15. *E. linnaeoides.*

Leaves suborbicular, petiolate or sessile; midrib and veins obscure .. 16. *E. nummularifolium.*

Leaves ovate; midrib and veins obscure .. .. var. *ovata.*

Leaves all sessile; midrib and veins prominent .. .. .. var. *pedunculare.*

Leaves suborbicular, acutely toothed; midrib and veins prominent .. var. *caespitosa.*

Leaves orbicular-oblong, coriaceous, purple beneath .. .. .. 17. *E. purpuratum.*

IV. DERMATOPHYLLAE. Small species, more or less suffruticose. Leaves coriaceous or rigid. Flowers few, terminal or subterminal.

† *Peduncles short, or capsules sessile.*

Erect, glabrous. Leaves linear-oblong or ovate-oblong. Stigma obcordate 18. *E. glabellum.*

Suberect. Leaves ovate-oblong. Stigma capitate .. .. .. 19. *E. Novae-Zelandiae.*

Erect, 2in.–5in. high. Leaves small, ovate or oblong. Stigma clavate .. 20. *E. Krulleanum.*

Stems very numerous. Leaves erect, dense, sessile .. .. .. 21. *E. erubescens.*

Stems stout, suberect or prostrate. Flowers $\frac{1}{2}$in.–$\frac{2}{3}$in. in diameter .. 22. *E. vernicosum.*

Stems numerous. Leaves toothed, erect. Capsule sessile .. .. 23. *E. pycnostachyum.*

Stems woody. Leaves tipped with a horny apiculus, red .. .. 24. *E. brevipes.*

Stems branched above. Leaves oblong, sharply toothed. Peduncles shorter than the leaves .. .. 25. *E. haloragifolium.*

Stems strict, black. Leaves deeply toothed. Capsule sessile .. .. 26. *E. melanocaulon.*

Stems very slender, branched above. Capsule shortly narrowed above .. 27. *E. polyclonum.*

Stems branched. Capsule sessile, narrowed into a short beak .. .. 28. *E. rostratum.*

Stems numerous, filiform. Leaves orbicular-ovate, $\frac{1}{8}$in.–$\frac{1}{3}$in. in diameter 29. *E. microphyllum.*

†† *Peduncles elongated in fruit.*

Stems numerous, wiry, ascending. Leaves broadly ovate. Flowers subterminal .. .. 30. *E. gracilipes.*

Stems stout. Leaves dense, obovate, coriaceous. Flowers red .. 31. *E. crassum.*

Stems branched, flaccid. Leaves distant, ovate. Flowers large, white 32. *E. macropus.*

1. **E. chionanthum,** *Haussk. in Oester. Bot. Zeitschr.* xxix. (1879) 149. Stems simple, ascending, 1ft.–2ft. high, puberulous above, decumbent at the

base, soboliferous in the early autumn. Leaves distant, sessile or subsessile, gradually diminishing in size upwards, ovate-oblong, obtuse, narrowed at the base, minutely denticulate, glabrous. Flowers 1–6, nearly ½in. in diameter, solitary in the axils of distant bracts. Pedicels puberulous. Calyx-segments acute or subacute, puberulous. Petals obcordate. Stigma capitate. Capsule pubescent or puberulous. Seeds smooth.—Haussk., Monog. Epiloh. 287, t. 22, f. 92 A, B.

NORTH and SOUTH Islands: in swamps, from Spirits Bay to Southland. Sea-level to 1,500ft. Nov. to Feb.

A very distinct species, characterised by its soboliferous habit, pale-green glabrous leaves, large flowers, and smooth seeds. I have never seen two fully-expanded flowers on the same stem. Originally discovered by Dr. Sinclair.

2. **E. junceum,** *Soland. in G. Forst., Prod. n.* 516. Stems erect or ascending, stout or slender, decumbent and woody at the base, 1ft.–2ft. high. Leaves sessile, pubescent, tomentose, coriaceous, alternate or opposite, oblong-lanceolate or linear-lanceolate, narrowed at the base, truncate at the apex or acute; margins denticulate or sinuate-toothed. Flowers usually numerous, terminal or subterminal, purple. Peduncles usually shorter than the leaves. Calyx-lobes acute. Stigma clavate. Fruiting peduncles longer or shorter than the leaves. Capsule glabrate, pubescent or pilose. Testa minutely papillose.—A. Cunn., Precurs. n. 551; Hook. f., Fl. N.Z. i. 60; Handbk. 80; Benth., Fl. Austr. iii. 304; Haussk., Monog. Epilob. 289.

Var. **cinereum.** Stems decumbent and woody at base, often excessively branched. Leaves crowded, narrow-linear, entire, denticulate or toothed, acute, often mucronate, grey with appressed pubescence. Flowers small. Capsule very slender, 1½in.–2in. long, pubescent. Flowering peduncles slender, exceeding the leaves. Seeds oblong-ovoid, papillose.—Haussk., Monog. Epilob. 289. *E. cinereum*, A. Rich., Fl. N.Z. 320; A. Cunn., Precurs. n. 544. *E. virgatum* and *E. confertium*, A. Cunn., Precurs. n. 547 and 549.

Var. **hirtigerum.** Stems strict, erect, pubescent or tomentose. Leaves uniform, mostly alternate, usually appressed to the stem, dense, linear, sparingly denticulate or sinuate-toothed, obtuse or mucronulate. Flowers erect in the uppermost axils. Stigma very small. Capsule strict, villous. Seeds densely papillose.—*E. hirtigerum*, A. Cunn., Precurs. n. 546; Haussk., Monog. Epilob. 291.

NORTH and SOUTH Islands: from the Three Kings Islands and the North Cape to Southland. Sea-level to 3,000ft. Nov. to Feb. Also in Australia.

An extremely variable plant, of which it would be easy to make numerous varieties, scarcely any of which could be separated by hard-and-fast lines.

3. **E. pallidiflorum,** *Soland.* ex *A. Cunn., Precurs. n.* 550. Stems slightly decumbent and woody at the base, emitting numerous slender stolons, ultimately erect, 1ft.–3ft. high, simple or branched. Leaves opposite or the uppermost alternate, sessile or subsessile, 1in.–3in. long, glabrous, often semi-amplexicaul, linear or lanceolate-oblong, acute, remotely toothed or entire. Flowers large, nodding in bud, white, ¾in. in diameter, forming an elongated leafy raceme. Peduncles always shorter than the leaves. Calyx-lobes long, acute, half the length of the corolla, pubescent. Stigma shortly clavate. Capsule 2in.–3in. long, hoary. Seeds minutely papillose.—Hook. f., Fl. N.Z. i. 61; Handbk. 81; Benth., Fl. Austr. iii. 305; Haussk., Monog. Epilob. 292. *E. macranthum*, Hook., Ic. Pl. t. 297.

Var. **stricta.** Stem simple, erect. Leaves narrow. Flowers small, purple.

NORTH and SOUTH Islands: in watery places, from the North Cape to Ruapuke Island, Foveaux Strait. Dec. to Feb. Also in Australia.

The largest New Zealand species.

4. **E. Billardierianum,** *Ser. in DC., Prod.* iii. 41. Stems decumbent and woody at the base, giving off numerous autumnal stolons, erect, 6in.–30in. high, simple or branched, red, bifariously pubescent. Leaves 1in.–2in. long, alternate and opposite, sessile or rarely subsessile, ovate-oblong or linear-oblong, obtuse, narrowed to the base or sometimes rounded, diminishing in size upwards, closely unevenly denticulate. Flowers numerous, reddish, small, crowded in the axils of the upper leaves, erect. Calyx-lobes ovate-lanceolate, apiculate. Stigma shortly clavate. Fruiting peduncle usually exceeding the leaves. Capsule pubescent or pilose. Seeds minutely papillose.—Haussk., Monog. Epilob. 293. *E. tetragonum*, Hook. f., Fl. N.Z. i. 60; Handbk. 80; Benth., Fl. Austr. iii. 305. *E. rubricaule*, Banks and Sol. MSS.!

Var. **simplex.** Stems simple, often numerous. Leaves narrow-linear-oblong.

NORTH and SOUTH Islands: from the North Cape to Stewart Island; CHATHAM Islands. Ascends to 2,000ft. Dec. to Feb. Also in Australia.

The robust habit, red stems, and numerous rosy flowers serve to distinguish this species at sight. It is remarkable that it does not seem to have been observed by the Cunninghams.

5. **E. pubens,** *A. Rich., Fl.* 329, *t.* 36, *f.* 1. Stems shortly decumbent and woody at base, developing small winter rosettes, ascending, 6in.–24in. high, much-branched above, slender, pubescent or puberulous. Leaves usually alternate, membranous, pubescent above and below; petioles very slender, ovate-oblong, obtuse, denticulate or distantly toothed. Flowers numerous in the axils of the upper leaves, small, white or pink, erect. Peduncles pubescent, always shorter than the leaves. Calyx-lobes lanceolate, acute, pubescent. Stigma clavate. Capsule 1in.–2in. long, pubescent. Seeds papillose.—A. Cunn., Precurs. n. 543; Hook. f., Fl. N.Z. i. 61; Handbk. 80; Haussk., Monog. Epilob. 295. *E. petiolatum*, A. Cunn. in Herb. n. 137.

Var. **minor.** Leaves smaller, oblong-obtuse.—Haussk. *l.c.*

NORTH and SOUTH Islands: from the North Cape to Stewart Island; CHATHAM Islands. Sea-level to 3,000ft. Nov. to Jan. Also in Australia.

6. **E. perplexum,** *n. s.* Stems slightly decumbent and woody at the base, emitting one or two stolons, ascending, 9in.–18in. high, rigid, terete, bifariously pubescent. Leaves opposite, distant, broadly ovate, obtuse or rarely acute, abruptly narrowed into the very short petiole, remotely toothed, glabrous or pubescent beneath. Flowers few, white, nodding in bud, ⅓in.–½in. in diameter, solitary in the axils of the upper leaves or terminating short branchlets. Peduncles and calyx-tube pubescent or puberulous. Calyx-lobes ovate-lanceolate, acute or subacute. Style thickened upwards; stigma broadly oblong, emarginate. Fruiting peduncles equalling or shortly exceeding the leaves. Capsule 2in. long, pubescent. Seeds scarcely papillose.

NORTH Island: East Cape, *J. B. Lee!* Ruahine Range, *Petrie!* Tararua Range, *T. P. Arnold.* SOUTH Island: Canterbury, *Cockayne!* Kelly's Creek, Westland, 2,500ft., *Petrie!*

This plant approaches *E. chloraefolium*, but is more robust, never branched at the base, and the capsules are longer; it also differs in the emarginate stigma. Mr. Petrie refers it to *E. Gunnianum*, Haussk. (*E. Billardierianum*, Hook. f., Fl. Tasm. i. t. 21); but that species has oblong-lanceolate leaves, larger purple flowers, shorter peduncles, and distinctly papillose seeds: there is no evidence that it has been found in New Zealand.

7. **E. confertifolium,** *Hook. f., Fl. Antarc.* i. 10. Stems prostrate, rooting at the nodes, ascending at the tips, sparingly branched; branches 1in.–4in. long. Leaves opposite, often imbricating, ¼in.–½in. long, rather fleshy, glabrous or glabrate, linear-oblong, oblong, or obovate-oblong, obtuse, shining, entire or with few minute teeth, sessile or narrowed at the base into a short broad sheathing petiole. Flowers few, in the axils of the upper leaves, small, red. Calyx-lobes lanceolate, acute. Petals shortly exceeding the sepals. Stigma shortly clavate, scarcely emarginate. Fruiting peduncles equalling or exceeding the leaves. Capsule ½in.–¾in. long, strict, glabrous. Seeds papillose.—Hook., Ic. Pl. t. 685; Hook. f., Handbk. 78; Benth., Fl. Austr. iii. 304; Haussk., Monog. Epilob. 296.

SOUTH Island: in alpine districts, from Nelson to Southland. 1,000ft. to 5,500ft. AUCKLAND and CAMPBELL Islands, *Hook. f.!* ANTIPODES Island, *T. K.!* Sea-level to 2,000ft. Jan., Feb. Also in Tasmania.

8. **E. Tasmanicum,** *Haussk., Monog. Epilob.* 296, *t.* 20, *f.* 84. Stems prostrate, rooting at the nodes, ascending at the tips, 1in.–4in. high. Leaves opposite, shortly petioled, spreading, about ⅓in. long, pale, oblong-ovate, almost entire or remotely toothed, obtuse, glabrous, rather fleshy. Flowers in the upper axils, few, erect, white or red. Calyx-lobes lanceolate. Fruiting peduncles about as long as the leaves, erect. Capsule (immature), ¾in.–1in. long, glabrous. Seeds not seen.

SOUTH Island: Rotoiti, Nelson. Mountains above Lake Harris, Otago, 3,000ft. to 4,000ft., *T. K.* Jan., Feb. Also in Tasmania.

Distinguished from *E. confertifolium* by the pale colour, slender spreading habit, by the distant leaves with obvious petioles, which are never sheathing or imbricate, and by the longer stigma. The capsules on my specimens are immature.

9. **E. tenuipes,** *Hook. f., Fl. N.Z.* i. 59. Stems short, 1in.–3in. long, prostrate, emitting short stolons, ascending at the tips, slender. Leaves mostly alternate, ascending, narrow-linear-oblong, the lower obtuse, the upper acute, narrowed at the base, remotely denticulate or entire, glabrous. Flowers white, erect, small, terminal and solitary, or rarely 2–3 in the upper axils. Flowering peduncles much shorter than the leaves. Calyx-lobes lanceolate, shortly acuminate, glabrous. Stigma clavate. Fruiting peduncles 2in.–3in. long, very slender. Capsule very slender, 1in. long. Seeds glabrous.—Fl. Tasm. i. 116; Haussk., Monog. Epilob. 297, t. xx. f. 83. *E. confertifolium, β tenuipes*, Hook. f., Handbk. 297.

NORTH Island: on boggy ground in alpine situations; local. Ruahine Mountains, *A. Hamilton!* Head of Wairarapa Valley, *Colenso* (Handbk.). SOUTH Island: Nelson Mountains, *Travers*. Broken River basin, Canterbury, *T. K.* 2,000ft.–5,500ft. Dec. Also in Tasmania.

Distinguished from *E. confertifolium* by the narrow-linear erect subimbricate reddish leaves, elongated solitary fruiting-peduncles, acuminate sepals, and smooth seeds. Its flowering season is remarkably brief.

10. **E. Hectori,** *Haussk., Monog. Epilob.* 298, *t.* xiv. *f.* 82A. Stems very slender, 3in.–6in. high, decumbent and much branched from the base, ascending. Leaves usually pubescent, mostly opposite, spreading, sessile or very shortly petioled, shortly oblong or linear-oblong, about ¼in.–⅓in. long, obtuse, entire or remotely denticulate, usually glabrous. Flowers numerous, in the upper axils, small, white. Peduncles very slender, exceeding the leaves. Calyx-lobes ovate-lanceolate, acute, shorter than the corolla, pubescent or pilose. Stigma clavate. Capsule ½in.–1in. long, purplish, obscurely tetragonous, abruptly contracted just below the apex, pubescent, pilose or rarely glabrous. Seeds smooth.—*E. atriplicifolium,* A. Cunn., Precurs. n. 542?

NORTH and SOUTH Islands: from the Bay of Islands to Southland. Probably common, although apparently local, but often confused with *E. alsinoides*. Sea-level to 2,800ft. or more. Dec. to Feb.

Mr. Petrie sends a variety with larger flowers, the corolla fully one-third longer than the calyx.

11. **E. alsinoides,** *A. Cunn., Precurs. n.* 540. Stems very slender, decumbent at the base, ascending or erect, 5in.–10in. long, glabrous or pubescent. Leaves usually opposite, very shortly petioled, ¼in.–½in. long, suborbicular or orbicular-ovate, obtuse, quite entire or remotely denticulate, glabrous. Flowers few, in the upper axils, small, erect. Peduncles slender. Calyx ovate-lanceolate, acute, equalling the corolla. Stigma broadly clavate. Fruiting-peduncles long and slender. Capsule 1in.–1½in. long, puberulous or pubescent. Seeds papillose.—Hook. f., Fl. N.Z. i. 59; Handbk. 78; Haussk., Monog. Epilob. 299, t. xxiii. f. 97. *E. thymifolium,* R. Cunn. in A. Cunn. Precurs. n. 539; Haussk., Monog. Epilob. 297.

NORTH and SOUTH Islands: Mangonui to Southland; STEWART Island; CHATHAM Islands; ANTIPODES Island: *T. K.* Sea-level to 2,000ft. Nov. to Feb.

Professor Haussknecht considers Cunningham's *E. thymifolium* a distinct species. An examination of one of the original specimens, for which I am indebted to Sir Joseph Hooker, does not enable me to confirm this view.

12. **E. chloraefolium,** *Haussk. in Oestr. Bot. Zeitsch.* xxix. (1879) 118. Stems 9in.–15in. high, slender, usually much branched below, decumbent at the base and emitting suberect stolons. Leaves opposite, shortly petioled, ascending, glabrate or glabrous, ovate truncate or rounded, or slightly cordate at the base, obtuse, obscurely remotely denticulate. Calyx-lobes ovate-lanceolate, glabrous, shorter than the petals. Stigma broadly oblong, rounded at the apex. Peduncles exceeding the leaves. Capsule about 1½in. long, pubescent. Seeds minutely papillose.—Monog. Epilob. 299, t. xix. f. 81A. *E. caespitosum,* T. Kirk, MSS.

SOUTH Island: mountains of Nelson, *Sinclair!* Spencer Mountains, *T. K.* Mount Torlesse, *Haast!* Broken River basin, *T. K.* Upper Waimakariri, *Cockayne! Cheeseman!* Kelly's Hill, Westland, *Cockayne!* Otago, *Hector* and *Buchanan*. 2,500ft. to 4,300ft.

13. **E. rotundifolium,** *G. Forst., Prod. n.* 161. Stems decumbent at the base, emitting one or more stolons, ascending, erect or prostrate, herbaceous, 3in.–15in. high, pubescent or glabrous. Leaves mostly in opposite pairs or the

upper alternate, rather fleshy, shortly petiolate, orbicular or orbicular-oblong, often purple beneath, glabrous or puberulous, unequally toothed; teeth short, acute. Flowers in the upper axils. Calyx-lobes oblong-lanceolate, subacute or apiculate, glabrous or puberulous, shorter than the corolla. Stigma narrow-clavate. Fruiting-peduncle equalling or exceeding the capsule, very slender. Capsule glabrate or pubescent. Seeds densely papillose.—DC., Prod. iii. 43; A. Rich., Fl. N.Z. 326; A. Cunn., Precurs. n. 538; Hook. f., Fl. N.Z. i. 58; Handbk. 78; Haussk., Monog. Epilob. 300. *E. flaccidum*, Banks and Sol. MSS.

NORTH and SOUTH Islands: from the North Cape to STEWART Island; CHATHAM Islands. Oct. to Feb.

In the winter state the stem is invariably creeping, and rooting at the nodes. Early in October the apex is prolonged into an erect or suberect stem, and the old creeping stem perishes. When growing in swamps it is invariably creeping, and resembles *E. linnaeoides*, all the leaves being opposite; but the flowers are developed from the upper axils, as in the type.

14. **E. insulare,** *Haussk., Monog.* 300. Stems slender, weak, prostrate or suberect, pubescent or puberulous, simple or sparingly branched, emitting stolons from the base. Leaves opposite, distant, shortly petioled, broadly ovate or oblong-ovate, obtuse or rounded at the tips, entire or remotely sinuate-toothed or denticulate, glabrous or glabrate. Flowers few, in the upper axils, small, erect, white. Calyx-lobes oblong, apiculate, one-third shorter than the corolla. Stigma clavate. Fruiting-peduncle 1in.–4in. long, slender. Capsule 1in.–2in. long, pubescent or glabrate. Seeds smooth.—*E. distans*, Petrie, MSS.

NORTH Island: East Cape, *Colenso*, n. 386, Herb. Kew (Haussknecht). SOUTH Island: Waimea River, Nelson, *Monro*, n. 76. New Brighton, Canterbury, *Petrie!* Lumsden to Dunedin, Kaitangata, and Catlin's River, &c., Otago, *Petrie!* In lowland boggy situations.

15. **E. linnaeoides,** *Hook. f., Fl. Antarc.* i. 10, *t.* 6. Herbaceous, slender, prostrate, creeping, simple or branched, 2in.–6in. long, perfectly glabrous. Leaves all opposite, flaccid and membranous, orbicular, denticulate or crose-denticulate. Flowers numerous or few, solitary in the middle or upper axils, small, erect, white or pink. Calyx-lobes lanceolate, apiculate, nearly equalling the corolla. Stigma clavate, rounded. Fruiting-peduncles, &c., 2in.–4in. long. Capsule ¾in.–2in. long, glabrous. Seeds papillose.—Fl. N.Z. i. 58; Handbk. 58; Haussk., Monog. Epilob. 301.

NORTH and SOUTH Islands: Ruahine Range, &c., *Colenso*. Tararua Range, *Buchanan*. More frequent in the Southern Alps. STEWART Island, AUCKLAND and CAMPBELL Islands, ANTIPODES Island, *T. K.* MACQUARIE Island, *A. Hamilton!* Sea-level to 4,000ft. Nov. to Feb.

This can only be distinguished from *E. rotundifolium* by its invariably prostrate habit, smaller size, uniform leaves, less numerous flowers, and more densely papillose seeds. Intermediate forms are common on Stewart Island.

16. **E. nummularifolium,** *R. Cunn.* ex *A. Cunn., Precurs. n.* 535. Stems cæspitose, slender, prostrate, creeping and rooting at the nodes, filiform, 3in.–7in. long, sparingly branched, ascending at the tips, pubescent or puberulous. Leaves opposite, in distant pairs, suborbicular, rounded at the apex, very shortly petioled or nearly sessile, rather fleshy or almost membranous, minutely

remotely denticulate or erose-denticulate; midrib and veins obscure. Flowers few, distant, erect. Calyx-lobes ovate-lanceolate, equalling the corolla. Stigma clavate. Fruiting-peduncles 3in.–4in. long, slender. Capsule 1in.–1½in. long, pubescent or glabrous, drooping, ultimately erect. Seeds papillose.—Hook. f., Fl. N.Z. i. 57; Handbk. 77; Haussk., Monog. Epilob. 302. *E. pedunculare,* A. Cunn., Precurs. n. 536. *E. pendulum,* Banks and Sol. MSS.

NORTH and SOUTH Islands: from the Three Kings Islands and the North Cape to STEWART Island; CHATHAM Islands. Ascends to 3,000ft. Oct. to March. Also on mountains in New Guinea, *Mueller.*

Var. **ovatum.** Leaves all ovate, truncate at base, obtuse, submembranous. Peduncles few, 1in.–2in. long, slender. On moist rocks, Papakura.

Subsp. **nertericides** (*sp.*), A. Cunn., Precurs. n. 541. Leaves distichous, ⅛in.–¼in. in diameter, dense, suborbicular, sessile or subsessile, coriaceous, erose or denticulate, rugose; midrib and veins usually prominent; margins revolute. Peduncles long or short, glabrous or pubescent.—*E. nummularifolium, β nertericides,* Hook. f., Fl. N.Z. i. 57; Handbk. 77. *E. pedunculare,* Haussk., Monog. Epilob. 303, t. xxiii. f. 96A. From the Three Kings Islands to STEWART Island; CHATHAM Islands, AUCKLAND Islands, 2,000ft., *T. K.* MACQUARIE Island, *A. Hamilton!*

Var. **minimum.** Stems ½in.–1in. long. Leaves opposite, close-set, about $\frac{1}{10}$in. in diameter, glabrous, orbicular-ovate, coriaceous, rugose, sessile or subsessile; midrib and veins prominent; margins revolute. Capsule ⅓in.–½in. long, erect, exceeding the peduncle, stout, spindle-shaped, deeply furrowed, glabrous. Seeds narrowed at the base and shortly curved, minutely papillose. Preservation Inlet and the Bluff Hill, *T. K.*

Var. **caespitosum.** Sparingly branched, puberulous. Leaves sessile or subsessile, less than ⅛in. in diameter, suborbicular, minutely denticulate; midrib and veins prominent. Flowers and capsule not seen. Bay of Islands and Hokianga, A. Cunn. in Herb. Kew. *E. caespitosum,* Haussk., Monog. Epilob. 301, t. xx. f. 85.

After many years' study of these plants in the living state, I find it quite impossible to separate the different forms as species, even with the aid of Professor Haussknecht's excellent descriptions and beautiful drawings. They pass into each other by the most insensible gradations.

17. **E. purpuratum,** *Hook., Handbk. N.Z. Fl.* 77. Stems prostrate, creeping, and rooting at the nodes, 1in.–3in. long, purplish-black, perfectly glabrous. Leaves all opposite, ⅓in.–½in. long including the petiole, orbicular-oblong, rounded at the apex, entire or nearly so, coriaceous. Flowers not seen. Peduncles slender, purplish-black, 2in. long, exceeding the slender capsule. Seeds papillose.—Haussk., Monog. Epilob. 303.

SOUTH Island: alps of Otago, 6,000ft., *Hector* and *Buchanan!*

I have only seen a single specimen of this rare plant, which has not been observed since its discovery more than thirty years ago.

18. **E. glabellum,** *G. Forst., Prod. n.* 160. Stems tetragonous, decumbent, and suffruticose at base, emitting stolons, simple or branched, erect or spreading, purplish, glabrous or bi- or quadrifariously pubescent between the nodes and on the margins of the short petioles. Leaves opposite or the upper alternate, shining, membranous or rather fleshy, glabrous, ovate or ovate-oblong, obtuse, entire, denticulate, coarsely toothed or sinuate-toothed. Flowers axillary, solitary in the upper axils, erect, white or pink. Calyx-lobes ovate-lanceolate, acute or apiculate, glabrous. Corolla shortly exceeding the calyx. Stigma obcordate. Fruiting peduncles rarely exceeding the leaves. Capsule slender, glabrous. Seeds minutely papillose.—Willd., Sp. Pl. ii. 316; Hook. f., Fl. N.Z. i. 59; Handbk. 79.

NORTH Island: Bay of Islands, *Colenso*, n. 105, Herb. Kew. More frequent southwards; Mount Egmont, Tararua Range, &c. SOUTH Island: Queen Charlotte Sound to Otago. Sea-level to 4,000ft. Dec. to Feb.

Some forms of *E. erubescens* (Haussk.) approach this very closely, but are easily distinguished by the rounded stigmas. This species is represented in Cunningham's Herbarium by *E. cinereum*.

19. **E. Novae-Zelandiae,** *Haussk., Monog. Epilob.* 305, *t.* 20, *f.* 86, A, B. Stems shortly creeping, decumbent and branched from the base, emitting short stolons, ascending, 2in.–6in. high, bifariously pubescent. Leaves mostly opposite, the upper alternate, linear-oblong or ovate-oblong, ¾in.–1in. long, obtuse, narrowed into a short petiole, remotely denticulate, glabrous. Flowers numerous, small, white, solitary in the axils of the upper leaves. Calyx-lobes ovate-lanceolate, glabrous. Stigma capitate. Fruiting-peduncle longer or shorter than the leaves. Capsule ¾in.–1½in. long, attenuated below, shortly narrowed towards the apex. Seeds more or less papillose.—*E. aggregata,* T. Kirk, MS. *E. elegans,* Petrie, MS.

NORTH Island: Bay of Islands, *Colenso*. No. 103, Herb. Kew, *Wilkes*. Herb. Benth., *Haussknecht*. SOUTH Island: Nelson: Rotoiti; Waiau-ua Valley; Mount Captain Range, *T. K.* Canterbury, *Sinclair* and *Haast*. Broken River basin and Upper Waimakariri, *T. K.* Otago, *Lyall*. Eweburn Creek and Naseby, *Petrie!* 1,800ft. to 4,500ft. Nov. to Jan.

20. **E. Krulleanum,** *Haussk., Monog. Epilob.* 305, *t.* xxiii. *f.* 95. Stems numerous, decumbent and woody at base, erect, strict, 2in.–6in. high, wiry, tetragonous, bifariously pubescent or glabrous. Leaves mostly opposite, approximate, ovate or oblong-ovate, obtuse, narrowed into short petioles, entire or denticulate, coriaceous. Flowers in the upper axils, erect. Calyx-lobes ovate-lanceolate, acute, shorter than the corolla. Stigma clavate. Fruiting-peduncles rarely exceeding the leaves. Capsule strict, ¾in.–1¼in. long.

SOUTH Island: Nelson; Amuri, Hanmer Plains, *T. K.* Canterbury, *Krull*, *Haast* (*Haussknecht*), 1,800ft. to 3,000ft. Dec., Jan.

Remarkable for its stout woody base, strict erect habit, and entire lower leaves.

21. **E. erubescens,** *Haussk. in Oestr. Bot. Zeitschr.* xxix. (1879) 150. Root stout, woody. Stems numerous, simple, erect, crowded and woody at the base, emitting short stolons, bi- or quadrifariously pubescent, otherwise perfectly glabrous, purplish. Leaves opposite, dense, sessile or very shortly petioled, coriaceous, strict, erect, ovate-oblong, obtuse, cuneate at the base, obscurely obtusely denticulate or sinuate. Flowers in the upper axils, numerous, erect, white or pink. Calyx-lobes lanceolate, acute or apiculate. Stigma small, shortly clavate. Fruiting-peduncles shorter than the leaves. Capsule 1in.–1¼in. long, curved, glabrate or glabrous. Seeds densely papillose.—Monog. Epilob. 306, t. xxiii. f. 98 A, B.

NORTH Island: Hikurangi, East Cape, *S. Dodgshun!* SOUTH Island: more frequent in mountain districts. Nelson to Otago, but local. Frequent in Westland, *T. K.* 1,500ft. to 6,000ft. Dec. to Feb.

Perhaps too closely related to *E. glabellum*, from which it is distinguished at sight by the numerous simple stems and crowded leaves, which, although often of a purplish hue when dry, never assume the deep-red tint so characteristic of *E. glabellum*.

Var. **suberecta.** Stems crowded, simple, decumbent at the base and woody; stolons more numerous. Leaves densely crowded, rather broader than in the type, often subimbricate; truncate, apiculate, or retuse when young. Rimutaka Range, *T. K.*

22. **E. vernicosum,** *Cheesem. in Trans. N.Z.I.* xxviii. (1895) 535. Shortly suffruticose. Rootstock stout. Stems numerous, 4in.–6in. long, decumbent at the base, or almost prostrate, terete, bifariously pubescent. Leaves opposite, or the upper alternate, dense, ascending, rarely imbricating, shortly petiolate, coriaceous, shining, oblong, linear-oblong or almost ovate, usually obtuse, ⅓in.–⅔in. long, obscurely denticulate or sinuate-toothed, glabrous. Flowers 2–5, in the upper axils, erect, ¼in.–⅓in. in diameter, white or pink. Calyx-lobes lanceolate, acute or apiculate, glabrous, shorter than the petals. Style very long, slender; stigma often exserted, capitate, unequally rounded. Fruiting-peduncles shorter than the leaves, slender. Capsule 1in.–1¼in. long, slender, faintly grooved. Seeds smooth, very small.

SOUTH Island: Nelson: Mount Arthur plateau, Mount Peel, *Cheeseman! Gibbs!* Otago: Clutha Valley, Te Anau, *Petrie!* 3,000ft. to 5,000ft. Dec., Jan.

A beautiful plant, with large showy flowers, and the most minute seeds of any New Zealand species.

23. **E. pycnostachyum,** *Haussk. in Oestr. Bot. Zeitschr.* xxix. (1879) 150. Stems numerous, 2in.–6in. high, ascending from the apex of a subterranean woody prostrate or inclined rhizome, quadrifariously pubescent. Leaves opposite or alternate, sessile or shortly petiolate, ascending, flaccid, oblong, cuneate at the base, acute or obtuse, glabrous or glabrate, ½in.–¾in. long; margins with 4–6 subacute teeth. Flowers in the upper axils, erect, large, white. Calyx-lobes lanceolate, acute, glabrous, much shorter than the corolla. Stigma clavate. Capsule sessile or subsessile, equalling the leaves, shortly narrowed at the apex. Seeds papillose.—Monog. Epilob. 306, t. xxi. f. 88. *E. Gilliesianum*, T. Kirk, MS.

SOUTH Island: Nelson, *Monro.* Mount Captain Range, Waiau-ua Valley, *T. K.* Clarence Valley, *Travers!* Canterbury: Whitcomb's Pass, *Haast!* Arthur's Pass, *Cheeseman.* Otago: Lake district, *Hector* and *Buchanan!* 2,000ft. to 5,000ft. Jan., Feb.

As the plant grows on shingle-slips the rootstock is often of great length and much branched. The leaves are blotched with velvety-brown patches, which form a pleasing contrast with the pure-white flowers. Mr. Cheeseman's specimens from Mount Arthur have broader spreading more obtuse leaves on long petioles.

24. **E. brevipes,** *Hook. f., Fl. N.Z.* ii. 328. Stems numerous, from a woody rootstock, branched, prostrate, shining, purple, tips ascending, 6in.–15in. high, naked below, densely foliaceous, perfectly glabrous. Leaves all opposite, coriaceous, shining, purplish-red; petioles ¾in.–1in. long, spreading, elliptical-lanceolate, with a thickened horny triangular apiculus, remotely denticulate. Flowers erect, sessile in the upper axils, white or pink. Calyx-lobes lanceolate, thickened at the apex, nearly equalling the corolla. Stigma clavate. Fruiting-peduncles very short. Capsule exceeding the leaves. Seeds minutely papillose. —Handbk. 78; Haussk., Monog. Epilob. 307, t. xxi. f. 89.

SOUTH Island: Marlborough: Upper Awatere, *Monro, T. K.* Taylor's Pass, *Spencer!* Kaikoura Mountains, *McDonald.* Nelson: Gorge of the Conway, *Cockayne!* Canterbury: Mount Torlesse, *Enys* and *Kirk.* 2,000ft. to 4,000ft. Dec., Jan.

A very handsome plant, apparently of extreme rarity. The valves of the old capsules are persistent for two or three years.

25. **E. haloragifolium,** *A. Cunn., Precurs.* 552. Suffruticose. Stems few, decumbent at the base, ascending or erect, simple or branched above, finely pubescent. Leaves all opposite or the upper alternate, distant, membranous, sessile or very shortly petioled, narrow-oblong or ovate-oblong, narrowed at base or almost cuneate, with few short acute teeth, obtuse, ¼in.–½in. long, ⅛in.–⅓in. broad, often blotched with red or grey. Flowers 1–6, in the upper axils, small, erect, white or pink. Calyx-lobes ovate, apiculate, rather shorter than the petals. Stigma narrow-clavate. Peduncles always shorter than the leaves. Capsule 1in.–1½in. long, densely pubescent.—*E. pictum*, Petrie in Trans. N.Z.I. xxviii. (1895) 538.

NORTH Island: Bay of Islands, *A. Cunn.* SOUTH Island: Upper Waimakariri, *T. K.* Hokitika, *Tipler!* Valleys of Central Otago, Mount St. Bathan's, &c., *Petrie!* Sea-level to 3,000ft.

Some of my specimens have a single terminal flower, thus exactly agreeing with Cunningham's description. Mr. Petrie's fine specimens have 3–6 flowers on the strongest shoots, and solitary flowers on the branches. I agree with Professor Haussknecht's identification of *E. pictum* with *E. haloragifolium*.

26. **E. melanocaulon,** *Hook., Ic. Pl. t.* 813. Rootstock woody, much branched from the base. Stems numerous, slender, simple, strict, erect or ascending, black, shining, wiry, bi- or quadrifariously pubescent. Leaves opposite or alternate, ⅛in.–⅜in. long, sessile or subsessile, rigid, linear-oblong, deeply obtusely toothed or almost lobed, obtuse, truncate or apiculate at the apex, cuneate at base, glabrous. Flowers small, erect, sessile, white or pink. Calyx-lobes ovate-lanceolate, acute, shorter than the corolla. Stigma clavate, rounded. Capsule sessile or subsessile, ¾in.–⅞in. long, glabrous. Seeds papillose.—Hook. f., Fl. N.Z. i. 60; Handbk. 80; Haussk., Monog. 307.

NORTH Island: Ruahine Range and mountains about Lake Taupo, *Colenso*, Herb. Kew. (*Haussk.*) SOUTH Island: common in mountain districts from Nelson to Southland. 1,000ft. to 3,500ft. Dec. to Feb.

Easily distinguished by the numerous slender simple black stems and deeply-toothed rigid erect leaves. North Island specimens have only been collected by Colenso.

27. **E. polyclonum,** *Haussk. in Oestr. Bot. Zeitsch.* xxix. (1879) 150. Suffruticose, strict, erect or decumbent at the base, very slender, branched, 6in.–10in. high, terete, black or dark-purple, glabrous or faintly bifariously pubescent. Leaves opposite or the upper alternate, rigid, distant, sessile or subsessile, ovate-oblong, ⅛in.–¼in. long, cuneate at base, obtuse or shortly acute. Flowers in the upper axils, very small. Calyx-lobes lanceolate-acute, equalling the corolla. Stigma clavate. Fruiting-peduncle slender, longer or shorter than the leaves. Capsule very slender, less than 1in. long, shortly narrowed below the apex. Seeds papillose.—Monog. Epilob. 308, t. xx. f. 87.

SOUTH Island: Canterbury, *Travers!* Otago, *Hector* and *Buchanan* in Herb. Kew; *Petrie!* 5,000ft. to 6,000ft.

Very near to *E. melanocaulon*, of which it should perhaps be considered a variety.

28. **E. rostratum,** *Cheesm. in Trans. N.Z.I.* xxviii. (1895) 534. Suffruticose, branched from the base. Stems 2in.–5in. high, wiry, simple or

branched, terete, erect or spreading, grey or whitish with close short pubescence. Leaves opposite or the upper alternate, ⅛in.–½in. long, sessile or subsessile, rigid, coriaceous, glabrous or puberulous near the base, narrow-oblong, obtuse or shortly mucronate, coarsely toothed. Flowers small, erect in the upper axils. Calyx-lobes narrow, ovate-lanceolate, minutely apiculate, pubescent, nearly equalling the petals. Style short; stigma linear-clavate, minutely emarginate. Capsule ½in.–¾in. long, sessile or subsessile, densely pubescent, curved, sharply narrowed below the apex, grooved. Seeds minutely papillose.

SOUTH Island: Canterbury: Upper Waimakariri; shingly beds of streams near Lakes Tekapo and Pukaki, *Cheeseman!* Otago: Naseby, Black's, *Petrie!* 1,000ft. to 2,500ft. Dec., Jan.

Closely related to *E. Hectori*, but distinguished by the woody habit, strongly-toothed leaves, and curved sessile capsules, which are sharply narrowed towards the apex.

29. **E. microphyllum,** *A. Rich., Fl.* 325, *t.* 36. Suffruticose. Stems numerous, very strict, erect, wiry, much branched from the base, black, bifariously pubescent. Leaves opposite or the upper alternate, ⅛in.–⅙in. long, glabrous, orbicular-ovate or oblong, sessile or very shortly petioled, obtuse, coriaceous, obscurely denticulate or rarely quite entire. Flowers usually few, in the upper axils, erect, very small, white or pink. Calyx-lobes ovate-lanceolate, puberulous, nearly equalling the corolla. Stigma clavate. Fruiting-peduncle exceeding the leaves, very slender. Capsule about ¾in. long, deeply grooved, purplish-black, pubescent, or minutely pilose between the angles. Seeds smooth.—A. Cunn., Precurs. n. 537; Hook. f., Fl. N.Z. i. 59; Handbk. 78. *E. nummularifolium*, var. γ *brevipes*, Hook. f., Handbk. 77 *ex* Haussk.

NORTH Island: East Coast and Cape Palliser, *Colenso*. Orongorongo River, *T. K.* SOUTH Island: common in shingly river-beds in mountain districts from Nelson to Southland. Sea-level to 2,000ft. Dec. to Feb.

30. **E. gracilipes,** *T. Kirk in Trans. N.Z.I.* xxvii. (1894) 351. Suffruticose. Stems numerous, decumbent, wiry, glabrous, reddish. Leaves distant, opposite, shortly petiolate, broadly ovate, obtuse or subacute, ¼in.–⅔in. long, obscurely denticulate, glabrous, purple beneath. Flowers 1–3, solitary in the upper axils, small, white. Calyx-lobes oblong, apiculate, glabrous, not equalling the corolla. Stigma clavate, oblique. Peduncles 1½in.–2in. long, very slender. Capsule equalling the peduncle, narrowed at the base, glabrous. Seeds oblong, acute at base, minutely papillose.

SOUTH Island: Canterbury: Broken River basin, *Enys* and *T. K.* Craigieburn Mountains, *L. Cockayne!* 2,000ft. to 4,000ft. Dec. to Feb.

Mr. Cockayne, to whom I am indebted for beautiful specimens, assures me that this elegant species keeps its character under cultivation. Its closest ally is *E. nummularifolium.*

31. **E. crassum,** *Hook. f., Fl. N.Z.* ii. 328. Perfectly glabrous. Stems woody at the base, prostrate, creeping and emitting roots from the nodes, 2in.–5in. long, rather stout, red, densely foliaceous. Leaves opposite, crowded, obovate, oblong-spathulate, rarely oblong, obtuse or rounded at the tips, 1in.–1¼in. long including the petiole, coriaceous, shining, obscurely denticulate,

glabrous. Flowers few, in the upper axils, large, red or pink-purple. Calyx-lobes lanceolate, acute, much shorter than the corolla. Fruiting peduncle erect, exceeding the leaves. Capsule strict, erect, rigid, 1in. long. Seeds minutely papillose.—Handbk. 78; Haussk., Monog. 309, t. xxii. f. 93A.

SOUTH Island: Nelson: Wairau Mountains, *Travers!* Summit of Mount Captain, Amuri, *T. K.* Marlborough: Upper Awatere, *Monro, Sinclair!* Otago, *Buchanan!* Kurow Mountains, *Petrie.* 3,000ft. to 6,000ft. Jan.

A handsome, rare, and local species, not easily mistaken for any other.

32. **E. macropus,** *Hook., Ic. Pl. t.* 812. Suffruticose. Stems numerous, simple or branched, decumbent or prostrate, ascending, purple, bifariously pubescent, 3in.–9in. long, sometimes emitting stolons from the base. Leaves all opposite, distant, shortly petioled, ovate ovate-oblong or lanceolate, obtuse or subacute, obscurely toothed, ¼in.–¾in. long, glabrous. Flowers few, solitary, axillary, erect, white, ⅓in.–½in. in diameter. Calyx-lobes lanceolate, acute, glabrous. Stigma shortly clavate, emarginate. Fruiting-peduncles 1in.–4in. long or more. Capsule 1in.–2in. long, erect, glabrous. Seeds minutely punctate.—Hook. f., Fl. N.Z. i. 58; Handbk. 78; Haussk., Monog. t. xxii. f. 91A.

NORTH Island: Ruahine Range, *W. E. Andrew!* Tararua Range, Wainuiomata, *Buchanan.* SOUTH Island: Nelson to Southland. 2,000ft. to 3,500ft. Common by mountain streams, amongst wet shingle, &c. Dec. to March.

The distant ovate leaves, purple stems, glabrous habit, and large flowers distinguish this from all other species, but small-leaved forms may be mistaken for *E. pubens.*

E. nanum, Col. in Trans. N.Z.I. xxvi. (1893) 315. Glabrous, 2½in. high. Leaves opposite, linear-lanceolate, with 1 or 2 small teeth; petiole very short, stout. Flowers solitary, subterminal, small. Ovary pubescent with curved hairs.

High land near Tongariro, North Island.

It is not possible to discover the affinities of this plant from the description given. Mr. Colenso states that he had only a solitary unexpanded flower for examination.

E. (*sp.*). Specimens of a plant from Campbell Island (*Lt. Rathouis*) and Clinton Valley (*Petrie*) appear distinct, but are too imperfect for determination. Stems 5in. long, decumbent or prostrate, emitting short stolons. Leaves all opposite, 1in. long, spreading, shortly petiolate, ovate-oblong, narrowed at both ends, obscurely toothed. Flowers not seen. Capsules few, straight, terminal, 1½in. long; peduncles shorter than the leaves. Seeds narrowed below, minutely papillose. Should it prove distinct it might be named *E. Petriei.*

Hybrid forms are occasionally found, and are usually treated by collectors as varieties of the species which they most nearly resemble. Professor Haussknecht describes the following:—

E. junceum × *pubens*, Monog. 291.
E. hirtigerum × *junceum*, Monog. 292.
E. Billardierianum × *junceum*, Monog. 294.

I have not seen specimens fully agreeing with the descriptions of these forms, and feel very doubtful as to their permanence, but must admit that the results of hybridisation in this and other variable genera have not been investigated by local botanists.

*OENOTHERA, Linn.

Calyx-tube 4-angled; limb cylindrical, 4-lobed, deciduous. Petals 4. Stamens 8. Ovary 4-celled; style 1; stigma capitate, entire or 4-lobed; ovules numerous. Capsule linear, 4-valved, 4-celled, rarely 1-celled, dehiscing from the apex into 4 valves, the seeds remaining on the axis, rarely indehiscent. Seeds many or few, not bearded. Herbs, rarely suffruticose, with alternate exstipulate leaves. Flowers solitary, axillary, or in leafy spikes or racemes.

Capsule cylindrical or tetragonous.

Leaves ovate-lanceolate .. .. * *O. biennis.*
Stems simple. Leaves all radical. Bracts foliaceous .. .. * *O. odorata.*

Capsule ovate-clavate.

tems branched, leafy .. .. .. * *O. tetraptera.*

* **O. biennis,** *L., Sp. Pl.* 346. A rather stout erect biennial, 2ft.–4ft. high, pubescent or hairy. Leaves shortly petiolate, lanceolate, ovate-lanceolate or oblong, 2in.–4in. long or more, acute, denticulate. Flowers nocturnal, erect, forming a leafy spike. Calyx-lobes exceeding the ovary. Petals exceeding the stamens, yellow, large. Capsule coriaceous, stout, pubescent. Seeds naked.

NORTH Island: sparingly established in Auckland, Hawke's Bay, Taranaki, and Wellington, but local. SOUTH Island: near Christchurch. *Evening primrose.* Jan., Feb. North America.

* **O. odorata,** *Jacq., Ic. Pl. Rar.* iii. 3, *t.* 456. Perennial. Rootstock stout. Leaves all radical, 3in.–6in. long, ½in.–¾in. broad, linear-lanceolate, acute, narrowed at base, obscurely toothed. Stems several, 2ft.–3ft. high, slender, simple; bracts numerous, linear-lanceolate, the upper pubescent, often coloured. Calyx-tube exceeding ovary, equalling the broad ovate-lanceolate deflexed sepals. Stigma 4-armed. Flowers large, nocturnal. Capsule 4-valved.

NORTH Island: abundantly naturalised on sandy soil, especially in places near the sea. Jan. to March.

* **O. tetraptera,** *Cav., Ic.* iii. 40, *t.* 279. Stem branched from the base, pilose or pubescent. Leaves 1½in.–2in. long, shortly petioled, lanceolate or rhomboid-lanceolate, with or without one or two large coarse teeth or pinnatifid at the base. Sepals lanceolate-acuminate, shorter than the calyx-tube. Capsule ovate-clavate, 8-ribbed, pilose.

NORTH Island: Auckland Domain, naturalised, *Cheeseman!* Mexico.

2. FUCHSIA, Linn.

Calyx-tube ovoid; limb of 4 spreading lobes. Petals 4, rarely 0, often small, inserted at the mouth of the tube, contorted in bud. Stamens 8 in 2 series, inserted with the petals. Ovary inferior, usually 4-celled; style slender; stigma capitate, entire or obscurely 4-lobed; ovules numerous. Fruit a berry, 4-celled. Seeds numerous. Shrubs or trees, rarely suffruticose, with opposite or alternate leaves. Flowers axillary, solitary, racemose or cymose, usually pendulous, hermaphrodite or rarely polygamous, trimorphic in the following species.

SPECIES, about 60, all natives of Mexico and South America, except the New Zealand species, which are endemic.

Flowers pendulous. Petals small or minute.

Leaves lanceolate or ovate-lanceolate, silvery beneath .. 1. *F. excorticata.*
Leaves ovate or orbicular-ovate, not silvery .. .. .. 2. *F. Colensoi.*

Flowers erect.

Petals 0 .. .. .. 3. *F. procumbens.*

1. **F. excorticata,** *L. f., Supp.* 217. A shrub or tree, 10ft.–15ft. high; trunk 6in.–36in. in diameter, clothed with brown papery bark. Leaves alternate, 1½in.–4in. long, lanceolate or ovate-lanceolate, acute or acuminate, entire or obscurely toothed, silvery beneath; petioles short. Flowers axillary, solitary;

¾in.–1in. long, trimorphic, pendulous. Calyx globose at the base, then suddenly constricted and expanded into a funnel-shaped tube with longitudinal ridges; segments 4, acuminate, spreading. Stamens exserted and, like the style, variable in length. Fruit a pendulous black or purple berry.—Link el Otto, Abb. t. 46; Lindl. in Bot. Reg. t. 857; DC., Prod. iii. 39; A. Cunn., Precurs. n. 533; Hook. f., Fl. N.Z. i. 56; Handbk. 75; T. Kirk, Forest Fl. N.Z. t. 36, 36A. *Skinnera excorticata*, Forst., Char. Gen. 58, t. 29; G. Forst., Prod. n. 163; A. Rich., Fl. N.Z. 331. *Agapanthus calycinus*, Banks and Sol. MSS.

From the North Cape to STEWART Island. Sea-level to 3,000ft. *Kotukutuku. Kohutuhutu. Konini* (the fruit only). Aug. to Dec.

The long-styled form of this and the next species has stamens with very short filaments and usually abortive anthers; the mid-styled form has a shorter style and perfect anthers, on longer filaments; while in the short-styled form the style is still shorter and the filaments longer. The dry pollen-grains are of a deep-blue colour, and are bound together with viscid threads. (See Trans. N.Z.I. xxv. (1892) t. xix.)

2. **F. Colensoi,** *Hook. f., Handbk.* 728. A small erect or prostrate shrub with slender branchlets. Leaves alternate, ovate or orbicular-ovate, rounded or cordate at the base, very membranous when dry, obscurely toothed; petioles very slender, longer or shorter than the blade. Flowers as in *F. excorticata*, but the calyx-tube is wider at the mouth, and the petals are minute.

From the Lower Waikato southward to STEWART Island; most plentiful in the SOUTH Island. Ascends to 1,500ft. Oct. to Feb.

In some places this species is less than 1ft. in height; in others it produces unbranched flexuous subscandent shoots 8ft.–9ft. long.

3. **F. procumbens,** *R. Cunn.* ex *A. Cunn., Precurs. n.* 534. Stems much branched, extremely slender, prostrate, 6in.–18in. long. Leaves alternate, rounded-ovate or cordate, obscurely toothed, ¼in.–½in. long, shorter than the slender petioles. Flowers axillary, solitary, ½in.–¾in. long. Peduncles erect. Calyx-tube without raised ridges; lobes reflexed. Petals 0. Stamens equal. Style varying in length; stigma capitate or 4-lobed. Berry large, clavate, pale-red, glaucous.—Hook., Ic. Pl. t. 421; Hook. f., Fl. N.Z. i. 57; Handbk. 76. *F. Kirkii*, Hook. f. in Hook. Ic. Pl. t. 1083.

NORTH Island: Auckland: in sandy or rocky places near high-water mark. Cape Maria and the North Cape; Matauri, *R. Cunningham.* Whangururu, *T. K.* Cape Colville Peninsula, *J. Adams!* Great Barrier Island, *T. K.* Nov. to Feb.

This appears to be the only species with erect flowers, and the only apetalous species with the calyx-tube destitute of external longitudinal ridges. The style may be shorter than the calyx-tube (*F. Kirkii*, Hook. f.) or equalling the stamens, or much longer than the stamens, but the stamens are of uniform length in all the forms.

ORDER XXXI.—PASSIFLOREAE.

Calyx-tube short or elongated; lobes 4 or 5, valvate or imbricate in bud. Petals alternating with the calyx-lobes, inserted at the base of the calyx, sometimes with one or more series of filaments at their base. Stamens as many or twice as many as the petals, and inserted with them; filaments often adnate with the stalk of the ovary. Ovary superior, usually stipitate with 3–5 parietal placentas, each with numerous ovules; styles 3–5, cohering at the base; stigmas terminal. Fruit coriaceous or succulent, 3–5-valved or indehiscent. Seeds numerous, often arillate; embryo straight; endosperm fleshy; cotyledons foliaceous. Usually climbers, with entire or palmatifid stipulate leaves and hermaphrodite or unisexual solitary or racemose axillary flowers.

A large tropical and subtropical family of wide distribution.

1. PASSIFLORA, Linn.

Calyx-tube short; lobes 4 or 5. Petals as many or rarely 0, sometimes resembling the sepals. One or several series of filaments form a corona within the petals. Stamens 4 or 5, adnate with the stalk of the ovary; anthers versatile. Styles 2 or 3; stigmas capitate, large. Fruit succulent, indehiscent or imperfectly 3-valved. Climbers, with stipulate simple or palmatifid leaves and axillary tendrils. Flowers dioecious or hermaphrodite.

A large genus, comprising about 125 species, widely distributed. The only New Zealand species is endemic, and constitutes the section *Tetrapathaea*. Name, an adaptation of *flos passionis*, applied to the flower from the supposed resemblance of its parts to the instruments of the crucifixion.

1. **P. tetrandra,** *Banks and Sol.* ex *DC., Prod.* iii. 323. A glabrous climber, ascending the loftiest trees. Stem sometimes 50ft. long, 3in.–4in. in diameter. Branchlets slender. Leaves alternate, 1in.–4in. long, lanceolate or ovate-lanceolate, acuminate, glossy. Tendrils very slender, elongated. Flowers in 2–4-flowered cymes or solitary, articulated with the short pedicels, unisexual, tetramerous. Petals as long as the sepals, but membranous. Coronal filaments yellow, shorter than the petals. Ovary stipitate, usually with adnate staminodia at its base; styles 2 or rarely 3. Fruit subglobose, 1in. or more in diameter, bright orange-coloured. Seeds compressed, wrinkled, black.—A. Cunn., Precurs. n. 524; Hook. f., Fl. N.Z. i. 73; Handbk. 81. *Tetrapathaea australis*, Raoul., Choix. t. 27.

NORTH and SOUTH Islands: Mangonui to Banks Peninsula. Ascends to 2,800ft. *Kohia.* Dec., Jan.

The seeds afford a pure oil, valued by watchmakers, armourers, &c.

ORDER XXXII.—CUCURBITACEAE.

Calyx-tube superior, produced above the ovary; limb campanulate or tubular, 5-lobed or toothed. Petals 5, free or coherent, often confluent below with the calyx-tube. Male: stamens 3–5, inserted below the petals; filaments free or coherent; anthers often confluent, forming a wavy or curved mass. Female: ovary inferior, usually 1-celled when young, with 3 or more placentas which are produced inwards until they meet in the axis and form a 3 or more celled ovary, or with 1 placenta and remaining 1-celled; style 3-fid or 3-partite; stigmas 3, entire or lobed; cells 1- or many-ovuled. Fruit succulent or coriaceous, usually indehiscent, few- or many-seeded. Seeds usually flat; testa coriaceous or bony; endosperm 0; cotyledons large; radicle short. Weak prostrate or climbing herbs, rarely woody. Leaves exstipulate, alternate, usually palmately veined and lobed or divided. Flowers unisexual or hermaphrodite.

A large order, comprising about 60 genera and 500 species, chiefly found in tropical countries, and including the rock-melon, cucumber, luffa, water-melon, pumpkin, &c.

1. SICYOS, Linn.

Male flowers racemed. Calyx campanulate with 5 minute subulate teeth. Corolla deeply 5-lobed. Filaments 3–5, coherent into a short column; anthers confluent. Female flowers panicled. Ovary inferior, 1-celled; ovule 1, pendulous. Fruit small, coriaceous, indehiscent, 1-seeded, clothed with barbed spines. Annual creeping or climbing plants. Leaves angular. Tendrils branched. Flowers monoecious.

A small tropical and subtropical genus. Etym. An old Greek name for the cucumber.

1. **S. angulata,** *L., Sp. Pl.* 1013. Stems prostrate or climbing, 2ft.–10ft. in length or more, nearly glabrous or more or less scabrid. Leaves broadly ovate-cordate or subreniform, palmately 5–7-lobed, the central lobe the longest, usually acute, toothed. Tendrils petiolate, branched. Male flowers racemose, on long peduncles. Female flowers sessile or on very short peduncles. Petals equalling the sepals. Fruits ovoid, spinous, compressed, 1-seeded.—G. Forst., Prod. n. 368; A. Rich., Fl. N.Z. 323; Hook. f., Fl. N.Z. i. 72; Handbk. 82; Benth., Fl. Austr. iii. 32. *S. australis,* Endl., Prod. Fl. Norf. n. 134; A. Cunn., Precurs. n. 360. *S. fretensis,* Hook. f. in Lond. Journ. Bot. vi. 473.

From the Kermadec and Three Kings Islands southward to Hawke's Bay; Great and Little Barrier Islands. SOUTH Island: Queen Charlotte Sound, *Banks* and *Sol. Mawhai.* Dec. to March. Also in North and South America, Polynesia, Australia, Lord Howe Island, and Norfolk Island.

On the outlying islands of New Zealand this plant has broadly-rounded reniform leaves, with a deep sinus; on the mainland the middle lobe is produced into a long acute or acuminate point.

Order XXXIII.—FICOIDEAE.

Calyx persistent, 4- or many-lobed or rarely divided to the base. Petals 4 or 5 or indefinite and narrow, rarely 0, perigynous or epigynous, rarely hypogynous. Filaments free or united at the base. Ovary inferior or superior, 3–5-celled or more, rarely 2-celled; styles as many as the ovary-cells, free or connate; ovules solitary or numerous. Fruit capsular or fleshy or drupaceous, opening in as many or twice as many valves as cells. Seeds with mealy endosperm, usually compressed; embryo curved or annular, succulent. Erect, prostrate, or creeping herbs, rarely suffruticose, with opposite or alternate exstipulate leaves and regular unisexual or hermaphrodite flowers.

A large order, widely dispersed through tropical and subtropical regions. Genera, about 24. Species, nearly 500.

1. Mesembryanthemum. Leaves angular. Petals numerous.
2. Tetragonia. Leaves flat, petioled. Petals 0.

1. MESEMBRYANTHEMUM, Linn.

Calyx-tube adnate to the ovary; limb usually 5-lobed, persistent. Petals numerous, in one or more series. Stamens in several series. Ovary inferior, with 5 or more cells, rarely few; ovules numerous; styles 4 or more, free or connate at the base. Capsule wholly included in the calyx-tube, flat above and dehiscing loculicidally. Seeds numerous, minute; testa crustaceous. Usually creeping or prostrate, rarely erect, herbs, or suffruticose. Leaves opposite, usually succulent, exstipulate, often angular. Flowers terminal or axillary.

A large genus, the species of which are abundant in South Africa. A few are found on the coasts of Chili, California, the Pacific Islands, Australia, and New Zealand. ETYM. From the Greek, signifying *midday* and *a flower*, in allusion to the flowers of many species expanding at midday.

Leaves above 1in. in length. Peduncles exceeding the leaves .. .. 1. *M. aequilaterale.*
Leaves less than 1in. in length. Peduncles shorter than the leaves .. .. 2. *M. australe.*
Leaves 3in. long, acute. Flowers large, terminal, yellow .. .. .. * *M. edule.*

1. **M. aequilaterale,** *Haw., Misc. Nat.* 77. Stems prostrate or rarely suberect, woody at base. Leaves opposite, stem-clasping, 1½in.–3in. long, succulent, linear, triquetrous, sometimes compressed laterally, acute. Peduncles 1in.–3in. long, winged, thickened upwards. Calyx-tube narrow-turbinate; lobes 5, two much longer than the others. Corolla 1in. in diameter or more. Styles 6–10. Ovary 6–10-celled.—DC., Prod. iii. 429; Hook. f., Fl. Tasm. i. 146; Benth., Fl. Austr. iii. 324.

NORTH Island: littoral. Napier (identified in the absence of flowers): Castlepoint, *T. K.* Dec. to Feb. Also in Australia, Chili, and California.

2. **M. australe,** *Sol.* ex *G. Forst. Prod. n.* 523. Stems creeping and rooting at the nodes, 1ft.–5ft. long. Leaves opposite, connate at the base, triquetrous, flat above, keeled or convex beneath, acute, fleshy, ¾in.–1½in. long. Peduncles solitary, axillary or terminal, stout, usually shorter than the leaves, which are often crowded in short axillary shoots. Calyx-tube turbinate, two of the lobes longer than the others and fleshy. Petals spreading, ¾in.–1in. in diameter when fully open. Styles 5–8. Ovary 5–8-celled.—Ait. Hort. Kew. ed. 1, ii. 187; A. Cunn., Precurs. n. 522; Hook. f., Fl. N.Z. i. 76; Handbk. 83; Benth., Fl. Austr. iii. 324.

From the KERMADEC Islands to Southland: common on all the coasts. CHATHAM Islands. *Pigs' faces.* Nov. to March. Also in Australia and the South Pacific, Lord Howe Island and Norfolk Island.

* **M. edule,** *L., Syst. ed.* x. 1060. Stems stout, spreading, prostrate or suberect, angular. Leaves opposite, fleshy, about 3in. long, shortly connate at base, linear, triquetrous, concave above; keel serrulate. Flowers large, terminal, yellow, sessile, or peduncle short and thick. Calyx-tube turbinate, the two larger lobes 1½in.–2in. long, triquetrous, fleshy. Styles 8–10. Ovary 8–10-celled.

NORTH Island: often planted to fix blown sand, and has become naturalised in many localities. *Hottentot fig.* Dec. to Feb. Cape of Good Hope.

2. TETRAGONIA, Linn.

Calyx-tube adnate with the ovary at the base; lobes 3, 4, or 5. Petals 0. Stamens few or many, perigynous, free. Ovary inferior, 2–8-celled; styles

2–8; ovules solitary, pendulous. Fruit indehiscent, globose or obconic, often angular or horned; epicarp fleshy; endocarp bony; testa membranous. Herbaceous or suffruticose, with erect or trailing stems and alternate petiolate rather succulent flat leaves. Embryo curved; radicle superior. Flowers unisexual or hermaphrodite, axillary.

SPECIES, about 25, chiefly distributed in South Africa, Japan, South America, Pacific islands, and New Zealand. ETYM. From the Greek, signifying *four* and *an angle*, in allusion to the angular calyx-tube.

Styles 3 or more. Fruit with 4 or more horns ... 1. *T. expansa.*
Styles 2 or rarely 3. Fruit globose, unarmed ... 2. *T. trigyna.*

1. **T. expansa,** *Murr. in Comm. Gotting.* vi. (1783) 13. A succulent erect or suberect herb, more or less branched, 1ft. high or more, glabrate or glabrous, more or less papillose. Leaves alternate, rhomboid-ovate or narrow-rhomboid, 1in.–3in. long, abruptly narrowed into the petiole, obtuse or acute. Flowers axillary, solitary or geminate, sessile or shortly peduncled. Calyx-tube broadly turbinate; lobes equalling the tube, obtuse. Stamens 12–16, irregularly inserted. Styles 3–8. Ovary half-inferior, 3–8-celled. Fruit furrowed, angular, usually with 2–4 erect spines or horns; endocarp woody.—A. Rich., Fl. N.Z. 320; DC., Prod. iii. 452; A. Cunn., Precurs. n. 523; Hook. f., Fl. N.Z. i. 77; Handbk. 84; Benth., Fl. Austr. iii. 326. *T. halimifolia*, G. Forst., Prod. n. 223.

KERMADEC Islands to STEWART Island: usually littoral. *New Zealand spinach.* Dec. to Feb. Also in temperate South America, Japan, Australia, Lord Howe Island, and Norfolk Island.

The spines vary greatly in number and length. In the specimens from Dog Island the calyx is broadly turbinate and obviously 8-ribbed. In Waikato specimens the calyx-tube is proliferous.

2. **T. trigyna,** *Banks and Sol.* ex *Hook. f., Fl. N.Z.* ii. 329. Stems suffruticose, 1ft.–8ft. long, subscandent or trailing. Leaves alternate, broadly ovate-rhomboid or ovate, obtuse, abruptly narrowed into the petiole, papillose. Peduncles axillary, filiform, rarely geminate. Flowers perfect or dioecious. Calyx-lobes broad, rounded. Stamens 12–20. Ovary 2- rarely 3-celled; cells 1-ovuled. Fruit red, subglobose; epicarp fleshy, at length woody, obscurely 4-lobed. Seeds 1–3.—Handbk. 84. *T. implexicoma,* Hook. f.; var. *Chathamica,* F. Muell., Veg. Chath. Isds. 12.

KERMADEC Islands to STEWART Island; CHATHAM Islands: littoral. *Ice-plant.* Nov. to Jan.

Distinguished from *T. expansa* by the broader leaves, dioecious flowers, and red globose unarmed fruits with 2 or 3 seeds. The Australian *T. implexicoma*, Hook. f., can only be regarded as a variety of *T. trigyna*, differing in the more acute leaves and longer peduncles.

ORDER XXXIV.—UMBELLIFERAE.

Calyx-tube adnate with the ovary; limb truncate or 5-toothed or lobed or obsolete. Petals 5, imbricate or valvate in bud, often with inflexed tips, inserted round a 2-lobed epigynous disk, the lobes often united with the base of the style, forming the stylopodium. Stamens 5, inserted with the petals; filaments incurved. Ovary inferior, 2-celled, or rarely 1-celled by absorption;

styles 2; stigmas small, terminal; ovules solitary, pendulous in each cell. Fruit of 2 dry carpels termed "mericarps," cohering by their inner face, and usually suspended from a persistent filiform axis called a "carpophore." Each mericarp is marked with 5 longitudinal primary ribs, and occasionally by 4 intermediate or secondary ribs. Oil-tubes, or vittae, are often found in the interspaces, but under the pericarp. Seeds pendulous; testa thin; embryo minute; radicle superior; endosperm hard. Erect or creeping herbs, with hollow stems and alternate simple or dissected leaves, having the petioles dilated into a sheathing base. Flowers small, in simple or compound lateral or terminal umbels. The bracts at the base of an umbel form the involucre, and those under a partial umbel or umbellule the involucel. The flowers are often dioecious or polygamous.

A large order, comprising about 160 genera and nearly 1,500 species, distributed through all lands except the coldest, and often presenting great difficulties to the student, owing to the close similarity between the flowers of different genera. Of the New Zealand genera, *Aciphylla* extends to Australia but not to the Antarctic islands. *Actinotus* is also found in Australia. *Azorella* and *Oreomyrrhis* are Australian, Andine, and Antarctic. The remaining genera are widely distributed. The order affords many valuable drugs.

I. Fruit with 1 or 3 ribs only on the face of each mericarp.

1. Hydrocotyle. Stems creeping. Umbels simple. Fruits laterally compressed. Oil-canals 0.
2. Azorella. Stems creeping, rarely erect. Fruits subquadrate. Oil-canals 0.
3. Eryngium. Stems erect or creeping. Leaves and involucres spinous. Umbels forming compact heads. Ribs 0. Oil-canals 0.
4. Actinotus. Leaves radical. Umbels simple. Styles 2. Fruits 1 seeded. Oil-canals 0.
* Bupleurum. Leaves divided or perfoliate. Oil-canals present or 0. Umbels compound.
* Conium. Stem spotted. Involucres of narrow-linear bracts. Calyx-teeth 0. Oil-canals 0.
5. Apium. Fruit with 5 obtuse ribs alternating with oil-canals on each mericarp.
* Ammi. Involucral leaves pinnate or pinnatifid. Fruits laterally compressed, ovate-oblong. Carpels 5 ribbed. Oil canals present.
* Carum. Fruits with 5 filiform ribs alternating with oil-tubes.
6. Oreomyrrhis. Pubescent. Fruits oblong, narrowed upwards. Carpels nearly terete. Ribs 5.
* Scandix. Stylopodia produced into a long beak. Oil-canals 0. Leaves simple, pinnate or decompound; segments flat, capillary or acicular.
* Foeniculum. Leaves excessively divided into capillary segments; ribs prominent.
7. Grantzia. Stem creeping. Leaves simple, fistular, transversely jointed. Umbels simple.
8. Ligusticum. Leaves pinnate or decompound. Umbels compound. Carpels with 3 or 5 narrow equal wings on each face. Oil-canals present.
9. Aciphylla. Leaves divided into rigid acicular segments. Umbels in erect spikes or panicles. Fruit linear or oblong, with 3 or 5 wings on each face.
10. Angelica. Erect or climbing. Carpels with 2 broad lateral wings.
* Peucedanum. Erect. Leaves usually pinnate. Involucres sometimes 0. Fruit ovoid or oblong, dorsally compressed, lateral primary ridges forming wings.

II. Carpels with slender primary ribs armed with bristles; the secondary ribs most prominent, armed with barbed or hooked prickles.

11. Daucus. Calyx-teeth 0. Fruits dorsally compressed. Secondary ribs with a row of barbed prickles.
* Caucalis. Calyx-teeth obvious. Fruits laterally compressed. Secondary ribs setose or prickly.

1. HYDROCOTYLE, Linn.

Calyx-teeth inconspicuous. Petals entire, valvate or rarely imbricate. Fruits without vittae, orbicular or suborbicular, flattened laterally. Mericarps placed edge to edge, usually with a prominent rib, rarely 2 or 3 on each face. Seed straight, laterally compressed. Herbs, with slender creeping prostrate stems often matted and rooting at the nodes, orbicular or reniform solitary or fascicled minutely stipulate leaves, and simple umbels of small involucrate hermaphrodite or rarely unisexual flowers.

SPECIES, about 80, generally distributed through the temperate and tropical regions. Of the New Zealand species, one is distributed through Asia, Africa, North and South America, and Australia, another extends to North and South America but not to Australia, while two are restricted to New Zealand and Australia; the others are endemic.

ETYM. From the Greek, signifying *water* and *a flat cup*, in allusion to the cup-shaped leaves of certain species. The fruits vary considerably in the different stages of growth.

* *Leaves solitary, orbicular or reniform, 3–5–7-lobed or partite. Petals valvate. Stems prostrate, creeping except in 6, often matted.*

Leaves deeply divided. Peduncles elongated. Fruits on strict pedicels ..	1. *H. elongata.*
Umbels sessile or very shortly peduncled	2. *H. americana.*
Leaves 3–5-foliolate	3. *H. tripartita.*
Leaves 5–7-partite, acutely toothed. Peduncles slender. Umbels many-flowered	4. *H. dissecta.*
Leaves obscurely 3–7-lobed, glabrous. Carpels large, flat	5. *H. pterocarpa.*
Stems erect or suberect. Carpels turgid, faintly keeled	6. *H. robusta.*
Leaves obscurely 3–7-lobed, pubescent or hairy. Umbels 5–10-flowered ..	7. *H. Novae-Zelandiae.*
Leaves distinctly 5–7-lobed, hispid or tomentose. Umbels many-flowered	8. *H. moschata.*
Leaves nearly glabrous, $\frac{1}{4}$in. in diameter, 5–7-lobed. Umbels few-flowered	9. *H. microphylla.*

** *Stems rather stout. Leaves fascicled. Petals imbricate.*

Leaves broadly cordate. Umbels 2–3-flowered. Fruits large	10. *H. asiatica.*

1. **H. elongata,** *A. Cunn., Precurs n.* 495. Stems filiform, prostrate, weak, pilose or rarely glabrate, 6in.–12in. long. Leaves on long petioles, distant, deeply 3–7-lobed; lobes acutely toothed, glabrate or hairy on both surfaces. Stipules minute. Peduncles exceeding the leaves, very slender; bracts minute. Umbels 12–30-flowered, $\frac{1}{4}$in.–$\frac{1}{3}$in. in diameter. Pedicels strict, slender. Flowers small. Fruits ovate-orbicular, truncate at base, not emarginate, flattened. Carpels with 1 rib on each face.—Hook. f., Fl. N.Z. i. 84; Handbk. 85. *H. concinna*, Col. in Trans. N.Z.I. xvii. (1884) 239.

Var. **echinella.** Margins of the sinus often sharply toothed. Carpels sparingly pubescent or almost bristly; sometimes the pubescence is confined to the carpellary rib.—*H. echinella*, Col. in Trans. N.Z.I. xx. 191.

NORTH and SOUTH Islands Mangonui to Southland, but often local. Sea-level to 2,000ft. Nov. to Feb.

2. **H. americana,** *L., Sp. Pl.* 234. Perfectly glabrous. Stems 3in.–6in. long, much branched, flaccid, matted. Leaves orbicular, reniform, shining, with 5–7 shallow lobes, crenate, $\frac{1}{4}$in.–$\frac{1}{2}$in. in diameter; petioles 1in.–1$\frac{1}{2}$in. long. Umbels axillary or leaf-opposed, 2–6-flowered, sessile or very shortly peduncled. Flowers sessile or subsessile. Fruits glabrous or shortly hispid. Carpels with 1 rib on each face; margins acute.—Hook. f., Fl. N.Z. i. 84; Handbk. 85.

NORTH and SOUTH Islands; STEWART Island. Ascends to 2,000ft. Oct. to Feb. Also in North and South America.

This does not appear to be found north of the Great Barrier Island.

Var. **heteromeria.** Very similar to the type, but the petioles are 2in.–3in. long, the leaves are larger, with a wider sinus, and more distinctly lobed, sometimes glabrate. The umbels are more frequently peduncled, although both sessile and pedunculate umbels may be found on the same specimen. The carpels exactly as in the type.—*H. heteromeria*, A. Rich., Hydroc. 200; A. Cunn., Precurs. n. 499; Hook. f., Fl. N.Z. i. 82; Handbk. 86. *H. nitens*, Col. in Trans. N.Z.I. xxiii. 386; Handbk. 86. *H. glabrata*, Banks and Sol. MSS. NORTH Island: very local. Three Kings Islands and Bay of Islands southwards. SOUTH Island: Queen Charlotte Sound, *Banks* and *Sol.*

3. **H. tripartita,** *R. Br.* ex *A. Rich., Hydroc.* 69, *t.* 21; *t.* 61, *f.* 25. Glabrous. Stems filiform, 1in.–2in. long, often matted. Leaves ¼in.–⅓in. in diameter, 3–5-foliolate; leaflets cuneate, 2-3-lobed, toothed or entire, rarely petiolulate; petioles ½in.–2in. long. Stipules rather large, entire. Peduncles shorter than the leaves. Umbels 2–5-flowered. Flowers sessile or shortly pedicelled. Fruits rather turgid, slightly rounded, retuse. Carpels with 1 obscure rib on each face.—DC., Prod. iv. 65; Hook. f., Fl. N.Z. i. 83; Benth., Fl. Austr. iii. 341. *H. muscosa*, Hook. f., Handbk. 86. *H. hygrophila*, Petrie in Trans. N.Z.I. xxviii. (1895) 552! (name only).

NORTH Island: Matata, Bay of Plenty, *Petrie!* Hawke's Bay, *Colenso.* SOUTH Island: Nelson: Lake Guyon, *T. K.* Canterbury: Broken River basin, *T. K.* Otago: Dunedin and Bluff, *Aston!* Te Anau and Wanaka Lakes, Taieri Plain, &c., *Petrie!* STEWART Island, *T. K.* Sea-level to 3,000ft. Dec., Jan.

Specimens from Broken River and Stewart Island have larger and more fleshy leaves.

4. **H. dissecta,** *Hook. f., Fl. N.Z.* i. 84. Hispid or hispid-pilose. Stems slender, 5in.–8in. long. Leaves alternate or fascicled, ¾in.–1in. in diameter, orbicular, with a narrow sinus, 5–7-lobed nearly to the base; segments obovate-cuneate, acutely toothed or laciniate, hairy on both surfaces; petioles ½in.–1½in. long. Peduncles 1in.–3in. long. Flowers 20–30, subsessile. Fruits densely crowded, suborbicular, rounded at base, retuse, turgid. Carpels with 1 obscure rib on each face.—Handbk. 86.

NORTH Island: *Colenso.* Matakana and Hunua, Auckland, *T. K.* SOUTH Island: Catlin's River, *Petrie.* Oct. to Jan.

A remarkably rare and local lowland species. Its leaves approach those of *H. elongata*, while the carpels are nearer to some forms of *H. moschata*.

5. **H. pterocarpa,** *F. Muell. in Trans. Vict. Inst.* i. (1855) 216. Glabrous or with few scattered hairs. Stems weak, 6in.–18in. long, prostrate. Leaves ½in.–1in. in diameter, orbicular-reniform, with a narrow or closed sinus, shining, obscurely 3–7-lobed; lobes crenate; petioles 1in.–3in. long or more. Peduncles very slender, shorter than the petioles. Umbels 3–6-flowered. Pedicels short. Fruits large, flattened, broader than long, the margins forming a flat wing, broadly emarginate, cordate below, mottled. Carpels with 2 ribs on each side; margins acute.—Hook. f., Fl. Tasm. i. 153, t. 33; Handbk. 86; Benth., Fl. Austr. iii. 342.

NORTH Island: in marshes, Auckland, from Mangonui southwards. Wellington: Lower Wairarapa, *T. K.* SOUTH Island: Nelson. Cheviot, *Haast!* Feb. Also in Victoria and Tasmania.

6. **H. robusta,** *n. s.* Stems as thick as a crow-quill, decumbent, sub-erect, 6in.–9in. high. Leaves distant, hairy on both surfaces, broadly reniform with a narrow sinus, ¾in.–1in. in diameter; lobes 5–7, very shallow, minutely crenate, ciliate, very membranous, pale; petioles 1in.–2in. long, stout. Stipules broad, obtuse. Peduncles usually about ½in. long. Umbels 8–12-flowered. Pedicels short. Fruits ⅛in. broad, $\frac{1}{12}$in. long, emarginate, truncate or slightly cordate at base, very turgid. Carpels with an obscure groove on each face, broadly rounded at the base and faintly keeled, glabrous.

NORTH Island: on sandy beaches. Auckland: between Waiwera and Mahurangi. Great Barrier Island: Port Tryphena, Blind Bay, Haratoanga. Dec., Jan.

This species is advanced with some hesitation, as, although very different from *H. Novae-Zelandiae* in appearance, it is not easy to find good specific characters. In addition to the stout erect habit and large uniform leaves, it differs notably in the remarkably broad and turgid fruits, which are obscurely grooved, broadly rounded at the back, and faintly keeled.

7. **H. Novae-Zelandiae,** *DC., Prod.* iv. 67. Pilose or glabrate. Stems branched, filiform, intricate or matted. Leaves distant, ¼in.–¾in. in diameter, membranous, orbicular-reniform, usually with a wide sinus, obscurely 5–7-lobed or angled, obscurely crenate, glabrate or sparsely hairy on both surfaces; petioles ½in.–2in. long, slender. Stipules usually large, broad, entire. Peduncles shorter than the petioles. Umbels 5–12-flowered. Pedicels short; bracts acute. Fruits glabrous, $\frac{1}{12}$in. broad, $\frac{1}{14}$in. long, slightly compressed, rounded on the back. Carpels with 1 obscure rib or furrow on each face.—A. Cunn., Precurs. n. 497; Hook. f., Fl. N.Z. i. 83; Handbk. 86. *H. dichondraefolia,* A. Cunn., Precurs. n. 498. *H. intermixta,* Col. in Trans. N.Z.I. xvii. (1885) 240. *H. alsophila,* Col. *l.c.* xviii. 261. *H. involucrata,* Col. *l.c.* xix. 262. *H. amaena,* Col. *l.c.* xxi. 83. *H. pilosa,* Banks and Sol. MSS.

Var. **lobulata.** Leaves very membranous, hairy on both surfaces. Peduncles exceeding the petioles. Carpels with a deep furrow on each face, giving the fruit a 4-lobed appearance, turgid, rounded at the back, mottled.

Var. **montana.** Rhizomes stout, extensively creeping, matted below; stems short. Leaves orbicular, with a narrow or closed sinus, coriaceous, glabrate; lobes rounded, very shallow. Peduncles shorter than the petioles. Fruits retuse, cordate at base, slightly compressed, with an obscure groove on each face; back rounded but thin.

From the Three Kings Islands and North Cape to STEWART Island. Var. *montana* ascends to 4,000ft. Var. *lobulata,* Waitemata. Nov. to March.

This species is extremely variable in the form of the leaf, the width of the sinus, and the teeth, but the lobes are always shallow. The hairs on the petiole and peduncle are usually reversed. In some forms the carpels are faintly keeled, especially in the young state. I have not seen authenticated specimens of Mr. Colenso's plants.

8. **H. moschata,** *G. Forst., Prod. n.* 135. Pilose or hispid, rarely glabrescent. Stems tufted or spreading, slender, 2in.–12in. long. Leaves ¼in.–1in. long, broad, reniform or orbicular, with a wide sinus, 5–7-lobed; lobes shallow, acutely toothed; teeth often minute, usually hispid on both surfaces; petioles ½in.–2in. long. Peduncles longer or shorter than the petioles. Umbels 10–40-flowered. Pedicels usually 0. Fruits densely crowded, minute, rather turgid. Carpels with 1 obvious rib on each face, acute at the back, sometimes faintly reticulate.—A. Rich., Hydroc. 66, t. 60, f. 24; A.

Cunn., Precurs. n. 501; Hook. f., Fl. N.Z. i. 83; Handbk. 87. *H. compacta*, A. Rich., Hydroc. 61; A. Cunn., Precurs. n. 500. *H. colorata*, Col. in Trans. N.Z.I. xviii. (1886) 260. *H. capitata*, Banks and Sol. MSS.

Var. **laciniata.** Leaves $\frac{1}{8}$in.–$\frac{3}{8}$in. in diameter, more deeply cut; teeth more acute. Peduncles longer or shorter than the leaves. Umbels 5–20-flowered.—*H. sibthorpioides*, Col. in Trans. N.Z.I. xxi. 83 (not of Lam.).

From the KERMADEC Islands and the North Cape to Southland; CHATHAM Islands. Sea-level to 2,000ft. Nov. to March.

The sparingly hispid or pilose leaves and acute carpels distinguish this from *H. Novae-Zelandiae* and *H. microphylla*.

9. **H. microphylla,** *A. Cunn., Precurs. n.* 496. Glabrous or with few hairs near the apex of petiole and peduncle. Stems slender or rather stout, 1in.–4in. long. Leaves $\frac{1}{12}$in.–$\frac{1}{3}$in. in diameter, orbicular, with a closed or narrow sinus and 5–7 shallow lobes, minutely crenate; petiole $\frac{1}{4}$in.–$\frac{3}{4}$in. long. Stipules large for the size of the plant. Peduncles shorter or longer than the petioles. Flowers 2–6, rarely more, sessile or subsessile. Carpels rounded on the back, with an obscure rib on each face, slightly turgid, glabrous.—Hook. f., Fl. N.Z. i. 84; Handbk. 87.

From Mangonui, NORTH Island, southwards to STEWART Island. Dec., Jan.

Distinguished by the glabrous leaves, few-flowered umbels, and carpels rounded at the back, except when young. In some specimens the carpels closely approach those of small forms of *H. muscosa*, but the leaves are always glabrous.

10. **H. asiatica,** *L., Sp. Pl.* 234. Stems creeping, rather stout. Leaves orbicular or oblong-reniform or cordate, rarely truncate at base, almost entire or repand, toothed, glabrous, $\frac{1}{4}$in.–1in. long; petioles $\frac{1}{2}$in.–4in. long. Peduncles $\frac{1}{2}$in.–2in. long. Umbels 2–4-flowered, rarely 1-flowered. Involucral leaves large, oblong or ovate, glabrous or pilose. Petals imbricate. Fruits broadly truncate at apex, turgid. Carpels with about 3 principal ribs, reticulated.—A. Cunn., Precurs. n. 502; Hook. f., Fl. N.Z. i. 82; Handbk. 86; Benth., Fl. Austr. iii. 346. *H. cordifolia*, Hook., Ic. Pl. t. 303. *H. uniflora*, Col. in Trans. N.Z.I. xvii. (1885) 239. *H. indivisa*, Banks and Sol. MSS.

From the Three Kings Islands and the North Cape to STEWART Island. Sea-level to 2,500ft. Oct. to March. Also in Asia, Africa, North and South America, and Australia.

2. AZORELLA, Lamk.

Calyx-teeth prominent, usually small, acute. Petals 5, imbricate in bud. Disk confluent with the styles, thick, depressed. Fruit transversely subquadrate, scarcely broader than thick or slightly compressed laterally. Carpels angular with 5 ribs, the lateral not close to the constricted commissure. Vittae 0. Carpophore simple. Tufted herbs with imbricate or subimbricate leaves forming dense masses, or slender stoloniferous herbs with radical simple or divided leaves. Umbels simple or irregularly compound, with free or coherent involucral bracts.

SPECIES, about 40, found in Andine South America, Australia, New Zealand, and the Antarctic regions. The Macquarie-Island plant is generally distributed through the Antarctic islands and extreme South America; the other New Zealand species are endemic. The student should be careful to examine mature fruits.

A. FRAGOSA. STEMS PULVINATE.

Leaves all cauline, appressed .. .. 1. *A. Selago.*

B. SCHIZEILEMA. LEAVES ALL RADICAL OR FASCICLED AT THE NODES OF LEAFY FLOWERING BRANCHES.

* *Leaves simple.*

Leaves entire or obscurely crenate-lobed .. .. .. 2. *A. exigua.*

Leaves reniform. Stipules entire. Pedicels shorter than the fruits 3. *A. reniformis.*

Leaves reniform. Stipules ciliate. Pedicels exceeding the fruits .. 4. *A. Haastii.*

** *Leaves 3–5-foliolate.*

1. Leaflets sessile; segments cuneate.

Leaflets coriaceous, 5-toothed or -lobed .. .. .. 5. *A. Roughii.*

Leaves crowded at the nodes of creeping scions, very coriaceous .. 6. *A. hydrocotyloides.*

Leaves membranous, pale-green .. .. 7. *A. pallida.*

Rhizomes matted. Leaflets minute, ovate-cuneate .. .. 8. *A. nitens.*

2. Leaflets petiolulate.

Umbels 2–8-flowered .. .. 9. *A. trifoliolata.*

Umbels 20–30-flowered .. .. .. 10. *A. elegans.*

1. **A. Selago,** *Hook. f., Fl. Antarc.* ii. 284, *t.* 99. Stems densely tufted, 1in.–5in. high or more, glabrous. Leaves alternate, with a broad sheathing membranous base; blade cartilaginous, appressed, 3–5-partite for half its length; the segments spreading, oblong, acute or apiculate, with several stiff bristles on the upper surface; marginal nerve stout, recurved. Umbels 3-flowered, sunk amongst the terminal leaves. Peduncle short. Involucral bracts linear, subacute. Calyx-teeth acute. Styles elongated. Fruits slightly compressed. Primary ribs 5.

MACQUARIE Island, *Fraser, Scott! Hamilton!* Also in Terra del Fuego, Port Famine, Hermite Island, Kerguelen's Land, Marion Island, Yong Island, and the Crozets.

The description of the flowers and fruit is taken from Hooker, the specimens from Macquarie Island being sterile. This singular plant forms large amorphous cushion-like masses, sufficiently compact to bear the weight of a man with but little injury; the upper surface consists of the living plant, while the base is composed of decomposing leaves and roots.

2. **A. exigua,** *Benth. and Hook. f. in Gen. Pl.* i. 875. Leaves densely crowded at the apex of a stout rootstock, ½in. long including the stout petiole, glabrous; blade orbicular-ovate, cordate, obscurely 3-lobed, coriaceous, minutely papillose; margins recurved; petiole dilated and sheathing at the base. Scape 3–5-flowered, usually shorter than the petiole; involucral leaves linear, obtuse, coherent at the base. Calyx-teeth minute, acute. Fruits $\frac{1}{12}$in.–$\frac{1}{10}$in. long, almost tetragonous, rounded at the angles. Carpels obscurely 5-ribbed on each face.—*Pozoa exigua,* Handbk. 88; Buch. in Trans. N.Z.I. xiv. (1881) t. 26, f. 2.

SOUTH Island: Otago: Black Peak, 6,000ft., *Hector* and *Buchanan!* Hector Mountains and Mount Cardrona, 5,000ft., *Petrie!*

A singular species, not easily mistaken for any other.

3. **A. reniformis,** *Benth. and Hook. f.* l.c. Rootstock slender, crowned with leaves and few leafy flowering-branches, rarely exceeding 2in.–3in. long, perfectly glabrous. Leaves ¼in.–¾in. in diameter, coriaceous or membranous, orbicular or reniform, crenate-lobed; petioles 1in.–2in. long; stipules acute or

shortly acuminate, quite entire. Peduncles shorter than the leaves. Umbels 3–8-flowered. Calyx-teeth acute. Pedicels shorter than the flowers. Fruits tetragonous; mericarp 5-ribbed.—*Pozoa reniformis*, Hook. f., Fl. Antarc. i. 15, t. 11.

AUCKLAND Islands, *J. D. Hooker!* ADAMS Island, *T. K.* CAMPBELL Island, *T. K.* 800ft. to 1,700ft. Dec., Jan.

A. Haastii has been mistaken for this species by myself on the Spencer Mountains, by Petrie in the Nevis Valley, and apparently by Cheeseman on Mount Peel. The stipules are invariably entire.

4. **A. Haastii**, *Benth. and Hook. f.* l.c. Rootstock slender or stout, crowned with radical leaves and prostrate or creeping leafy flowering-branches. Leaves glabrous or sparingly setose above or below, orbicular or reniform, ¼in.–1½in. broad, usually with an open sinus, more or less coriaceous, glossy, broadly crenate-lobed; lobes shallow; petiole 1in.–6in. or more; stipules ciliate or almost laciniate. Primary umbels 1–3 or more, overtopped by 1–3 secondary umbels. Peduncles ½in.–3in. long, in the axils of shortly-petiolate leaves. Involucral leaves linear-oblong, united at the base. Flowers 3–50. Pedicels exceeding the tetragonous fruits. Carpels 5-ribbed.—*Pozoa Haastii*, Hook. f., Handbk. 88.

NORTH Island: Ruabine Range, *A. Hamilton!* SOUTH Island: Nelson to Southland. 2,000ft. to 5,000ft. Not unfrequent. Dec., Jan.

Much too closely related to *A. reniformis*, from which it can only be distinguished by the ciliated or laciniate stipules, pedicels longer than the fruits, and more numerous flowers.

5. **A. Roughii**, *Benth. and Hook f.* l.c. Habit of *A. Haastii*. Leaves orbicular, ¾in.–1½in. in diameter, 3–5-foliolate or -partite, coriaceous, glabrous, shining; leaflets broadly obcuneate, 5-toothed or -lobed, sessile; petioles 2in.–6in. long; stipules acute or laciniate. Stolons exceeding the leaves. Peduncles 1–4, ½in.–3in. long, developed in the axils of lobed or divided leaves. Primary umbels overtopped by 1 or more secondary umbels rising from the base of the next below it. Umbels 5–20-flowered; involucral leaves free, obtuse, entire or toothed below the middle. Calyx-tube ovate, obtuse. Stylopodia large; styles elongated. Fruits somewhat rounded, obviously 5-ribbed.—*Pozoa Roughii*, Hook. f., Handbk. 89.

SOUTH Island: Marlborough: Mount Stokes, *MacMahon!* Nelson: Dun Mountain, *Rough! Sinclair!* Wooded Peak, Ben Nevis, &c., *Gibbs!* Raglan Mountains and above the Wairau Gorge, *Cheeseman.* Fowler's Pass, Amuri, *T. K.* 3,000ft. to 5,000ft. Dec., Jan.

Most nearly related to *A. Haastii*, but distinguished by the divided leaves, more numerous umbels, obtuse calyx-lobes, and longer styles.

6. **A. hydrocotyloides**, *Benth. and Hook. f.* l.c. Root stout, often 1ft. long or more. Stems stout, tufted, creeping and rooting at the nodes, often forming small rather compact mounds. Leaves crowded at the rootstock and nodes, ⅓in.–¾in. in diameter, orbicular or orbicular-reniform, thick and coriaceous, 3–5-foliolate or -partite; leaflets sessile, broadly obovate-cuneate, 3–5-lobed or obtusely toothed; petiole ½in.–2in. long; stipules entire, acute or ciliate. Peduncles solitary or 2–4 at the apex of a short leafy flowering-

branch. Umbels 3–15-flowered. Pedicels usually exceeding the fruits. Involucral leaves obtuse. Fruits tetragonous. Carpels 5-ribbed.—*Pozoa hydrocotyloides*, Hook. f., Handbk. 88.

SOUTH Island: Canterbury: Rangitata River, *Sinclair!* Near the source of the Kowai, *Haast!* Mount Torlesse, *T. K.* Leith Hill and Mount Enys, *Enys!* Otago: Mount St. Bathan's, Kurow Mountains, *Petrie!* 2,000ft. to 4,500ft. Dec., Jan.

Easily distinguished by the peculiar habit and excessively coriaceous leaves.

7. **A. pallida.** Glabrous, flaccid. Rootstock emitting creeping stolons. Leaves crowded at the apex, ½in.–¾in. in diameter, 3-foliolate or rarely 3-partite; leaflets sessile, cuneate or obovate-cuneate, 3–6-lobed at the tips, shining, pale-green; petioles 1in.–3in. long; stipules lacerate. Peduncles usually shorter than the leaves, bearing a single terminal umbel in the axil of a petioled 3-lobed leaf; frequently secondary or tertiary umbels are developed in like manner, on very short peduncles. Umbels 4–12-flowered. Involucral leaves linear, obtuse. Flowers very small; calyx-teeth minute. Pedicels exceeding the fruit. Fruits slightly narrowed above, obtusely 4-angled; ribs 5. —*Pozoa pallida*, T. Kirk in Trans. N.Z.I. x. (1877) 419.

SOUTH Island: rather local. Nelson: Rotoiti; Spencer Mountains and Upper Wairau Valley, *T. K.* Mount Arthur, *Cheeseman.* Canterbury: Pukunui Creek, *T. K.* 2,000ft. to 4,000ft. Dec., Jan.

8. **A. nitens,** *Petrie in Trans. N.Z.I.* xxv. (1892) 270. Rhizomes very slender, creeping, matted. Leaves few, solitary, 3-foliolate or 3-partite; leaflets $\frac{1}{16}$in.–¼in. long, shortly petiolulate or sessile, ovate-cuneate below, entire or obscurely 1–3-toothed, obtuse or subacute, perfectly glabrous; petioles ½in.–1½in. long; stipules broadly ovate, acute. Peduncles axillary, on short 1–3-leaved scapes, solitary or 2 or 3 together, very short. Flowers solitary or in 2–3-flowered umbels. Involucral leaves 3, narrow-linear. Calyx-teeth minute, acute. Fruits equalling the pedicels, shortly tetragonous, $\frac{1}{16}$in. long, turgid; mericarp indistinctly 5-ribbed.—*A. pusilla*, T. Kirk, MSS.

SOUTH Island: Nelson: shores of Lake Guyon, &c., Amuri (1875), *T. K.* Canterbury: Broken River basin, *Enys* and *Kirk* (1876). Otago: Clinton River and Te Anau Lake, *Petrie!* 700ft. to 3,000ft. Dec., Jan.

Easily recognised by its diminutive size, matted rhizomes, and small fruits with rounded angles. It resembles depauperated states of *Hydrocotyle muscosa*.

9. **A. trifoliolata,** *Benth. and Hook.* l.c. A very slender usually glabrous species. Stems filiform, creeping and rooting at the nodes. Leaves 2–5, fascicled at the nodes, very membranous, 3-foliolate, rarely 3-partite; leaflets shortly petiolate or sessile, obovate-cuneate, obcuneate or flabellate, unequally crenate or 3–4-lobed or toothed, apiculate, rarely laciniate, glabrous or with scattered hairs above or below; petioles 1in.–4in. long; stipules small, ciliate or laciniate. Peduncles ¾in.–1in. long. Umbels 1–4, 2–8-flowered. Involucral leaves free. Calyx-lobes acute. Fruits tetragonous, shortly rounded at both ends; mericarps distinctly 5-ribbed. Pedicels shorter

than the fruits.—*Pozoa trifoliolata*, Hook. f., Fl. N.Z. i. 85, t. 18; Handbk. 87. *P. microdonta*, Col. in Trans. N.Z.I. xxiii. (1890) 387. *Hydrocotyle trifolia*, Banks and Sol. MSS.!

NORTH and SOUTH Islands: Hawke's Bay to Foveaux Strait. Sea-level to 2,600ft. Dec. to Feb.

10. **A. elegans,** *Col. in Trans. N.Z.I.* xxiii. (1890) 386. Densely tufted. Stems simple, erect, slender, succulent, glabrous, 3in. high. Leaves radical, 2–3-foliolate, 1in.–1½in. in diameter; leaflets orbicular, membranous, obscurely 3–4-lobed, crenate, petiolulate; stipules large, fimbriate. Peduncles 4–5 lines long. Umbels 2–3, 20–30-flowered. Involucral leaves linear, obtuse. Calyx-teeth subobovate. Immature fruit faintly ribbed.

SOUTH Island: "Sealy Range, 6,000ft.; the entire plant forming a big bunch or rosette with many flowers."—*H. Suter.*

Not having seen specimens of this, I have copied Colenso's description in a slightly condensed form. It may be an alpine variety of the preceding.

3. ERYNGIUM, Linn.

Calyx-lobes 5, acute, with pungent tips. Petals erect, deeply lobed; margins reduplicate or recurved, with a laciniate process from between the lobes. Margin of disk thickened. Ovary densely clothed with scales. Fruits subterete; ribs inconspicuous or 0; vittae 0. Herbs, with rigid prickly or spinous leaves, involucres, and involucels, the umbels reduced to globose or oblong bracts or heads.

The genus comprises about 150 species, distributed through warm and temperate countries. The only New Zealand species extends to Australia. Name of uncertain origin.

1. **E. vesiculosum,** *Labill, Pl. Nov. Holl.* i. 73, *t.* 98. Root stout. Radical leaves rosulate, lanceolate, oblong or spathulate, narrowed into a flat petiole, 1in.–3in. long or more, pinnatifid or deeply toothed; teeth spinescent. Stems 2in.–5in. long, prostrate but never rooting at the nodes. Peduncles ½in.–1in. long or more, axillary; heads globose or ovoid. Outer involucral bracts rigid, stellate, pungent. Flowers mixed with the projecting bracts of the involucels.—DC., Prod. iv. 92; Hook f., Fl. N.Z. i. 85; Handbk. 90; Benth., Fl. Austr. iii. 369.

NORTH and SOUTH Islands: on sandy beaches. East Cape and Poverty Bay to Oamaru on the east coast and Okarito on the west. Dec., Jan. Also a native of Australia.

4. ACTINOTUS, Labill.

Calyx-limb 5-lobed, rarely 0. Petals 5 or 0. Stamens 5, rarely 2. Ovary 1-celled, 1-ovuled; styles 2, free or coherent at the base. Fruit of a single carpel, crowned by the calyx-limb when present, dorsally compressed; ribs obvious or obscure; vittae 0. Annual or perennial tufted herbs. Leaves crenate-toothed, rarely entire or divided. Umbels simple, or the flowers capitate. Involucres deeply divided.

A small genus, restricted to Australia and New Zealand.

1. **A. Novae-Zelandiae,** *Petrie in Trans. N.Z.I.* xiii. (1881) 324. A minute tufted herb, forming small hoary patches. Stems ¼in.–1in. long. Leaves $\frac{1}{16}$in.–⅛in. in length, ciliate at the apex, entire, oblong or oblong-spathulate, narrowed into a canaliculate sheathing villous petiole. Peduncles ⅛in.–¾in. long, glabrate or villous. Involucral leaves 5 or more. Flowers 2–6. Calyx-limb obscure or 0. Petals 0. Stamens 2. Carpels compressed, convex on the outer face, obscurely 3-ribbed, 1-seeded. Stylopodia coherent nearly to the apex; stigmas short, divergent.—*A. suffocata*, Rodway in Bot. Notes (Tasm.) 2. *Hemiphues suffocata*, Hook. f. in Lond. Journ. Bot. vi. (1847) 471. *H. bellidioides*, var. *suffocata*, Fl. Tasm. i. 158, t. 36. *Actinotus bellidioides*, Benth., Fl. Austr. iii. 369.

SOUTH Island: Nelson: Heaphy River, *J. Dall!* Mount Rochfort, *Rev. F. H. Spencer!* Otago: Blue Mountains, *Petrie!* Longwood Range, *T. K.* STEWART Island, *Petrie* and *Thomson*, *T. K.* Sea-level to 3,500ft. Nov., Dec. Also in Tasmania.

This plant varies to a great extent in the degree of hairiness, some leaves being glabrous with the exception of a minute pencil of hairs at the tip. I have only seen a solitary head of perfect flowers. *Actinotus* was originally discovered in New Zealand by the Rev. F. H. Spencer, although first published by Mr. Petrie. Mr. Rodway was the first to point out its distinctive characters.

* BUPLEURUM, Linn.

Calyx-teeth 0. Petals rounded with a retuse inflexed point. Disk-lobes dilated. Fruit laterally compressed, subdidymous; commissure broad; carpellary ridges 5, filiform, winged, or 0; vittae 0. Styles short, reflexed. Seed flat or concave, deeply grooved in front. Perennial or annual herbs or rarely shrubs, with entire leaves, few- or many-rayed umbels. Bracts leafy or small or 0.

* **B. rotundifolium,** *L.* Annual. Stem 12in.–18in. high, fistular, branched above. Leaves oblong or oval, perfoliate, apiculate. Umbels small; rays many. General involucre 0; partial involucre of 3–5 ovate connivent bracteoles exceeding the rays.

NORTH Island: occasionally seen in cultivated land near Auckland and Wellington, but scarcely established. *Thorough-wax.* Dec., Jan. Europe, Western Asia.

* CONIUM, Linn.

Calyx-teeth 0. Petals obcordate, inflexed at the apex. Disk-lobes depressed. Fruit ovoid, laterally compressed, glabrous with obvious wavy acute ribs; lateral ribs marginal; interstices striate; commissure constricted; vittae 0. Seed deeply grooved in front. Styles short, reflexed. Biennial herbs, with erect branched stems and 3-pinnate leaves and terminal or axillary umbels.

* **C. maculatum,** *L.* Stems 2ft.–5ft. high, fistular, spotted with purple, foetid. Leaves large, deltoid, 3-pinnate, flaccid; lower leaflets stalked, pinnatifid, lanceolate or ovate-oblong; segments acute. Umbels on short terminal or axillary peduncles; rays numerous. Partial involucres unilateral, attenuate, shorter than the umbels.

Highly poisonous. Easily distinguished by the crenate ridge of the fruit, the glaucous spotted stem, and foetid odour.

NORTH and SOUTH Islands: naturalised in many localities from Auckland to Akaroa; often abundant. *Hemlock.* Oct., Nov. Europe, North Africa.

5. APIUM, Linn.

Calyx-teeth 0. Petals entire, acute or with a short involute point. Disk-lobes depressed or conical. Fruit roundish or ovoid, laterally compressed, didymous; commissure constricted; primary ridges 5, filiform, equal, obtuse; vittae solitary between the ridges. Seed subterete. Carpophore simple; styles divergent. Annual or perennial herbs, glabrous. Leaves pinnately or ternately divided. Umbels compound, terminal or lateral. Bracts and bracteoles many or few.

A small genus, distributed through the temperate or warmer regions, extending to Fuegia.

Leaves pinnate; leaflets toothed, cuneate below ..	* *A. graveolens.*
Leaves 3-foliolate or 2–3-pinnate. Umbels sessile ..	.. 1. *A. prostratum.*
Leaves dissected	* *A. leptophyllum.*

* **A. graveolens,** *L.* Biennial. Root fusiform. Stems erect, 2ft.–4ft. high or more, crowded or furrowed. Leaves pinnate or 3-foliolate; leaflets obovate or rhomboid, cuneate below, toothed or lobed, glossy. Umbels on slender peduncles or sessile, often leaf-opposed, usually with 1 or 2 ternately-divided leaves beneath. Bracteoles 0. Flowers small. Styles divergent. Fruits rounded; ribs prominent.

NORTH and SOUTH Islands: sparingly naturalised in many localities between Auckland and Akaroa; attaining over 6ft. in height in the Pelorus Sound. *Celery.* Dec., Jan. Europe, South Africa, West Asia.

1. **A. prostratum,** *Labill, Relat.* i. 141; *Pl. Nov. Holl.* i. 176, *t.* 103. Root stout. Stems prostrate or suberect, 6in.–24in. long, stout, grooved. Leaves excessively variable, 2–3-pinnate; leaflets sessile or petioled, variously lobed or cut, membranous or subcoriaceous. Umbels compound, axillary or leaf-opposed, sessile; rays 3–12, ½in.–1½in. long; pedicels ¼in. long. Involucral bracts 0. Carpels ovoid. Primary ribs prominent; vittae obscure.—*A. australe,* Thou., Fl. Trist. d'Acugn. 43; Hook. f., Fl. N.Z. i. 86; Handbk. 90; Benth., Fl. Austr. i. 372. *Petroselinum prostratum,* DC., Prod. iv. 102; A. Rich., Fl. Nouv.-Zel. 278; A. Cunn., Precurs. n. 503. *Apium decumbens, α sapidum,* Banks and Sol. MSS.

Var. **filiforme** (*sp.*), Hook., Ic. Pl. t. 819. Stems slender, prostrate. Leaves 3-foliolate; leaflets lobed or incised, rarely pinnate. Umbels smaller; rays few, rarely 1.—*A. filiforme,* Hook. f., Fl. N.Z. i. 87; Handbk. 90. *Petroselinum filiforme,* A. Rich., Fl. Nouv.-Zel. 278; A. Cunn., Precurs. n. 504. *Apium decumbens, β tenellum,* Banks and Sol. MSS.

From the KERMADEC Islands to STEWART Island: common, littoral. Var. *filiforme* is occasionally found inland. *Wild celery.* Dec. to Feb. Also in Australia, South Pacific Islands, South America, South Africa, Tristan d'Acunha, and St. Paul's Island.

* **A. leptophyllum,** *F. Muell.* ex *Benth., Fl. Austr.* iii. 372. Stems usually slender, suberect or spreading, 6in.–12in. high. Leaves 1in.–1½in. long; petioles slender, sheathing, 3-pinnate; the leaflets ternately dissected into many narrow-linear-acute or mucronate segments. Umbels leaf-opposed, compound; rays 1–3. Involucral bracts 0. Pedicels short. Flowers white. Styles very short. Ribs turgid, with a single oil-tube under each furrow.

NORTH Island: naturalised in various localities from Mangonui to Wellington, but local. SOUTH Island: Nelson. Nov. to March. Australia, North and South America.

* AMMI, Tourn.

Calyx-teeth 0. Petals emarginate; lobes unequal, irregular. Fruits laterally compressed, ovate-oblong. Carpels 5-ribbed; oil-canals 5. Annual or biennial herbs, with erect or suberect stems, pinnate or pinnatisect leaves, compressed umbels, and pinnate or pinnatifid involucres.

* **A. majus,** *L.* Glabrous. Lower leaves pinnate; pinnules lanceolate, serrulate, apiculate; upper leaves pinnatisect; involucral leaves 3-fid. Flowers white.

NORTH Island: naturalised in the Auckland Domain, *Cheeseman.*

* CARUM, Linn.

Calyx-teeth small or 0. Petals obcordate, usually with the tip inflexed. Stylopodium conical. Fruit ovate or oblong, glabrous or hispid, compressed laterally; carpophore 2-fid; carpellary ridges 5, filiform; vittae usually solitary, rarely 1–3 together. Glabrous herbs. Roots fusiform or tuberous. Leaves pinnate or 2–3-pinnate. Bracts and bracteoles few or 0. Flowers perfect, dioecious or polygamous.

Leaves dissected; segments filiform ... * *C. Carui.*
Leaves 2–3-pinnate; segments broad ... * *C. Petroselinum.*

* **C. Carui,** *L.* Root fusiform, biennial. Stem erect, slender, 1ft.–2ft. high, branched, glabrous. Leaves 2-pinnate; leaflets cut into opposite filiform segments. Umbels irregular. General involucre reduced to 1 leaf or 0; partial involucres 0. Vittae prominent. Styles spreading. Carpels aromatic.

SOUTH Island: near Dunedin, *A. Hamilton!* Local. *Carraway.* Dec. Europe.

* **C. Petroselinum,** *Benth.* Root fusiform, biennial. Stem erect, branched, solid. Leaves deltoid, 2- or 3-pinnate, shining; segments cuneate at the base, crenate or toothed; cauline leaves often with entire linear segments. Umbels regular, flat. Bracts of general involucre often divided; bracteoles filiform. Flowers yellow.

Not infrequent from Auckland to Otago, but closely eaten down by sheep. *Parsley.* Dec., Jan.

6. OREOMYRRHIS, Endl.

Calyx-teeth 0. Petals shortly inflexed at the tip, imbricate in bud. Disk continuous with the base of the styles. Fruit linear or ovate-oblong, narrowed above, slightly compressed laterally. Carpels with 5 obtuse prominent ridges and a vitta under each furrow. Seed nearly terete. Perennial silky hairy or glabrous scapigerous herbs, with pinnate or 2–3-pinnate leaves and simple pedunculate umbels. Involucral bracts small, linear or ovate.

A small genus, comprising about 6 species, chiefly Andine, ranging from Mexico to Fuegia, with a single species extending to Australia and New Zealand. Etym. From the Greek, in reference to the montane habitat of most of the species, and to *Myrrhis*, a closely-related plant.

1. **O. andicola,** *Endl., Gen. Pl.* 787. Solitary or densely tufted, nearly glabrous or hairy or silky, slender or rigid. Radical leaves few or many, 1in.–6in. long, linear-oblong, pinnate or 2-pinnate; leaflets sessile or petioled, broadly oblong, pinnatifid or incised; segments acute or obtuse. Scapes simple or branched, 2in.–18in. long. Umbels 2–30-flowered. Involucral leaves 6–8, ovate or linear. Flowers sessile in the involucres or nearly so. Pedicels lengthening in fruit, ¼in.–2in. long, unequal. Fruit linear or ovate-oblong, glabrous, pubescent, or almost hispid. Carpels glabrous or pubescent.—Benth., Fl. Austr. iii. 377; F. Muell., 2nd Cens. Austr. Pl. 108. *O. eriopoda,* Hook. f., Fl. Tasm. i. 162. *O. argentea,* Hook. f., *l.c.,* and in Hook. Ic. Pl. t. 300. *Caldasia andicola,* Lag. *ex* DC., Prod. iv. 229. *Myrrhis andicola,* H. B. and K., Nov. Gen. et Sp. v. 13, t. 419.

NORTH and SOUTH Islands: from the East Cape to the Bluff. CHATHAM Islands. Sea-level to 4,500ft. Nov. to Feb. Also in Australia and America, where it ascends to nearly 15,000ft., descending to sea-level on the Falkland Islands.

The amount of hairiness is extremely variable, but I follow Bentham in uniting the different species described in the Handbook under this. The following are the principal forms observed in New Zealand, but they pass into each other so insensibly that it is impossible to draw permanent lines of separation. The hairs on the scape and pedicels usually point downward.

Var. **Colensoi** (*sp.*), Hook. f., Fl. N.Z. i. 92. Glabrate or hairy. Leaves all radical, pinnate or 2-pinnate; leaflets incised or pinnatifid. Scapes simple, naked. Involucral leaves 6–8, usually ovate.—Handbk. 91. *O. Haastii*, Hook. f. *l.c.* 91, is a form with pedicels shorter than the tomentose fruits.

Var. **ramosa** (*sp.*), Hook. f., Handbk. 91. Stems 12in.–24in. high, very slender, much branched, glabrescent or hairy; hairs spreading. Leaves usually pinnate; leaflets membranous, distant, lobed incised or partite or again pinnate, obtuse or subacute. Peduncles axillary, longer or shorter than the leaves, sometimes 1ft. long, 2–8-flowered. Pedicels often 2in. long. Involucral leaves usually linear-acute. Fruits glabrous or almost tomentose.

Var. **apiculata.** Stems branched, almost capillary. Peduncles short. Leaves glabrous, deltoid-ovate, ternately divided; leaflets mostly 3-foliolate; segments petioled, lobed and toothed, apiculate.—*Ligusticum trifoliolatum*, Hook. f., Handbk. 97? SOUTH Island: Invercargill, *W. S. Hamilton!* Better specimens of this curious plant are badly wanted; it may prove a distinct species.

Var. **rigida.** Stems stout and branched at the very base only, 6in.–8in. high. Leaves twice pinnate; segments acute, pubescent, hairy or woolly. Peduncles depressed, 6in.–8in. long. Pedicels stout. Involucral leaves ovate, oblong or linear. Fruits linear, downy or glabrous.—*O. andicola*, Hook. f., Fl. Antarc. ii. 288, t. 101.

* SCANDIX, Tourn.

Calyx-teeth minute or 0. Petals obovate, with a short inflexed point or 0. Disk dilated. Fruit compressed or contracted at the sides, almost cylindrical, produced into a long beak; carpophore simple or 2-fid. Carpels subterete; primary ridges filiform, secondary 0; vittae solitary in the interspaces or 0. Seed deeply furrowed in fruit. Annual herbs, with pinnate decompound leaves and simple or compound umbels. Flowers unequal, polygamous.

* **S. Pecten-veneris,** *L.* Stem erect, 1ft. high or more, branched, pubescent, spreading. Leaves 3-pinnate; segments short, linear. Umbels axillary or terminal, small; rays few, short. Bracteoles often divided and longer than the pedicels. Fruit rough, dorsally compressed; edges finely setose; beak three times longer than the fruit.

NORTH and SOUTH Islands: cultivated and waste land, but often local. *Shepherd's needle.* Dec. to Feb. Europe.

* FOENICULUM, Tourn.

Calyx-teeth 0. Petals entire, with a broad inflexed obtuse lobe. Stylopodium large, conical. Fruit oblong or ovate; commissure broad. Carpels subterete; primary ridges 5, obtuse, stout; vittae solitary in the interspaces. Seed flat or concave in front, furrowed. Erect glabrous annual or perennial herbs, with decompound leaves and compound umbels of yellow flowers. General and partial involucres 0.

* **F. vulgare,** *Mill.* An aromatic perennial, stout, erect, 4ft.–6ft. high, terete. Leaves spreading, dissected into innumerable filiform or capillary channelled segments. Umbels large; rays numerous. Flowers yellow.

Naturalised throughout the colony, especially in places near the sea. *Fennel.* Feb., March. Europe.

7. CRANTZIA, Nutt.

Flowers minute. Calyx minutely 5-toothed. Petals acute, imbricate in the bud. Fruit ovoid-globose, slightly compressed laterally. Carpels with 5 corky ribs separated by narrow furrows, a vitta under each furrow and 2 on the commissure. A small creeping herb, with solitary or tufted erect cylindrical leaves springing from the nodes, and simple few-flowered axillary umbels.

A monotypic genus, extending to Australia, extra-tropical North and South America, and the Falkland Islands. Named in honour of *Henry John Crantz*, an Austrian botanist.

1. **C. lineata,** *Nutt., Gen. N. Am. Pl.* i. 177. Rhizomes slender, rooting at the nodes. Leaves ½in.–5in. long, linear-fistular, internally septate, obtuse, sometimes compressed. Peduncles axillary, solitary, filiform, 2–6-flowered. Involucral leaves minute. Pedicels 1–2 lines long, capillary, spreading. Disk confluent with the conical base of the style. Fruits 1–2 lines long.—Hook. f., Fl. Antarc. ii. 287, t. 100; Fl. N.Z. i. 87; Handbk. 89. *Hydrocotyle lineata*, Monog. Hydr. 77, f. 38.

From the North Cape to STEWART Island; CHATHAM Islands. In wet places, especially near the sea; rarely in running water. Ascends to fully 2,000ft. Nov. to Feb.

The leaves are usually compressed when growing in elevated situations.

8. LIGUSTICUM, Linn.

Calyx-teeth usually obsolete or minute, sometimes unequal. Petals inflexed at the tip. Stylopodium conical. Fruit linear-oblong, elliptic-oblong, or ovate-oblong. Carpels dorsally compressed, each with 5 equal winged ridges, or one or both carpels with 3 winged ridges; vittae often obscure. Glabrous or rarely pilose or silky erect perennial herbs, from 1in.–4ft. high, often robust, aromatic or strong-smelling, with pinnate or decompound leaves, the rhachis jointed at its juncture with the leaflets. Umbels polygamous, compound or rarely simple, often panicled. Flowers white or red; female sometimes on very short pedicels. Herbs, often of large stature, with pinnate or decompound leaves.

Much uncertainty exists as to the limits of this genus. Bentham restricts it to species of the Northern Hemisphere in which vittae are well developed, and unites those of the Southern Hemisphere with *Aciphylla*, in *Genera Plantarum*. I prefer to follow Hooker in placing them under *Ligusticum*, although it might be better to restore the genus *Anisotome* for their reception. All the New Zealand species of this section are endemic. Two or three others are found in South America, and one in Tasmania. Name from *Liguria*, the home of the officinal species.

* *Leaves 2–3-pinnate or decompound.*

† Robust species 2ft.–4ft. high.

Leaflets decurrent, with pungent lobes .. .. 1. *L. latifolium.*
Leaves decompound; segments subulate .. .. .. 2. *L. antipodum.*
Sheath of petiole produced into a hooded ligule .. .. .. 3. *L. acutifolium.*
Leaflets contracted at the base; lobes obtuse .. .. 4. *L. intermedium.*
Leaflets cut into narrow-linear lobes .. .. 5. *L. Lyallii.*
Leaves decompound; lobes hair-pointed .. .. 6. *L. Haastii.*

†† Slender species rarely exceeding 1ft. in height.

Leaves ovate-lanceolate, acute, 3-pinnate. Umbels few or many .. 7. *L. dissectum.*
Leaves small, 2–3-pinnate; segments almost capillary. Umbels few, small .. .. 8. *L. politum.*
Leaves 2-pinnate; lobes very narrow-linear, piliferous .. .. .. 9. *L. brevistyle.*
Erect, very slender. Leaflets cut into linear-cuneate or almost filiform segments. Umbels compound .. .. 10. *L filifolium.*

Small. Leaves broadly deltoid. Scape simple	.. 11. *L. deltoideum*.
Stems spreading, slender. Leaflets distant. Umbels 6–10-flowered	.. 12. *L. patulum*.
Leaves and umbels all radical, fleshy, erect	.. 13. *L. carnosulum*.
** *Leaves pinnate or 3-foliolate.*	
Leaves oblong, obtuse, pinnate, or leaflets pinnatifid	.. 14. *L. piliferum*.
Rhachis broad. Leaflets toothed or pinnatifid, piliferous or woolly	.. 15. *L. aromaticum*.
Leaflets irregularly pinnatifid, not piliferous	16. *L. decipiens*.
Densely tufted, palmately 3–5-lobed; lobes piliferous ..	.. 17. *L. imbricatum*.
Leaflets in 2–3 pairs, often involute, fan-shaped, entire..	.. 18. *L. flabellatum*.
Leaflets in 3–6 pairs, glaucous, sharply toothed ..	.. 19. *L. Enysii*.
Leaves 3-foliolate or pinnate; leaflets petioled..	.. 20. *L.? trifoliolatum*.

1. **L. latifolium,** *Hook. f., Handbk.* 94. A noble plant, 3ft.–5ft. high or more. Stem 3in.–4in. in diameter at the base, grooved. Radical leaves 1ft.–2ft. long or more; petioles ¾in.–1in. in diameter, sheathing at the base; sheath shortly ligulate*; blade ovate, excessively coriaceous, 2-pinnate; primary divisions sessile, 2in.–4in. long, linear-oblong; segments obliquely cuneate below, with broad winged bases, unequally 3–5-lobed; lobes acuminate, with acute points and thickened margins. Bracts large, 2in. broad, concave base with foliaceous tips. Umbels numerous, 2in.–3in. in diameter, mostly polygamous, but some entirely male. Involucral leaves numerous as long as the rays, linear, acute. Flowers reddish. Calyx unequally 5-toothed. Petals not inflexed. Fruit ⅛in. long. Carpels with 5 primary ridges, rarely 4 or 3, and a single vitta under each furrow.—*Anisotome antarctica*, Hook. f., Fl. Antarc. i. 16, *t.* 8. *Calosciadium latifolium*, Endl. *ex* Walp. Ann. ii. 702.

Var. **angustatum.** Ultimate segments of the leaves ⅛in. in diameter or less, acicular points longer.

AUCKLAND and CAMPBELL Islands. Dec., Jan.

The numerous crowded umbels often form large masses 6in.–10in. in diameter.

2. **L. antipodum,** *Homb. and Jacq.* ex *Dcne, Bot. Voy. Astrol. et Zél.* 63, *t.* 3. Stem stout, 3ft. high, deeply furrowed. Leaves 1ft.–2ft. long; petioles stout, sheath shortly ligulate at the apex; blade oblong or broadly oblong, 2–3-pinnate, coriaceous, dissected into countless narrow-linear subulate segments $\frac{1}{12}$in. broad, with acicular points. Umbels mostly unisexual, smaller and less numerous than in *L. latifolium*, compound. Bracts narrower, inflated. Involucral leaves longer, acuminate. Calyx unequally 5-toothed. Petals inflexed at the tip. Fruit ¼in. long. Styles short, narrow-oblong. One carpel with 5 wings, the other 3-winged.—Hook. f., Handbk. 94. *Anisotome antipoda*, Hook. f., Fl. Antarc. i. 17, t. 9, 10. *Calosciadium antipodum*, Endl. *ex* Walp., Ann. ii. 702.

AUCKLAND and CAMPBELL Islands, *Hook. f.* ANTIPODES Island, *T. K.* Dec., Jan.

Readily distinguished from *L. latifolium*, to which it is closely allied, by the dissected leaves with linear acicular segments.

* The term is here applied to a membranous process arising from the sheathing base of the petiole in certain species.

3. **L. acutifolium,** *T. Kirk in Journ. Bot.* (1891) 237. Stems stout, deeply furrowed, 3ft.–5ft. high. Rootstock as thick as a man's wrist. Leaves spreading, 2ft. long or more, 6in.–9in. broad, oblong or broadly ovate-oblong, 3-pinnate, coriaceous; segments large, acute, sharply toothed; petiole 1ft. long or more, finely grooved, with the upper part of the sheath ligulate for half its length. Flowers not seen. Fruiting-umbels 2in.–2½in. in diameter, compound, dense; rays numerous, about 1in. long. Fruits $\frac{3}{16}$in. long, exceeding the pedicels. One carpel 5-ribbed, the other 3-ribbed. Involucral bracts apparently 0.

The SNARES, *T. K.* Dec.

Distinguished from *L. intermedium* by the more sharply toothed leaves, the ligulate leaf-sheath, the broader fruits, and the absence of milky juice.

4. **L. intermedium,** *Hook. f., Handbk.* 94. Rather stout, 6in.–24in. high. Leaves 4in.–18in. long or more, charged with white milky juice, 4in.–6in. broad; petioles with long narrow sheathing bases; blade coriaceous, oblong, 2–3-pinnate; primary divisions in 5–8 pairs, 2in.–4in. long; leaflets ½in.–1in. long, broad, sessile or shortly petioled, deltoid-ovate, often cuneate at the base, unequally lobed or toothed or cut into crowded linear-obtuse or subacute segments $\frac{1}{14}$in.–$\frac{1}{10}$in. broad. Umbels polygamous or unisexual, few or many, 1½in.–2½in. in diameter, compound. Involucral leaves linear, acute, shorter than the rays; secondary involucres obtuse, exceeding the winged pedicels. Flowers white. Calyx-teeth minute, unequal. Petals slightly inflexed at the tip. Fruits longer or shorter than the pedicels, linear-oblong. One carpel with 5 ridges, the other with 4, and 1 vitta under each furrow. Seed grooved.—*Anisotome intermedia*, Hook f., Fl. N.Z. i. 94.

SOUTH Island: littoral. North shore of Foveaux Strait from the Nuggets to Preservation Inlet, and northward to Martin's Bay. STEWART Island and islands of Foveaux Strait. Dec., Jan.

Var. **oblongifolium.** Leaves linear-oblong, 2in. broad, 3-pinnate; segments narrow-linear, crowded, subacute. Umbels very numerous, almost corymbose. STEWART Island: inland base of the Ruggedy Range.

5. **L. Lyallii,** *Hook. f., Handbk.* 95. Similar to the preceding but stouter, 1½ft.–2ft. high, purplish, obscurely grooved. Stem 1in.–2in. thick. Leaves linear-oblong, 2–3-pinnate; leaflets 8–10 pairs, linear-oblong; pinnules crowded, 1in. long, obovate-cuneate, cut to the base into linear-obtuse lobes, $\frac{1}{12}$in. broad, 1-nerved; petiole as thick as the little finger, with a narrow sheathing base. Flowers not seen. Fruiting-umbels compound. Fruit ¼in. long, exceeding its pedicel, linear-oblong.—*Anisotome Lyallii*, Hook. f., Fl. N.Z. i. 88.

SOUTH Island: Port Preservation, sounds of the West Coast, *Hector ex* Handbk.

I have copied the description from the Handbook, as the plant appears to me only a form of *L. intermedium* with excessively divided leaves, a character of small importance in the genus; but I hesitate to unite it with that species, not having had the opportunity of examining the type specimen. Some pinnules kindly sent by Sir Joseph Hooker agree closely with *L. intermedium* from Preservation Inlet.

6. **L. Haastii,** *F. Muell.* ex *Hook. f., Handbk. N.Z. Fl.* 95. Dioecious. Stem 1ft.–2ft. high, sparingly branched, grooved. Leaves 6in.–18in. long,

narrow-oblong or ovate-oblong, 2–4-pinnate; leaflets $\frac{1}{4}$in.–$\frac{3}{4}$in. long, cut into crowded narrow flaccid linear hair-pointed lobes $\frac{1}{30}$in.–$\frac{1}{24}$in. broad, 1-nerved; petiole long; sheath narrow, purple. Umbels numerous on spreading peduncles. Umbels few or many, 1in.–3in. in diameter; compound rays $\frac{3}{4}$in.–$1\frac{1}{2}$in. long. Involucral leaves filiform, shorter than the rays. Fruit ovoid-oblong, $\frac{3}{16}$in. long. Carpels compound, 5-winged; styles very slender.

SOUTH Island: not uncommon in mountain districts from Pelorus Sound to the lake district of Otago and Southland. 1,500ft. to 4,500ft. Dec., Jan.

Distinguished from all its congeners by the long membranous decompound leaves.

7. **L. dissectum,** *n. s.* Erect, 6in.–16in. high, rather stout. Stem striated. Radical leaves 3in.–12in. long, $1\frac{1}{2}$in.–$2\frac{1}{2}$in. broad, ovate-lanceolate or oblong-lanceolate, coriaceous, acute, 2–3-pinnate; leaflets in 4–9 pairs, petioled, 1in.–2in. long or more, often imbricating, pinnately divided into crowded linear rigid lobes $\frac{1}{20}$in.–$\frac{1}{16}$in. broad, tipped with short bristles; petiole exceeding the blade; sheath narrow. Rhachis with an obvious midrib. Umbels few or many, compound, paniculate, 1in.–3in. in diameter; bracts of involucre linear-acuminate; primary rays $\frac{1}{2}$in.–3in. long, secondary shorter. Pedicels winged, shorter than the linear-oblong fruit. Flowers white, dioecious or polygamous. Calyx-tube acute. Carpels 5-winged.

NORTH Island: Mount Holdsworth and other peaks of the Tararua Range, *T. P. Arnold! Buchanan!* 3,500ft. to 6,000ft. Jan.

Related to *L. piliferum*, from which it is separated by the acute more-divided leaves, longer lobes, and more linear fruits.

8. **L. politum,** *n. s.* Tufted. Stem erect, polished, 2in.–6in. high. Leaves ovate-oblong, green, membranous or rarely stiff, but not rigid, 2in.–4in. long, 2- rarely 3-pinnate; petiole about as long as the blade, with a narrow sheath tipped at the mouth by a pair of simple or divided leaflets; leaflets in 4–6 pairs, divided into narrow-linear hair-tipped segments not exceeding $\frac{1}{40}$in. wide or almost capillary. Panicles subterminal. Umbels compound, $\frac{1}{4}$in.–1in. in diameter. Male: 4 or 5, crowded each in the axil of a broad dilated sometimes leafy bract; rays 3–6; involucral bracts linear-acuminate; pedicels very short. Flowers white. Female umbels smaller, with narrow bracts and longer peduncles. Calyx-teeth minute. Fruit (immature) oblong. Carpels 5-ribbed or one 4-ribbed. Stylopodia conical; styles rather stout.

SOUTH Island: Ben Nevis, Mount Starveall, and Mount Lunar, *Gibbs! Bryant! Kingsley!* Dec., Jan.

A small delicate species, with the habit of *Aciphylla Monroi*, but smaller in all its parts; never truly coriaceous or pungent, although sometimes stiff; quite unlike any other species. Better material is required for a good diagnosis.

9. **L. brevistyle,** *Hook. f., Handbk.* 95. Dioecious, erect, 8in.–15in. high, slender. Radical leaves 4in.–6in. long, scarcely sheathing, linear-oblong, 2-pinnate; leaflets about $\frac{1}{2}$in. long, cleft to the rhachis into 3–5 almost filiform hair-pointed segments $\frac{1}{6}$in.–$\frac{1}{3}$in. long. Umbels 1–3, compound; rays unequal,

$\frac{1}{2}$in.–1in. long or more. Involucral bracts shorter than the rays. Calyx-teeth 0 in the female flower. Fruits on very short pedicels, oblong, $\frac{1}{8}$in.–$\frac{1}{5}$in. long. Styles very short.

SOUTH Island: Canterbury: Lake Hawea and Waitaki, *Haast!* Otago: Lake district, *Hector* and *Buchanan!* Kurow and Mount Ida Ranges, Nevis Valley, &c., *Petrie!* 800ft. to 1,600ft. Nov., Dec.

10. **L. filifolium,** *Hook. f., Handbk.* 95. A very slender dioecious species, 4in.–12in. high or more. Stems simple or branched above, with a small leaf at the base of each branch. Radical leaves 3in.–10in. long, on long and slender petioles, 2–3-pinnate; leaflets ternately divided into narrow-linear flat acute segments varying in width from almost capillary or filiform to $\frac{1}{8}$in. broad, the broadest cuneate at the base and 2–5-lobed or toothed. Umbels few, compound, narrow; rays unequal. Involucral leaves short, lanceolate or ovate, acute; rays unequal, mixed with pedicels. Pedicels unequal, $\frac{1}{20}$in.–$\frac{1}{2}$in. long. Flowers minute. Calyx-teeth 0 in the female flowers. Fruit compressed, linear-oblong, $\frac{1}{4}$in. long. Carpels with 5, rarely 4, prominent ridges; vittae obscure.

SOUTH Island: mountains of Marlborough, Nelson, Westland, and Canterbury. 2,000ft. to 3,500ft. Dec., Jan.

11. **L. deltoideum,** *Cheesem. in Trans. N.Z.I.* xiv. (1881) 299. A small slender aromatic species, 2in.–6in. high. Rootstock stout, covered with chaffy scales above. Leaves all radical, 2in.–4in. long including the petiole, broadly deltoid, 2-pinnate; secondary leaflets ternately divided into acute or acuminate linear-cuneate spreading lobes, not hair-pointed. Scapes shorter than the leaves, usually naked. Umbels $\frac{1}{2}$in.–1in. in diameter, few-flowered. Pedicels very short. Ripe fruit not seen.

SOUTH Island: Mount Arthur Plateau, Nelson, 4,500ft., *Cheeseman!*

In the absence of ripe fruit this may be only a variety of *L. filifolium*, some forms of which have deltoid leaves, differing only in the rather broader segments.

12. **L. patulum,** *n. s.* Stems very slender, sparingly branched, inclined, spreading, 6in.–12in. high or more. Radical leaves 3in.–6in. long or more, pinnate or 2-pinnate or pinnatifid; leaflets in 6–7 distant pairs, petioled, laxly divided into very narrow-linear toothed or tridentate acute lobes, rarely entire. Rhachis very slender, scarcely jointed. Cauline leaves smaller, with narrower segments. Umbels simple, on very slender axillary or terminal peduncles, 6–10-flowered. Involucral bracts linear-ovate, shorter than the longest pedicels. Pedicels unequal. Calyx-teeth extremely minute, acute. Petals inflexed at the tips. Fruit linear-oblong. Styles very slender.

SOUTH Island: Canterbury: on limestone, Burke's Pass, *J. B. Armstrong!* Otago, *Buchanan!*

The grey tint of the leaves, spreading stems, and few-flowered umbels render this species easily recognised. I have only seen two specimens, both in poor condition.

13. **L. carnosulum,** *Hook. f., Handbk.* 96. Leaves and umbels all radical; petioles and peduncles tapering downwards, 1in.–6in. long or more, the former tumid, $\frac{1}{4}$in. in diameter; the rhachis jointed at the junction and ter-

nately divided at the apex into petioled leaflets, which are 2–3-ternately or -pinnately cut into glaucous linear subulate fleshy lobes $\frac{1}{4}$in.–$\frac{2}{3}$in. long. Umbels compound, glaucous; rays 12–20 or more, $\frac{1}{3}$in.–1in. long; umbellules simple, 5–10-flowered. Flowers on short stout pedicels, dioecious or polygamous, overtopped by the secondary involucral leaves, which are divided into 3–5 linear segments at the apex. Calyx-teeth acute, prominent. Stylopodia conical; style slender. Fruit oblong. Carpels with 4 obscure scarcely acute ridges.

SOUTH Island: on shingle-slips, mountains of Nelson and Canterbury. Wairau Gorge, *Cheeseman.* Mount Captain, Amuri, *T. K.* Mount Torlesse, *Haast!* Leith Hill, *Enys!* 5,000ft. to 6,500ft. Jan.

One of the most remarkable plants in the order; now very rare, being everywhere cropped by sheep.

14. **L. piliferum,** *Hook. f., Handbk.* 96. Dioecious, erect, robust. Stems 10in.–20in. high, sparingly branched above. Leaves 3in.–12in. long, linear-oblong, very coriaceous, glaucous, red-purple, pinnate; leaflets in 8–12 pairs, sessile, often imbricating, ovate orbicular-ovate or deltoid-ovate, entire with few rounded apiculate lobes or teeth or 2–3-lobed to the base, the margins lobulate and thickened, tipped with a stout bristle; petiole and rhachis $\frac{1}{4}$in.–$\frac{1}{3}$in. broad, deeply striated; sheath long and narrow. Umbels few, on stout peduncles, compound, 2in.–3in. in diameter; rays $\frac{3}{4}$in.–1$\frac{1}{2}$in. long, unequal; involucral bracts linear-oblong, shorter than the rays. Flowers rather small, white. Calyx-teeth obscure. Fruit ovate-oblong. Carpels 3-winged or one 5-winged. Styles slender.

Var. **pinnatifidum.** Leaves green, less coriaceous, longer, pinnatifidly divided into crowded narrow-linear bristle-pointed segments.

SOUTH Island: mountains of Marlborough, Nelson, Canterbury, and western Otago, 2,800ft. to 4,000ft.

In some small-leaved specimens the segments are almost capillary, $\frac{1}{8}$in.–$\frac{1}{4}$in. long, shorter than the hair-like points.

15. **L. aromaticum,** *Hook. f., Handbk. N.Z. Fl.* 96. Dioecious, aromatic. Scapes 1in.–10in. high, simple or sparingly branched. Leaves all radical, spreading, coriaceous, shining, 1in.–6in. long, $\frac{1}{2}$in.–1in. broad, linear-oblong, pinnate; leaflets sessile, 6–12 pairs, $\frac{1}{6}$in.–$\frac{1}{2}$in. long, ovate deltoid-ovate or orbicular, toothed, incised, pinnatifid or even pinnate; lobes and teeth piliferous; petiole stout; sheath short, broad. Umbels small, $\frac{1}{2}$in.–1$\frac{1}{2}$in. broad, open or compact; rays unequal, slender, $\frac{1}{4}$in.–2in. long. Involucral bracts very short, linear-acute, subulate. Flowers white. Calyx-teeth inconspicuous. Style slender. Fruit on short pedicels or almost sessile, linear-oblong, $\frac{1}{10}$in.–$\frac{1}{7}$in. long. Carpels 5-winged.—*Anisotome aromatica*, Hook. f., Fl. N.Z. i. 99.

NORTH and SOUTH Islands: very frequent in mountain districts from the East Cape to Southland. 2,000ft. to 6,000ft. Nov. to Jan.

Var. **incisum.** Larger. 12in.–20in. high. Leaves more membranous; leaflets fan shaped or rhomboid, 3-partite nearly to the base; segments deeply incised; lobules toothed, spreading, hair-pointed. In limestone districts, Broken River, Canterbury, &c., *Enys* and *Kirk.*

Var. **lanuginosum.** Leaf-segments and teeth tipped with copious woolly snow-white hairs, which completely hide the upper surface of the leaf. Peduncles stout. Umbels compact. Mountains above Lake Tekapo, *Cheeseman!* Hector Mountains, Mount Pisa, Mount Cardrona, &c., Otago, *Petrie!*

16. **L. decipiens.** Root stout. Scape 3in.–6in. high, very slender, simple. Leaves usually radical, spreading, 3in.–6in. long, linear-oblong, pinnate; leaflets in 5–10 pairs, ¼in.–½in. long, flaccid, unequally pinnatifid; lobes linear or often reduced to mere teeth, acute or subacute. Scape naked or with 1 or rarely 2 cauline leaves, sometimes equalling or overtopping the scape. Flowers minute. Umbels ½in.–1½in. in diameter, compound; rays unequal. Involucral leaves dilated at the base. Calyx-teeth acute. Styles slender, recurved. Fruit narrow-ovate-oblong, not cordate at the base in my specimens; wings narrow, coriaceous. Pedicels shorter than the carpels.—*Angelica decipiens*, Hook. f., Handbk. 98. *Aciphylla decipiens*, Benth. and Hook. f., Gen. Pl. i. 916.

NORTH Island: "I have an imperfect specimen, apparently of this plant, from Colenso," *Hooker f.* SOUTH Island: not uncommon in the Southern Alps from Nelson to Otago. 2,000ft. to 6,000ft. Nov., Dec.

Closely resembling *Ligusticum aromaticum*, which is often mistaken for it. Perhaps most readily recognised by the deeply-pinnatifid leaflets with wide sinuses. The wings are usually very narrow.

17. **L. imbricatum,** *Hook. f., Handbk.* 97. Dioecious, forming large flat or convex green patches. Stems stout, much branched, clothed with densely-imbricating coriaceous shining leaves. Leaves ¼in.–⅔in. long; petioles short, broadly sheathing, the sheath produced upwards into a hooded ligule; leaflets in 4–7 pairs, sessile, palmate, 4–5-lobed, each lobe tipped by a stout bristle twice the length of the lobe. Rhachis flattened, broad. Umbels small, simple or compound, hidden amongst the apical leaves. Calyx-teeth acute. Fruit orbicular-ovoid. Carpels with 5 prominent wings.

SOUTH Island: on high mountain peaks from Marlborough to Southland. 4,500ft. to 6,500ft. Jan., Feb.

In early leafy shoots the produced portion of the sheath is not developed, while the bristle-points are very short and slender. The broad fruits distinguish it from all forms of *L. aromaticum*.

18. **L. flabellatum,** *n. s.* Polygamous, ½in.–1½in. high. Rootstock stout, penetrating the rock to a considerable depth. Leaves ½in.–2in. long, pinnate, linear, very coriaceous; leaflets in 1–3 pairs, often reduced to a single leaflet, sessile, fan-shaped or rounded-rhomboid, entire or minutely sinuate-crenate; margins usually involute; sheath very short and broad. Rhachis obscurely jointed. Scape decumbent, shorter than the leaves, with a small inflated bract above the middle. Umbels small; rays 3 or 4. General involucre 0; partial involucre of connate bracts open on one side; pedicels unequal. Flowers minute, white. Calyx-teeth extremely minute, acute. Fruit almost orbicular. Stylopodia conical at base. Carpels 4-winged or one 5-winged; vittae 4–5.

STEWART Island: in crevices of syenitic rocks near the South Cape. Jan.

Distinguished from all other species by the 3-lobed partial involucres and the almost orbicular carpels. One of the rarest plants in the colony. First observed by Dr. Lyall.

19. **L. Enysii,** *T. Kirk in Trans. N.Z.I.* ix. (1876) 548. Dioecious, depressed, 2in.–3in. high, solitary. Leaves all radical, few, spreading, when fresh excessively thick and glaucous, linear-oblong, pinnate; leaflets in 3–6

pairs, sessile, ovate or ovate-acuminate, sharply toothed or lobed, not piliferous. Stem decumbent, 2in.–4in. high, simple or with a single branch. Umbels of 3–5 unequal spreading rays. Involucre cup-shaped, consisting of 2 broadly-ovate connate apiculate bracts. Umbellules 3–6-flowered. Pedicels equalling or exceeding the ovate fruits. Carpels with 5 obscure ridges.

SOUTH Island: on limestone. Trelissick basin, Canterbury, *Enys* and *Kirk*. Naseby, Otago, *Petrie!* 1,800ft. to 2,500ft. Jan.

Distinguished from all other species by the entire cup-shaped involucres.

20. **L. (?) trifoliolatum,** *Hook. f., Handbk.* 97. Small, glabrous. Stem slender, 6in. high, sparingly divided above. Leaves 3-foliolate or pinnate; leaflets in 1 or 2 distant pairs, $1\frac{1}{2}$in. long, on slender petioles, rhombeo-orbicular, cuneate at the base, the rounded tip crenate, glaucous below, reticulate, lowermost sometimes lobed or 3-fid; petioles slender; sheath short, broad. Umbels small, few-flowered; rays short or long, unequal, slender. Involucral bracts very short. Flowers white. Styles slender. Fruit unknown.

SOUTH Island: Canterbury: water-courses by the Kowhai River, 2,000ft. to 3,000ft., *Haast*.

"A curious little species, at once known by the few petioled leaflets. It is probably 2-pinnate or 2-ternately pinnate. I have only two specimens, and in the absence of fruit am not certain of its genus."

Not having seen specimens, I have copied this description from the Handbook. The absence of fruit renders it impossible to identify the original plant from description only, but it is most likely identical with the plant described in this work as *Oreomyrrhis andicola*, Endl., var. *apiculata*.

[A remarkable plant, of which I have very imperfect specimens, has leaves 2–3-pinnate, 5in.–8in. long and 1in.–$1\frac{1}{4}$in. broad. The leaflets, 2in.–3in. long, are cut into strict erect apiculate segments, $\frac{1}{12}$in. broad, not coriaceous. As my specimens were collected in the South Island by Dr. Sinclair, I suggest that it should be named *L. Sinclairii* should better material prove it to be distinct.]

9. ACIPHYLLA, Forst.

Calyx-teeth small, often unequal. Petals incurved, rarely inflexed at the apex. Stylopodia depressed in the male flowers, conical in the female. Fruit oblong or linear-oblong. Carpels sometimes dorsally compressed, each with 5 or 4 narrowly-winged ridges or one carpel 3-winged. Vittae 1 beneath each furrow, usually obscure. Erect rigid glabrous perennials. Leaves pinnate or 2–3-pinnate, with dagger-like segments and short spines at the apex of the sheath. The rhachis jointed at the insertion of the leaflets. Leaves rarely reduced to a petiole. Umbels compound, usually in the axils of divided spinescent bracteate leaves. Flowers unisexual, forming an elongated dense inflorescence, or rarely paniculate; the females often with males intermixed. Involucral bracts linear.

In addition to the New Zealand species, which are endemic and distributed from the East Cape to Stewart Island, two species are found in Australia. The genus is chiefly distinguished from *Ligusticum* by its coriaceous spinescent leaves and bracts and singular habit. The male inflorescence of *A. Monroi* and *A. Lyallii*, &c., cannot be distinguished from *Ligusticum*.

ETYM. From the Greek, signifying *sharp* and *a leaf*.

* *Male and female umbels in the axils of spinous bracts, forming a stout erect leafy raceme.*

2ft.–8ft. high.	Leaf-segments $\frac{1}{2}$in.–$\frac{3}{4}$in. broad, pungent, brownish-yellow	1. *A. Colensoi.*
2ft.–4ft. high.	Leaf-segments $\frac{1}{4}$in. broad, pungent, grey	2. *A. squarrosa*
1ft.–2ft. high.	Leaf-segments pungent, $\frac{1}{8}$in. broad, red	3. *A. crenulata.*
$\frac{1}{2}$ft.–3ft. high.	Leaf-segments transversely jointed, $\frac{1}{10}$in.–$\frac{1}{8}$in. broad ..	4. *A. Traversii.*
6in.–9in. high.	Leaf-segments $\frac{1}{4}$in.–$\frac{1}{2}$in. long, pungent, subulate ..	5. *A. Hookeri.*

** *Male umbels paniculate; female dense, contracted; scape naked below.*

Leaves pinnate, shining, polished .. .. 6. *A. Lyallii.*
Leaves entire or 2-3-foliolate; segments ⅜in. broad, apiculate .. .. 7. *A. Kirkii.*
Leaves 3-5-foliolate; segments ⅛in. broad, pungent .. .. .. 8. *A. Traillii.*

*** *Male and female umbels paniculate.*

6in.–18in. high. Leaves pinnate .. .. 9. *A. Monroi.*

**** *Umbels terminal.*

3in.–4in. high. Leaves excessively coriaceous, imbricating, 3-fid .. .. 10. *A. Dobsoni.*
3in.–4in. high. Leaves entire .. .. 11. *A. simplex.*

***** *Leaves flaccid.*

Male and female umbels forming an erect raceme. Carpels much compressed 12. *A. Dieffenbachii.*

1. **A. Colensoi,** *Hook. f., Handbk.* 92. Erect, glaucous when young. Stem 2ft.–4ft. high, 2in.–3in. in diameter, furrowed. Leaves 1ft.–2ft. long, with a pair of simple or divided spines 1in.–4in. long at the mouth of the sheath, pinnate or rarely the leaflets 2-pinnate; segments excessively coriaceous, thick and rigid, narrow-linear, 5in.–9in. long, ½in.–¾in. broad or more, acuminate, forming a mass of bayonet-like spikes 2ft.–3ft. across; margins rough, scarcely serrulate. Umbels in the axils of dilated 3–5-partite spinous bracts. Male: 2in.–6in. long; umbellules numerous, whorled; calyx-teeth minute, very acute. Female: less crowded; peduncles 2in.–3in. long, with 3–5 short simple umbels at the base; rays unequal, ½in.–¾in. long, with a few pedicellate flowers in their axils; pedicels short, slender; calyx-teeth acute. Fruit $\frac{5}{20}$in. long, broadly oblong. Carpels usually with 4 wings, or one 3-winged with 2 intermediate ridges; vittae 5 on the commissural face and 1 beneath each furrow; styles slender, recurved.—Lindsay, Contrib. N.Z. Bot. 49, t. 1. *A. squarrosa,* Forst., β Hook. f., Fl. N.Z. i. 87.

Var. **conspicua.** Segments slender, acuminate, with a broad orange or red midrib, submembranous or cartilaginous; margins perfectly even.

NORTH and SOUTH Islands: from Mount Hikurangi, East Cape, to Southland. Ascends to 4,500ft.; descends to 500ft., rarely to sea-level. Ward Island, Port Nicholson, *Hector.* Var. *conspicua:* Ruahine Mountains, *W. F. Howlett!* Whangapeka, Nelson, *Kingsley!* *Taramea. Spaniard. Bayonet-grass.* Dec., Jan.

Var. **maxima.** Stem 4ft.–9ft. high, 2in.–4in. in diameter. Leaves 1½ft.–4ft. long; segments ¾in. broad or more, extremely rigid and pungent, nerveless, forming a ring of spikes 4ft.–6ft. in diameter. Male umbels 6in.–18in. long; calyx-teeth inconspicuous. Female 3in.–4in. long. Fruits linear-oblong, $\frac{4}{10}$in. long, $\frac{3}{20}$in. broad. Carpels, one with 4 wings, the other with 3. Umbellules more lax and pedicels longer than in the type.

SOUTH Island: frequent in the mountains of Nelson and Canterbury. *Greater-spaniard.* Probably a distinct species.

Distinguished from all other species by its robust habit, broad leaf-segments, and large fruit. It yields a semi-transparent gum-resin. The spines at the mouth of the sheath are sometimes produced into leaflets.

2. **A. squarrosa,** *Forst., Char. Gen.* 136, *t.* 38. Erect, 2ft.–5ft. high, grey or glaucous. Stem 2in.–4in. in diameter at base, deeply grooved. Radical leaves 1ft.–2½ft. long, spreading, 2- rarely 3-pinnate; leaflets in 3–4 pairs, divided into rigid pungent crowded narrow-linear segments, 6in.–12in. long or more, $\frac{1}{16}$in.–⅛in. broad. Scape leafy at the base. Umbels compound,

irregular, crowded in the axils of dilated linear-oblong bracts, with 5-partite or rarely pinnate spines. Male: primary peduncle 1in.–3in. long, with 1 or more simple umbels at the base; rays 6–12, very unequal. Involucral bracts narrow-linear, acute; flowers densely crowded. Female: smaller and shorter; rays about 3, with few flowers on very short pedicels; bracts shorter and broader. Fruit broadly oblong. Carpels, one 4-winged, the other 3-winged.—Hook., Ic. Pl. t. 607, 608; Hook. f., Fl. N.Z. i. 87; Handbk. 92. *Ligusticum Aciphylla*, Spreng. in Schult. Syst. Veg. vi. 551; DC., Prod. iv. 159; A. Rich., Fl. Nov.-Zel. 274; A. Cunn., Precurs. n. 505. *Laserpitium spinosissimum*, Banks and Sol. MSS. and Icon.

NORTH and SOUTH Islands: from the East Cape to Southland. Sea-level to 3,300ft. *Kuri-kuri. Spear-grass.* Oct. to Jan.

Var. **flaccida.** Smaller and more slender. Leaves softer, 3-pinnate; sheath broader, sometimes 2in. wide or more; segments narrower, more crowded. Bracts spreading, never refracted. Ruahine Mountains, *Howlett!*

The spines at the mouth of the sheath are usually developed into pinnate leaflets.

3. **A. crenulata,** *J. B. Armst. in Trans. N.Z.I.* xiii. (1880) 336. Erect, 1ft.–2ft. high, similar to *A. squarrosa* but smaller, green or red, shining when fresh. Stem furrowed. Leaves 6in.–12in. long, pinnate; leaflets in 1–3 pairs, spreading, $\frac{1}{12}$in.–$\frac{3}{16}$in. broad, pungent when fresh; midrib red; sheath of petiole short and broad, with 2 spines at the mouth. Male umbels as in *A. squarrosa*, but smaller, spreading; peduncles and rays slender; spines of bracts not refracted, lowest sometimes nearly equalling the scape, 3-partite. Female partly hidden in the sheath of the bracts, shorter and more compact; rays 1–3, unequal, with pedicellate flowers in the axils, 2–6-flowered. Fruit short, oblong. Carpels 4-winged or one 3-winged; vittae obscure; styles very short.

SOUTH Island: Canterbury: source of the Rakaia, *J. B. A.!* Waimakariri glaciers, *Enys* and *Kirk*. Browning's Pass, *Haast!* 2,800ft. to 4,500ft. Dec., Jan.

This plant appears to be partially confused with *A. Lyallii* in the Handbook, and has been mistaken by local botanists for that species, but is easily distinguished by its leafy scape, longer leaf-segments, and 4-ribbed carpels.

4. **A. Traversii,** *Hook. f., Handbk.* 729. Slender or robust, 6in.–30in. high. Stem 2in. in diameter, furrowed in large specimens. Leaves 4in.–30in. long, equally or unequally pinnate; leaflets in 1–3 pairs, 2in.–12in. long, $\frac{1}{8}$in.–$\frac{1}{4}$in. broad, narrow-linear, acute, coriaceous, pungent, striated and transversely articulated; petioles sometimes 1ft. long, with two short spines at the mouth of the sheath. Umbels very numerous, usually solitary in the axils of opposite or verticillate dilated bracts, each tipped with a simple or 3-fid subpungent leaflet. Flowers polygamous. Male: 1in.–5in. long, irregular; rays unequal, $\frac{1}{4}$in.–1in. long; involucral bracts very small, linear, acute; calyx-teeth minute, acute. Female: fewer, on shorter peduncles, usually intermixed with male; rays very unequal, with a few pedicellate flowers in their axils; pedicels shorter than the narrow-linear oblong fruit. Carpels narrow-linear, 3-winged, or one 4-winged with 2 intermediate ridges; vittae 3 on the commissure.—*Gingidium Traversii*, F. Muell., Veg. Chath. Isds. 18.

CHATHAM Islands, *Captain Gilbert Mair! H. H. Travers! F. A. D. Cox!* *Taramea.* Nov., Dec.

Easily distinguished from all other species by the narrow-linear 3-winged carpels and the transversely-articulated leaf-segments, which, although pungent, are much less rigid than those of *A. Colensoi.*

5. **A. Hookeri,** *n. s.* Erect. Scape 5in.–9in. high. Leaves 3in.–6in. long, rigid, curved outwards for the upper half of their length, 2-pinnate; leaflets in 2–5 pairs, rather crowded, ½in.–1½in. long, divided into rigid almost terete grooved spreading abruptly-acuminate pungent segments ⅛in.–½in. long; rhachis concave above; petiole exceeding the blade, almost flaccid; sheath narrow, dilated at the base, with 2 spines at its mouth. Male scapes leafy below; umbels in the axils of the leafy pinnate spinous bracts, 1in.–3in. long or more; primary peduncles equalling the dilated portion of the bract, solitary or with 1–3 shorter peduncles in the same axil; rays unequal, often racemose, 5–10-flowered; flowers minute; calyx-teeth nearly obsolete; involucral bracts as long as the slender pedicels. Female: umbels and fruit not seen.

SOUTH Island: Heaphy River, Nelson, *J. Dall!*

One of the most remarkable species in this singular genus, easily recognised by the very short pungent segments. Rarely the leaves are reduced to 1 or 2 pairs of entire acicular leaflets. A plant of which imperfect specimens were collected by Dr. Gaze on mountains near Westport may be identical, but the segments are rigid.

6. **A. Lyallii,** *Hook. f., Handbk.* 92. The entire plant polished and shining, yellow. Scape 6in.–12in. high, ¼in.–⅜in. in diameter, naked below, deeply grooved. Leaves numerous, 3in.–5in. long, pinnate; leaflets in 1–4 pairs, coriaceous, rigid, $\frac{1}{15}$in.–$\frac{1}{10}$in. broad, pungent; petiole-sheath narrow, with 2 short spines at its mouth. Male umbels compound, paniculate; peduncles 1in.–3in. long in the axils of 3–5-partite or entire spinous bracts as long as the peduncles; rays 6–10, subequal, slender; pedicels short. Female umbels restricted to the upper part of the scape, almost hidden in the base of the polished tumid bracts. Carpels narrow-oblong, 5-winged, ⅙in. long.—*A. montana,* J. F. Armst. in Trans. N.Z.I. iv. (1871) 290. *A. Hectori,* Buch. in Trans. N.Z.I. xiv. (1881) 346, t. 27 (in part).

SOUTH Island: Nelson: Mount Arthur, *Cheeseman.* Canterbury: Rangitata Range and Ashburnam Glaciers, *Sinclair* and *Haast!* Otago: Dusky Sound, *Lyall.* Lake district, *Hector* and *Buchanan.* 3,000ft. to 5,000ft.

In the leaves and male inflorescence this species approaches *A. Monroi,* but the leaf segments are more acute. The female inflorescence is very different.

7. **A. Kirkii,** *Buch. in Trans. N.Z.I.* xix. (1886) 214, *t.* xxii. Scape erect, 8in.–12in. high, rather stout, naked below, grooved. Leaves all radical, brown, 5in.–9in. long, almost ⅜in. broad, quite entire or 2–3-foliolate, very thick and coriaceous, striated, subacute, abruptly apiculate, pungent; petiole-sheath short and narrow, jointed at its junction with the blade. Male umbels with long 5-partite bracts. Female inflorescence contracted, 2in.–3in. long; umbels in the axils of coriaceous simple or 2–3-fid coriaceous pungent bracts, 5–10-flowered, sometimes with 2 or 3 pedicellate flowers at the base; peduncles

shorter than the pedicels or 0; apical umbels with almost sessile flowers. Carpels 5-winged or rarely 4-winged, with intermediate ridges; vittae 1 under each furrow, 4 on the commissural face; stylopodia much thickened but scarcely conical; styles very short and stout.

SOUTH Island: Otago: Mount Alta, *Buchanan!* Hector Mountains; on a hill opposite Mount Aspiring, *Petrie!* 5,000ft. to 6,000ft.

A very distinct species, never polished or shining. Buchanan's description is imperfect and his drawing somewhat misleading, being partly a restoration. I am indebted to Mr. Petrie for my specimens, but have not seen male flowers.

8. **A. Traillii,** *T. Kirk in Trans. N.Z.I.* xvi. (1883) 371. Tufted, erect. Scape 3in.–6in. high. Leaves rather flaccid when fresh, coriaceous, pungent when dry, 2in.–4in. long, simple or 3- rarely 5-foliolate; leaflets linear, $\frac{1}{12}$in. broad, with a stout marginal nerve; petiole short, denticulate at the mouth. Male umbels on short peduncles or almost sessile, distant or crowded; bracts spinous, simple or 3-partite; pedicels very short; calyx-teeth obsolete. Female umbels 5–10, crowded; bracts shorter, with a broad tumid sheath enclosing the umbel; umbel of a single ray or with from 1–3 flowers at the base of the short peduncle; involucral leaves minute. Fruit sessile or nearly so. Carpels (immature) apparently with 5 narrow wings; styles very short.

SOUTH Island: Otago: Mount Ida Ranges, *Petrie!* STEWART Island: Mounts Anglem and Rakiahua, 2,000ft.–3,000ft., *T. K.* Dec., Jan.

Nearly allied to *A. Lyallii.* Petrie's specimens have the umbels almost sessile, usually forming a compact spiciform inflorescence.

9. **A. Monroi,** *Hook. f., Handbk.* 93. Scape erect, 6in.–18in. high, less rigid than *A. Lyallii,* smooth and shining. Leaves 3in.–8in. long, pinnate; leaflets in 2–6 pairs, $\frac{1}{2}$in.–2in. long, $\frac{1}{12}$in.–$\frac{1}{8}$in. broad, linear, subacute or pungent; midrib obscure; sheath with 2 short spines at the mouth. Male umbels panicled as in *A. Lyallii,* spreading, each in the axil of a flaccid 3–5-partite linear-oblong bract. Female umbels 2–5, compound, paniculate, subterminal; rays 4–8; pedicels exceeding the carpels; involucral bracts narrow-linear, obtuse or acute. Fruit oblong. Carpels 5-ribbed or one 3-ribbed; vittae obscure.

SOUTH Island: mountain districts of Marlborough, Nelson, Westland, Canterbury, and Otago. 3,000ft. to 6,500ft. Dec., Jan.

The most plentiful of all the mountain species except *A. Colensoi.*

10. **A. Dobsoni,** *Hook. f., Handbk.* 93. A robust plant, forming compact patches 3in.–4in. high, yellow-brown. Leaves all radical, densely imbricated, excessively thick and coriaceous, 1in.–2$\frac{1}{2}$in. long, consisting of a sheath carrying 3 short linear jointed subequal pungent segments, keeled at the back, $\frac{1}{4}$in.–$\frac{1}{3}$in. long. "Flowering-stem as thick as the little finger, terete, striate, bearing at the top 2 small leaves like the radical, and 5 pedunculate densely capitate globular umbels. Peduncles unequal, $\frac{3}{4}$in.–1$\frac{1}{2}$in. long, stout, grooved. Umbels (or heads) 1in. in diameter, compound, both peduncles and pedicels very short and thick. Fruits densely packed, mixed with short subulate invo-

lucral leaves, linear-oblong, ⅛in. long. Calyx-teeth rather large, unequal. Carpels usually with 5 narrow wings."

SOUTH Island: Canterbury: Mount Dobson, *Dobson* and *Haast*, *Cheeseman!* Mountains above Lake Hawea, *Haast!* Mountain above Lake Ohau, *Buchanan!* Otago: Mount St. Bathans, *Petrie!* 6,000ft.

I have not seen the flowers or fruit of this rare and singular plant. The leaf-segments develop strong nerves parallel with the midrib; these are intersected by lateral veins dividing the surface into narrow rectilinear interspaces.

11. **A. simplex,** *Petrie in Trans. N.Z.I.* xxii. (1889) 440. Habit of the preceding, and differing only in the leaves, which are less coriaceous, entire, 1½in.–2in. long, dilated at the base and closely imbricating, narrowed upwards, deeply concave, usually with a distinct midrib and thickened margins; tips free, spreading, obtuse, but with a strong mucro; the upper surface divided into rectilinear interspaces. Flowers as in the preceding. Fruits not seen.

SOUTH Island: Otago: Mount Pisa, Mount Cardrona, and the Hector Mountains, 5,000ft. to 6,000ft., *Petrie!* Feb.

A singular plant, probably not specifically distinct from *A. Dobsoni.*

12. **A. Dieffenbachii.** Erect, 2ft.–3ft. high, glabrous, grey. Stem robust, 1in. or more in diameter, furrowed. Leaves all radical, flaccid, 1½ft.–2ft. long, 4in.–7in. broad, 3–4-pinnatisect, broadly oblong or oblong in outline, cuneate at base; leaflets in 4–5 pairs, ascending; segments 2in.–3in. long, 1 line broad, striate, mucronate; petiole exceeding the blade, the sheath auricled at the mouth. Umbels compound, solitary in the axils of dilated sheathing pinnatisect bracts; sheath auricled at the mouth, acute in the male, obtuse in the female. Male: peduncle 2in.–4in. long or more; rays numerous, verticillate or terminal, very slender; calyx-teeth inconspicuous, acute. Female: peduncle stout, 3in.–5in. long, sometimes with a solitary ray at the base; rays about 6; pedicels 8–12, shorter than the fruit; involucral leaves linear, subulate, acute. Fruit broadly oblong, not cordate at the base, much compressed, ⅜in. long, ¾in. broad. Styles very short. Carpels with 3 closely-parallel ridges and 3 broad cartilaginous wings, or 1 carpel with 2 wings; vittae 1 under each furrow and 2 on the commissural face. Seed furrowed. — *Gingidium Dieffenbachii,* F. Muell., Veg. Chath. Isds. 17, t. 1. *Ligusticum Dieffenbachii,* Hook. f., Handbk. 729. *Angelica Dieffenbachii,* Index Kew. i. 133.

CHATHAM Islands, *H. Travers! Cox!* Extremely rare.

This is a somewhat aberrant member of *Aciphylla*, differing in the flaccid habit and the structure of the fruit, while it diverges still further from *Ligusticum* and *Angelica*, to which it has been referred. It will probably form the type of a new genus. My friend Mr. Cox has kindly favoured me with the only fruiting specimen that has been collected.

10. ANGELICA, Linn.

Calyx-teeth minute or unequal. Petals incurved at the apex. Fruit usually cordate at the base, compressed dorsally; 3 primary ridges slender but prominent, the lateral 2 forming broad membranous wings. Stylopodia depressed; styles slender. Erect biennial or perennial herbs, rarely subscandent, and almost suffruticose, with pinnate or rarely 2–3-pinnate leaves and compound umbels of unisexual or polygamous white flowers.

A small genus, most frequent in northern, temperate, and subarctic regions: apparently restricted to New Zealand in the Southern Hemisphere. The subscandent species are aberrant in the order. SPECIES, about 25.

* *Herbaceous, erect. Leaves radical.*

Leaves pinnate, erect, 1ft.–1½ft. high .. .. .. 1. *A. Gingidium.*

** *Suffruticose, subscandent.*

Leaves pinnately 5-foliolate, shining .. .. 2. *A. rosaefolia.*
Leaves distant, unifoliolate, small .. .. 3. *A. geniculata.*

1. **A. Gingidium,** *Hook. f., Handbk.* 97. Root stout, fleshy, strong-smelling. Stem slender or stout, striate, 1ft.–1½ft. high, sparingly branched above. Leaves linear-oblong in outline, pinnate, 6in.–12in. long; leaflets in 5–10 pairs, sessile, obliquely ovate-oblong, often subimbricate, rarely lobed, crenate or serrate, nerveless; petiole often longer than the blade; rhachis jointed at the insertion of the leaflets. Umbels compound, 1in.–3in. in diameter; general involucre 0; rays 8–15, spreading, slender; partial involucre of few short linear bracts, exceeding the pedicels. Flowers white. Fruit ovate-oblong, cordate at base, $\frac{1}{16}$in. long; wings membranous.—*Anisotome Gingidium,* Hook. f., Fl. N.Z. i. 89. *Ligusticum Gingidium,* G. Forst., Prod. n. 140; A. Rich., Fl. N.Z. 272; A. Cunn., Precurs. n. 506. *Gingidium montanum,* Forst., Gen. t. 21. *Ligusticum anisatum,* Banks and Sol. MSS.

NORTH and SOUTH Islands: Tararua Range to Otago. Sea-level to 3,000ft. *Aniseed.* Nov., Dec.

Formerly abundant, but has been almost extirpated by cattle in many localities.

2. **A. rosaefolia,** *Hook., Ic. Pl. t.* 581. Suffruticose. Stem 2ft.–5ft. long, clothed with the sheaths of old leaves below, stout, much branched, scrambling over rocks. Leaves petioled, 2in.–5in. long, pinnately 5- rarely 3-foliolate; leaflets sessile, 1in.–2in. long, with 2 glands at the base of each, obliquely ovate-lanceolate or oblong-acute, acutely serrate, shining; petiole one-third the length of the leaf; sheath persistent, produced upwards into 2 sub-acute or obtuse lobes. Umbels terminal or axillary, compound, spreading; rays 8–12, subequal, exceeding the involucral bracts. Flowers white. Calyx-teeth acute. Style slender, straight. Fruit exceeding the pedicels, ovate, cordate; wings broad, membranous.—Hook. f., Handbk. 98; Raoul, Choix. 47. *Anisotome rosaefolia,* Hook. f., Fl. N.Z. i. 90. *Ligusticum aromaticum,* Banks and Sol. MSS.

THREE KINGS Islands, *Cheeseman.* Bay of Islands, *Banks* and *Sol.* MSS. Whangarei, *T. K.* Great and Little Barrier Islands, Kawau Island, *T. K.* Between Kaipara Harbour and Port Waikato, *Cheeseman.* Manukau Harbour, *T. K.* Mercury Bay and East Cape, *Banks* and *Sol!* Hawke's Bay, *A. Hamilton!* Upper Rangitikei, *Buchanan.* SOUTH Island: Banks Peninsula, *Raoul. Koheriki. Koheripa.* Oct., Nov. Chiefly littoral. Ascending to 2,000ft. in the Urewera Country, *E. Best!*

The leaves are beautifully reticulated with pellucid veins, and the rhachis is obscurely jointed.

3. **A. geniculata,** *Hook. f., Handbk.* 98. Suffruticose. Stems 2ft.–4ft. long, much branched, slender, subscandent or scrambling over shrubs; internodes 1in.–3in. long. Leaves in young state 3-partite or 3-lobed, when mature 1-foliolate; sheath narrow-linear, with two obtuse lobes at the mouth; blade orbicular-ovate or obscurely rhomboid or deltoid-ovate, rarely oblong, obtuse or rounded, minutely crenate. Peduncles short. Umbels axillary or terminal; rays 2–4, unequal, usually 3–5-flowered; pedicels shorter than the ovary; involucral bracts very short, acute. Flowers white. Petals inflexed at the tips. Calyx-teeth acute. Styles slender. Fruits ovate-oblong, cordate at the base, spreading. Carpels with 3 rarely 4 filiform ridges and 2 membranous wings; vittae 1 beneath each furrow and 2 on the commissural face.—*Anisotome geniculata*, Hook. f., Fl. N.Z. i. 90, t. 19 (*Eustylis*). *Peucedanum geniculatum*, G. Forst., Prod. n. 136; A. Rich., Fl. Nouv.-Zel. 272; A. Cunn., Precurs. n. 507. *Bowlesia geniculata*, Schultes, Syst. Veg. vi. 364.

NORTH Island: local. East Cape and interior, *Colenso*. Near Paikakariki, *H. B. Kirk!* Port Nicholson, *Buchanan!* SOUTH Island: Queen Charlotte Sound, *Forster*. Banks Peninsula, *Raoul!* East coast of Canterbury and Otago, *J. F. Armst.! Buchanan!* Jan., Feb.

I have never seen this plant growing far from the sea, and doubt its occurrence inland.

* PEUCEDANUM, Linn.

Calyx inconspicuous or 0. Stylopodium small; margin expanded or undulate. Petals rounded or obovate, with an inflexed point. Fruit oblong, broadly oblong, or ovoid, compressed dorsally; primary ridges 3, equidistant, two marginal dilated into wings, intermediate filiform. Vittae solitary in the interspaces. Biennial or perennial herbs, with pinnately or ternately divided leaves. Umbels compound; bracts and bracteoles many or few or 0. Flowers often polygamous.

* **P. sativum,** *Benth. and Hook. f., Gen. Pl.* i. 920. A stout erect biennial. Root fusiform. Stem furrowed, fistular. Leaves pinnate; leaflets sessile, ovate-oblong, coarsely toothed or serrate, shining above, downy beneath. Umbels large; rays many, long. Flowers yellow. General and partial involucres 0. Calyx-teeth 0. Fruit oblong; wings rather broad; styles very short.

Sparingly naturalised from Auckland to Otago, especially on calcareous soils. *Parsnip.* Jan., Feb. Europe, Siberia.

11. DAUCUS, Linn.

Calyx-teeth inconspicuous. Petals emarginate, inflexed at the tips, slightly imbricate in the bud. Disk small. Fruit ovoid or oblong, slightly dorsally compressed. Carpels with 5 slender primary bristly ribs and 4 prominent winged secondary ribs with glochidiate bristles; vittae under the secondary ribs and 2 on the commissural face. Erect bristly annual or perennial herbs, with pinnate or 2–3-pinnate leaves and narrow segments. Umbels of white flowers with pinnatifid or pinnate bracts, terminal, compound.

A genus comprising about 30 species, chiefly found in the temperate regions of the Northern Hemisphere.

Umbels large, compact .. * *D. Carota.*

Umbels small, lax .. 1. *D. brachiatus.*

* **D. Carota**, *L.*, *Sp. Pl.* 242. Biennial, more or less hispid. Root fusiform. Stem erect, furrowed, solid, 1ft.–2ft. high. Leaves oblong or narrow-oblong, 2-pinnate; leaflets inciso-dentate; segments narrow, acute or lanceolate. Umbels large, concave; rays 1in.–2in. long. Bracts pinnatifid; bracteoles lanceolate. Fruit with primary or secondary ridges carrying a row of usually hooked bristles.

Naturalised from Auckland to Otago, but local in many districts. *Carrot.* Dec. to Feb. Europe.

1. **D. brachiatus**, *Sieber.* ex *DC.*, *Prod.* iv. 214. Annual or biennial, 6in.–18in. high. Stems slender, strict, branched, hirsute or glabrate. Leaves 2–3-pinnate; leaflets short, narrow, incised or pinnatifid. Umbels axillary or terminal, compound; rays 3–5, unequal, with or without a few pedicellate flowers in the axils. Involucral bracts small, entire, pinnatifid or pinnate. Flowers red, minute. Fruits $\frac{1}{8}$in.–$\frac{1}{10}$in. long. Carpels with the secondary ridges bearing a row of barbed purple spines; primary ridges bearing a double row of short colourless bristles.—Hook. f., Fl. N.Z. i. 61; Handbk. 99; Benth., Fl. Austr. iii. 376. *Scandix glochidiata*, Labill, Pl. Nouv. Holl. i 75, t. 102.

Common in lowland districts from the Three Kings Islands and North Cape to Southland; CHATHAM Islands. Oct. to Dec.

* CAUCALIS, Linn.

Calyx-teeth 5, acute, rarely 0. Petals unequal, obcordate, with the apex inflexed. Stylopodium conical, thickened. Fruit oblong or ovate, compressed, laterally constricted at the commissure; primary ridges filiform, prominent, and with the secondary ridges bearing 1 or 3 rows of shortly-hooked spines or bristles; vittae solitary between the secondary ridges. Annual herbs, with more or less hispid erect or rarely decumbent stems and terminal or axillary umbels. Flowers small, often dioecious or polygamous.

* **C. nodosa**, *Scop.*, *Fl. Carn.*, *ed.* 2, i. 192. Annual. Stems diffuse or prostrate, solid, branching at the base only, 1ft.–2ft. long, hispid; hairs retrorse. Leaves rather distant, pinnate or 2-pinnate; leaflets small, narrow-pinnate. Umbels small, subglobose, sessile, often leaf-opposed. Involucre 0. Outer carpels with hooked bristles; inner carpels tubercled.

NORTH and SOUTH Islands: naturalised in many localities, but rarely abundant. *Hedge-parsley.* Sept. to Nov. Europe, West Africa, West Asia to India.

Order XXXV.—ARALIACEAE.

Calyx-tube adnate to the ovary; limb persistent, forming a raised ring or cup about the apex, truncate or 5- rarely 3-toothed. Petals 5, rarely 4, coriaceous, valvate or rarely imbricate, deciduous, inserted round an epigynous disk. Stamens as many as the petals (rarely more) and inserted with them; anthers versatile. Ovary inferior, 2 or more celled or rarely 1-celled with 1 pendulous anatropous ovule in each cell. Styles as many as cells, erect, with recurved stigmatiferous tips, or coherent into a cone, or reduced to a small protuberance, with as many inconspicuous stigmas as cells. Fruit drupaceous, indehiscent, the epicarp usually succulent, 2- or many-celled; cells 1-seeded. Seeds with copious endosperm. Embryo minute; radicle superior. Shrubs or trees, rarely climbers or herbs, with alternate simple or compound shining leaves, stipulate or exstipulate, the petiole often dilated at the base, the expansion forming an interpetiolar stipule. Flowers often unisexual or polygamous, in simple or highly compound umbels or racemes or panicles. Bracts small, often 0.

Distributed through the temperate and especially the tropical regions of both hemispheres. GENERA, about 40. SPECIES, about 375. All the New Zealand species are endemic. The foliage of many species exhibits a large amount of polymorphism.

* *Herbs. Petals imbricate in bud.*

1. STILBOCARPA. Petals obovate, obtuse. Styles 3-4. Drupe spherical, 3-4-celled.
2. ARALIA. Petals linear-acute. Styles 2. Drupe spherical, 2-celled.

** *Shrubs or trees. Petals valvate in bud. Stamens as many as the petals.*

3. PANAX. Leaves simple or digitate. Flowers jointed to the pedicels. Styles distinct, recurved at the apex.
4. MERYTA. Leaves entire, large. Flowers paniculate.
5. SCHEFFLERA. Umbels small, in a large racemose panicle.
6. PSEUDOPANAX. Leaves simple or digitate. Flowers 5- or 4-merous. Styles united into a cone; stigma discoid, small.

*** *Shrub. Petals valvate in bud.*

* HEDERA. Climbing. Leaves entire, palmatilobed or ovate.

1. STILBOCARPA, A. Gray.

Calyx-tube 3–4-grooved; limb entire. Petals 5, imbricate. Male flower: stamens 5; styles 0; lobes of disk flat. Female: stamens usually 0; lobes of disk 3 or 4, surrounding a cavity sunk in the apex of the ovary; styles 3 or 4, recurved; ovary broadly turbinate, 3–4-celled. Fruit globose, axis hollow, 3–4-furrowed, 3–4-celled, each cell containing a single nut. A herb, with large orbicular-reniform leaves and huge masses of unisexual or polygamous flowers.

The only species; endemic.

1. **S. polaris,** *A. Gray, Bot. U.S. Expl. Exped.* i. 714. Stems short, given off from a thick fleshy annulate rhizome, strong-smelling when bruised. Leaves 6in.–12in. broad or more, almost fleshy, bristly on both surfaces, orbicular-reniform, many-lobed, strongly toothed; petiole 12in.–18in. long, almost terete; sheath semiamplexicaul, produced upwards into a leafy lobed or laciniate ciliate ligule. Umbels compound, terminal and axillary, 4in.–9in. in diameter. Flowers ¾in. in diameter, yellow, waxy, shining, with purple centres. Pedicels articulated. Petals of the female 5, oblong-obovate; of the male smaller, dull. Lower involucral bracts foliaceous, pedicellate, lobed. Fruit as large as a peppercorn, shining.—Denc. and Planch., Voy. au Pole Sud, t. 2, Dicot. *Aralia polaris,* Hook. f., Fl. Antarc. i. 21; Hook., Ic. Pl. t. 477.

AUCKLAND, CAMPBELL, ANTIPODES, and MACQUARIE Islands. *Punui.* Dec., Jan.

2. ARALIA, Tourn.

Calyx-teeth minute or obsolete. Petals 5, more or less imbricate in bud. Stamens 5. Styles 2–5, free or slightly connate at base. Fruit a 2–5-celled drupe; cells 1-seeded. Herbs, shrubs, or trees, with unisexual or polygamous, umbellate, racemose or paniculate flowers on jointed pedicels and simple pinnate, digitate, or decompound leaves.

The genus differs from *Stilbocarpa* in the absence of the cavity in the axis of the fruit. It comprises about 40 species, distributed through the temperate regions; most plentiful in the Northern Hemisphere. Name of uncertain origin.

1. **A. Lyallii,** *T. Kirk in Trans. N.Z.I.* xvii. (1884) 293, *t.* 17. A stout herb, 1ft.–3ft. high. Stems ¾in.–1in. in diameter, horizontal, emitting strong arcuate stolons. Leaves radical, crowded, more or less clothed with rather soft bristles beneath, usually glabrous and shining above, orbicular-reniform, lobed, deeply toothed, slightly coriaceous; petioles terete, fistulose, with thin walls, pilose or pubescent. Umbels 3in.–12in. in diameter. Flowers unisexual or polygamous, purplish-red. Calyx-teeth 0. Petals 4, linear, acute. Male: stamens 5. Female: ovary 2-celled. Stylopodia 2, forming a fleshy indented disk; styles free. Fruit spherical, 2-celled, black, shining; cells 1-seeded.—*Stilbocarpa Lyallii,* J. B. Armst. in Trans. N.Z.I. xiii. (1880) 336.

Var. **robusta.** More robust than the typical form, less pubescent, never pilose. Stolons 0. Teeth of leaves more strongly mucronate. Petioles stouter, plano-convex, solid or nearly so. Flowers smaller. Petals shorter, yellowish.

SOUTH Island: Coal Island, Preservation Inlet. STEWART Island: Herekopere and adjacent islands, Ruapuke, Green and Centre Islands, Foveaux Strait. Var. *robusta,* The SNARES, *T. K. Punui.* Dec. to Feb.

In "Index Kewensis" it is doubtfully suggested that this plant might be referred to *Pseudopanax;* and Baron von Mueller thought that it might be referred to *Panax,* notwithstanding its imbricate aestivation; but it seems more correctly placed under *Aralia.*

3. PANAX, Linn.

Calyx-limb forming a raised border; teeth 0 or inconspicuous. Petals 5, valvate. Stamens 5. Disk broad. Ovary 2–4-celled; styles connate at the base; tips free or recurved. Fruit coriaceous or fleshy, 2–5-celled, 1-seeded. Evergreen shrubs or trees, usually glabrous, with simple digitate or pinnate leaves. Flowers unisexual or polygamous, umbellate, racemose or paniculate.

SPECIES, about 45, widely distributed in tropical and temperate countries. Much uncertainty still exists as to the limitations of many species. All the New Zealand forms are endemic, and several exhibit polymorphism in the foliage to a remarkable extent.

ETYM. From the Greek, signifying *everything* and *a remedy,* in reference to the alleged virtues of the ginseng, which was formerly referred to this genus.

* *Leaves always simple.*

Leaves in young plants linear, 6in. long; in old plants lanceolate, 2in.–3in. long .. 1. *P. lineare.*

** *Leaves in old plants 1-foliolate; in young plants usually 3–5-foliolate.*

Leaflets lanceolate, serrate. Stipules 0. Styles 2 .. .. 2. *P. simplex.*

Leaflets oblong, entire. Stipules 0. Styles 3 or 4 .. .. 3. *P. Edgerleyi.*

Leaflets small, rounded. Umbels small. Stipules minute. Styles 2 .. .. 4. *P. anomalum.*

*** *Leaves in old plants 3–5- or 7-foliolate.*

a. Petiole with a 2-lobed sheathing stipule. Umbels twice compound.

Leaves 3–5-foliolate; leaflets sessile, usually acute .. .. .. 5. *P. Colensoi.*

Leaves 3–7-foliolate; leaflets petiolate, usually obtuse .. .. 6. *P. arboreum.*

b. Petiole slender, not sheathing.

Leaves 3–5-foliolate. Fruit urceolate. Styles 3–5 .. .. .. 7. *P. discolorum.*

Leaves 3–7-foliolate. Fruit compressed. Styles 2 .. .. .. 8. *P. Sinclairii.*

1. **P. lineare,** *Hook. f., Fl. N.Z.* i. 93. A small sparingly-branched dioecious shrub, 5ft.–8ft. high. Branchlets short, stout, often crowded. Leaves dimorphic; in the young state linear, 6in.–9in. long, ¼in.–⅓in. wide, narrowed into a short stout petiole, coriaceous, rigid, acute, midrib stout, margins distantly and obscurely serrate; mature leaves crowded at the tips of the branchlets, 1½in.–3½in. long, linear-lanceolate, excessively coriaceous, acute or obtuse, obscurely serrate; petiole ¼in. long, stout, mixed with 3-fid or simple coriaceous subulate scales. Stipules minute. Umbels small, terminal. Peduncles very short or 0. Male: compound; rays bracteolate, 4–8, unequal; pedicels short; petals 5; stamens 5. Female: umbels compound; rays 1–4-flowered; pedicels short; petals 0; ovary 3- rarely 5-celled; styles 3–5-celled, connate at the base; tips recurved. Fruit urceolate, 3- or 5-celled and seeded.—Handbk. 101; T. Kirk in Trans. N.Z.I. ix. (1876) 492.

SOUTH Island: Southern Alps, not common; from Nelson to Southland. 2,800ft. to 4,000ft. Jan., Feb.

The linear erect or spreading leaves pass very gradually into those of the mature state.

2. **P. simplex,** *G. Forst., Prod. n.* 399. A shrub or small tree, 5ft.–20ft. high. Leaves polymorphic: in the young state (1), all ovate or broadly ovate, serrate on long slender petioles; (2), 5-foliolate on slender peduncles; leaflets petiolulate, linear, lobed or pinnatipartite: both these are replaced by 3-foliolate leaves with lanceolate sessile leaflets, and these again by unifoliolate oblong or obovate-lanceolate leaves 2in.–5in. long, subacute acute or acuminate, serrate or dentate or almost entire, rarely opposite. Flowers in compound axillary or terminal umbels shorter than the leaves; the terminal umbellule female, the lateral male. Male flowers: petals 5; stamens 5. Female: petals and stamens 0; ovary 2-celled; styles 2, free, recurved. Fruit compressed, 2-seeded.—DC., Prod. iv. 253; A. Rich., Fl. N.Z. 280, t. 31; A. Cunn., Precurs. n. 509; Hook. f., Fl. Antarc. i. 18, t. 12; Fl. N.Z. i. 93; Handbk. 100; T. Kirk, Forest Fl. N.Z. t. 106, 107.

Var. **quercifolium.** Mature leaves 1-foliolate, 3in.–5in. long, coriaceous, narrow-lanceolate, deeply pinnatifid or lobulate, acute. Female umbels simple, few-flowered.—T. Kirk, Forest Fl. N.Z. t. 106, f. 2.

Var. **parvum.** Leaves when mature 1-foliolate, acute or subacute, crenate or sharply serrate, ¾in.–1in. long. Umbels few-flowered. Approaches *P. anomalum* in general appearance, but the leaves are more acute.

NORTH and SOUTH Islands: from Tairua, Thames Goldfield, southwards to STEWART Island and the AUCKLAND Islands. Sea-level to 4,000ft. Var. *quercifolium*, Waimakariri, *Enys!* Var. *parvum*, Stewart Island, *T. K.* *Haumakoroa. Haumangaroa.* Nov. to Jan.

3. **P. Edgerleyi,** *Hook. f., Fl. N.Z.* i. 94. A dioecious tree, 20ft.–40ft. high. Leaves dimorphic: in young plants on long slender petioles, 3–5-foliolate; leaflets oblong-lanceolate or lanceolate, acute, pinnatipartite or pinnatifid or lobed, purple beneath: on mature plants 1-foliolate, 3in.–9in. long, obovate or oblong-lanceolate, acute acuminate or rarely obtuse, membranous, shining; petioles slender, 1in.–4in. long, jointed to the blade. Umbels compound, axillary or developed below the leaves, 1in.–2in. long, racemose or in slender

racemose panicles. Pedicels short. Calyx-teeth minute. Styles 3 or 4, connate at the base; tips free, recurved. Ovary 3–4-celled. Fruit spherical, small, 3–4-seeded.—Handbk. 101; T. Kirk, Forest Fl. N.Z. t. 44, 45.

Var. **serratum.** Leaves of the mature state with serrated or lobulate margins.—T. Kirk, Forest Fl. t. 45 A.

NORTH and SOUTH Islands: from the Bay of Islands and Hokianga southwards to STEWART Island. Var. *serratum*, Stewart Island, *T. K.* Sea-level to 2,000ft. *Raukawa*. Jan., Feb.

The leaves afford a delicious perfume, used by the Maoris for scenting oil. Some forms approach *P. simplex*, but may always be distinguished by the more numerous styles.

4. **P. anomalum,** *Hook in Lond. Journ. Bot.* ii. (1843) 422, *t.* 12. A bushy shrub 6ft.–12ft. high, with crowded branches divaricating at right angles, setose or hispid. Leaves dimorphic: in the young state 3-foliolate; petioles slender, winged; leaflets with stipellae at the base, jointed to the petiole, ovate, sharply toothed or crenate: in the mature state rarely pinnatifid. Leaves 1-foliolate, with stipellae at the base and apex of the flattened petiole, orbicular or oblong-orbicular, rounded at the tip, subcoriaceous, obscurely crenate; petiole $\frac{1}{8}$in. long. Flowers minute, in small axillary simple umbels or corymbs, 2–10-flowered. Umbels rarely fascicled. Peduncles much shorter than the pedicels. Calyx-teeth acute. Male flowers: stamens 5; petals 5. Female: ovary 2-celled; styles 2, free. Fruit $\frac{1}{8}$in.–$\frac{1}{6}$in. in diameter, compressed, mottled; styles 2, recurved.—Hook. f., Fl. N.Z. i. 93; Handbk. 101; Raoul, Enum. 40.

Var. **microphyllum** (*sp.*), Col. in Trans. N.Z.I. xvi. 328, xvii. 240. Branches more slender, spreading irregularly or divaricating, but not at right angles. Leaves solitary or fascicled, lanceolate or obovate, narrowed below, rounded at apex, sinuate-crenate. Flowers and fruit as in the type.

NORTH and SOUTH Islands: from Whangaroa North to Southland. Ascends to 2,000ft. *Wawa-paku*. Dec., Jan.

A singular plant. In some seasons two or three late flowers are produced at the base of the umbels after the fruit is fully formed. The typical plant is found from the Waikato northwards, and passes into the common form by insensible gradations.

5. **P. Colensoi,** *Hook. f., Fl. N.Z.* i. 94, *t.* 21. A dioecious shrub or small tree, 4ft.–10ft. high, with stout branches. Leaves on stout petioles, 3in.–9in. long, with a 2-lobed sheath, 3–5-foliolate; leaflets sessile or shortly petioled, 2in.–6in. long, thick and coriaceous, oblong-lanceolate or obovate-cuneate, coarsely serrate, shortly abruptly acuminate; veins usually indistinct. Umbels terminal, similar to *P. arboreum*, but smaller, and with fewer primary rays. Pedicels more slender. Fruit nearly orbicular, compressed, 2-celled, 2-seeded. Styles 2, connate at the base; tips short, stout, recurved.—Handbk. 102.

Var. **montanum.** Spreading, slender. Leaves smaller; leaflets usually oblong-lanceolate, crenate at base, scarcely coriaceous. Umbels of 3–4 primary rays and as many secondary, very lax. Flowers small. Male: calyx-teeth 0. Female with minute obtuse teeth.

NORTH Island: from the Thames Goldfield southwards to STEWART Island. Descends to sea-level on Stewart Island; ascends to 4,500ft. in the Ruahine and Tararua Ranges. Var. *montanum*: SOUTH Island: Nelson to Banks Peninsula.

Nearly related to *P. arboreum*, but distinguished by the 3–5-foliolate leaves, with sessile or subsessile usually obovate leaflets and indistinct veins.

6. **P. arboreum,** *G. Forst., Prod. n.* 398. A much-branched dioecious shrub or small tree, 12ft.–20ft. high. Branches stout. Leaves digitate, on long petioles, 4in.–9in. long, with a 2-lobed sheath, 5–7-foliolate; leaflets 4in.–7in. long, coriaceous, shining, oblong or obovate-oblong, obtuse or acute, serrate or sinuate; petioles ½in.–1in. long; veins distinct. Umbels very large, terminal, compound. Peduncle short, stout. Primary rays 8–12, about 4in. long; secondary rays 6–10, ½in.–1in. long, each bearing a 10- or 12-flowered umbel. Pedicels very short. Male: calyx-teeth obsolete; petals 5; stamens 5. Female: calyx-teeth minute; ovary 2-celled; styles 2; tips free, recurved. Fruits urceolate, compressed, 2-celled and -seeded.—DC., Prod. iv. 253; A. Rich., Fl. N.Z. 281; A. Cunn., Precurs. n. 510; Hook., Lond. Journ. Bot. ii. (1843) 421, t. 11; Hook. f., Fl. N.Z. i. 94; Handbk. 102.

Var. **laetum.** Leaflets larger, 7in.–10in. long, 3in.–4in. broad, broadly ovate-lanceolate or obovate, shortly abruptly acuminate, dentate or coarsely serrate.

KERMADEC Islands and North Cape southwards to Dunedin. Common by riversides, &c., in lowland districts; ascending to 1,500ft. *Whauwhau-paku.* Dec., Jan.

7. **P. discolorum,** *T. Kirk in Trans. N.Z.I.* iii. (1871) 178. A much-branched dioecious shrub, 6ft.–20ft. high. Leaves shining, 1–3-foliolate, on slender petioles ¼in.–2in. long; leaflets 2in.–3in. long, lanceolate or ovate-lanceolate, narrowed at the base, acute or acuminate, upper part of the blade with large sharp teeth. Umbels terminal: male of 5–10 very slender rays of racemose flowers on a slender peduncle; pedicels equalling the unexpanded buds; calyx-teeth 0: female of 5–10 subequal 1–3-flowered strict rays; calyx-limb minutely 5-toothed; ovary 3–5-celled. Fruit urceolate; seeds 3–5. Styles 3–5, connate at the base; tips very short, scarcely recurved.

NORTH Island: Auckland, local. Whangaroa North; Great Omaha; Thames Goldfield; Great and Little Barrier Islands. Sea-level to 2,800ft. Nov.

Easily recognised by the racemose male flowers and the peculiar bronzed hue of the foliage.

8. **P. Sinclairii,** *Hook. f., Handbk.* 103. A shrub or small tree, 20ft. high or more. Leaves on slender petioles ½in.–1½in. long, 3–5-foliolate; leaflets sessile or petiolulate, 1in.–2in. long, coriaceous, dull, obovate or oblong-lanceolate, acute, the upper part of the blade sharply serrate; veins obscure. Umbels small, axillary or terminal, ½in.–1½in. long; rays 2–4. Pedicels short. Female flowers: calyx-teeth minute, acute; ovary 2-celled. Fruit orbicular, ovate, compressed, 2-celled, 2-seeded. Styles 2, short, recurved.

NORTH Island: Auckland District, very local. Karioi Mountain (Raglan) and Pirongia, *Cheeseman.* East Cape, *Sinclair.* Ruahine Range, *Colenso.* Opepe, Taupo, *T. K.* Mount Egmont, *Buchanan.* 1,000ft. to 4,000ft.

This species is closely related to *P. simplex*, but the fruits are narrower and the leaves are never 1-foliolate.

4. MERYTA, Forst.

Dioecious. Male flowers: calyx-teeth 3–5, minute, acute; petals 4–5, valvate; stamens 4–5, filaments longer than the petals. Female flowers: calyx-limb obsolete; petals 4–5, very small, valvate; ovary 5–∞-celled; styles

5, short, connate at base, tips spreading. Berry ovoid or oblong, 5–∞-celled. Seeds compressed. Small trees, with stout resinous branches and very large alternate coriaceous leaves. Flowers in terminal panicles, sessile.

A small genus, comprising about 16 species, occurring in islands of the South Pacific, 10 in New Caledonia, 2 in Norfolk Island. The New Zealand species is endemic.

1. **M. Sinclairii,** *Seeman in Bonplandia* x. (1862) 295. A handsome dioecious tree, 12ft.–25ft. high, with stout brittle branches. Leaves with stout marginal nerves, alternate, on stout petioles 4in.–14in. long; blades 9in.–20in. long, 4in.–10in. broad, very coriaceous, shining, obovate-oblong or oblong, sometimes contracted below the middle; veins prominent; margin slightly waved. Panicles terminal, erect; branches stout, jointed to the rhachis. Male flowers in fours with an ovate bract at the base of each fascicle, and a pair of bractlets at the base of each flower; petals ligulate, flexuous; stamens 4, inserted beneath a corrugated glandular disk; filaments slender, exserted. Female flowers solitary or crowded; petals 5–6, alternating with staminodia; ovary 3–6-celled; styles 3–6, tips spreading, short. Fruit black, shining, ¾in. long, oblong, 3–6-celled, each cell containing 1 compressed bony seed.—Hook. f., Handbk. 104; T. Kirk, Forest Fl. N.Z. t. 121. *Botryodendrum Sinclairii*, Hook. f., Fl. N.Z. i. 97.

THREE KINGS Islands, *Cheeseman*. Taranga Islands, near Whangarei, *T. K.* Reported from the Poor Knights Island. *Puka*. Dec.

One of the rarest trees in the world.

5. SCHEFFLERA, Forst.

Flowers polygamous, not jointed to the pedicel. Calyx-limb minutely 5-toothed. Petals 5, valvate. Stamens 5. Disk with a waved margin. Ovary 5–10-celled. Styles 5–10, connate to above their middle; tips spreading, free. Fruit nearly globose, fleshy, 5–10-celled, 5–10-ribbed. Pyrenes 5–10, compressed. Shrubs or small trees, with digitate leaves and sheathing petioles. Umbels few-flowered, arranged in paniculate racemes.

A genus of about 20 species, chiefly natives of New Caledonia, with one from Lifo, one from Fiji, and the present endemic in New Zealand.

1. **S. digitata,** *Forst., Char. Gen.* 46, *t.* 23. A shrub or small tree, 10ft.–20ft. high or more. Branches stout. Leaves 5–10-foliolate; petioles 4in.–7in. long, sheathing at the base; leaflets petioled, membranous, 3in.–7in. long, oblong-lanceolate, shortly acuminate, finely serrate. Umbels 3–10-flowered, ¼in.–½in. in diameter, racemose. Peduncles ½in. long. Pedicels ¼in. long. Flowers small. Calyx-teeth obsolete. Fruit purplish-black, $\frac{1}{12}$in.–$\frac{1}{10}$in. in diameter, furrowed when dry. In the young state the leaflets are often irregularly pinnatifid or partite.—Hook. f., Handbk. 103. *S. Cunninghamii*, Miq. in Linnaea xviii. (1844) 89. *Aralia Schefflera*, Spreng., Pl. Pugill. i. 28; DC., Prod. iv. 258; A. Rich., Fl. N.Z. 283; A. Cunn., Precurs. n. 513; Hook. f.,

Fl. N.Z. i. 95, t. 22; A. Gray, Bot. U.S. Expl. Exped. 715. *A. polygama*, Banks and Sol. MSS.

From the North Cape to STEWART Island. Ascends to 2,800ft. *Patete. Kotete. Patata.* Feb., March.

6. PSEUDOPANAX, C. Koch.

Calyx-teeth obsolete or inconspicuous. Petals: male 5, valvate, inflexed at the tips; female 0. Stamens 5; anthers ovate or oblong. Ovary 5-celled; styles connate at the base, forming a short cone or column; stigmas sessile. Fruit subglobose, fleshy, ribbed when dry. Pyrenes laterally compressed, bony. Glabrous shrubs or trees. Leaves usually polymorphic; foliage simple or digitate, coriaceous, often coarsely toothed. Flowers in paniculate or simple dioecious umbels or racemes. Bracts caducous or 0.

As now constituted the genus is restricted to New Zealand, and is specially remarkable for the large amount of metamorphism exhibited by the foliage of certain species. The flowers are not jointed to the pedicels.

Leaves all 3-5-foliolate; petioles and peduncles stout .. .. .. 1. *P. Lessonii.*
Leaves mostly 1-foliolate, ovate, rarely 3-foliolate; petioles and peduncles slender .. .. 2. *P. Gilliesii.*
Leaves of unbranched state deflexed, with short distant teeth. Fruit small. Styles conical .. .. 3. *P. crassifolium.*
Leaves of unbranched state deflexed, with lobulate hooked teeth. Fruit large. Styles truncate .. .. 4. *P. ferox.*
Leaves of unbranched state erect. Female umbels simple. Fruit large, globose .. .. 5. *P. Chathamicum.*

1. **P. Lessonii,** *C. Koch in C. Koch and Fint Wochenschrift* ii. (1859) 336. A much-branched shrub or small tree, 6ft.–20ft. high. Branches stout. Leaves 3–5-foliolate; leaflets sessile, 1in.–4in. long, obovate or oblong-lanceolate, subacute or obtuse, thick and coriaceous, glossy, entire or sinuate-serrate; veins indistinct; petioles 2in.–6in. long. Umbels terminal, sessile or on short stout peduncles. Male flowers racemose; primary rays 5–8, 1in.–8in. long, each carrying 3–10 terminal racemes, often with lateral racemes below; pedicels exceeding the flowers, slender. Female: primary rays shorter; secondary rays ½in.–¾in. long, 3–5-flowered; pedicels exceeding the ripe fruit; styles 5, short, truncate. Fruit broadly oblong, 5-celled, 5-seeded.—*Cussonia Lessonii*, A. Rich., Fl. N.Z. 285, t. 32; A. Cunn., Precurs. n. 511; Raoul, Enum. 46. *Aralia Lessonii*, Hook. f., Fl. N.Z. i. 96. *A. trifolia*, Banks and Sol. MSS. *Panax Lessonii*, DC., Prod. iv. 253; Hook. f., Handbk. 102. *Hedera Lessonii*, A. Gray, Bot. U.S. Expl. Exped. 719.

NORTH Island: chiefly littoral: Three Kings Islands and North Cape southwards to Poverty Bay.

2. **P. Gilliesii,** *n. s.* A shrub or small tree, 10ft.–15ft. high. Branches slender. Bark whitish. Leaves chiefly 1-foliolate, coriaceous, but not thick; blade 1½in.–2½in. long, 1½in. broad, ovate acute or acuminate, acutely serrate; petioles very slender, shorter than the blade, rarely 3-foliolate; leaflets 3in. long, lanceolate, coarsely toothed, narrowed below. Flowers not seen. Fruiting-

umbels simple, terminal. Peduncles very short or 0. Rays 3–6, 3in.–4in. long, with racemose flowers near the tips. Pedicels exceeding the fruits or 1in. long, 3–4-flowered, with pedicels usually shorter than the fruits, slender. Fruits almost globose, 5-celled. Seeds 2–5. Calyx-teeth minute. Styles 5, connate, truncate.

NORTH Island: Auckland District: Whangaroa North, inland (1868), *Gillies* and *Kirk*.

Related to *P. Lessonii*, but more slender in all its parts, with mostly unifoliolate leaves, smaller umbels, and almost globose fruits.

3. **P. crassifolium,** *C. Koch in Koch and Fint Wochenschrift,* ii. (1859) 336. A round-headed tree, with spreading or straggling branches, 20ft.–60ft. high; trunk 10in.–20in. in diameter. Leaves polymorphic.

Var. **unifoliolatum.** Leaves all simple: (1), in the seedling state membranous, rhomboid or rhomboid-acuminate, sharply toothed, lobed or deeply incised; (2), in the unbranched state shortly petioled, simple, coriaceous, linear, 1ft.–3½ft. long, ¼in. wide, deflexed, with few distant acute teeth; (3), erect, 6in.–8in. long, with shorter broader leaves on longer petioles, often coarsely toothed, obtuse, narrowed below; (4), mature state excessively coriaceous, oblong, narrowed at the ends, teeth few, small. Umbels terminal, compound. Peduncles short or 0. Primary rays 6–10, unequal, with caducous bracts at the base, spreading or compact; terminal rays 5–6. Flowers shortly pedicelled, racemose or umbellate. Male: calyx-teeth minute; stamens 5, on slender filaments. Female: calyx-teeth 0; ovary 5-celled; styles 5, conical. Fruit globose. Seeds 5.

Var. **trifoliolatum.** Leaves of the third stage 3–5-foliolate; leaflets sessile, narrow-linear, 6in.–12in. long, ½in.–¾in. broad, sessile or shortly petioled, acute or obtuse, coarsely or distantly toothed. Other stages as in the type.—T. Kirk, Forest Fl. N.Z. 59, t. 38, 38A, 38B, 38C, 38D. *Aralia crassifolium*, Banks and Sol. MSS.; A. Cunn., Precurs. n. 514; Hook., Ic. Pl. t. 583, 584; Hook. f., Fl. N.Z. i. 96. *A. heterophylla*, A. Cunn. MSS. *Panax crassifolium*, Decn. and Planch. in Rev. Hort. 1854, 105; Hook. f., Handbk. 101; T. Kirk in Trans. N.Z.I. x., xxxiii. App. *P. longissimum*, Hook. f., Handbk. 102; Buch. in Trans. N.Z.I. ix. (1876) 530, t. 20. *P. coriaceum*, Regel, Gartenflora, 1859, 45. *Hedera crassifolia*, A. Gray, Bot. U.S. Expl. Exped. 719. *Xylophylla heterophylla*, Banks and Sol. MSS.

From the North Cape to STEWART Island. Var. *unifoliolatum* common in the Auckland District; rare elsewhere. Sea-level to 1,800ft. *Horoeka*. *Hohoeka*. Feb., March.

An extremely variable and interesting plant, of which a more detailed account may be found in the "Forest Flora of New Zealand."

4. **P. ferox,** *T. Kirk, Forest Fl. N.Z.* 35, *t.* 23, 24, 25, 26. A small tree, 15ft.–20ft. high. Leaves trimorphic: (1), in the seedling state narrow-linear-lanceolate, acute; (2), in the unbranched state 12in.–18in. long, ½in. broad, on short stout petioles, deflexed, narrow-linear, slightly expanded at the tips, excessively thick and coriaceous, lobulate-dentate; teeth acute, hooked; (3), mature state 3in.–5in. long, ½in.–⅝in. broad, linear-obovate, apiculate, very thick and rigid. Flowers in terminal umbels: male of from 6–10 slender racemes; pedicels shorter than the flowers; petals 4; stamens 4: female umbel compact; rays 6–9, 1in. long, 1–3-flowered; ovary 5-celled. Fruit ovoid, glaucous, 5-seeded. Styles connate into a column, truncate.—*Panax ferox*, T. Kirk in Trans. N.Z.I. x. App. xxxiv. *P. crassifolium*, Buch., Trans. N.Z.I. ix. 529, t. 20.

NORTH Island: local. Auckland: Kauaeoruruwahine Forest, *T. K.* East Cape, *Bishop Williams!* SOUTH Island: Nelson: Wairoa, *Hector* and *Kirk*. Moutere, Matukituki, *T. K.* Canterbury: Lake Forsyth, *T. K.* Near Dunedin, *Buchanan!* Ascends to 1,000ft.

Differentiated from *P. crassifolium* by the acute lobulate teeth of the deflexed leaves, the slender racemes of the male flowers, and the larger fruits with truncate style-columns.

5. **P. Chathamicum,** *n. s.* A small tree, 20ft.–25ft. high. Branches stout. Leaves never deflexed, all simple, dimorphic; in the unbranched state 2in.–6in. long, ¾in.–1¼in. broad, lanceolate or oblong, narrowed into the short petiole, usually acute, coarsely or finely toothed above, membranous or sub-coriaceous; in the mature state 5in.–7in. long, coriaceous, linear-obovate or oblanceolate, gradually narrowed into the short winged petiole, subacute, obtuse, or truncate at the apex, obscurely sinuate-dentate or with 2–3 coarse teeth at the apex. Umbels terminal, sessile: male very large, of 6–10 primary rays, each with 5–8 slender secondary rays 2in.–3in. long, carrying crowded racemose flowers often mixed with small umbels; calyx-limb obsolete; petals 4; stamens 4: female umbels sessile, simple; rays 3–7, very slender, subequal, 2in.–4in. long, forming terminal 6–10-fruited umbels, with or without a few racemose fruits below. Fruits globose when fully ripe, large, 5-celled, 5-seeded. Styles forming a short truncate column.

CHATHAM Islands: *Enys!* (March, 1891), *Cox!* *Hoho.* Feb.

Distinguished from *P. crassifolium* and *P. ferox* by the absence of deflexed leaves, the simple female umbels, and globose fruits with truncate styles.

I am greatly indebted to my friend Mr. Cox for specimens of this curious species, and greatly regret that my diagnosis is still imperfect.

[Two specimens of what appears to be another species may be briefly described: Branchlets stout. Leaves excessively coriaceous, quite entire, 3in.–4in. long, 1in. broad or more, linear-oblong, gradually narrowed into a short petiole, acute or apiculate. Male umbels simple; rays 5–8, 1in.–1½in. long; pedicels 1–5, exceeding the flowers. Petals 4, inflexed at the tips. Stamens 4. Should better specimens prove this to be distinct it might be named *P. apiculatum*. CHATHAM Islands: *Cox!*]

*HEDERA, Linn.

Calyx entire or 4–5-toothed, superior. Petals and stamens 5–10, free. Ovary 5-celled; styles 5–10, connate or connivent; stigmas terminal. Berry subglobose, 5-celled, 5-seeded. Seed invested by a thin cartilaginous endocarp. Endosperm fleshy, ruminate. Shrubs, climbing by clasping rootlets. Leaves alternate, exstipulate, simple, lobed. Umbels simple. Flowers polygamous.

***H. Helix,** *L., Sp. Pl.* 202. A climbing shrub, ascending the loftiest trees. Trunk sometimes 1ft. in diameter. Lower leaves 1in.–4in. broad, 5-lobed, cordate; of flowering-branches ovate or ovate-oblong, acute, entire. Umbels simple or sub-racemose, erect, finely pubescent or downy; bracts minute. Berry black.

NORTH Island: on the sites of deserted gardens in several localities, but not naturalised. *Ivy.* April, May. Europe.

ORDER XXXVI.—CORNACEAE.

Calyx-tube adnate to the ovary; teeth 4 or 5, or 0. Petals 4 or 5, rarely 0, valvate. Stamens as many or twice as many as the petals, inserted at the base of an epigynous disk. Ovary inferior, 1–2-celled, with 1 anatropous ovule in each cell; style very short or connate at the base, with recurved stigmas or stigma capitate; ovules usually solitary, 1 in each cell, pendulous or 3 from the apex of a column in a 1-celled ovary. Fruit an indehiscent drupe. Testa thin; endosperm fleshy; embryo minute. Shrubs or trees, rarely herbs, with opposite or alternate entire exstipulate leaves and regular hermaphrodite or unisexual flowers in axillary or terminal cymes or panicles.

GENERA, 12. SPECIES, 80. Sparingly distributed through the temperate regions of both hemispheres, but most frequent in the northern.

1. COROKIA. Leaves white with silky pubescence beneath. Flowers perfect.

2. GRISELINIA. Leaves glabrous. Flowers unisexual.

1. COROKIA, A. Cunn.

Calyx-tube turbinate, 5-toothed, valvate in bud. Petals 5, valvate, a small scale at the base of each, silky on the outer surface. Stamens 5; filaments short. Ovary 1–2-celled; ovules solitary; style short; stigma ovoid or almost capitate, 2-lobed. Drupe broadly oblong, ovoid or globose, ¼in.–½in. long, crowned with the persistent calyx-teeth, 1-celled, 1-seeded. Endosperm fleshy; embryo elongate. Evergreen shrubs, with straight or tortuous branches and black bark. Leaves silky, pubescent below. Flowers yellow, in axillary or terminal cymes, racemes, or panicles.

The genus comprises 3 species, endemic in New Zealand and the Chatham Islands. NAME, a modification of the Maori *korokia*.

Flowers in slender terminal panicles .. .. 1. *C. buddleioides.*
Flowers in axillary racemes .. .. .. 2. *C. macrocarpa.*
Flowers in small axillary cymes or fascicles .. 3. *C. Cotoneaster.*

1. **C. buddleioides,** *A. Cunn., Precurs. n.* 579. A much-branched shrub, 6ft.–12ft. high. Branches slender. Leaves alternate, narrow-lanceolate, acuminate, 3in.–6in. long, ½in. broad; petioles short, silky, pubescent beneath. Panicles very slender, terminal, leafy at the base. Flowers numerous, yellow. Pedicels very slender, silky, longer than the petioles. Drupe ovoid, orange-red, ¼in. long.—Hook., Ic. Pl. t. 424; Hook. f., Fl. N.Z. i. 98; Handbk. 106.

NORTH Island: Auckland: Mangonui to the East Cape. Ascends to 2,000ft. *Korokia-taranga.* Dec.

2. **C. macrocarpa,** *n. s.* Erect, with spreading rather robust branches, 15ft.–20ft. high. Leaves oblong-lanceolate or broadly oblong-lanceolate, 1in.–2½in. long, ¾in.–1¼in. broad, acute subacute or apiculate, coriaceous, narrowed into a short petiole, densely silky, tomentose beneath. Flowers in axillary racemes, shorter than the leaves. Rhachis, pedicels, and calyx silky, tomentose. Pedicels shorter than the petioles. Petals linear-acute. Drupe large, fully ⅗in. long, broadly oblong, red.—*C. buddleioides*, var. *β*, Hook. f., Fl. N.Z. i. 98; F. Muell., Veg. Chath. Isds. 16.

CHATHAM Islands, *Dieffenbach, Gilbert Mair! H. Travers! Cox!* *Whakataka. Hokotaka.*

Mr. Cox sends a sterile shoot with oblong obtuse membranous leaves 6in. long and nearly 2in. broad, evidently grown in the shade.

3. **C. Cotoneaster,** *Raoul, Ann. Sc. Nat.* ii. (1844) 120; *Choix.* 22, *t.* 20. A much-branched shrub, 3ft.–8ft. high, with tortuous black interlacing branches. Leaves alternate or fascicled, ¼in.–1in. long, orbicular, obovate, or oblong-ovate, emarginate or obcordate, abruptly narrowed into the short flat

petiole. Flowers axillary or terminal, solitary or fascicled. Petals broader. Drupe globose, ¼in. in diameter, red.—Hook. f., Fl. N.Z. i. 98; Handbk. 106.

From the Three Kings Islands and North Cape to the mouth of the Waiau and the Bluff, but often local. Ascends to 2,600ft. Dec., Jan.

Vigorous young shoots rarely produce acutely toothed or lobed leaves exceeding 1½in. in length.

2. GRISELINIA, Forst.

Flowers dioecious. Male: calyx-teeth 5, minute or 0; petals 5, valvate; stamens 5; disk fleshy, pentagonous. Female: calyx-teeth adnate with the ovary, ovoid or turbinate, 5-toothed; petals 5 or 0; stamens 0; ovary 1–2-celled; styles 3, very short, connate at the base, stigmatic tips, subulate, recurved; ovules solitary. Fruit a 1-celled 1-seeded drupe, crowned with the minute calyx-teeth. Cotyledons divaricating. Evergreen shrubs or large trees. Branches scarred at the nodes. Leaves very oblique at the base or nearly symmetrical, glossy, the petiole shortly sheathing and jointed to the branch. Panicles axillary or subterminal.

Besides the New Zealand species, which are endemic, 3, or perhaps 4, others are found in Chili.

Leaves large, very unequal. Female flowers apetalous .. .. 1. *G. lucida.*
Leaves smaller, often nearly symmetrical. Female flowers polypetalous.. .. 2. *G. littoralis.*

1. **G. lucida,** *G. Forst., Prod. n.* 401. A shrub or small tree, 3ft.–30ft. high, often epiphytic. Bark furrowed. Leaves all obliquely ovate or oblong, rounded at the apex, very unequal at the base, or almost cordate, glossy, 4in.–6in. long; petiole short, stout. Flowers in subterminal or axillary much-branched panicles, 3in.–6in. long. Rhachis and pedicels pubescent; pedicels jointed to the rhachis. Female flowers minute, apetalous. Drupe fleshy, blackish-purple, 1-celled, 1-seeded.—Willd., Sp. Pl. iv. 1128; A. Cunn., Precurs. n. 261; Raoul, Choix. 46; Hook. f., Fl. N.Z. i. 98; Handbk. 105; T. Kirk, Forest Fl. N.Z. t. 41. *Scopolia lucida*, Forst., Char. Gen. t. 70. *Lissophyllum lucidum,* Banks and Sol. MSS.

NORTH and SOUTH Islands: from Spirits Bay to Dusky Sound and Foveaux Strait. *Puka. Poukater.* Oct., Nov.

2. **G. littoralis,** *Raoul, Choix.* 12, *t.* 19. A round-headed tree, 40ft.–60ft. high, with a short trunk 2ft.–5ft. in diameter. Bark roughly furrowed. Leaves almost symmetrical at the base, ovate or oblong-ovate, narrow, rounded at the apex, 1½in.–4in. long, yellowish-green; veins obscure; petiole slender. Panicles axillary, ½in.–2in. long, finely pubescent; branches very short or wanting. Female flowers: calyx-teeth obsolete; petals 5, narrower and longer than in the male. Fruit smaller than in *G. lucida.* Panicle sometimes reduced to a raceme.—Hook. f., Handbk. 105; T. Kirk, Forest Fl. N.Z. 69, t. 42. *Pukateria littoralis,* Raoul in Ann. Sc. Nat. ii. (1844) 121.

NORTH and SOUTH Islands: from the Cape Colville Peninsula southward to Stewart Island. Ascends to 3,000ft. *Kapuka. Papauma. Tapatapauma.* Oct., Nov.

Timber extremely durable, although not of large dimensions.

Division II.—**GAMOPETALAE.**

Order XXXVII.—CAPRIFOLIACEAE.

Calyx-limb 4–5-lobed or -toothed, adnate with the ovary. Corolla inserted round the epigynous disk, regular or irregular, tubular, campanulate or rotate, with 4–5 imbricate teeth or lobes. Stamens usually as many as the corolla-lobes, inserted on the tube, equal or unequal. Ovary inferior, 2–5-celled, rarely 1-celled, with 1 or more pendulous ovules in each cell. Style filiform; stigma entire or divided. Fruit a 1–5-celled berry. Seeds 1 or more in each cell. Endosperm fleshy; embryo axile; radicle superior. Trees or shrubs, often climbing, rarely herbs. Leaves usually exstipulate, opposite or rarely alternate, simple or rarely pinnate.

A small order, most frequent in the northern temperate region, with a few southern species. Represented in Australia by *Sambucus* alone, and in New Zealand by the endemic genus *Alseuosmia*.

* Sambucus. Leaves pinnate. Corolla rotate. Flowers in terminal corymbose cymes.

1. Alseuosmia. Leaves simple. Corolla tubular. Flowers axillary.

* SAMBUCUS, Linn.

Calyx-teeth 3–5, minute or obsolete. Corolla with a very short tube, 3–5-partite, rotate or campanulate. Stamens 3–5, inserted at the base of the corolla. Ovary 3–5-celled; style short, or stigma sessile, 3–5-lobed. Fruit a small berry-like juicy drupe containing 3–5 pyrenes. Trees, shrubs, or herbs, usually strong-smelling. Leaves opposite, pinnate. Flowers in large terminal compound cymes.

* **S. nigra,** *L., Sp. Pl.* 269. A shrub or small tree, 10ft.–15ft. high, with corky lenticellate bark; branches with thick pith. Leaflets in 2–4 pairs, ovate-oblong or lanceolate, acute, serrate. Cymes 5-rayed, 5in.–8in. in diameter, flat-topped. Berry small, black, shining.

NORTH and SOUTH Islands: frequently naturalised; sometimes forming a large part of the undergrowth in woods. *Elder.* Oct., Nov. Europe.

1. ALSEUOSMIA, A. Cunn.

Calyx-teeth 4 or 5, narrow, deciduous. Corolla tubular or narrowly funnel-shaped, with 4–5 spreading lobes; margins inflexed, sometimes toothed or lobulate. Stamens 4–5, inserted near the mouth of the corolla; filaments short; anthers oblong. Ovary 2-celled; style filiform; stigma clavate or sub-clavate; ovules 2-seriate, several in each cell, pendulous. Berry ovoid, crimson, 2-celled, several-seeded, crowned with the epigynous disk. Seeds angular; testa bony. Shrubs, with alternate subcoriaceous polymorphous leaves and axillary solitary or fascicled drooping flowers on short bracteolate pedicels.

Small tufts of minute red hairs are found in the axils of the leaves and bracteoles. The genus is endemic in New Zealand, and the species are so excessively variable that it is impossible to separate them by sharply-defined characters. The flowers are deliciously fragrant.

Leaves 3in.–7in. long, broad. Flowers 1in.–2in. long .. .. 1. *A. macrophylla.*
Leaves 1in.–3in. long, oblong or obovate. Flowers ¾in.–1in. long .. 2. *A. quercifolia.*
Leaves ½in.–1in. long, ovate or rhomboid. Flowers ½in.–¾in. long .. 3. *A. Banksii.*
Leaves narrow-linear or linear .. .. 4. *A. linariifolia.*

1. **A. macrophylla**, *A. Cunn., Precurs. n.* 487. A glabrous erect or spreading shrub, 4ft.–10ft. high. Leaves 3in.–7in. long, oblong or linear-oblong or oblong-obovate, narrowed into the short slender petiole, entire or sinuate-serrate or distantly toothed, subcoriaceous, often glossy. Flowers solitary or in 1–3-flowered fascicles, pendulous. Corolla 1½in. long or more, crimson; pedicels very slender; bracteoles acute. Calyx-lobes linear-subulate, acute. Corolla-lobes usually 5, with fimbriate or minutely-toothed margins. Berry crimson, ⅓in. long.—Hook. f., Fl. N.Z. i. 102, t. 23; Handbk. 109.

NORTH Island: from Spirits' Bay to Cook Strait, but rare and local south of Taupo. SOUTH Island: Marlborough: very rare, *J. Rutland!* Nelson: Collingwood, *J. Dall!* Oct., Nov. Ascends to 3,000ft. A beautiful plant.

2. **A. quercifolia**, *A. Cunn., Precurs. n.* 493. Erect, slender, sparingly branched, or branches rarely dichotomous. Leaves 1in.–4in. long, variable even on the same branch, lanceolate-oblong or obovate-lanceolate or ovate-lanceolate, narrowed into a very slender petiole, obtuse or rarely acute, entire or obscurely or distantly lobed or deeply sinuate-lobed, almost membranous. Flowers solitary, or in 3–6-flowered fascicles, ⅓in.–¾in. long, very slender; corolla-tube red, lobes subacute, greenish; stigma often exserted.—Hook. f., Fl. N.Z. i. 102; Handbk. 109. *A. Ilex*, A. Cunn., Precurs. n. 492.

NORTH Island: From Mangonui southwards, but often local. SOUTH Island: Mount Peter, *J. MacMahon!* Oct., Dec.

Var. **puella** (*sp.*), Col. in Trans. N.Z.I. xvii. 241. Erect, 1ft.–2ft. high. Leaves 1in.–1½in. long, broadly ovate, narrowed below, crenate-serrate. Flowers few, pendulous, very fragrant. Norsewood, Hawke's Bay, *Colenso!*

Var. **glauca.** Erect, spreading, 3ft.–5ft. high; branches slender. Leaves oblong-obovate, 2in.–3in. long, ¾in.–1in. broad, glaucous beneath, entire. Flowers few, restricted to the upper axils. Marlborough: Pelorus Sound, *MacMahon!* Mount Stokes, *T. K.* Rai Valley, *Rutland!* Ascends to 2,800ft.

3. **A. Banksii**, *A. Cunn., Precurs. n.* 493. A small bush with slender spreading or straggling branches, 6in.–24in. high; young shoots puberulous. Leaves ½in.–1in. long, varying from broadly oblong-obovate to ovate-orbicular or almost rhomboid; margins waved or irregularly toothed or almost lobed, especially in the upper part of the leaf; petiole very slender, short. Flowers solitary, ¼in.–½in. long, axillary or terminal, greenish-yellow. Berry almost spherical; seeds 4–8.—Hook. f., Fl. N.Z. i. 102, t. 24; Handbk. 110. *A. palaeiformis* and *A. atriplicifolia*, A. Cunn., Precurs. n. 490, 491.

NORTH Island: from Mangonui and Hokianga to the Waitemata, but rare and local. Oct. to Dec.

Specimens less than 3in. high may sometimes be found with a short greenish-yellow flower in each axil.

4. **A. linariifolia**, *A. Cunn., Precurs. n.* 487. An erect bushy shrub, 1ft.–4ft. high, with numerous slender branches; young shoots clothed with minute spreading pubescence. Leaves crowded, 1in.–2in. long, $\frac{1}{16}$in.–½in. wide, linear or linear-lanceolate, acute, entire or sinuate-toothed or -lobed; petioles very short. Flowers drooping, solitary or in 2–4-flowered fascicles, ½in.–¾in. long,

red; lobes of corolla more or less toothed and fimbriate.—Hook. f., Fl. N.Z. i. 103, t. 25; Handbk. 110. *A. ligustrifolia*, A. Cunn., *l.c.* 488. *A. Hookeria*, Col., Excurs. North Isd. 84.

NORTH Island: from Whangaroa North to Manukau Harbour. Local. Oct.

The crowded linear leaves give this pretty species a different aspect from any other. The corolla-tube is broader than those of *A. quercifolia* or *A. Banksii*.

ORDER XXXVIII.—RUBIACEAE.

Calyx-tube adnate to the ovary, 4- or 5-toothed or -lobed or 0. Corolla rotate, tubular, funnel-shaped or campanulate, inserted round an epigynous disk; lobes usually 4 or 5, rarely 3, valvate or imbricate in bud. Stamens alternating with the corolla-lobes and inserted on its tube. Ovary inferior, 2–4-celled; ovules 1 or more in each cell; styles 1 or 2; stigma 2–3-lobed or capitate, lobes often linear. Fruit an indehiscent berry or nut, or a capsule usually with 2 or more cells or nuts; cells 1-seeded. Seeds with a very thin testa. Embryo straight, rather small, with flat cotyledons; endosperm fleshy or horny. Trees, shrubs, or herbs, rarely scandent, with simple entire opposite leaves and interpetiolar or whorled stipules. Flowers hermaphrodite or unisexual, axillary or terminal, cymose, often solitary.

A very large order, widely distributed through all regions. *Cinchona* and *Coffea* are amongst its most important products. Of the New Zealand genera, *Coprosma* and *Nertera* are restricted to the extra-tropical or mountain regions of the Southern Hemisphere, extending to the verge of the antarctic circle. *Galium* and *Asperula* have a wide range in both hemispheres. GENERA, 340. SPECIES, 4,300.

Tribe I. COFFEEAE.—Leaves opposite or rarely whorled. Ovules usually solitary in the cells. Cells of the fruit always 1-seeded. Fruit a berry-like drupe, 2-seeded.

1. COPROSMA. Trees or shrubs, sometimes suffruticose, creeping. Flowers unisexual.
2. NERTERA. Slender creeping herbs. Flowers hermaphrodite.

II. STELLATEAE.—Leaves verticillate. Ovary 2-celled, with 1 ovule in each cell. Fruit of 2 dry indehiscent 1-seeded carpels.

3. ASPERULA. Calyx-teeth 0. Corolla funnel-shaped, 4-lobed.
4. GALIUM. Calyx-teeth 0. Corolla rotate, 4-lobed.

* SHERARDIA. Calyx-teeth 4–6. Flowers in small involucrate heads.

1. COPROSMA, Forst.

Flowers often inconspicuous, dioecious. Calyx-limb 4–5-toothed or -lobed, often 0 in the male; tube ovoid in the female. Corolla-tube funnel-shaped or campanulate, 4–5-lobed: lobes valvate in the bud. Stamens usually 4–5, inserted at the base of the corolla-tube; filaments capillary. Anthers exserted, apiculate. Ovary 2-celled, rarely 3–4-celled, with 1 erect ovule in each cell. Styles divided to the base into 2, 3, or 4 linear stigmatic papillose lobes. Drupe rounded or ovoid, with 2 plano-convex 1-seeded pyrenes. Trees and shrubs, rarely creeping. Leaves usually small, with interpetiolar stipules, often intolerably foetid when bruised. Flowers in trichotomous or simple axillary cymes or fascicles, or solitary.

A genus comprising about 50 species, most abundant in New Zealand, but extending to temperate Australia, the Pacific Islands, and the highest mountains of Borneo. Some of the New

Zealand species form almost impenetrable thickets as far south as Campbell Island. Many of the species are difficult to discriminate, owing to their great variability. The student should endeavour to obtain, in addition to male flowers, female flowers and fruits taken from the same plant, as the fruit is often of great importance in the determination of species. The calyx is absent in the male flower of many species, but is invariably present in the female; in the former the corolla is most frequently funnel-shaped, in the latter it is usually tubular. In many of the small-leaved species the male, or both male and female flowers are seated in a kind of involucel, which usually is unequally 4-lobed, and is easily mistaken for a calyx, especially in the male flower, which is often destitute of this organ; one or more pairs of connate stipules are frequently developed below the involucel. The New Zealand species have been admirably monographed by Mr. T. F. Cheeseman in Trans. N.Z.I. xix., and in many cases I have gladly availed myself of the results of his work. Without doubt additions will be made to the genus, especially from alpine and subalpine districts.

NAME, from the Greek, in reference to the nauseating odour of certain species.

I. ERECT SHRUBS OR TREES. LEAVES EXCEEDING 1IN. IN LENGTH. FLOWERS IN DENSE FASCICLES.

* *Peduncles 3in.–4in. long (except in* C. macrocarpa), *trichotomously divided; fascicles dense.*

Leaves subcoriaceous, 4in.–7in. long. Peduncles 1in.–1½in. long .. 1. *C. macrocarpa.*

Leaves membranous, 3in.–8in. long.. .. .. 2. *C. grandifolia.*

Leaves coriaceous, 2in.–5in. long .. .. 3. *C. lucida.*

** *Peduncles short. Flowers numerous, in dense fascicles, or rarely few.*

Leaves fleshy, glossy, obtuse or retuse .. .. 4. *C. Baueri.*

Leaves thin. Branchlets and young leaves puberulous .. .. 5. *C. petiolata.*

Leaves coriaceous; margins serrulate. Fascicles 1–3-flowered.. .. 6. *C. serrulata.*

Leaves coriaceous, entire, oblong or elliptical .. .. 7. *C. robusta.*

Leaves coriaceous, linear. Fascicles 3–12-flowered, axillary .. .. 8. *C. Cunninghamii.*

Leaves membranous, acute or acuminate. Fascicles 3–5-flowered .. 9. *C. acutifolia.*

Leaves membranous, acute. Fascicles 1–8-flowered .. .. .. 10. *C. tenuifolia.*

Leaves coriaceous, orbicular-spathulate. Fascicles dense .. .. 11. *C. arborea.*

II. ERECT OR PROSTRATE SHRUBS. FEMALE FLOWERS USUALLY SOLITARY. MALE FLOWERS SOLITARY OR IN FEW-FLOWERED FASCICLES. LEAVES LESS THAN 1IN. LONG (EXCEPT IN C. FOETIDISSIMA).

A. FLOWERS terminal on arrested lateral branchlets, or apparently axillary from the branchlets being scarcely developed.

* *Leaves spathulate.*

Leaves abruptly narrowed into winged petioles .. .. 12. *C. spathulata.*

** *Leaves orbicular, linear-oblong or obovate.*

a. *Twigs densely pubescent, except in* C. tenuicaulis *and occasionally in* C. rhamnoides:—

Branches divaricating. Leaves membranous, orbicular-cuspidate, ⅛in.–1in. in diameter. Male and female flowers in small fascicles .. 13. *C. rotundifolia.*

Branchlets fastigiate. Leaves membranous, orbicular-spathulate, ⅛in.–¼in. in diameter. Male flowers geminate or fascicled .. .. 14. *C. areolata.*

Branchlets slender, flexuous, glabrous or glabrate. Leaves orbicular, ⅛in.–½in. in diameter. Flowers solitary or geminate .. .. 15. *C. tenuicaulis.*

Erect or prostrate. Branches dense or open, often interlaced. Leaves orbicular, ovate, narrow-oblong, or linear. Flowers solitary or fascicled .. .. 16. *C. rhamnoides.*

b. *Twigs puberulous or glabrate:*—

Decumbent or prostrate. Branchlets puberulous. Leaves narrow-obovate, ½in. long, ¼in. wide .. .. 17. *C. ramulosa.*

Erect. Branchlets puberulous. Leaves obovate. Drupe turbinate .. 18. *C. turbinata.*

Erect. Branchlets puberulous or glabrate. Leaves linear-oblong. Drupe obconic, compressed .. .. 19. *C. obconica.*

c. *Branchlets densely pubescent:—*

Leaves oblong to obovate, ciliate .. .. 20. *C. ciliata.*

Leaves obovate or linear-oblong. Flowers all solitary .. .. 21. *C. parviflora.*

d. *Branchlets glabrous or glabrate, except in* C. Kirkii. *Leaves orbicular to obovate, or linear:—*

Branchlets rigid, divaricating, often interlaced. Leaves orbicular, coriaceous. Flowers solitary or geminate .. .. 22. *C. crassifolia.*

Branchlets divaricating. Bark reddish. Leaves orbicular or oblong-spathulate, $\frac{1}{4}$in.–$\frac{3}{4}$in. long, glabrous. Stigmas very slender .. 23. *C. rigida.*

Branchlets ascending. Leaves obovate, $\frac{1}{2}$in.–1in. long, pubescent. Stigmas robust .. .. 24. *C. Buchanani.*

Branchlets divaricating. Leaves round, oblong, or orbicular, $\frac{1}{4}$in.–$\frac{2}{3}$in. long, membranous .. .. 25. *C. rubra.*

Branchlets slender, intricate. Leaves ovate- or elliptic-spathulate, membranous, $\frac{1}{5}$in.–$\frac{1}{3}$in. long .. .. 26. *C. virescens.*

Branches flexuous, interlaced. Leaves narrow-linear, $\frac{1}{4}$in.–$\frac{1}{3}$in. long, $\frac{1}{20}$in. wide, often fascicled. Male flowers solitary or fascicled .. 27. *C. acerosa.*

Twigs strigose, pubescent. Leaves $\frac{3}{4}$in. long, $\frac{1}{16}$in. broad. Female flowers solitary. Drupe globose, $\frac{1}{10}$in. in diameter .. .. 28. *C. margarita.*

Branches divaricating. Leaves linear-oblong, $\frac{1}{4}$in.–$\frac{1}{2}$in. long, $\frac{1}{10}$in. wide. Male flowers solitary or fascicled .. .. 29. *C. propinqua.*

Decumbent or prostrate. Twigs pubescent. Leaves $\frac{3}{8}$in.–$\frac{3}{4}$in. long, fascicled. Female flowers in 3–5-flowered fascicles on abortive branchlets .. .. 30. *C. Kirkii.*

B. Flowers usually terminating leafy branches; females solitary; males solitary or rarely fascicled.

Erect. Branchlets slender. Leaves linear-lanceolate; petioles short, glabrous. Male flowers in 3–5-flowered fascicles .. .. 31. *C. linariifolia.*

Erect. Branchlets stout, setose. Leaves linear-lanceolate, sessile, sparingly ciliate .. .. 32. *C. Solandri.*

Branchlets very slender. Leaves oblong, ovate, or obovate, membranous, distant. Male flowers solitary or in 2–3-flowered fascicles .. 33. *C. foetidissima.*

Erect or prostrate. Leaves linear, obovate, or oblong. Peduncle short, decurved .. .. 34. *C. Colensoi.*

Prostrate. Leaves coriaceous, retuse or emarginate, with thickened erosulate margins .. .. 35. *C. retusa.*

Erect. Branches rigid. Leaves numerous, coriaceous, linear or cuneate-oblong or obovate-oblong, $\frac{1}{4}$in.–$\frac{1}{2}$in. long .. .. .. 36. *C. cuneata.*

Erect. Branches very slender. Leaves linear or linear-lanceolate, $\frac{1}{4}$in.–$\frac{1}{3}$in. long, membranous .. .. 37. *C. microcarpa.*

Erect or prostrate. Leaves $\frac{1}{4}$in.–$\frac{1}{3}$in. long, linear-lanceolate, coriaceous, concave .. .. 38. *C. depressa.*

III. Stems prostrate and rooting. Leaves small. Flowers solitary, terminal.

Leaves glabrous, narrow-oblong or obovate. Male corolla tubular .. 39. *C. repens.*

Leaves linear-oblong or obovate, hairy. Male corolla shorter, narrowed below .. .. 40. *C. Petriei.*

1. **C. macrocarpa,** *Cheesem. in Trans. N.Z.I.* xx. (1887) 147. A robust glabrous shrub, 5ft.–12ft. high. Bark dark-brown. Leaves subcoriaceous, glossy-green, 4in.–7in. long, $1\frac{1}{2}$in.–$3\frac{1}{2}$in. broad, broadly ovate-oblong or elliptic-oblong, obtuse, subacute or apiculate, slightly thickened at the margins, rather suddenly narrowed into a short petiole; veins conspicuous. Stipules large, loosely sheathing, truncate. Flowers not seen. Fruit very large (immature), $\frac{1}{2}$in.–1in. long, sessile or on short pedicels, in 3–7-fruited axillary fascicles or

short racemes, broadly oblong, rounded at both ends, or sometimes nearly orbicular.

THREE KINGS Islands: *Cheeseman!* 1887.

A noble species, with fruit twice the size of that of *C. grandifolia*. Both leaves and fruit show a near relationship to *C. grandifolia*, especially in the venation of the former, but the arrangement of the fruits is nearer that of *C. robusta*. My thanks are due to Mr. Cheeseman for the only specimens I have seen of this fine plant.

2. **C. grandifolia,** *Hook. f., Fl. N.Z.* i. 104. A sparingly-branched shrub, 6ft.–15ft. high. Branches naked below. Bark brown. Leaves membranous, not shining, 4in.–8in. long, 1½in.–3in. broad, broadly oblong, obovate-oblong, acute or rarely acuminate, narrowed at the base; petioles slender, ¾in.–1in. long. Peduncles 1in.–3in. long, trichotomously branched. Flowers sessile, in terminal or lateral 2-bracteate fascicles; male fascicles very dense. Calyx minute in both, but the corolla of the male is funnel-shaped, of the female tubular, very narrow. Drupe ⅓in. long, broadly oblong, rounded at both ends, orange-red.—Handbk. 112. *C. autumnalis,* Col. in Trans. N.Z.I. xix. (1886) 263. *Ronabea australis,* A. Rich., Fl. N.Z. 265. *Pelaphia grandifolia,* Banks and Sol. MS.

NORTH and SOUTH Islands: from the Three Kings Islands and the North Cape to Awatere and the Valley of the Buller. Ascends to 2,800ft. *Menono. Kanono. Raurekau.* April to June.

3. **C. lucida,** *Forst., Char. Gen.* 138. Glabrous, erect, 2ft.–15ft. high. Leaves 2in.–5in. long, obovate oblong-obovate or lanceolate-obovate, acute subacute or apiculate, narrowed into the short petiole, coriaceous, shining. Peduncles 1in.–2in. long, trichotomously divided. Calyx minutely 4–5-toothed both in male and female. Corolla 4–5-lobed; male broadly tubular; female narrower and shorter. Stamens 4 or 5. Styles three or four times longer than the corolla, filiform. Drupe ⅓in. long, orange-red.—DC., Prod. iv. 378; A. Rich., Fl. N.Z. 262; A. Cunn., Precurs. n. 470; Hook. f., Fl. N.Z. i. 104; Handbk. 112. *Pelaphia laurifolia,* Banks and Sol. MS.

Var. **obovata.** Erect, branches slender, strict. Leaves 1½in.–3in. long, obovate or oblong-obovate, obtuse, subcoriaceous, not shining. Peduncles shorter, with fewer flowers.

Throughout the colony, from the Three Kings to Stewart Island. Ascends to fully 3,000ft. Var. *obovata*: Great Barrier Island and Cape Colville ranges. *Karamu. Karangu.* Sept. to Nov.

Easily distinguished from *C. grandifolia* by the coriaceous obovate shining leaves and smaller flowers.

4. **C. Baueri,** *Endlich., Iconog. Gen. Pl. t.* iii. A shrub or small round-headed tree, from 1ft.–20ft. high, glabrous except the young shoots, which are sometimes minutely pubescent. Branchlets stout, terete, or obscurely tetragonous. Leaves almost fleshy, broadly oblong or ovate or nearly obovate, narrowed into the short slender petiole, rounded or retuse at the tip; margins often recurved, glossy. Stipules broad, acute, minutely toothed. Peduncles axillary, about as long as the petioles. Male in dense terminal heads; calyx very small, minutely 4-toothed; corolla campanulate, usually 4-lobed. Female:

peduncle exceeding the petioles in fruit, simple or trichotomous; heads 3–5-flowered; calyx-limb truncate or minutely 4-toothed; corolla short, tubular, 4-toothed. Drupe broadly ovoid, ¼in.–⅓in. long, orange-yellow.—*C. Baueriana*, Hook. f., Fl. N.Z. i. 106; Handbk. 112; Cheesem., Trans. N.Z.I. xix. 232; T. Kirk, Forest Fl. N.Z. t. 62. *C. lucida*, Endl., Prod. Fl. Ins. Norf. n. 117. *C. retusa*, Hook. f. in Lond. Journ. Bot. iii. 415. *Pelaphia retusa*, Banks and Sol. MS.

NORTH and SOUTH Islands: on sea-cliffs. From the KERMADEC Islands, the Three Kings Islands, and the North Cape to Marlborough and Greymouth. CHATHAM Islands, *H. H. Travers!* Also in LORD HOWE'S Island and NORFOLK Island. *Naupata. Angiangi. Mamangi.* Sept. to Nov.

The stipules of this species and some others secrete a watery fluid, which serves to protect the buds.

? Var. **oblongifolia.** Branches very stout, rigid, densely pubescent or almost piloso. Leaves glabrous or glabrate, linear-oblong-spathulate, about 1in. long, ¼in.–⅓in. wide, rounded at the tips, gradually narrowed into very short petioles, rather fleshy; midrib pubescent. Flowers and fruit not seen. NORTH Island: Tapotopoto Bay, *Gillies* and *Kirk*.

5. **C. petiolata,** *Hook. f. in Journ. Linn. Soc. Bot.* i. (1857) 128. A shrub or small tree, 10ft.–30ft. high, with spreading branches. Young shoots and petioles pubescent. Leaves glabrous and pubescent, about 2in. long, 1in. broad, broadly elliptical-oblong or oblong-spathulate, rounded above, narrowed into a slender petiole, almost coriaceous; petiole and midrib pubescent. Stipules deltoid, short. Peduncles and flowers as in *C. Baueri*, but the latter rather smaller.—*C. Baueri*, F. Muell., Fram. Phyt. Fragm. Austr. ix. 69 (not of Endl.).

KERMADEC Islands, *Cheeseman*. NORTH Island: maritime rocks south of Castlepoint, *Colenso*. CHATHAM Islands, *Cox!* Also on LORD HOWE'S Island and NORFOLK Island. *Karamu.*

I have not seen the flowers of this species, which differs from *C. Baueri* in the narrower, thin, almost coriaceous leaves and short pubescent petioles and branchlets. Mr. Cox states that on the Chatham Islands it forms a large portion of the forest, sometimes with a trunk 2ft. in diameter, and often hollow. My attempts to find Colenso's plant near Castlepoint have been futile.

6. **C. serrulata,** *Hook. f., MS.* ex *Buch. in Trans. N.Z.I.* iii. (1870) 212. A sparingly-branched glabrous shrub, 1ft.–3ft. high, rarely more. Branches stout, erect or inclined. Bark white when old. Leaves ¾in.–2in. long or more, ½in.–1½in. broad, coriaceous, shining, varying from broadly obovate to oblong-obovate or orbicular-spathulate, rounded at the apex, obtuse or apiculate, narrowed into a very short broad petiole, minutely serrulate. Stipules very large, persistent, acute or apiculate, with toothed or ciliated margins. Male flowers in 3–6-flowered axillary fascicles; calyx 0; corolla 4–5-lobed, campanulate; stamens 4–5. Female: solitary or in 2–3-flowered fascicles; calyx-teeth minute; corolla long and tubular; lobes 3–5, very short. Drupe ⅓in. long, broadly oblong.—T. Kirk in Trans. N.Z.I. x. App. xxxv.; Cheesem., *l.c.* xix. 231.

SOUTH Island: not uncommon in the Alps from Mount Arthur to mountains above Dusky Sound, but local in some districts. 2,000ft. to 4,500ft. Nov., Dec.

Distinguished from all other species by the apparently serrulate leaves.

7. **C. robusta,** *Raoul in Ann. Sc. Nat.* ii. (1844) 121; *Choix de Pl.* 23, *t.* 21. A stout erect glabrous shrub, 2ft.–12ft. high, with pale-brown shining bark. Leaves numerous, close-set, coriaceous, glossy, 1in.–5in. long, ⅓in.–2in. broad, elliptic-oblong or lanceolate, acute subacute or obtuse, narrowed into a short petiole; margins often faintly recurved. Stipules shortly deltoid, acute. Peduncles short, stout, simple or branched. Flowers densely capitate. Male: calyx minute, truncate or minutely 4–5-toothed; corolla campanulate, 4–5-lobed. Female: smaller; corolla tubular, 4–5-lobed; lobes narrow, often spreading. Drupes crowded, ⅓in. long, deep-red or golden-yellow.—Hook. f., Fl. N.Z. i. 105; Handbk. 113. *C. coffaeoides,* Col. in Trans. N.Z.I. xxi. (1888) 87! *Pelaphia laeta,* Banks and Sol. MS.

Var. **angustata.** Leaves linear-oblong-lanceolate, 1in.–3in. long, ¼in.–½in. broad, less coriaceous. Not uncommon.

Var. **parva.** Leaves ½in.–¾in. long, oblong, more membranous. SOUTH Island: Queen Charlotte Sound, *Banks* and *Sol.!* Kaikoura Mountains, *T. K.*

From the Three Kings Islands to Southland; Chatham Islands: *Travers! Cox!* Ascends to 2,300ft. Common. *Karamuramu. Kakaraumu. Karamu.* Sept. to Nov.

Some forms approach *C. Cunninghamii* in the foliage, but are easily distinguished by the more numerous flowers and red or yellow fruits.

8. **C. Cunninghamii,** *Hook. f., Handbk.* 113. A large shrub or small tree, 6ft.–15ft. high. Branches ascending. Bark pale-brown. Leaves erect, distant or crowded, ½in.–2in. long, ⅛in.–¼in. broad, linear or linear-lanceolate, flat, acute or subacute, narrowed into the short petiole, scarcely coriaceous. Flowers sessile in 3–10-flowered axillary glomerules or terminating short abortive branchlets. Male: calyx minute, 4-toothed, truncate; corolla campanulate, 4–5-lobed. Female: smaller and less numerous; calyx-limb 4–5-toothed; corolla tubular, narrow, with 4–5 acute usually spreading lobes; styles very long and slender. Drupe subglobose, ⅙in. long, translucent.—Cheesem. in Trans. N.Z.I. xix. 234. *C. foetidissima,* A. Cunn., Precurs. n. 471 (in part)!

NORTH Island: not infrequent in lowland districts, but rather local. SOUTH Island: in many localities in Marlborough and Nelson; less frequent in Westland, Canterbury, and Otago. CHATHAM Islands, *Dieffenbach. Mingimingi.* July, Aug.

Distinguished from all the larger species by the linear leaves and translucent fruits.

9. **C. acutifolia,** *Hook. f. in Journ. Linn. Soc. Bot.* (1857) 128. A glabrous shrub or small tree, with slender branches. Leaves 1in.–2½in. long, ½in.–1in. broad, membranous, ovate or ovate-lanceolate or oblong-ovate, acute or shortly acuminate, narrowed into the slender petiole. Stipules short, broad, deciduous. Peduncles exceeding the petioles, very slender, simple or trichotomously branched. Male in 3-flowered fascicles; calyx minutely 4-toothed; corolla-tube broadly funnel-shaped, with 4–5 spreading lobes; stamens 4–5. Female: peduncles simple or branched, usually with a terminal 3-flowered fascicle and 2 lateral flowers; calyx shortly 4-toothed; corolla tubular, limb 4–5-lobed, spreading; styles rather stout. Drupes (immature) ⅜in. long, narrowed at both ends.—Handbk. 114; Cheesem., *l.c.* 235!

SUNDAY Island, KERMADEC Group, *McGillivray, Cheeseman!* July, Aug.

The membranous acute or acuminate leaves distinguish this plant from its nearest allies. I am indebted to Mr. Cheeseman for specimens.

10. **C. tenuifolia,** *Cheesem. in Trans. N.Z.I.* xviii. (1885) 315. A sparingly-branched shrub, 8ft.–18ft. high. Bark pale or white. Leaves membranous or rarely subcoriaceous, glabrous or rarely the midribs and petioles hairy below, 1½in.–4in. long, ⅓in.–1⅓in. broad, ovate oblong-ovate lanceolate oblong-lanceolate or elliptic-lanceolate, acute or acuminate; petioles very slender. Stipules broadly deltoid, ciliate when young. Male flowers sessile on 2–4-flowered axillary or terminal fascicles, with a cupular involucel at the base; calyx 0; corolla broadly funnel-shaped; limb spreading or shortly recurved; filaments elongated, pendulous. Female not seen. Fruits ovoid or oblong, about ⅛in. long, sessile in fascicles of 3, 8, or more on short lateral branches.

NORTH Island: East Cape, *Banks* and *Sol.*, *Bishop Williams! Lee!* Urewera Country, *E. Best!* Ruahine Range, *Colenso, T. K.* Near Woodville; Karioi, Ruapehu, Waimarino, and upper portions of the Whanganui and Rangitikei Valleys as far south as Hunterville, common, *T. K.* Karioi and Mount Pirongia, Auckland; Mount Egmont Range and Stratford, Taranaki; *Cheeseman!*

This species has a close general resemblance to *C. robusta*, but is distinguished by the almost membranous pale-green leaves and small glomerules of male flowers.

11. **C. arborea,** *T. Kirk in Trans. N.Z.I.* x. (1877) 420. A small narrow- or round-headed tree, 15ft.–30ft. high.; trunk 6in.–18in. in diameter. Branchlets slender, puberulous near the tips. Leaves 1in.–2in. long, ½in.–1in. broad, ovate-spathulate or almost orbicular-spathulate, obtuse or rounded at the apex, rarely retuse, abruptly narrowed into winged petioles, ⅛in.–¾in. long, usually reddish below. Stipules short, deltoid, ciliolate. Flowers sessile, forming densely-crowded spherical heads at the tips of arrested axillary shoots, or less frequently of larger branchlets. Male: calyx 4–5-lobed, lobes ciliate; corolla very short, broadly campanulate, cleft into 4–5 acute lobes; stamens 4–5. Females in 4–10-flowered glomerules; flowers shorter and smaller. Drupes forming dense closely-packed heads, broadly oblong or almost globose, ¼in. in diameter, translucent, at length black.—Cheesem., *l.c.* xix. 236; T. Kirk, Forest Fl. N.Z. t. 132.

NORTH Island: from the North Cape district to the Hauraki Gulf and Lower Waikato: sea-level to 1,300ft. Oct., Nov.

In habit, colour, and general appearance this fine species resembles *Myrsine Urvillei.* With the exception of *C. petiolata*, it is the tallest species of the genus, and one of the most distinct.

12. **C. spathulata,** *A. Cunn., Precurs. n.* 479. A sparingly-branched shrub, 1ft.–5ft. high, rarely more. Young branches puberulous. Bark pale. Leaves usually distant, coriaceous, glossy, ¼in.–1½in. long, orbicular or broadly ovate, rounded retuse or emarginate, abruptly contracted into a narrow winged petiole longer or shorter than the blade; margins recurved; veins few, obscure. Stipules cuspidate, deciduous. Flowers sessile, axillary, solitary or in 2–3-flowered fascicles. Male: drooping, seated in an unequally 4-lobed involucel;

calyx deeply 4–5-lobed; corolla campanulate, deeply 4–5-cleft, lobes spreading or revolute; stamens 4. Female: usually solitary, smaller and narrower than the male; calyx deeply 4-toothed, teeth acute; corolla tubular, deeply 4-cleft, lobes narrow, acute, revolute. Drupe ovoid, ¼in. long, slightly narrowed at both ends, black.—Hook. f., Fl. N.Z. i. 106; Handbk. 114; Cheesem. in Trans. N.Z.I. xix. 237. *Pelaphioides rotundifolia*, Banks and Sol. MS.

NORTH Island: Auckland: not infrequent in lowland forests from the North Cape to the Upper Waikato. Aug., Sept.

Allied to *C. arborea*, but distinguished by its smaller size, straggling habit, pendulous male flowers, and solitary black fruit.

13. **C. rotundifolia**, *A. Cunn., Precurs. n.* 473. A shrub, 4ft.–12ft. high or more, with few slender spreading branches, often divaricating at right angles. Young shoots pubescent or villous near the tips. Bark pale- or ashy-brown. Leaves in distant pairs, ¼in.–1in. long, orbicular or broadly oblong or ovate-oblong, abruptly acute or rarely obtuse, membranous, pubescent on both surfaces, margins ciliate, suddenly contracted into a slender villous petiole usually shorter than the blade. Stipules small, deciduous. Flowers axillary, sessile, fascicled or solitary. Male: calyx 0; corolla deeply 4–5-cleft, lobes spreading. Female: calyx-limb minutely 4–5-toothed; corolla tubular, narrow, shortly 3–4-lobed. Drupe subglobose, usually didymous, broader than long, ⅙in. broad, red.—Hook. f., Fl. N.Z. i. 108; Handbk. 114; Raoul, Choix 46; Cheesem., Trans. N.Z.I. xix. 237 *C. rufescens*, Col., Trans. N.Z.I. xviii. 261.

NORTH and SOUTH Islands: common in dark forests from Mangonui and Hokianga to Southland. Ascends to 2,000ft. Sept., Oct.

Specimens with small leaves are sometimes mistaken for *C. areolata;* but the spreading habit and broader drupes enable it to be recognised without difficulty. The leaves are often blotched, and more or less deciduous.

14. **C. areolata**, *Cheesem. in Trans. N.Z.I.* xviii. (1885) 315. An erect shrub or small tree, 6ft.–15ft. high. Branches rather slender, strict, divaricating in young plants, fastigiate when old. Young shoots pubescent or villous. Leaves ¼in.–⅔in. long, membranous and thin, often hairy beneath, orbicular-spathulate ovate-spathulate or elliptic-spathulate, acute or apiculate, abruptly narrowed into short pubescent or hairy petioles; veins forming large areoles. Stipules short, broad, acute. Flowers axillary, solitary or in 2–4-flowered fascicles. Males: ⅛in. long, involucellate; calyx 0; corolla deeply 4–5-lobed, lobes spreading. Female. solitary or geminate, $\frac{1}{10}$in. long; calyx obscurely 4-toothed; corolla narrow, funnel-shaped, shortly 4-lobed, lobes scarcely spreading. Drupe spherical, $\frac{1}{10}$in.–⅛in. in diameter, black, shining when mature.—*C. multiflora*, Col. in Trans. N.Z.I. xxi. 84. *C. pallida*, T. Kirk MS.

NORTH and SOUTH Islands: in lowland woods, from Mangonui to Southland. Sept., Oct. Ascends to 1,250ft.

This species resembles *Melicytus micranthus* in habit and general appearance. Distinguished from most other species of *Coprosma* by the fastigiate habit, small flat acute leaves, and spherical drupes, which are never didymous.

15. **C. tenuicaulis,** *Hook. f., Fl. N.Z.* i. 106. A much-branched shrub, 3ft.–8ft. high. Branches slender, often interlaced. Bark purplish-brown, glabrous or glabrate. Young branches puberulous. Leaves slightly coriaceous, ¼in.–½in. long, orbicular-spathulate or ovate-spathulate, rounded at the apex, obtuse or subacute, abruptly narrowed into a short petiole; veins reticulated in large areoles. Flowers axillary, solitary or in 2–3-flowered fascicles, involucellate. Male: calyx 0; corolla ⅛in.–⅙in. long, broadly funnel-shaped, 4–5-lobed, lobes short. Female: shorter than the male; calyx-limb truncate; corolla narrow, funnel-shaped, 4–5-lobed, lobes acute. Drupe globose or rarely depressed, ⅛in. in diameter, black, shining.—Handbk. 115; Cheesem., Trans. N.Z.I. xix. 239.

NORTH Island: Auckland: in swampy lowland forests and open turfy swamps. Sept., Oct.

Distinguished from *C. areolata* by the more obtuse and almost coriaceous leaves, dark bark, and more slender twigs.

16. **C. rhamnoides,** *A. Cunn., Precurs. n.* 474. Erect, 2ft.–6ft. high, or sometimes prostrate and creeping over rocks. Branches usually very numerous, divaricating, dense, rigid or almost spinous, and interlaced when growing in exposed situations. Young shoots pubescent or villous. Bark reddish-brown, uneven. Leaves coriaceous or almost membranous, ¼in.–½in. long, ⅛in.–⅜in. broad, orbicular or depressed-orbicular, broadly ovate, oblong or linear, rounded, acute or retuse, suddenly narrowed into a very short petiole, often puberulous beneath; veins not evidently reticulate in the coriaceous forms. Stipules minute. Flowers involucellate, axillary, solitary or in 2–3-flowered fascicles. Male: calyx 0; corolla $\frac{1}{10}$in. long, 4–5-partite, lobes spreading or recurved. Female smaller than the male; calyx-limb truncate or minutely 4-toothed; corolla narrow, funnel-shaped, deeply 4-cleft, lobes linear, often revolute. Drupe globose, bright-red, ultimately black, ⅛in.–⅓in. in diameter.—Hook. f., Fl. N.Z. i. 107; Handbk. 116. *C. gracilis*, A. Cunn., Precurs. n. 475. ? *C. divaricata*, A. Cunn., *l.c.* n. 476 (*not* of Hook. f.). *C. concinna*, Col. in Trans. N.Z.I. xvi. (1886) 330.

From the North Cape to STEWART Island: common. Sea-level to fully 3,000ft. Aug. to Oct.

An extremely variable plant, the variations being chiefly caused by situation, shelter, or exposure. Cheeseman distinguishes two principal forms: *vera*, with orbicular or broadly ovate obtuse often coriaceous leaves (to this must be referred *C. orbiculata*, Col. in Trans. N.Z.I. xxii. 465); and *C. concinna*, Col., *l.c.* xvi. 330, a form with thinner subacute leaves.

Var. **divaricata.** Leaves submembranous, broadly ovate oblong-ovate or linear-oblong, acute or subacute.—*C. divaricata*, A. Cunn. (not of Hook. f). When linear or lanceolate leaves are mixed with the larger it is *C. heterophylla*, Col., *l.c.* xviii. 263.

All the forms, however, are connected by the most minute gradations. In a remarkable subvariety of the typical form from Mount Manaia the leaves are broader than long.

17. **C. ramulosa,** *Petrie in Trans. N.Z.I.* xxvii. (1894) 405. Much branched, slender prostrate or decumbent. Bark pale-brown or grey. Twigs usually pubescent. Leaves ⅓in. long, ⅛in. wide or less, narrow-obovate, rounded at the apex, narrowed into a short broad petiole or sessile; veins indistinct.

Stipules deltoid, acute. Flowers not seen. Drupes solitary, sessile or nearly sessile on the younger lateral shoots, deep-red.—*C. pubens*, Petrie, *l.c.* xxvi. 267 (not of A. Gray).

SOUTH Island: not uncommon in the mountains of Westland and Canterbury. Kelly's Hill, *Petrie!* Arthur's Pass; mountains above Bealey, *T. K.*

The dried fruit cannot be distinguished from that of *C. rhamnoides*, of which it is probably a mountain form.

18. **C. turbinata,** *Col. in Trans. N.Z.I.* xxiv. (1891) 389. A much-branched shrub, 8ft.–9ft. high, with shining red-brown bark. Branches suberect; branchlets slender, puberulous at the tips. Leaves obovate, 2–3 lines long, rather distant, very obtuse, subcoriaceous; veins obscure. Stipules blunt, glabrous. Flowers numerous, solitary, or 2–3 together. Male: involucel glabrous, lobes 4, two longer than the others, obtuse; corolla 4-cleft, segments acute, recurved. Female not seen. Fruits lateral, solitary, turbinate, 1–1½ lines long, yellow when ripe. Peduncles short, stoutish.

NORTH Island: woods south of Dannevirke, Hawke's Bay, *Colenso.*

I have not seen specimens of this plant, which appears to be related to *C. rhamnoides*, differing essentially in the turbinate fruit.

19. **C. obconica,** *n. s.* Erect, 4ft.–5ft. high or more. Primary branches divaricating; branchlets numerous, dense, interlaced. Bark pale. Twigs pubescent or puberulous. Stipules narrow, acute, pubescent, fugacious. Leaves mostly fascicled on very short arrested branchlets, sessile or minutely petioled, oblong or narrow-oblong, ⅛in.–¼in. long, $\frac{1}{16}$in.–$\frac{1}{12}$in. broad, obtuse or minutely apiculate, coriaceous, glabrous; margins thickened and recurved; veins obscure. Flowers involucellate, terminating short abortive branchlets, solitary or rarely geminate. Peduncles decurved, much shorter than the corolla. Male: calyx shortly funnel-shaped, teeth deltoid; corolla broadly funnel-shaped, 4-cleft for half its length, segments ovate, recurved. Female: calyx 0; corolla tubular, slightly ventricose at base, teeth short, straight; styles very long. Drupes sessile or on very short decurved pedicels, broadly obconic or obcordate, compressed laterally, $\frac{3}{16}$in.–⅛in. broad; margins rounded, yellowish-white, almost translucent.

SOUTH Island: Nelson: Wairoa Gorge, *Bryant* and *Kirk.* Aug.

The drupes of this curious plant are broader than long, and cannot be mistaken for those of any other species. I am greatly indebted to my friend W. H. Bryant, who has kindly forwarded better specimens than those we collected together.

20. **C. ciliata,** *Hook. f., Fl. Antarc.* i. 22. A spreading shrub, 4ft.–10ft. high, forming thickets. Branches lax or dense, stout or slender. Bark very pale or nearly white, furrowed. Young shoots villous with rigid hairs. Leaves tufted on short arrested branchlets, ⅓in.–⅔in. long, pubescent beneath, oblong, linear-oblong or slightly obovate, rounded at the apex or subacute, flat, membranous, narrowed into a short petiole; margins ciliated; petioles pubescent or hairy; veins obscure, not reticulated. Stipules short, broad, acute, villous.

Flowers not seen. Drupe subglobose, $\frac{3}{16}$in. in diameter, black.—Handbk. 115.

AUCKLAND and CAMPBELL Islands. ANTIPODES Island, *T. K.*

This species forms almost impenetrable thickets, especially on Campbell Island. The form with linear-oblong leaves is characterized by more robust branches than the type. It is to be regretted that only a solitary drupe has been seen, while the flowers are quite unknown.

21. **C. parviflora,** *Hook. f., Fl. N.Z.* i. 107. Erect, rigid, densely branched, and leafy, 4ft.–14ft. high. Bark pale-brown, nearly smooth. Young branches densely pubescent or villous, often divaricating. Leaves $\frac{1}{4}$in.–$\frac{3}{4}$in. long, coriaceous, mostly close-set, fascicled on short lateral branches, obovate linear-obovate or linear-oblong, rounded at the tip or rarely subacute; margins slightly recurved or flat, narrowed into the short petiole; veins obscure. Stipules broad, pubescent or villous. Flowers involucellate, solitary or in 2–4-flowered fascicles. Male: calyx 0; corolla $\frac{1}{10}$in. long, broadly campanulate, 4–5-partite. Female: calyx-limb 4–5-toothed, teeth minute; corolla $\frac{1}{12}$in. long, narrow, tubular, 4-cleft. Drupe spherical, whitish or violet-tinged, translucent, ultimately black.—Handbk. 116; Cheesem., Trans. N.Z.I. xix. 241. *C. myrtillifolia*, Hook. f., Fl. Antarc. i. 21; Fl. N.Z. i. 108.

From the North Cape to STEWART Island: common. AUCKLAND Islands, *Hook. f.* CAMPBELL Island, *T. K.* Sea-level to 4,000ft. Nov. to Jan.

An extremely variable plant, the variations depending largely upon soil, situation, and elevation. Luxuriant lowland specimens sometimes have subverticillate branches with crowded pubescent or villous branchlets arranged in the same plane. Mountain specimens, on the other hand, frequently have short stout rigid divaricating branches, the bark villous or shaggy, and the coriaceous leaves less than $\frac{1}{4}$in. long. A form from the Amuri has perfectly glabrous plum-coloured bark with rigid decurved branchlets, and leaves with ciliate margins, but it is not possible to separate any of these as permanent varieties. The fruits appear to vary greatly in colour, being sometimes of a deep rich violet. I have never seen fruits absolutely white.

22. **C. crassifolia,** *Col., Excurs. North Isl.* 75. A shrub, 4ft.–12ft. high, with stiff or rigid opposite divaricating branches. Bark pale-brown or reddish, twisted in young branches, longitudinally fissured in old specimens. Young branchlets glabrous or rarely puberulous. Leaves $\frac{1}{3}$in.–$\frac{2}{3}$in. long, glabrous, broadly oblong, ovate, orbicular or obovate, thick and coriaceous, rounded at the apex or retuse, abruptly narrowed into a short petiole; margins thickened; veins obscure. Flowers involucellate, solitary or in 2–3-flowered fascicles, sessile, axillary, or terminating very short arrested branchlets. Male: calyx 0; corolla $\frac{1}{3}$in. long, campanulate, deeply 4-partite, lobes broad; stamens 4. Female: calyx truncate or minutely 4-toothed; corolla tubular, narrow, $\frac{1}{8}$in. long, 4-cleft, lobes linear, acute, revolute. Drupe broadly oblong, yellow, $\frac{1}{4}$in. in diameter.—*C. divaricata*, Hook. f., Fl. N.Z. i. 107 (*in part*); Cheesem. in Trans. N.Z.I. xix. 242. *C. aurantiaca*, Col., *l.c.* xxii. (1889) 464.

NORTH Island: widely distributed, but very local. Auckland: Hokianga, Kaipara, and Great Omaha, *T. K.* Whangarei and sandhills between Helensville and the West Coast, *Cheeseman!* Head of the Manukau Harbour; Hawke's Bay; *Colenso*. Wellington: Karori, *T. K.* SOUTH Island: Nelson: Maitai Valley and other places, *Cheeseman!* Otago: near the sea and inland, *Petrie!*

Originally discovered by Colenso in 1842. Best distinguished by the rigid divaricating branches, few coriaceous leaves with thickened margins, and yellow drupes.

23. **C. rigida,** *Cheesem. in Trans. N.Z.I.* xix. 243. An erect shrub, with spreading branches, 5ft.–15ft. high. Branches divaricating but rarely at right angles, often interlaced; young branches puberulous. Bark red or reddish-brown, often furrowed. Leaves $\frac{1}{4}$in.–$\frac{3}{4}$in. long, mostly on short arrested branchlets, coriaceous or nearly membranous, obovate or oblong-spathulate, rounded or retuse, narrowed into the short petiole; veins obscurely reticulated. Stipules loosely sheathing, deltoid, glabrous. Flowers involucellate, solitary or in 2–3-flowered fascicles. Male: calyx 0; corolla campanulate, deeply 4–5-cleft. Female: calyx-limb minutely 4–5-toothed; corolla shorter than in the male, tubular, 3–5-lobed, lobes spreading but not revolute. Drupe broadly oblong, orange-yellow, $\frac{1}{6}$in. long.—*C. divaricata*, Hook. f. in Fl. N.Z. i. 107 (in part).

NORTH and SOUTH Islands: not uncommon in swampy lowland forests, but often local; most plentiful in the North Island. Sept., Oct.

Nearly related to the preceding, from which it is distinguished by the ascending branches, red bark, more membranous leaves, and broader fruit.

24. **C. Buchanani,** *T. Kirk in Trans. N.Z.I.* xxiv. (1891) 424. A much-branched shrub, 5ft.–10ft. high. Branches opposite, ascending. Bark reddish-grey, papery. Young branchlets puberulous or pubescent. Leaves distant, $\frac{1}{2}$in.–1in. long, puberulous and minutely ciliate when young, subcoriaceous, obovate or oblong-ovate, rounded or minutely apiculate, narrowed into a short puberulous petiole; margins slightly thickened. Stipules broadly deltoid, puberulous or minutely ciliate. Male flowers not seen. Female: axillary, solitary, involucellate; calyx-limb 4–5-toothed, teeth very short and broad, acute; corolla narrow-campanulate, 4–5-cleft half-way down; segments acute, shortly recurved. Stigmas very robust, tapered towards the base. Fruit unknown.

NORTH Island: Cape Terawhiti, *T. K.* Oct.

In the absence of the male flowers and the fruit of this rare species its true affinities cannot be determined; it is, however, related to *C. rigida*, but differs in the more erect branches, in the larger pubescent more distant leaves, and especially in the stouter styles, which are curiously tapered at the base. Specimens sent to Kew were doubtfully referred to *C. petiolata*, Hook. f.

25. **C. rubra,** *Petrie in Trans. N.Z.I.* xvii. (1884) 269. Erect, forming a much-branched lax or close bush or shrub, 5ft.–10ft. high. Branches slender, divaricating or ascending. Bark usually reddish-brown. Young branchlets puberulous. Leaves membranous or subcoriaceous, $\frac{1}{2}$in.–1in. long, $\frac{1}{5}$in.–$\frac{1}{2}$in. wide, oblong or broadly oblong-obovate or almost orbicular, subacute, rounded or minutely apiculate, suddenly narrowed into long or short petioles, which are often puberulous or ciliolate; veins obscurely reticulate. Flowers axillary, solitary or in 2–4-flowered fascicles, on very short abortive branchlets, sessile or shortly pedicellate, involucellate. Male: calyx 0; corolla $\frac{1}{3}$in. long, campanulate, 4-partite; segments acute. Female: calyx-limb minutely 4-toothed; corolla narrow, tubular, 4-lobed. Drupe oblong, $\frac{1}{4}$in. long, yellowish-white, translucent.—Cheesem., *l.c.* xix. 243. *C. divaricata*, var. *latifolia*, Hook. f., Fl. N.Z. i. 107. *C. lentissima*, Col. in Trans. N.Z.I. xxii. 165.

Var. **pendula** (*sp.*), Col., *l.c.* xxi. 84. Branches very slender, distant, divaricating at right angles. Leaves distant, broadly ovate-orbicular, abruptly narrowed into the short petiole, subcoriaceous.

NORTH Island: Hawke's Bay, *Colenso!* SOUTH Island: Nelson: Wairoa Gorge, *Bryant* and *Kirk*. Otago: East Coast, Dunedin, and Catlin's River, *Petrie!* Var. *pendula*, Hawke's Bay, *Colenso!*

Closely related to *C. crassifolia*, Col., var. *pendula*, forming a connecting-link between the two. Best distinguished by the thin membranous leaves, which are destitute of a thickened margin, and the smaller flowers.

26. **C. virescens,** *Petrie, l.c.* xi. 426. A glabrous much-branched bush or small shrub, 5ft.–10ft. high. Branches very slender, flexuose, divaricating, interlaced. Bark pale-brown or white, slightly furrowed. Leaves $\frac{1}{8}$in.–$\frac{1}{3}$in. long, membranous, green, spathulate, obtuse or subacute, margins often slightly sinuate, narrowed into a short slender petiole. Stipules minute, subacute, ciliolate. Flowers solitary or in 2–4-flowered fascicles, involucellate. Male: calyx 0; corolla $\frac{1}{4}$in. long, campanulate, 4-partite nearly to the base; stamens 4; anthers broadly oblong, short. Female: calyx-limb obscurely 4-toothed; corolla shorter than in the male, tubular, deeply 4-toothed, teeth acute, spreading, but not revolute. Drupe oblong, yellowish-white, translucent, $\frac{1}{4}$in. long.—Cheesem., *l.c.* xix. 244.

NORTH Island: Wellington: Wairarapa, *Colenso* in Herb. Kew. No. 333, on the authority of Mr. N. E. Brown. SOUTH Island: Marlborough: Pelorus Sound, *J. Rutland!* Nelson: Wairoa Gorge, *Bryant* and *Kirk*. Canterbury: Lake Forsyth, *T. K.* Otago: widely spread from Otepopo to the lake district and Catlin's River, but local, *Petrie!* Ascends to 1,500ft. Originally discovered by Colenso.

Most nearly related to *C. acerosa*, but easily recognised by the slender strict branches, pale bark, and very small spathulate leaves with uneven margins. The female flowers are erect and closely parallel with the branchlets, often presenting a curious appearance.

27. **C. acerosa,** *A. Cunn., Precurs. n.* 477. Prostrate or suberect, 1ft.–4ft. high, much branched. Branches straight or flexuous or zigzag, often interlaced. Young branchlets puberulous. Bark yellowish-brown or dark-brown, often twisted. Leaves in opposite pairs or 2–5-leaved fascicles, uniform, $\frac{1}{4}$in.–$\frac{3}{8}$in. long, $\frac{1}{24}$in.–$\frac{1}{16}$in. broad, very narrow-linear, erecto-patent, obtuse or subacute; veins obscure. Stipules shortly sheathing, puberulous or ciliolate. Flowers apparently axillary but really terminating minute branchlets, involucellate. Male: solitary or in 3–4-flowered fascicles; calyx 0; corolla campanulate, 4-cleft to below the middle, $\frac{1}{6}$in. long; stamens 4. Female: solitary; calyx-limb minutely 4-toothed; corolla $\frac{1}{12}$in.–$\frac{1}{10}$in. long, narrow, tubular, shortly 4-lobed. Drupe spherical or shortly oblong, $\frac{1}{6}$in.–$\frac{1}{4}$in. long, white or pale-blue, translucent.—Hook. f., Fl. N.Z. i. 109; Handbk. 118; Raoul, Enum. Pl. N.Z. 46; Cheesem. in Trans. N.Z.I. xix. 244. *Pelaphia acerosa*, Banks and Sol. MSS.

Two prevailing forms may be distinguished, though closely connected by intermediates, as follows:—

(1) **arenaria.** Branches numerous, with yellowish-brown bark, slender, flexuous, spreading, and closely interlaced. Leaves numerous, very narrow-linear, close-set. Drupe rarely exceeding $\frac{1}{6}$in. long, white, translucent.

(2) **brunnea.** Branches fewer, stouter, with dark-brown bark, often twisted, divaricating. Leaves more distant, often stiff and rather broader. Drupes larger, ¼in. long, subglobose or broadly oblong, translucent, pale-blue.

Mr. Cheeseman describes the male flowers of the first as usually solitary: I find them usually fascicled. The larger involucellate leaves often differ but little from the ordinary cauline form.

NORTH and SOUTH Islands; STEWART Island; CHATHAM Islands. Form (1) *arenaria*, common on blown sand all round the coasts, rarely inland; (2) *brunnea*, in river-valleys, &c.; rare in the North Island, but common in the South, where it ascends to 4,000ft. Very different in appearance from *arenaria*, but cannot be separated.

28. **C. margarita,** *Col. in Trans. N.Z.I.* xxviii. (1895) 594. Bark purplish. Branches very slender, erect, drooping when in fruit, strigosely pubescent. Leaves few, linear, ¾in. long, $\frac{1}{12}$in. broad, acute or subacute, slightly falcate, narrowed at base; petiole very short. Stipules small, deltoid-acuminate, pilose. Male flowers not seen. Female: solitary, terminal on short branchlets, involucellate; peduncles 1 line long; calyx-teeth minute; corolla funnel-shaped, ¼in. long, pale-yellow, lobes subacute, spreading; stigmas stout, obtuse. Drupes globular, $\frac{1}{10}$in. in diameter, white, translucent.

NORTH Island: Ruahine Mountains, east side, *A. Olsen.*

Closely related to *C. propinqua*, of which it may be a variety, but differs in the strigosely pubescent twigs, acute leaves, larger flowers, and small fruit. I have not seen specimens.

29. **C. propinqua,** *A. Cunn., Precurs. n.* 472. A much-branched shrub or small tree, 6ft.–20ft. high, forming extensive thickets. Branches divaricating. Bark brown or greyish. Young branchlets puberulous. Leaves mostly in fascicled pairs on short arrested branchlets, ⅓in.–½in. long, $\frac{1}{16}$in.–⅛in. broad, narrow-linear-oblong, mostly obtuse, rarely subacute, gradually narrowed into a short petiole or sessile, slightly coriaceous; veins obscure. Stipules truncate, glabrous. Flowers solitary or in 2–5-flowered fascicles; the fascicles involucrate, the flowers separately involucellate. Male flowers usually fascicled; calyx 0; corolla ⅛in.–⅙in. long, broadly campanulate, 4–5-lobed or -partite. Female: calyx-limb 4-toothed or erose; corolla $\frac{1}{12}$in.–$\frac{1}{10}$in. long, tubular, shortly 3–4-lobed. Styles short. Drupe ⅓in. long, usually globose, yellow, at length black, or rarely white or of a bluish tinge.—Raoul, Enum. 46; Hook. f., Fl. N.Z. i. 109; Handbk. 116; Cheesem., Trans. N.Z.I. xix. 245. *C. alba*, Col., *l.c.* xxiv. (1891) 388. *Pelaphia parvifolia,* Banks and Sol. MSS.

NORTH and SOUTH Islands: from Hokianga and Mangonui to Foveaux Strait. STEWART Island; CHATHAM Islands. Common in open swamps, swampy forests, and by river-sides. Ascends to 1,500ft. *Mingimingi.* Sept., Oct.

Some forms of this closely approach *C. Cunninghamii*, but may be recognised by the dark bark, involucrate fascicles of male flowers, and the drupes always opaque.

30. **C. Kirkii,** *Cheesem. in Trans. N.Z.I.* xxix. (1896) 391. A procumbent shrub, forming rounded masses 1ft.–4ft. high, and the same in diameter. Branches rather stout, often interlaced, subterete or 4-angled when young, with short whitish pubescence. Leaves in opposite pairs or fascicles, close-set, ½in.–¾in. long, linear or narrow-linear-oblong or narrow-linear-obovate, obtuse or subacute, gradually narrowed into a very short petiole, flat; veins indistinct above; midrib evident below. Stipules very short, broad, ciliate. Flowers on

short lateral branches. Males not seen. Females in 3–6-flowered fascicles; calyx-teeth 4, minute; corolla narrow-campanulate, 4-lobed. Immature drupes oblong.—*Plagianthus linariifolius*, Buch. in Trans. N.Z.I. xvi. t. 34, f. 1.

NORTH Island: Auckland: Tapotopoto Bay, April, 1867, *T. K.* Whangakea: coast between Tom Bowline's Bay and Hooper's Point, near Ahipara, *Cheeseman!* South Head of Hokianga Harbour, *T. K.* Near Tauranga, *T. K.* Taranaki: near Opunake, *T. K.*

This is distinguished from all other narrow-leaved species by its peculiar habit, pubescent branchlets, and fascicled leaves. Its nearer affinities, so far as can be determined at present, appear to be with *C. propinqua* and *C. Cunninghamii*. Some of the Opunake specimens have broader leaves, and may be different.

31. **C. linariifolia,** *Hook. f., Handbk.* 118. A much-branched shrub or small tree, 4ft.–20ft. high. Branches slender, spreading, but not divaricating. Bark greyish-brown. Branchlets puberulous. Leaves never fascicled, ½in.–1½in. long, ⅛in.–⅓in. broad, linear-lanceolate or oblong-lanceolate, suddenly narrowed into a short slender petiole, flat, almost membranous; veins obscure. Upper stipules connate, produced into rather long puberulous sheaths, ciliolate at the mouth. Flowers at the tips of terminal or lateral branchlets involucellate. Male: in 2–5-flowered involucrate fascicles; calyx 0; corolla ⅕in.–¼in. long, 4–5-lobed for half its length, lobes revolute. Female: solitary; calyx-limb 4–5-lobed, lobes linear-oblong, erect; corolla $\frac{1}{10}$in.–⅛in. long, tubular, lobes short, acuminate. Drupe oblong, broad, ⅓in.–¼in. long, pale and translucent, ultimately black. Calyx-lobes persistent.—Cheesem. in Trans. N.Z.I. xix. 246; T. Kirk, Forest Fl. N.Z. t. 95. *C. propinqua*, var. γ, Hook. f., Fl. N.Z. i. 109.

NORTH and SOUTH Islands: from the upper part of the Thames Valley southward to Otago. Sea-level to fully 2,500ft. Oct.

This species approaches *C. Cunninghamii* and *C. propinqua*, but is distinguished from both by the slender graceful habit, acute leaves, and long calyx-lobes of the female flowers.

32. **C. Solandri,** *T. Kirk in Trans. N.Z.I.* xxix. (1896). Apparently a much-branched shrub. Branches as thick as a goose-quill, rigid, obscurely tetragonous; branchlets numerous, short, erect. Bark pale or whitish, setose. Leaves sessile, very coriaceous, laxly imbricating, linear-lanceolate, about ⅓in. long, $\frac{1}{10}$in. broad, acute or apiculate, sparingly ciliate, erect; midrib sunken on both surfaces. Stipules loosely sheathing, setose, ciliate. Drupes solitary, terminal, seated in an involucel formed of 2 reduced leaves with their stipules, globose-ovoid, $\frac{4}{16}$in.–$\frac{5}{16}$in. long, crowned with the acute connivent ciliate calyx-lobes.

NORTH Island: East Cape district. From the Banksian Herbarium, British Museum.

This fine species is related to *C. linariifolia* and *C. Colensoi*, from both of which it differs in the stout tetragonous branchlets and sessile erect imbricating coriaceous leaves. Its affinities, however, cannot be fully determined in the absence of flowers.

33. **C. foetidissima,** *Forst., Char. Gen.* 138. Usually a slender sparingly-branched or twiggy shrub, 6ft.–10ft. high, but occasionally forming a small tree 20ft. high, rarely with trunk 1ft. or more in diameter. Branches slender, ascending, flexuous, interlaced; young branchlets puberulous. Bark pale-red

or greyish. Leaves almost membranous, variable, $\frac{3}{4}$in.–$2\frac{1}{2}$in. long, $\frac{1}{4}$in.–$\frac{3}{4}$in. broad, oblong or linear- or obovate-oblong or rounded-ovate, obtuse acute or retuse, suddenly narrowed into long or short petioles; midrib distinct; veins usually obscure. Stipules short, cuspidate, puberulous or ciliate. Flowers sessile, terminal or rarely axillary, solitary or rarely geminate, involucellate. Male: calyx 0, or the limb minutely 4-toothed; corolla $\frac{1}{2}$in.–$\frac{3}{4}$in. long, broadly funnel-shaped, 4–5 to 8- or 10-lobed, rarely 6-lobed, lobes one-third the length of the tube, acute; stamens as many as the lobes. Female: erect; calyx-limb obscurely toothed or entire; corolla shorter than the male, 3–5-lobed. Drupe $\frac{1}{4}$in.–$\frac{1}{3}$in. long, yellowish passing into red, rarely white and translucent.—DC., Prod. iv. 578; A. Rich., Fl. Nouv.-Zel. 261; Hook. f., Fl. Antarc. i. 20, t. 13; Fl. N.Z. i. 105; Handbk. 116; A. Cunn., Precurs. n. 471 (in part). *C. affinis*, Hook. f., Fl. Antarc. i. 21, t. 14. *C. pusilla*, G. Forst., Prod. n. 513. *C. repens*, A. Rich., Fl. N.Z. 264; A. Cunn., Precurs. n. 478 (not of Hook. f.).

NORTH and SOUTH Islands: from the southern part of the Thames Goldfields and the East Cape to Foveaux Strait. CHATHAM Islands, *Cox!* STEWART Island; AUCKLAND and CAMPBELL Islands. Sea-level to 5,000ft. *Karamu. Hupiro.* Aug. to Oct.

Easily recognised at sight by the slender twigs, distant leaves, and terminal fuchsia-like flowers. The whole plant has a pungent and intensely nauseating odour.

34. **C. Colensoi**, *Hook. f., Handbk.* 117. A small erect or rarely prostrate glabrous slender shrub, 2ft.–5ft. high. Bark pale-brown or whitish. Young branchlets usually puberulous. Leaves varying greatly in form and texture, $\frac{1}{4}$in.–$1\frac{1}{2}$in. long, $\frac{1}{3}$in.–$\frac{3}{4}$in. broad, opposite or fascicled, yellowish-green when dry, submembranous or coriaceous, linear-lanceolate or oblong, obovate to ovate, broadly oblong or obovate, obtuse retuse or emarginate, narrowed into slender petioles; margins recurved in the coriaceous forms; veins obscure. Stipules small, acute, mostly deciduous. Flowers solitary on short decurved terminal peduncles, involucellate, the two longer lobes of the involucel being foliaceous and minutely ciliate or erose at the apex. Male: calyx 0; corolla broadly campanulate, $\frac{1}{4}$in. long, 4-lobed. Female: calyx-limb minutely 4-toothed; corolla $\frac{1}{6}$in.–$\frac{1}{5}$in. long, tubular, shortly 4-lobed, lobes revolute; stigmas 3–4 times longer than the corolla, capillary. Drupe oblong, $\frac{1}{6}$in.–$\frac{1}{4}$in. long, crowned with the persistent calyx and involucel, red, ultimately black.—Cheesem., Trans. N.Z.I. xix. 248. *C. myrtillifolia β linearis*, Hook. f., Fl. N.Z. i. 108. *C. Banksii*, Petrie MS.

NORTH Island: Auckland: Mount Wynyard and other peaks of the Cape Colville Range, *T. K.* Thames Goldfields and Te Aroha, *J. Adams.* Karioi and Pirongia Mountains, *Cheeseman!* Waikaremoana, *Colenso, Bishop Williams!* Wellington: mountains near Cook Strait, *Colenso.* Tararua Range, *Budden!* SOUTH Island: Marlborough: Mount Stokes, *T. K.* Westland: Paparoa Range, *Helms!* Otago: West Coast sounds and Riverton, *T. K.* STEWART Island: common throughout. Descends to the sea-level at Riverton and on Stewart Island. Ascends to 3,500ft. and upwards. Oct., Nov.

A very distinct species, differentiated from most others by the large involucels and solitary flowers on decurved terminal peduncles. The broad-leaved forms might be mistaken for *C. spathulata.*

35. **C. retusa**, *Petrie in Trans. N.Z.I.* xxvi. (1893) 268 (not of Hook. f. in Lond. Journ. Bot. iii. (1844) 415). A procumbent much-branched shrub,

Branches short and often rather stout. Bark pale or whitish on old branches and longitudinally furrowed, usually with 2 opposite lines of pubescence interrupted at the nodes. Leaves $\frac{1}{4}$in.–$\frac{1}{2}$in. long, obcordate or retuse, cuneate, narrowed into a short stout petiole, coriaceous; margin thickened and usually erosulate, flat or slightly concave above but sometimes keeled; midrib stout; nerves obscure. Stipules broad, acutely 3-toothed, pubescent or ciliate. Flowers solitary, terminal. Corolla-segments almost fleshy. Male within an involucel consisting of a pair of 3-toothed stipules and 2 coriaceous bracts; calyx 0; corolla 4-cleft for fully half its length, limb spreading. Female: involucel as in the male; calyx campanulate, with 4 short subulate teeth; corolla cleft nearly to the base, limb tubular or spreading. Styles very stout. "Drupes orange, ovoid, $\frac{3}{8}$in. long."

SOUTH Island: Westland: Kelly's Hill. Southland: Clinton Saddle, Lake Te Anau, *Petrie!* Longwood Range, *T. K.* 2,500ft. to 3,500ft.

Its nearest allies are *C. Colensoi* and *C. foetidissima*, but it is abundantly distinct from either. Its pungent odour is almost as nauseating as that of the latter species.

36. **C. cuneata,** *Hook. f., Fl. Antarc. i.* 21, *t.* 15. A stout or spreading much-branched shrub, 1ft.–12ft. high. Branches woody, rigid, excessively leafy; young branchlets puberulous. Bark whitish or greyish-brown to almost black. Leaves yellowish-green when fresh, crowded, mostly fascicled on short arrested branches, $\frac{1}{5}$in.–$\frac{2}{3}$in. long, $\frac{1}{10}$in.–$\frac{1}{3}$in. broad, linear- or oblong-obovate, obovate-lanceolate or cuneate-oblong, retuse, subacute or obtuse, patent or recurved, very rigid, coriaceous, often shining; midrib deeply sunk above, narrowed at the base, but scarcely petioled. Stipules short, broad, fimbriate or ciliate. Flowers solitary, terminal, sessile, involucellate. Male: lobes of the involucel linear, almost equal; calyx 0; corolla $\frac{1}{4}$in. long, campanulate, 4–5-lobed. Female: calyx-limb 4–5-lobed, lobes unequal; corolla $\frac{1}{5}$in. long, deeply 4-lobed. Drupe globose, $\frac{1}{8}$in.–$\frac{1}{6}$in. in diameter, red.—Hook. f., Fl. N.Z. i. 110; Handbk. 117; Cheesem. in Trans. N.Z.I. xix. 249.

NORTH Island: Mount Hikurangi, Ruahine Mountains, and Lake Taupo, *Colenso*. Ngauruhoe, *T. K.* Mount Egmont, *Dieffenbach, Cheeseman!* SOUTH Island: plentiful in mountain districts from Cook Strait to Foveaux Strait. 2,000ft. to 5,000ft. STEWART Island: Mount Anglem, 3,300ft., *T. K.* AUCKLAND and CAMPBELL Islands, *Hook. f.* ANTIPODES Island, *T. K.* Descending to sea-level. Nov., Dec.

When growing above the forest-line specimens are excessively branched, the branches rigid and densely matted, the leaves excessively coriaceous, with recurved margins. I have seen no South Island specimens with leaves as broad as those collected by Cheeseman on Mount Egmont.

37. **C. microcarpa,** *Hook. f., Fl. N.Z.* i. 110 *and* ii. 331. A leafy shrub, 1ft.–10ft. high. Branches slender, close-set, divaricating, pubescent, leafy. Bark grey. Leaves in pairs on short slender lateral branchlets, $\frac{1}{4}$in.–$\frac{1}{3}$in. long, $\frac{1}{15}$in.–$\frac{1}{12}$in. broad, linear or linear-lanceolate, acute, flat, veinless, dark-brown when dry, not coriaceous. Stipules short, ciliate. Flowers minute. Male: calyx cup-shaped, 4-toothed; corolla broadly bell-shaped, $\frac{1}{5}$in. in diameter, 4-partite, lobes narrow-acuminate, long. Female: calyx-limb short, tubular, 4-toothed; corolla $\frac{1}{12}$in. long, tubular or funnel-shaped, 4-cleft

quarter-way down. Drupe very small, globose, $\frac{1}{10}$in. in diameter.—Handbk. 118.

NORTH Island: tops of the Ruahine Mountains, *Colenso.* Perhaps a variety of *C. cuneata.*

I have ventured to copy the above from the Handbook, not having seen any plant fully agreeing with it. A plant which is most probably identical, however, but of which flowers and fruit have not been obtained, differs in the following trivial particulars: Branchlets often glabrous, sometimes rather stout. Leaves rarely exceeding $\frac{1}{4}$in. in length, submembranous or very coriaceous, acute or obtuse; petiole minute. A single imperfect tubular corolla, doubtless female, has 4 short acute lobes, scarcely spreading. Hab. North Island: Ruahine Range, *T. K.*; Taupo, *G. Mair!* South Island: Ahaura Plain, Nelson, *T. K.*; high ranges north of Oxford Forest, Canterbury, *T. K.*; Otago, *Buchanan!*

38. **C. depressa,** *Col.* ex *Hook. f., Fl. N.Z.* i. 110. A much-branched bush, 1ft.–4ft. high. Branches often prostrate or trailing. Bark often greyish-brown or black. Young branchlets puberulous. Leaves mostly in opposite fascicles, $\frac{1}{8}$in.–$\frac{1}{4}$in. long, $\frac{1}{15}$in.–$\frac{1}{12}$in. broad, spreading or recurved, sessile or with a short stout petiole, linear-lanceolate, acute or obtuse, coriaceous, rigid, veinless but with the midrib obvious beneath in some forms, usually glabrous. Stipules short, broad, pubescent and ciliate. Flowers solitary, terminal, involucellate. Male: calyx 0; corolla campanulate, $\frac{1}{10}$in. long, 4-lobed. Female: calyx-limb 4-toothed; corolla $\frac{1}{10}$in. long, 4-lobed; styles short and stout. Drupe $\frac{1}{8}$in. long, globose, orange-yellow.—Handbk. 118; Cheesem. in Trans. N.Z.I. xix. 250.

NORTH Island: Lake Taupo, top of Ruahine and Hawke's Bay Ranges, *Colenso.* Ngauruhoe and Ruapehu, *T. K.* SOUTH Island: Mount Arthur, Nelson, *Cheeseman.* STEWART Island: near the summit of Mount Anglem, *T. K.* Identified in the absence of flowers. Dec., Jan.

In its ordinary form this approaches *C. cuneata*, but is a more slender plant, with fewer and more distant leaves.

39. **C. repens,** *Hook. f., Fl. Antarc.* i. 22, *t.* 16A. A small glabrous creeping matted species. Bark grey. Branches from 1in. to 2ft. in length, sometimes flaccid, densely leafy. Leaves close-set, rarely distant or fascicled, suberect or spreading, $\frac{1}{10}$in.–$\frac{1}{3}$in. long, linear-oblong or broadly oblong to linear or broadly obovate, rounded at the tips or subacute, narrowed into very short broad petioles, veinless, very coriaceous; margins thickened. Stipules broad, obtuse, usually glabrous. Flowers solitary, terminal. Male: erect, conspicuous; calyx-limb small, 4–8-toothed; corolla erect or slightly curved, tubular, 4–8-toothed or -lobed; stamens 4–8; filaments twice as long as the corolla, erect when first extruded. Female: less conspicuous, $\frac{1}{4}$in.–$\frac{1}{3}$in. long; calyx-limb 4–8-toothed; corolla $\frac{1}{2}$in.–1in. long, tubular, 4–8-lobed, lobes short; styles 2–4, rarely 3 or 5. Drupe globose, $\frac{1}{4}$in.–$\frac{1}{3}$in. in diameter, orange-yellow or red; pyrenes 2–4.—Hook. f., Fl. N.Z. i. 110; Handbk. 119; Cheesem. in Trans. N.Z.I. xix. 250.

Var. **pumila** (*sp.*), Hook f., Fl. Antarc. ii. 543, t. 16B. Forming small, matted, subherbaceous patches less than 1in. in height. Leaves $\frac{1}{10}$in.–$\frac{3}{16}$in. long, obtuse. Stipules minute. Male flowers: corolla slightly funnel-shaped, shortly 4-lobed. Female: corolla nearly as long as male, shortly 4-toothed. Drupe "orange-yellow."—Handbk. 119.

NORTH and SOUTH Islands: common in hilly and mountainous districts; descending to below 1,000ft. on the Longwood Range, ascending to 6,000ft. in the Southern Alps. STEWART

Island, *T. K.* AUCKLAND and CAMPBELL Islands, from sea-level to 2,000ft, *Hook. f.* ANTIPODES Island, *T. K.* MACQUARIE Island, *Scott! A. Hamilton!* Var. *pumila*, Amuri and STEWART Island, *T. K.*

This plant attains the extreme limit of ligneous vegetation in the Southern Hemisphere, and is the only woody plant found on Macquarie Island. All the fruits seen by me on the Auckland Islands were orange-yellow, but in "Flora Antarctica" they are represented by Hooker as red. On the other hand, I have not seen yellow fruits in either the North or South Island.

10. **C. Petriei,** *Cheesem. in Trans. N.Z.I.* xviii. (1885) 316. Stems creeping and densely matted or prostrate with distant short rigid branches. Leaves usually close-set, rarely fascicled, erect or spreading, $\frac{1}{16}$in.–$\frac{1}{4}$in. long, linear-obovate or linear-oblong, obtuse or acute, narrowed into very short petioles, veinless, glabrous or margins and both surfaces ciliate. Flowers solitary, terminal, involucellate. Male: calyx 0; corolla $\frac{1}{6}$in.–$\frac{1}{3}$in. long, tubular, campanulate or slightly funnel-shaped at the mouth, shortly 4-lobed. Female: $\frac{1}{10}$in.–$\frac{1}{8}$in. long; calyx-limb 3–5-toothed; corolla broadly tubular, deeply 4-lobed. Drupe $\frac{1}{6}$in.–$\frac{1}{3}$in. long, globose or broadly oblong, red or bluish, 2-seeded.

SOUTH Island: Mount Arthur, Nelson, *Cheeseman!* Lake Lyndon and Lake Pearson, *T. K.* Mountains near Lake Tekapo, Canterbury, *Cheeseman.* Abundant in the interior of Otago, *Petrie!* From coast-level to 6,000ft. Dec., Jan.

Too nearly related to *C. repens*, which it closely resembles, except when, from growing in dry situations, it assumes a rigid habit. Both campanulate and funnel-shaped corollas may be found on the same plant. The drupes are rather larger than those of typical *C. repens*, and often almost translucent.

2. NERTERA, Banks and Sol.

Calyx-limb truncate or obscurely 4-toothed. Corolla tubular or funnel-shaped, 4–5-lobed, valvate in the bud. Stamens 4 or 5, inserted at the base of the corolla-tube; filaments long; anthers exserted. Ovary 2-celled; cells 1-ovuled. Styles divided nearly to the base into 2 long filiform arms, papillose, hirsute. Fruit a red fleshy drupe with 2 one-seeded pyrenes. Slender, prostrate, or creeping perennial herbs, with minute interpetiolar stipules and solitary axillary or terminal usually sessile hermaphrodite flowers.

A small genus, of which species are found in New Zealand, Australia, Andine and antarctic America, the Indian Archipelago, and the Pacific Islands. Two of the New Zealand species are endemic, another is widely distributed, and the fourth extends to the Philippine Islands.

Glabrous.

Leaves glabrous, broadly ovate, green .. .. 1. *N. depressa.*
Leaves glabrous, narrow-ovate, acute, reddish .. .. .. 2. *N. Cunninghamii.*

Hispid or ciliate.

Leaves hairy or villous. Petiole exceeding the blade .. .. 3. *N. dichondraefolia.*
Leaves ciliate. Petiole shorter than the blade .. .. .. 4. *N. ciliata.*
Leaves, flowers, and fruit hispid .. .. 5. *N. setulosa.*

1. **N. depressa,** *Banks and Sol.* ex *G. Forst., Prod. n.* 501. A glabrous perennial, forming small or large patches. Stem rooting at the nodes. Leaves shortly petioled, broadly ovate or almost orbicular, obtuse or acute, truncate, rounded or almost cordate at the base, green. Flowers terminal. Calyx-limb minutely 4-toothed. Corolla funnel-shaped, minutely 4-lobed; lobes short.

Drupe depressed.—Gaert., Fruct. i. 26; Pet. Thouars, Fl. Trist. d'Acun. 42, t. 10; DC., Prod. iv. 451; A. Cunn., Precurs. n. 481; Hook. f., Fl. Antarc. i. 167; Fl. N.Z. i. 112; Handbk. 120; Benth., Fl. Austr. iii. 431. *N. repens*, Ruiz and Pav., Fl. i. 60, t. 90. *N. montana*, Col. in Trans. N.Z.I. xxviii. (1895) 595. *Coprosma nertera*, F. Muell., Fragm. ix. 186. *Cunina Sanfuentes*, Clos. in Gay Fl. Chili, iii. 203, t. 34. *Gamosia granatensis*, Mut. in L. fil. Supp. 29.

NORTH Island: Ruahine Range (*Olsen*), *Colenso* (if my identification be correct). SOUTH Island: from Cook Strait to Foveaux Strait. STEWART Island, *T. K.* AUCKLAND Islands, *Hook. f.* Sea-level to 3,000ft. Oct. to Jan.

Originally discovered by Banks and Solander in Queen Charlotte Sound. Distributed over the entire area of the genus except the East Indian Archipelago. I have not seen specimens of Colenso's plant.

2. **N. Cunninghamii**, *Hook. f., Fl. N.Z.* i. 112. More slender than *N. depressa*. Stems often red, filiform or almost axillary. Leaves narrow-ovate, rounded at the base, acute. Stipules acute. Calyx-limb 4-toothed. Corolla broadly funnel-shaped. Stamens usually erect. Drupe smaller than in *N. depressa*, spherical, crimson.—Handbk. 120. *N. papillosa*, Col. in Trans. N.Z.I. xxviii. 595.

NORTH Island: most plentiful in stony river-beds, but often local; from the Bay of Islands southward. Great Omaha, Te Whau, and Auckland, *T. K.* Taupo Plain, *Colenso*. Tongariro, *H. Hill*. Wellington, *T. K.* Oct. to Jan. Also in the Philippine Islands.

3. **N. dichondraefolia**, *Hook. f., Fl. N.Z.* i. 112, t. 28A. Stems 3in. to 2ft. long, slender, creeping, hairy or villous. Leaves dull-green, membranous, hairy and slightly hispid above, glabrate or glabrous below, broadly ovate, cordate or rounded at the base, acute or apiculate; petiole as long as the blade, slender. Stipules acute. Flowers and fruit similar to those of *N. depressa*, but rather larger.—*N. gracilis*, Raoul in Ann. Sc. Nat. III. ii. (1864) 121. *Geophila dichondraefolia*, A. Cunn., Precurs n. 482.

NORTH and SOUTH Islands: common in forests from Mangonui to the Bluff. STEWART Island. Ascends to 2,300ft. Oct. to Dec.

4. **N. ciliata**, *n. s.* Stems creeping and rooting at the nodes, rather stout, glabrous. Leaves smaller than in *N. dichondraefolia*, broadly ovate, subacute or acute, cuneate, rounded at the base, membranous, ciliate; petiole shorter than the blade. Stipules truncate. Flowers smaller than in *N. dichondraefolia*. Calyx-teeth 0. Corolla funnel-shaped, 4–5-lobed; lobes linear, shortly recurved. Stamens 4; filaments very short.

SOUTH Island: Bealey Gorge, 2,000ft., Jan., 1876, *Enys* and *Kirk*.

Closely allied to *N. dichondraefolia* and *N. setulosa*, but distinguished from both by the more robust stems. From *N. dichondraefolia* it is further distinguished by the ciliate leaves, petiole shorter than the blade, recurved linear corolla-lobes and truncate stipules; from *N. setulosa* by the glabrous corolla and drupe.

5. **N. setulosa**, *Hook. f., Fl. N.Z.* i. 112, *t.* 28B. Stems extremely slender, 2in.–5in. long, wiry. Branches spreading or ascending, hispid or

glabrate. Leaves $\frac{1}{4}$in.–$\frac{1}{2}$in. long, broadly ovate, oblong or narrow-obovate, rounded or obtuse, usually hispid with scattered white hairs below and ciliate, glabrous above; petiole shorter than the blade. Flowers as long as the leaves. Calyx-teeth 0. Corolla-tube funnel-shaped; teeth 4–5, short, not recurved. Ovary and corolla-tube hispid. Drupe small, hispid, often with very little pulp or nearly dry.—Handbk. 120. *N. pusilla*, Col. in Trans. N.Z.I. xvi. (1883) 331.

NORTH and SOUTH Islands: from Hawke's Bay to Otago, but rare and local in the North; more frequent in the South. STEWART Island, *T. K.* Oct., Nov.

Easily distinguished by its hispid leaves, corollas, and fruits: the fruits are often dry.

3. ASPERULA, Linn.

Calyx-limb 0. Corolla with a short but distinct tube and 4 spreading lobes, valvate in the bud. Anthers exserted. Ovary 2-celled; cells 1-ovuled. Fruit indehiscent, small, dry, usually 2-lobed. Herbs, with slender angular stems. Leaves usually in whorls of 4–8, of which 2–6 are stipules, which assume the size and appearance of true leaves so closely that they cannot be distinguished. Flowers solitary or cymose, axillary or terminal, rarely unisexual.

A small genus, generally distributed through the temperate regions of the earth, but not represented in America or South Africa. It differs from *Galium* in the shape of the corolla. The only New Zealand species is endemic. NAME, from *asper*, in reference to the hispid leaves and stems of many species.

Stems 1in.–4in. long. Flowers solitary or geminate .. .. 1. *A. perpusilla*.
Stems 6in.–15in. long. Flowers in axillary cymes .. .. 2. *A. fragrantissima*.

1. **A. perpusilla,** *Hook. f., Fl. N.Z.* i. 114. A very slender glabrous perennial. Stems 1in.–4in. long, branched, decumbent, filiform. Leaves 4 in a whorl, $\frac{1}{20}$in.–$\frac{1}{10}$in. long, lanceolate, acuminate, awned or obtuse, straight or curved. Flowers solitary, axillary or terminal, often unisexual: males usually shortly pedicelled; female sessile, corolla-tube 4–5-lobed, scarcely $\frac{1}{12}$in. in diameter. Stamens 4–5. Styles free at the tips and divergent.—Handbk. 121.

Var. **aristifera** (*sp.*), Col. in Trans. N.Z.I. xxi. (1888) 88. Leaves aristate. Flowers mostly pedicellate. Corolla-lobes 5. Stamens 5.

NORTH and SOUTH Islands: from the Lower Waikato to Foveaux Strait. STEWART Island. *T. K.* Plentiful in the South Island, rare and local in the North. Sea-level to 3,000ft. Dec., Jan.

2. **A. fragrantissima,** *J. B. Armst. in Trans. N.Z.I.* xiv. (1881) 359. Stems slender, much branched, creeping, 6in.–15in. long, forming broad dense patches 1ft.–3ft. across, glabrous or glandular-pubescent. Leaves in whorls of 4, sessile, linear-oblong, obtuse or subacute, awnless, glandular, dotted, slightly pubescent on both surfaces, flaccid when dry. "Flowers fragrant, very numerous, in axillary clusters of 3–8, on branched peduncles $\frac{1}{10}$in.–$\frac{1}{8}$in. long or more." Calyx short. "Corolla $\frac{1}{10}$in.–$\frac{1}{8}$in. in diameter, campanulate, split to below the middle into 4, rarely 5 or 3, rather broad obtuse lobes." "Styles 2, shorter than the stamens, united almost their entire length"; tips divergent. Stigmas unequal. Ovary glandular. Fruit not seen.

SOUTH Island; on dry banks, Fairlie Creek, Geraldine; also in Selwyn County; *J. F. Armstrong.* Dec.

I have not seen specimens, and fear that the founder of the species has been misled by some form of *Galium umbrosum.*

4. GALIUM, Linn.

Calyx-teeth 0. Corolla rotate, 4- rarely 3- or 5-partite; lobes spreading, valvate. Stamens 4, rarely 3; filaments short. Ovary 2-celled; cells 1-ovuled. Styles 2, connate at the base; stigmas capitate or simple. Fruit didymous, dry, of 2 indehiscent carpels. Slender weak annual or perennial herbs, with erect prostrate or climbing 4-angled stems and verticillate entire leaves, of which only 2 are true leaves, the remainder being stipules as in *Asperula.* Flowers in axillary or terminal cymes or panicles, white or yellow.

SPECIES, about 160, widely distributed. The New Zealand species are endemic.

NAME, from the Greek, in reference to some species being used to curdle milk.

Leaves 4 in a whorl, linear-lanceolate. Flowers in 2-3-flowered cymes .. .. 1. *G. tenuicaule.*
Leaves 4 in a whorl, oblong, mucronate. Flowers solitary .. .. .. 2. *G. umbrosum.*
Leaves 6-8 in a whorl, lanceolate, scabrid. Carpels hispid .. .. .. * *G. Aparine.*
Leaves 6 in a whorl, shortly mucronate. Carpels minutely tubercled .. .. * *G. parisiense.*

1. **G. tenuicaule,** *A. Cunn., Precurs. n.* 468. A weak slender annual, glabrous or slightly scabrid. Stems straggling, 4in. to 3ft. long. Leaves distant, 4 in a whorl, ¼in.–¾in. long, linear-lanceolate or oblong-lanceolate, acuminate or awned, gradually narrowed into a short petiole; margins and midrib faintly scabrid beneath. Cymes 1–3-flowered on slender peduncles, usually exceeding the leaves; pedicels short, decurved in fruit. Fruit of 2 minute globose carpels, glabrous.—Hook. f., Fl. N.Z. i. 113; Handbk. 120. *G. triloba,* Col. in Trans. N.Z.I. xx. (1887) 192.

NORTH and SOUTH Islands: in damp situations on the margins of woods and swamps, from the Lower Waikato to Southland. Ruapuke Island, *Mrs. A. W. Traill!* Ascends to 4,000ft. Jan. to March.

2. **G. umbrosum,** *Soland.* ex *G. Forst., Prod. n.* 500. Annual, erect or suberect or prostrate, sometimes rather stiff but often weak and straggling, 1in.–10in. long, usually glabrous or stem and leaves ciliate. Leaves 4 in a whorl, broadly oblong or elliptical-oblong, acuminate or shortly awned or apiculate, pellucid, dotted when seen between the eye and the light; petiole short. Peduncles short, exceeding the leaves. Flowers mostly solitary, rarely geminate, minute. Fruit of 2 globose carpels, somewhat rough and uneven.—Hook. f., Fl. N.Z. i. 113; Handbk. 120. *G. propinquum,* A. Cunn., Precurs. n. 469.

NORTH and SOUTH Islands: from the North Cape to Southland. Ruapuke Island, *Mrs. A. W. Traill!* Ascends to 4,000ft. Jan. to March.

The late Baron Von Mueller united this with *G. Gaudichaudi,* DC., but that is a much larger plant, with sessile leaves, flowers in threes, and perfectly glabrous carpels. In our plant the flowers are nearly always solitary or rarely geminate.

* **G. Aparine,** *L., Sp. Pl.* 108. Annual, 5ft.-8ft. long or more, weak, straggling. Angles of the stems and midribs and margins of the leaves rough with small recurved

prickles or minute asperities. Leaves 6–8 in a whorl, linear-lanceolate, 1in. long, hispid. Cymes 3- or more flowered; flowers small. Fruits usually purplish, clothed with hooked bristles.

NORTH and SOUTH Islands: naturalised in many localities. *Eriff. Goose-grass.* Jan. to March. Europe.

* **G. parisiense**, *L.*, *Sp. Pl.* 108, subsp. *anglicum*. Annual. Stems slender, spreading, 5in.–9in. high, glabrous. Leaves 5–7 in a whorl, ⅛in.–⅓in. long, scabrous, mucronate. Cymes few-flowered, axillary or terminal, paniculate. Flowers very small. Fruit glabrous, slightly tuberculate.

NORTH Island: Auckland: sparingly naturalised. Whangarei, *T. K.* Auckland, *Cheeseman!* SOUTH Island: Motueka, *Kingsley!*

* SHERARDIA, Dill.

Calyx 4–6, persistent. Corolla funnel-shaped, 4–5-lobed. Stamens 4–5. Style filiform, bifid; stigmas capitate. Fruit of 2 dry indehiscent 1-seeded carpels. An annual or biennial herb, with short slender stems and lanceolate acute leaves in whorls of 4–6. Flowers in small sessile or subsessile heads.

* **S. arvensis**, *L.*, *Sp. Pl.* 102. Annual. Stems numerous, prostrate, 4-angled, scabrid or clothed with recurved points. Leaves opposite below, verticillate above, oblong-lanceolate, sessile, margins recurved, acute. Flowers in terminal involucrate heads, small. Fruit hispid. Calyx-teeth persistent, ciliate.

NORTH and SOUTH Islands: naturalised in most districts. *Field-madder.* Dec. to Feb. Europe, &c.

Order—* VALERIANEAE.

Calyx-tube adherent with the ovary; limb lobed or pappose; lobes involute in bud. Corolla tubular or funnel-shaped, regular or gibbous or spurred at the base, unequally 3–5-lobed, imbricate in bud. Disk epigynous, inserted on the corolla-tube and fewer than its lobes. Anthers versatile. Style slender; stigmas 1–3. Ovary 3-celled, 2 cells empty, 1 with a solitary pendulous anatropous ovule. Fruit indehiscent, small, membranous or coriaceous, rarely woody, with 1 fertile cell: 2 may be sterile or confluent, absorbed or suppressed. Embryo large; endosperm 0. Herbs, rarely shrubs, with opposite exstipulate leaves and usually irregular flowers in dichotomous cymes.

* Centranthus. Calyx-limb pappose. Corolla-tube spurred. Stamen 1.

* Valerianella. Calyx-limb toothed or lobed. Corolla-tube obconic. Stamens 3.

* CENTRANTHUS, DC.

Calyx-limb annular, minutely crenulate. Corolla-tube elongated, narrow, with a longitudinal septum spurred at the base; limb 5-lobed. Stamen 1. Stigma capitate. Fruit membranous, indehiscent, 1-celled, 1-seeded, crowned with the calyx-limb expanded into a feathery pappus. Perennial calyx-limb glabrous. Herbs, with terminal unilateral paniculate cymes and bracteolate flowers.

* **C. ruber**, *DC.*, *Fl. Fr.* iv. 239. An erect herb, 2ft.–3ft. high, sometimes suffruticose. Young branches fistular. Leaves below 2in.–3in. long, petioled, lanceolate; upper leaves larger, ovate-lanceolate, sessile, entire or toothed at the base. Cymes elongated. Flowers crowded, red or white. Spur very short, acute.

NORTH and SOUTH Islands: on rocky banks; naturalised in many localities. *Red valerian.* Dec. to March. South Europe.

* VALERIANELLA, Tourn.

Calyx-limb toothed or 0. Corolla funnel-shaped, 5-lobed, not spurred. Stamens 3, rarely 2. Stigma simple or 3-fid. Fruit compressed, grooved, 3-celled, 2 of the cells empty and sometimes confluent, forming a single empty cell, the other 1-seeded. Annual or biennial dichotomously-branched herbs, with rather succulent delicate leaves and solitary or cymose heads of small bracteate flowers.

* **V. olitoria,** *Poll., Hist. Pl. Palat.* i. 30. Annual, erect, glabrous, much branched, 3in.–6in. high. Leaves entire or toothed, linear-oblong; cauline partly amplexicaul. Cymes capitate; bracts leafy, entire or toothed. Flowers minute, lilac. Fruit glabrous or pubescent, obliquely rhomboidal; empty cells at length confluent; fertile cell corky on the back.

NORTH Island: naturalised in grassy places, but local. Auckland: East Cape, &c. *Lamb's lettuce.* Sept., Oct. Europe, &c.

ORDER—* DIPSACEAE.

Calyx-tube adnate with the ovary; limb entire, lobed or ciliate. Corolla nearly regular, cylindric, obtusely 4–5-lobed; lobes imbricate in bud, the anterior larger and overlapping. Stamens 4, inserted on the corolla-tube; filaments often unequal. Ovary 1-celled, with 1 pendulous anatropous ovule. Style filiform; stigma simple. Fruit indehiscent, invested by the indurated involucel and often crowned by the calyx-limb. Seed pendulous; testa membranaceous; endosperm fleshy; embryo straight, with a short radicle and broad flat cotyledons. Coarse or stout, hairy or prickly, biennial or perennial herbs, with opposite exstipulate leaves and capitate involucrate heads of small bracteolate flowers.

* DIPSACUS. Erect, prickly, or hairy, with spinescent floral bracts.

* SCABIOSA. Erect or spreading, smooth or hairy. Floral bracts reduced to scales or 0.

* DIPSACUS, Linn.

Calyx-limb discoid or cupular, lobed. Corolla nearly regular, unequally 4-cleft. Stamens 4. Style slender; stigmas dilated. Stout biennial hairy or prickly herbs, with angular stems and large oblong or cylindric heads of small flowers having an involucre of large spreading bracts at the base, and each flower invested by a 4-angled involucel of exserted spinous floral bracts.

* **D. sylvestris,** *Mill., Gard. Dict. ed.* viii. *n.* 1. Erect, 2ft.–5ft. high. Stems stout, angular, prickly. Leaves lanceolate-oblong, sessile; cauline leaves 5in.–8in. long, the uppermost connate at the base. Heads 2in.–4in. long; bracts rigid, narrow-linear, exceeding the head. Floral bracts long, strict, subulate, ciliate. Involucel pubescent. Flowers purple. Calyx-limb caducous.

NORTH Island: plentifully naturalised in many localities, but rather local. *Teasel.* Feb. March. Europe.

* SCABIOSA, Linn.

Calyx-tube included in the angular truncate or lobed involucel, contracted at the mouth; limb cupular, with 4–10 teeth, each tooth a rigid bristle. Corolla oblique or 2-lipped, curved; lobes 4–5, obtuse. Stamens 4. Style filiform; stigma small, notched. Flowers in hemispherical or depressed heads, with the involucral bracts 1- or 2-seriate. Receptacle depressed or elongated, hairy or with scaly floral bracts, the outer flowers sometimes rayed and larger than the inner. Perennial herbs, with entire or pinnatifid leaves.

Branches spreading, hairy. Leaves pilose. Involucel 4-furrowed .. * *S. arvensis.*

Branches erect, glabrous. Involucel 8-furrowed .. .. .. * *S. maritima.*

* **S. arvensis,** *L., Sp. Pl.* 99. Stem rather stout, 2ft.–3ft. high, branched above, hairy. Radical leaves oblong-lanceolate, pilose, entire serrate or lobed; cauline pinnatifid. Involucral bracts shorter than the corolla. Receptacle hemispherical, hairy; involucel 4-furrowed. Calyx-teeth 8–16, reduced to bristles. Outer corollas 4-lipped; inner shorter, usually 4-lobed.

NORTH Island: Auckland, naturalised, *Cheeseman*. *Corn scabious*. Europe.

* **S. maritima,** *L., Cent. Pl.* ii. 8. Erect. Branches fastigiate, glabrous. Leaves 2in.–4in. long, oblong-spathulate, rounded at the apex, narrowed into broadly-winged petioles, coarsely serrate or almost dentate, often pinnatifid. Peduncles 5in.–12in. long. Involucral leaves exceeding the flowers. Receptacle elongated in fruit. Involucels 8-furrowed. Calyx-limb with 5 persistent bristles.—*S. atropurpurea*, L., Sp. Pl. 100.

NORTH Island: abundantly naturalised at Mangonui and in other localities. *Sweet scabious*. Jan. to April. Mediterranean.

ORDER XXXIX.—COMPOSITAE.

Flowers (or florets) minute, densely crowded on depressed or conical receptacles forming heads surrounded by an involucre of 1 or several series of erect bracts, the whole resembling a single flower. Receptacle naked or with chaffy scales, hairs, or bristles between the flowers. Calyx-tube adherent with the 1-celled ovary, the limb represented by a pappus or by a ring of scales at the apex or 0. Corollas all tubular, hermaphrodite, 5- rarely 4-toothed, valvate, forming discoid heads; or all hermaphrodite, ligulate, with a short tube and linear elongated blade: most frequently both kinds occur in one head, the central or disk-florets tubular and hermaphrodite or male only, and those of the circumference ligulate and female or neuter in 1 or 2 series, forming a ray (heads radiate). Stamens 5 or 4, inserted on the corolla-tube, the anthers usually united longitudinally, and forming a ring or sheath which surrounds the style (syngenesious), the connective shortly produced upwards, and the lobes obtuse or sometimes produced below into hair-like points or tails. Ovary inferior, 1-celled, 1-ovuled. Style filiform in the perfect flowers, usually divided into 2 short stigmatic arms. Fruit an achene or rarely a nut. Seed erect; endosperm 0; embryo straight or rarely curved; radicle inferior. Flower-heads with disk-florets only are termed "discoid"; with ligulate or ray-florets only, or with both, "radiate": in the latter case the head is said to be heterogamous; when the florets are of one kind only, homogamous. Herbs, shrubs, or rarely trees, with alternate or opposite exstipulate leaves. Flower-heads terminal, rarely axillary, solitary, paniculate, corymbose or cymose.

The largest order of flowering-plants, distributed through all regions and all countries, but less frequent in tropical Asia and Africa than in other large areas. GENERA, 800. SPECIES, 12,000.

Amongst the New Zealand genera *Pleurophyllum* is endemic in the antarctic islands; *Olearia* and *Celmisia, Brachycome* and *Craspedia* are restricted to New Zealand and Australia, but, with the exception of one *Celmisia* and one *Craspedia*, none of the species are identical.

The limitation of the principal genera of this large order is attended with unusual difficulty, and even now is in a very unsatisfactory condition. I have followed the arrangement of Bentham and Hook. f. in "Genera Plantarum" with a few trivial exceptions, although fully convinced that such genera as *Olearia, Pleurophyllum*, and *Celmisia* must ultimately be merged in *Aster*, while *Raoulia* stands in exactly the same relation to *Helichrysum*; but the present arrangement is preferred in this work on account of its convenience.

In the following synopsis of the tribes and genera stress has been laid upon the minute distinctive character afforded by the style-branches of the hermaphrodite florets, and the presence or absence of minute appendages or tails at the base of the lobes of the anther. It must, however, be remembered that it is only the styles of the hermaphrodite florets that are available for this purpose, those of the female florets being, with very few exceptions, uniform throughout the order.

Sub-order I.—CORYMBIFERAE.

Corolla tubular in all the perfect florets, 5- or rarely 3–4-toothed; ligulate in the outer or ray florets, which are pistillate only, or neither stamens nor pistils (neutral), or 0.

Tribe I. EUPATORIACEAE.—Heads discoid; florets all hermaphrodite, tubular. Anthers obtuse at the base. Style-arms obtuse, usually thickened upwards.

1. Ageratum. Pappus of 5 or more chaffy scales or bristles.

II. ASTEROIDEAE.—Heads heterogamous or dioecious, the ray-florets ligulate, the male hermaphrodite florets tubular and 4–5-toothed, or rarely the florets all hermaphrodite. Style-branches more or less flattened, usually produced into tips or appendages. Leaves alternate, rarely opposite.

Female florets usually ligulate, 1-seriate.

Pappus 0 or of very short scales or bristles.

2. Lagenophora. Achenes contracted into a short beak or boss. Pappus 0.

3. Brachycome. Achenes obtuse or truncate. Pappus obscure or 0.

Pappus of unequal rigid awns or spines.

* Calotis. Pappus of barbed rigid awns or spines.

Achenes terete or slightly flattened.

Involucral bracts usually with scarious margins.

4. Olearia. Heads paniculate or solitary and terminal or fascicled, rarely axillary. Shrubs or trees. Leaves alternate or rarely opposite.

5. Shawia. Heads panicled; floret 1, hermaphrodite. Shrub or tree, with alternate leaves.

6. Pleurophyllum. Herbs, with large radical leaves and erect racemose scapes. Heads large, radiate or apparently discoid.

7. Celmisia. Scapigerous herbs, mostly with radical leaves and solitary terminal heads.

Achenes much flattened. Ray-florets in 2 or more series.

8. Vittadinia. Heads solitary, terminal. Style-arms with subulate tips.

* Erigeron. Heads on naked peduncles. Style-arms with obtuse lanceolate tips.

9. Haastia. Involucral bracts in 2 series, woolly. Ray-florets in several series. Leaves imbricating, woolly or villous. Stems often pulvinate.

III. INULOIDEAE.—Heads heterogamous, discoid or rarely radiate; all the florets tubular or hermaphrodite, or the central florets male, or the outer female, or filiform or rarely ligulate or irregular, or the heads dioecious. Anthers usually sagittate at the base, with delicate hair-like tails. Style-branches narrow, mostly subterete, obtuse or truncate. Involucral bracts in several series, scarious, herbaceous, or rarely coriaceous. Florets all hermaphrodite or rarely a few sterile. Herbs or shrubs, with alternate entire leaves.

Female filiform florets numerous, in several series or in homogamous heads.

10. Gnaphalium. Heads small and clustered, rarely solitary. Female florets in several series. Pappus-bristles capillary.

11. Raoulia. Female florets in 1 or 2 rows or numerous. Involucral bracts soft or scarious, sometimes with radiating tips. Small tufted or pulvinate plants, with solitary terminal sessile heads.

12. Helichrysum. Achenes compressed or angular but not flat, and mostly perfect. Herbs or small shrubs.

13. Cassinia. Receptacle narrow, paleaceous. Pappus-bristles capillary. Florets all hermaphrodite, or a few sterile. Shrubs.

14. Craspedia. Heads small, sessile or subsessile, forming a dense globose or discoid capitulum. Herb.

Tribe IV. HELIANTHOIDEAE.—Heads heterogamous, radiate or rarely discoid; outer florets female or neuter, ligulate, rarely irregular or 0, or all the florets tubular, hermaphrodite or male; rarely the heads unisexual and anthers free. Receptacle with chaffy or rigid scales amongst the florets. Anthers usually obtuse at the base. Style-arms of the hermaphrodite florets truncate. Pappus of short scales or awns or 0. Herbs, with opposite or rarely alternate leaves.

* Xanthium. Heads unisexual. Female florets 2 together, forming with the involucre a 2–4-celled prickly burr. Male florets in globose heads. Herbs, with alternate leaves. Pappus 0.

15. Siegesbeckia. Involucral bracts in 2 series, the outer leafy and glandular, the inner and the receptacular scales enclosing the florets. Pappus 0.

16. Bidens. Involucral bracts in 2–3 series. Ray-florets, when present, neuter. Pappus of 2–4 rigid awns.

* Madia. Involucre angular. Ray-florets 1–2-seriate, fertile ligule short or 0. Disk-florets hermaphrodite, fertile or sterile. Achenes compressed, enclosed in the infolded involucral bracts.

V. ANTHEMIDEAE.—Heads heterogamous, radiate or discoid; the females ligulate or filiform or without corollas; the disk-florets hermaphrodite or male; or rarely all the florets tubular and homogamous. Anthers usually obtuse at the base. Style-arms usually truncate. Pappus 0 or represented by a raised border or by minute scales. Leaves alternate.

Receptacle with chaffy scales. Heads radiate.

* Achillea. Heads small, numerous. Rays short and broad. Achenes compressed, winged.

* Anthemis. Rays elongated. Achenes terete, ribbed or angled.

Receptacle naked. Rays white or yellow.

* Matricaria. Receptacle conical, often elongating. Rays female or 0.

* Chrysanthemum. Receptacle convex, flat. Rays female.

Heads discoid, or apparently discoid.

17. Cotula. Heads pedunculate. Achenes compressed or turgid, acute or truncate.

18. Centipeda. Heads axillary, sessile. Corollas of female florets minute, 4-toothed. Achenes trigonous or tetragonous.

* Soliva. Heads sessile. Achenes compressed, crowned by the indurated style. Pappus 0.

19. Abrotanella. Leaves densely crowded, rarely spreading. Style-arms short, truncate. Female florets slender.

Florets all tubular.

* Tanacetum. Heads corymbose. Receptacle broad. Achene crowned with an epigynous disk.

* Artemisia. Heads in elongated panicles. Receptacle narrow. Pappus 0.

VI. SENECIONIDEAE.—Heads heterogamous with the ray-florets ligulate or filiform, or homogamous with all the florets hermaphrodite and tubular. Receptacle usually naked. Involucral bracts herbaceous, usually in a single row, with or without a few bractlets at the base. Anthers usually obtuse at the base. Style-arms usually penicillate at the base, rarely produced into appendages. Achenes usually crowned with a setose pappus.

20. Erechtites. Herbs, with the outer female florets usually in 2–3 series, very slender, filiform.

21. Brachyglottis. Shrubs. Outer florets small, 2-lipped. Achene terete, papillose.

22. Senecio. Herbs or shrubs, with terminal inflorescence. Outer florets ligulate or tubular.

VII. CALENDULACEAE.—Heads heterogamous, radiate. Involucral bracts 1–2-seriate, or rarely 3–4-seriate, herbaceous. Rays in 3 series, female or rarely neuter, 3-dentate. Disk-florets hermaphrodite, rarely fertile, tubular. Style-arms complanate, truncate, concrete or thickened at the base. Herbs or shrubs, with alternate or rarely opposite leaves.

* CALENDULA. Herbs, with entire alternate leaves. Achenes with dilated margins, incurved, muricate at the base.

* OSTEOSPERMUM. Shrubs, with entire or toothed leaves. Ray-florets female, spreading. Achenes rounded, indurated, smooth.

Tribe VIII. ARCTOTIDEAE.—Heads heterogamous, radiate. Involucral bracts in several series, scarious or spinescent at the apex. Ray-florets spreading, sterile. Disk-florets hermaphrodite, fertile. Style-arms rounded at the tips, more or less concrete. Achenes thickened. Pappus paleaceous or 0.

* CRYPTOSTEMMA. Herbs, with radical pinnate leaves, white beneath. Achenes woolly. Pappus of short scales hidden amongst the wool.

IX. CYNAROIDEAE.—Heads discoid, usually homogamous, the florets all hermaphrodite, tubular, and regular, or nearly so, 5-merous. Involucral bracts in many series, imbricate, often mucronate or spinescent. Receptacle usually setose or paleaceous. Anthers sagittate at the base, with fimbriate or plumose tails. Style-arms short, obtuse or pointed, often erect. Achenes hard, shining, or rarely rugose or villous. Herbs, with alternate often spinous leaves.

* ARCTIUM. Coarse herbs, with entire leaves. Involucral scales hooked at the tip. Pappus of short bristles.

* CARDUUS. Leaves often spinous. Involucre usually spinous. Receptacle bristly. Pappus-hairs not plumose, equal.

* CNICUS. Similar to *Carduus*, but the pappus-hairs plumose.

* CYNARA. Involucral bracts large, ending in spreading leafy or spinous processes. Pappus-bristles plumose, unequal.

* SILYBUM. Leaves reticulated with white veins. Heads solitary; involucral bracts large, each with a long recurved subulate spine. Filaments glandular.

* CENTAUREA. Involucral bracts tipped with a rigid spine. Outer florets often dilated, sterile. Pappus bristly or 0.

SUB-ORDER II.—LIGULIFLORAE.

Florets all ligulate and hermaphrodite. Herbs with milky juice.

Tribe X. CICHORIACEAE.—Heads homogamous; all the florets hermaphrodite, with narrow tubular and ligulate 5-toothed corollas. Anthers sagittate at the base. Leaves alternate.

* CICHORIUM. Leaves chiefly radical. Stem and branches rigid, leafy. Florets blue. Pappus 0.

23. MICROSERIS. Leaves all radical. Pappus of flat scales tipped with simple or plumose bristles.

* TOLPIS. Involucre campanulate; bracts 2–3-seriate. Achenes subterete, ribbed, truncate. Pappus of 3–10 slender bristles.

* LAPSANA. Involucral bracts 1-seriate. Pappus 0.

24. PICRIS. Stems leafy, hispid. Outer involucral bracts many. Achenes terete; beak long or short. Pappus 2–∞-seriate, plumose.

25. CREPIS. Involucral bracts 1-seriate. Achenes terete; beak long, short, or 0. Pappus white, silky.

* HYPOCHAERIS. Involucral bracts bearing narrow bractlets. Receptacle sparingly paleaceous. Achenes beaked.

* LEONTODON. Involucral bracts 2–3-seriate, the outer few and small. Achenes terete or fusiform; beak short. Pappus of toothed scales or feathery bristles.

26. TARAXACUM. Scapes leafless. Achenes muricate, abruptly contracted into a long slender beak.

Achenes very flat. Erect leafy herbs.

* LACTUCA. Achenes contracted into a slender beak.

27. SONCHUS. Achenes grooved or ribbed, not beaked.

* TRAGOPOGON. Involucral bracts long, 1-seriate. Achenes with a long slender beak. Pappus-hairs plumose.

1. AGERATUM, Linn.

Flower-heads discoid; involucre hemispherical, cylindrical, or campanulate; involucral bracts imbricate, 2-seriate. Receptacle flat or nearly so; scales 0. Florets hermaphrodite, all tubular, 5-toothed. Anthers obtuse below. Style-arms obtuse, elongated. Achenes angular. Pappus of 5 or 10 bristles or chaffy scales, dilated at the base. Herbs, with opposite leaves and corymbose flower-heads.

SPECIES, about 23, chiefly American.

1. **A. conyzoides,** *L., Sp. Pl.* 839. Annual, erect, branching, 1ft.–2ft. high, more or less clothed with spreading hairs. Leaves opposite, petiolate, crenate. Flower-heads in dense terminal corymbs, pale-blue or white; involucral bracts striate, acute. Achenes black. Pappus composed of 5 lanceolate awned chaffy scales, often serrate below.—DC., Prod. v. 108; Hook., Exot. Fl. t. 15; Benth., Fl. Austr. iv. 462.

KERMADEC Islands, *Cheeseman*.

A common weed in warm countries. I have not seen specimens from the New Zealand area. *Wild heliotrope.*

2. LAGENOPHORA, Cass.

Involucre nearly hemispherical; bracts in about 2 series, broad or narrow, with scarious margins. Receptacle convex; scales 0. Ray-florets numerous in 1 series, female, ligulate or short and tubular, white or rarely purple. Disk-florets numerous, hermaphrodite, tubular; limb 5-toothed. Anthers obtuse below. Style-arms of the disk-florets long and slender, slightly flattened. Achenes compressed, contracted at the top, sometimes into a short beak. Pappus 0. Small perennial herbs, with radical leaves and simple scapes bearing solitary heads; rarely the scapes are modified into erect or suberect leafy stems.

SPECIES, about 16, distributed through New Zealand, Australia, India, China, and extra-tropical South America. All the New Zealand species are endemic.

ETYM. From the Greek, signifying *a flagon* and *to bear*, referring to the shape of the achenes.

* *Stems slender, usually creeping at base, more or less branched below. Leaves radical or cauline.*

eaves on slender petioles. Achenes with thickened margins, compressed .. 1. *L. Forsteri.*

Leaves on slender petioles, orbicular. Achenes slightly falcate .. .. 2. *L. petiolata.*

Leaves cauline, ovate. Achenes lanceolate .. .. 3. *L. Barkeri.*

Leaves cauline, oblong-spathulate. Achenes oblanceolate .. .. .. 4. *L. purpurea.*

** *Stems never creeping. Leaves all radical, hirsute or pubescent. Root-fibres fleshy in the last two.*

Leaves hirsute, oblong-obovate, lobed or pinnatifid; petioles slender. Scapes very slender .. .. 5. *L. pinnatifida.*

Leaves spreading, hirsute or tomentose. Scape slender .. .. .. 6. *L. lanata.*

Leaves ascending, glabrate or hirsute. Scape stout, strict .. .. .. * *L. emphysopus.*

1. **L. Forsteri,** *DC., Prod.* v. 307. Rootstock slender, usually branched, ¾in.–2in. long, all radical or stems leafy and decumbent. Leaves orbicular-oblong obovate or almost orbicular, obtuse, more or less crenate or lobed, narrowed into a slender petiole, glabrous glabrate or hirsute. Scape 1in.–6in. long, naked

or with 1–3 minute linear bracts, slender. Heads ¼in.–½in. in diameter; involucral bracts narrow-linear, acute, with scarious margins. Ray-florets numerous, revolute, white. Achenes compressed; margins thick, abruptly contracted into a short almost viscid beak.—A. Cunn., Precurs. n. 436; Hook. f., Fl. N.Z. i. 125; Handbk. 137. *Calendula pumila*, Forst., Prod. n. 305. *Microcalia australis*, A. Rich., Fl. N.Z. 231, t. 30. *Bellis geum*, Banks and Sol. MSS.

KERMADEC Islands; NORTH and SOUTH Islands, from the North Cape to Southland; STEWART Island; CHATHAM Islands. Sea-level to 2,800ft. *Papataniwhaniwha. Native daisy.* Oct. to Jan.

The leaves are usually pale-green and rather fleshy; the bracts are frequently erose or almost ciliate. This species often develops strong prostrate leafy branches.

Var. **minima.** 1in.–2in. high. Leaves very membranous, including the petiole ¼in.–1in. long, obovate or suborbicular, gradually narrowed into the slender petiole; teeth rounded at the apex, with two teeth or lobes on each side, rarely shortly pinnate at base, teeth mucronate. Scapes filiform. Heads ⅛in.–⅙in. in diameter. Achenes convex, shortly beaked, rounded at the margins. NORTH Island: Great Omaha and other places north of the Waitemata, *T. K.*

2. **L. petiolata,** *Hook. f., Fl. N.Z.* i. 125. A smaller and more slender plant than *L. Forsteri.* Leaves usually radical, spreading, ¾in.–2in. long including the slender petiole, very thin, obovate or suborbicular, obtuse, acutely toothed or crenate, hirsute or with scattered hairs on both surfaces; teeth minutely apiculate. Scape filiform, strict, 2in.–6in. long, pubescent or hirsute. Heads ¼in.–⅓in. in diameter; involucral scales in about 3 series, linear-acute, with scarious margins. Achenes obovate, slightly falcate, shortly beaked, sparingly hispid above; margins thickened.—Handbk. 137. *L. strangulata*, Col. in Trans. N.Z.I. xxii. (1889) 471.

NORTH Island: Te Aroha Mountain, *Adams.* Ruahine Range, *Colenso, W. E. Andrew!* Tararua Range. SOUTH Island: frequent in mountain districts, 1,000ft. to 4,000ft. STEWART Island, *T. K.* Dec., Jan.

This has been recorded both from the Kermadec and Auckland Islands, but it is to be feared that some error has occurred. The slender curved rather turgid achenes and the spreading suborbicular leaves distinguish this from other species.

3. **L. Barkeri,** *n. s.* Stems slender, erect, 6in.–9in. high or more, leafy, the leaves gradually diminishing in size upwards, and the peduncle longer or shorter than the leafy portion of the stem. Leaves erect or ascending, oblong-spathulate or narrow-oblong-spathulate, 1in.–1½in. long, scaberulous on both surfaces, acute or obtuse, distantly serrate or crenate-serrate; teeth apiculate. Head solitary, ⅓in.–½in. in diameter; involucral bracts in about 3 series, hyaline, acute. Disk-florets longer than in either of the preceding. Limb rather deeply 5-lobed; lobes acute, spreading. Achenes (immature) narrow-lanceolate, compressed, thin, glabrous; beak short.

SOUTH Island: Nelson: Amuri. Canterbury: by the Porter River, 1,800ft. to 3,000ft., *Enys* and *Kirk.*

Distinguished from either of the preceding by the scaberulous leaves, hyaline bracts, and linear compressed shortly-beaked achenes. Named in acknowledgment of S. D. Barker's valued assistance in botanical matters.

4. **L. purpurea,** *n. s.* Stems leafy below, naked above, erect, slender, grooved, 4in.–6in. high, pubescent or puberulous. Leaves including the petiole 1½in. long, membranous, ovate, radical and cauline, rather distant, truncate at

base, rounded at the apex, serrate or crenate-serrate, teeth apiculate, pubescent on both surfaces, ciliate, purple beneath. Heads ⅓in.–½in. in diameter; involucral bracts in about 3 rows, linear, acute, with scarious margins, keeled, midrib distinct, often tipped with purple. Achenes oblanceolate, compressed, with a rather long beak and thin margins.

SOUTH Island: Catlin's River, *T. K.*

Best distinguished by the long beaks and thin margins of the achenes. The leaves are brown above and usually purple beneath.

5. **L. pinnatifida,** *Hook. f., Fl. N.Z.* i. 126. Hirsute or pilose in all its parts. Leaves 1in.–4in. long, all radical, narrow-oblong-obovate or oblong-spathulate, obtuse, deeply crenate-lobed or pinnatifid, ciliate; petiole short. Scape slender, 3in.–10in. high. Heads ¼in.–½in. in diameter; involucral bracts linear, acute; midrib not prominent. Achenes (immature) compressed, straight, dimidiate; margin thin, slightly viscid.—Handbk. 137.

NORTH Island: Auckland, *Sinclair.* East Cape, *Colenso.* SOUTH Island: Nelson: Wairau Valley, &c., *T. K.*, *Cheeseman.* Marlborough, *Rough!* Canterbury, *Sinclair* and *Haast!* Otago, *Buchanan! Lindsay, Petrie!* Dec., Jan.

Easily recognised by the soft hirsute leaves and dimidiate achenes.

6. **L. lanata,** *A. Cunn., Precurs. n.* 126. Solitary or tufted. Root-fibres stout, fleshy. Leaves all radical, hirsute or tomentose, rather fleshy, 1in.–1½in. long including the petiole, obovate-spathulate or oblong, coarsely crenate-dentate, teeth rarely acute, obtuse at the apex, narrowed into a very short petiole. Scapes 2in.–6in. long, with 1 or 2 minute bracts, glabrous. Heads ¼in.–⅓in. in diameter; involucral bracts linear, obtuse or subacute, purple, with a very narrow scarious margin. Achenes compressed, slightly oblique, abruptly narrowed into a slender beak, glabrous or rarely papillose at the neck.—Hook. f., Fl. N.Z. i. 126; Handbk. 137. *Bellis pilosa,* Banks and Sol. MS.

NORTH Island: on dry clay hills from near Mangonui to the Auckland Isthmus, but rare and local. SOUTH Island: Queen Charlotte Sound, *Banks* and *Sol.!*

* **L. emphysopus,** *Hook. f., Fl. Tasm.* i. 189. Tufted or solitary. Root of stout fleshy fibres. Leaves all radical, 1½in.–3in. long, tufted, hirsute or pubescent, oblong or narrow-obovate, obtuse, narrowed at the base into a short broad petiole, obscurely toothed or crenate. Scapes stout, strict, hirsute, exceeding the leaves, constricted immediately beneath the head, naked or with 1 rarely 2 short bracts. Heads less than ¼in. in diameter; involucral bracts in 2 series, broadly oblong, with narrow scarious margins, rounded at the tips. Ray-florets scarcely exceeding the involucral scales, tubular or very shortly ligulate. Achenes of the disk abortive; of the ray compressed, glabrous, narrowed at both ends; beak very short.

NORTH Island: naturalised, hills near Paikakariki and Wellington. SOUTH Island: Taylor's Mistake, Banks Peninsula. Jan. to April. Australia.

3. BRACHYCOME, Cass.

Involucre hemispherical or nearly so; bracts 1- or 2-seriate, linear-oblong, with scarious margins. Receptacle convex or conical, papillose; scales 0. Ray-florets in 1 series, female, ligulate, revolute, short. Disk-florets numerous, tubular; limb 5-toothed. Anthers obtuse below. Style-arms with lanceolate or triangular tips, slightly flattened, papillose outside. Achenes compressed, flat, with winged margins, or thick and obtusely tetragonous ; beak 0. Pappus of short scales or bristles or 0.

All the New Zealand species are tufted perennial herbs, with either radical leaves and naked scapes or with ascending leafy stems, simple or branched.

Besides the New Zealand species, which are all endemic, about 40 species are found in Australia.

ETYM. From the Greek, signifying *short* and *hair*, in reference to the short pappus.

Scapes simple, naked.

Minute. Leaves narrow-linear, entire, ⅜in. long .. .. 1. *B. lineata.*
Leaves linear-oblong, pinnatifid, uniform .. .. 2. *B. pinnata.*
Leaves oblong or obovate-spathulate, entire, lobed or rarely pinnatifid .. 3. *B. Sinclairii.*

Stems branched, leafy.

Slender, 2in.–4in. high. Leaves 3–6-lobed .. .. 4. *B. odorata.*
Stout, excessively glandular. Leaves with broadly-winged petioles .. .. 5. *B. Thomsonii.*
Glabrous. Upper leaves reduced to bracts .. .. 6. *B. polita.*
Leaves linear-spathulate, entire, obscurely 3-nerved ; petioles sheathing .. 7. *B. simplicifolia.*

1. **B. lineata.** (¹) A minute glabrous species. Leaves all radical, erect or spreading, narrow-linear, narrowed towards the base, which is slightly expanded, ¼in.–½in. long, $\frac{1}{22}$in. broad, obtuse or subacute. Scapes about 1in. high in fruit, very slender. Heads $\frac{1}{18}$in.–$\frac{1}{16}$in. in diameter ; involucral bracts 7–10, short, broad, rounded-ovate, scarious and almost hyaline, ultimately with purple margins. Ray-florets few, exceeding the bracts ; tips revolute. Disk-florets very short, dilated above, 4-toothed. Achenes very short, compressed, cuneate or shortly obovate, not narrowed above.—*Lagenophora linearis*, Petrie in Trans. N.Z.I. xxv. (1892) 271.

SOUTH Island : in grassy places by Te Anau Lake, *Petrie !* Jan.

Mr. Petrie remarks that the scapes become elongated in fruit, and states that the achenes are narrowed at both ends. The achenes with which he has kindly favoured me do not show this.

2. **B. pinnata,** *Hook. f., Handbk.* 138. Rootstock branched, ascending. Whole plant glabrous or minutely glandular-pubescent. Leaves all radical, coriaceous, ½in.–1½in. long, narrow linear-oblong, ⅛in.–⅙in. broad, rarely broader above, closely pinnatifid ; segments suborbicular, entire, flat or concave beneath. Scape very slender, naked, glandular-pubescent above, 2in.–6in. long. Head ⅓in. in diameter ; involucral bracts oblong, obtuse, with very narrow scarious erose margins. Achenes obovate, with thickened margins, glabrous. Pappus 0 or inconspicuous.—*B. radicata*, var. β, Hook. f., Fl. N.Z. i. 127.

SOUTH Island : remarkably local ; near Burnham and in other dry localities on the Canterbury Plains, *T. K.* STEWART Island, *Lyall* in Handbk. Dec., Jan.

This species is included by Petrie in his catalogue of Otago plants (Trans. N.Z.I. xxviii., 1895, 560), but the specimens sent by him under this name are *B. Sinclairii.* The record for Stewart Island has not been verified, and is extremely doubtful. It seems this plant was discovered by Lyall

(¹) The blank existed in the MS. that was in the printer's hands at the time of the author's death.

on the Canterbury Plains (associated with *Iphigenia Novae-Zelandiae*), possibly in the locality where it was rediscovered by the writer in 1882. Stewart Island is a most unlikely habitat for a plant which flourishes in dry situations.

3. **B. Sinclairii,** *Hook. f., Handbk.* 137. Glandular-pubescent or glabrous. Rootstock short, simple or branched. Leaves radical, ¾in.–3in. long, oblong-spathulate or narrow-obovate-spathulate or linear, rounded at the tip, entire, toothed, lobed or pinnatifid, narrowed into a slender or rather broad petiole, submembranous or coriaceous. Scapes 1–6, slender or strict, naked. Heads ⅓in.–1in. in diameter; involucral bracts linear or linear-oblong, acute or subacute, green or with erose purple margins, minutely glandular-pubescent or glabrous. Achenes very small, cuneate; margins slightly thickened, usually glabrous.

Var. **montana.** Rather stout, usually glabrate or glabrous. Leaves almost fleshy, broadly obovate-spathulate, crenate, or almost lobulate. Scapes rather stout, naked. Heads large. Rays long.

NORTH Island: East Cape, Ruahine and Tararua Mountains, but very local. SOUTH Island: common in subalpine and alpine situations from Cook Strait to Southland. 2,000ft. to 6,000ft. *Native daisy.* Dec., Jan.

The different forms of leaf are not correlated with variations in the flowers and fruit, and appear to depend largely on soil, situation, and moisture. Entire leaves are most frequent in rather dry situations, while those growing in moist situations, especially at great altitudes, are fleshy or coriaceous, and deeply lobed or pinnatifid.

4. **B. odorata,** *Hook. f., Handbk.* 158. Branched from the base. Stems erect, leafy, 2in.–4in. high. Leaves few, obovate-spathulate, rounded at the apex, deeply unequally 3–6-lobed, ½in.–1in. long, narrowed into slender petioles, pubescent, subglandular. Peduncles terminal, slender, glandular. Heads ¼in.–⅓in. in diameter; involucral bracts oblong, glabrous. Rays very short. Achenes linear-clavate, glandular or nearly glabrous. Pappus inconspicuous.—*B. radicata*, Hook. f., Fl. N.Z. i. 127 (in part).

NORTH Island: Hawke's Bay: Kawaka, *H. Tryon!* Wellington: Patea Village, *Colenso. Ronin.*

A favourite plant of the Natives, on account of its perfume. I have only seen a single specimen.

5. **B. Thomsonii,** *T. Kirk in Trans. N.Z.I.* xvi. (1883) 372, *t.* 27. A rather coarse glandular pubescent herb, 4in.–12in. high or more. Rootstock stout. Stems rather stout, slightly decumbent, branched from the base, leafy, erect or spreading. Leaves 1in.–1½in. long, oblong-spathulate, narrowed into broad petioles, deeply crenate-lobed, rarely dentate. Heads ½in. in diameter, on stout terminal peduncles 3in.–6in. long or more, naked or with a solitary bract; involucral scales oblong or oblong-ovate, with purple tips; receptacle convex. Ray-florets numerous, female, white, spreading, papillose at base. Disk-florets tubular. Achenes broadly clavate, margins slightly thickened, excessively glandular. Pappus minute, bristly, more conspicuous in the achenes of the disk than in those of the ray.

STEWART Island: common in littoral situations. Dec., Jan.

Var. **minima.** Scapes 2in.–3in. high. Heads much smaller. Ray-florets 0. Ruapuke Island; Dog Island.

Var. **membranifolia.** Slender, usually glabrous except the upper part of the slender peduncle, which is usually pubescent or glandular. Leaves 1in.–3in. long, very membranous or rarely pubescent and ciliated; petioles very slender, sometimes 3in. long. Peduncles 6in.–9in. long, naked above. Involucral leaves crowded, glabrous. Achenes with thickened margins and inconspicuous pappus. Mount Arthur Plateau, *Cheeseman!* Dunedin, *Petrie!*

Var. **dubia.** Pubescent or glandular-pubescent. Rootstock rarely branched. Leaves 1in.–2in. long, nearly all radical, narrowed into slender petioles. Scape 3in.–5in. long, pubescent or glandular, usually naked or with a leaf or small bract above the base. Involucral bracts glandular-pubescent. Achenes (imperfect) glandular. Otago: cliffs near Cape Whanbrow, *T. K.* Near Green Island, *Petrie!* This may be a distinct species.

The typical form is distinguished by the branched leafy excessively glandular habit, the lobulate or almost pinnatifid leaves with broad petioles, and the broad glandular achenes. Professor Oliver considers that it is most nearly related to *B. diversifolia*, Fisch. and Mey.

6. **B. polita,** *n. s.* Stems erect, 8in.–12in. high or more, mostly simple; leafy at the base. Leaves 2in.–3in. long, linear-oblong-spathulate, deeply lobed or toothed, rounded at the apex, narrowed into slender petioles, glabrous or nearly so, reduced to 1 or 2 small bracts above. Peduncles long, terminal, thickened upwards, glandular. Heads ½in. in diameter; involucral bracts oblong, obtuse, slightly erose. Ray-florets short. Achenes clavate, much compressed, thickened at the margins, glandular. Pappus 0.

SOUTH Island: Arthur's Pass, *T. K.* 3,000ft. Dec.

The small heads and much compressed achenes with raised margins distinguish this from all the other species in this section.

7. **B. simplicifolia,** *J. B. Armst. in Trans. N.Z.I.* xiii. (1880) 338. Stems stout, branched, leafy, 3in.–4in. high. "Radical leaves 2in.–3in. long, linear-spathulate or linear, obscurely 3-nerved, obtuse, with broad membranous sheathing petioles and revolute margins, quite entire except the sheaths, which are somewhat shaggy, glandular-pubescent or glabrous. Scapes 1in.–3in. high, pubescent, 1–2-flowered." Cauline leaves "few, 1in. long, linear-spathulate, obtuse, subamplexicaul, glandular-pubescent, more distinctly nerved." Heads ½in. in diameter; involucral scales 12–16 in one series, linear, acute, 3-nerved. Ray-florets few or 0, short. Disk-florets tubular, 3–5-toothed. Pappus 0. Achene compressed, $\frac{1}{10}$in. long, glandular-pubescent, thickened at the tip. Receptacle very narrow, convex, papillose.

SOUTH Island: Nelson Provincial District, *C. W. Jennings.* Marlborough Provincial District, *J. B. Armstrong.*

Not having seen specimens, I have adopted the original description, omitting some unimportant details. Possibly *Lagenophora Barkeri* may be the plant intended, but its scapes are invariably 1-flowered.

* BELLIS, Linn.

Involucre campanulate; bracts in about 2 series, with scarious margins, herbaceous; receptacle conical. Ray-florets many, 1-seriate, female, ligulate. Disk-florets many, hermaphrodite, tubular; limb 4–5-toothed. Style-arms short, thick, the papillose tips conical. Achenes compressed, obovate, slightly hispid. Pappus 0. Perennial or annual herbs, scapigerous or with leafy stems. Flower-heads solitary. Rays white or purple.

* **B. perennis,** *L., Sp. Pl.* 886. Perennial. Rootstock short. Leaves radical, 1in.–2in. long or more, obovate-spathulate or oblong-spathulate, obtuse or

rounded at the apex, rather fleshy, crenate, glabrous or hairy; petiole broad. Ray-florets white, often tipped with pink. Disk-florets yellow.

Naturalised from the North Cape to Stewart Island. *Daisy.* Sept. to May. Europe.

* CALOTIS, R. Br.

Heads terminal, solitary. Involucre hemispherical; bracts nearly equal, 2-seriate, with scarious margins and a few narrow inner bracts. Receptacle flat or convex. Scales 0. Ray-florets female, few or many, ligulate, 1-seriate. Disk-florets numerous, apparently hermaphrodite but usually sterile, tubular; limb 5-toothed, papillose at the tips. Heads globose in fruit. Disk-achenes usually sterile. Achenes of the ray flat, obovate or oblong. Pappus of few barbed bristles, all short, or 1 or more elongating into rigid spreading spines or awns, usually accompanied by 2 or more truncate scales. Perennial or annual herbs, rarely woody at the base. Leaves entire, toothed, or pinnately divided. Flower-heads terminal, solitary; rays usually white, rarely blue, purple, or yellow.

* **C. lappulacea,** *Benth. in Hueg. Enum.* 60. Perennial; old specimens excessively branched and woody at the base, pubescent or hirsute, 6in.–12in. high. Branches numerous, erect or suberect. Lower leaves oblong, cuneate at the base, usually toothed or lobed, rarely pinnatifid; upper leaves linear, entire or toothed, small. Heads small, bright-yellow. Fruiting-heads ¼in.–½in. in diameter. Achenes finely muricate. Pappus of from 4–8 barbed rigid awns, hirsute at the base; 1–4 of them being much longer than the rest, which are very short.—*Glossogyne Hennedyi,* R. Brown in Trans. N.Z.I. xv. (1882) 259.

NORTH Island: naturalised, Poverty Bay, *Bishop Williams!* SOUTH Island: near Nelson, *Kingsley!* Banks Peninsula, *T. K.* Feb. to April. Australia.

4. OLEARIA, Moench.

Involucre broad or narrow, with few or many series of imbricating bracts, the margins usually dry or scarious. Receptacle pitted. Florets usually numerous, rarely solitary; ray-florets female, usually ligulate and spreading, rarely filiform or rays 0; disk-florets few or many, hermaphrodite, usually 5-lobed, tubular or abruptly contracted below. Anthers 5, with very short tails or rarely obtuse. Style-arms flattened, rarely with short appendages, papillose at the back. Achenes ribbed, terete or flattened. Pappus of 1 or more rows of unequal scabrid hairs, sometimes thickened at the points. Trees, shrubs, or undershrubs. Leaves alternate, rarely opposite or fascicled, usually more or less tomentose, rarely viscid. Heads varying greatly in size, solitary or paniculate, sessile or pedunculate, terminal or rarely axillary.

In addition to the New Zealand species, about 65 are restricted to Australia; 2 species are found on Lord Howe Island, but the genus is absent from Norfolk Island. In New Zealand it ranges from the North Cape to the Auckland Islands, attaining its largest dimensions on Stewart Island and the Snares. It includes some of the most beautiful and striking members of the flora. The late Baron von Mueller referred all the species of *Olearia* and *Celmisia* to *Aster*, but the disadvantages attending this course have decided me against its adoption. Bentham remarks, "There appear to be, indeed, better grounds for maintaining *Olearia* as distinct from *Aster* than for retaining *Erigeron*, which passes so gradually into it, and that again into *Conyza*; and if all these were united we should have a group quite unmanageable without dividing it into sections corresponding to the present genera, which would be, in fact, the present arrangement with all the evils consequent on the nominal change." Many species are extremely variable and difficult to define.

NAME, from *Olea*, the olive, on account of the resemblance of the leaves of some species to those of that tree.

I. Leaves alternate (except in Nos. 9 and 10).

A. *Heads 1in.–1½in. in diameter, solitary on terminal bracteate peduncles, or in simple racemes.*

* Heads radiate.

Leaves 1in.–2½in. long, linear, acutely toothed towards the tips. Rays purple	1. *O. semidentata.*
Leaves 1in.–3in. long, broadly lanceolate-serrate. Peduncle slender ..	2. *O. Chathamica.*
Leaves 2in.–4in. long, lanceolate, obtusely toothed. Bracts short, imbricating. Rays white	3. *O. operina.*
Leaves as in 3, but bracts leafy. Disk purple	4. *O. angustifolia.*
Leaves 4in.–6in. long, obovate-lanceolate. Heads racemed	5. *O. Traillii.*

** Heads discoid.

Leaves broadly oblong or obovate-lanceolate	6. *O. Colensoi.*
Leaves broadly ovate or orbicular-ovate	7. *O. Lyallii.*

B. *Heads on long naked pubescent peduncles.*

Leaves 4in.–6in. long, oblong to ovate, shining, smooth	8. *O. insignis.*

C. *Heads paniculate, ¼in.–½in. in diameter. Leaves 1½in.–5in. long, oblong to ovate.*

*** Leaves opposite.

Leaves oblong to broadly ovate. Heads discoid	9. *O. Traversii.*
Leaves oblanceolate, 2in.–3½in. long. Heads radiate	10. *O. Buchanani.*

**** Leaves alternate.

Leaves ovate-oblong, coriaceous. Florets 8–12	11. *O. furfuracea.*
Leaves ovate or broadly ovate, satiny below. Florets 16–20	12. *O. nitida.*
Leaves orbicular-cordate, shortly acuminate	var. *cordatifolia.*
Leaves ovate-oblong or ovate, strongly toothed. Florets 8–12	13. *O. macrodonta.*
Leaves linear, oblong-lanceolate, spinous	14. *O. ilicifolia.*
Leaves oblong to ovate-oblong, toothed, white with soft tomentum beneath. Florets 16–24. Achene glabrous	15. *O. Cunninghamii.*

D. *Leaves linear; veins forming a right angle with the midrib. Heads small, ⅛in.–⅛in. long.*

Leaves narrow-linear, 4in.–6in. long, ferruginous beneath	16. *O. lacunosa.*
Leaves linear, oblong-lanceolate, acuminate, 1in.–4in. long, white with appressed tomentum beneath	17. *O. excorticata.*
Leaves very narrow-linear, 5in.–6in. long, ¼in. broad	18. *O. alpina.*

E. *Leaves 1in.–2in. long. Panicles diffuse. Heads ¼in.–⅜in. in diameter.*

Leaves broadly oblong or elliptical-ovate, thick and coriaceous, white and silvery beneath	19. *O. Allomii.*

F. *Heads corymbose, rarely solitary. Leaves ¼in.–1½in. long (2in.–3in. long in No. 23).*

Leaves ¾in.–3in. long, elliptic, rigid. Corymbs rounded	20. *O. rigida.*
Leaves ¼in.–⅝in. long, obovate-oblong. Florets 12–20	21. *O. moschata.*
Leaves ½in.–1in. long, oblong-ovate. Florets 7–9	22. *O. Haastii.*
Leaves 1½in.–3in. long, lanceolate or oblong-lanceolate. Florets 4–6 ..	23. *O. oleifolia.*
Leaves ¾in.–1½in. long, narrow-oblong or ovoid, ferruginous beneath. Florets 6–10	24. *O. suavis.*
Leaves ¼in.–½in. long, suborbicular or orbicular-oblong. Heads solitary or axillary	25. *O. nummularifolia.*
Leaves convex above, boat-shaped	var. *cymbifolia.*

G. *Heads small, corymbose. Leaves 1in.–4in. long. Florets 3–5.*

Leaves 1½in.–2½in. long, broadly elliptical, undulate	26. *O. angulata.*
Leaves 3in.–4in. long, oblong or ovate-oblong	27. *O. albida.*
Leaves 1½in.–4in. long, broadly lanceolate. Ray-florets usually solitary ..	28. *O. avicenniaefolia.*

II. Flowers in globose axillary capitula.

Leaves elliptical-lanceolate, ¾in.–1½in. long .. .. 29. *O. fragrantissima.*

III. Leaves opposite or in opposite fascicles.

Leaves obovate, membranous. Heads on slender pedicels, fascicled. Florets 20–26 .. .. 30. *O. Hectori.*

Leaves linear-spathulate, narrowed at the tips. Heads on slender pedicels, fascicled. Florets 5–8 .. .. 31. *O. laxiflora.*

Leaves linear-spathulate, rounded at the tips. Heads on short pedicels. Florets 15–30 .. .. 32. *O. odorata.*

Leaves ⅛in.–⅓in. long, linear-obovate; margins recurved. Heads sessile or shortly pedicelled. Florets 7–12 .. .. 33. *O. virgata.*

Leaves ½in.–1½in. long, linear; margins recurved .. .. .. var. *lineata.*

Leaves linear-obovate, ⅛in.–⅓in. long. Heads solitary, terminal. Florets 18–22 .. .. 34. *O. Solandri.*

1. **O. semidentata,** *Decaisn.* ex *Hook. f., Fl. N.Z.* i. 115. A small sparingly-branched shrub, 1ft.–2ft. high. Branches slender, sparingly clothed with loose tomentum. Leaves close-set, thin, white with loose appressed tomentum beneath, spreading or ascending, 1in.–2½in. long, ¼in.–⅓in. broad, linear-lanceolate, acute, narrowed at the base, serrate or serrulate near the tips, coriaceous. Heads numerous, on very slender peduncles equalling or exceeding the leaves and clothed with distant linear bracts. Involucral leaves in 3 series, acute, the lower pubescent at the tips or ciliate. Achenes faintly striate, glabrous or faintly puberulous.—Handbk. 124; T. Kirk in Trans. N.Z.I. xxiii. (1890) 444; Buch. in Trans. N.Z.I. vii. (1874) 336, t. 14. *Eurybia semidentata,* F. Muell., Veg. Chath. Isds. 21.

CHATHAM Islands, *Dieffenbach, Bishop Williams! G. Mair! H. H. Travers! Cox! Hangatara.* Nov., Dec.

On margins of woods near the sea. Distinguished from all other species of this section by the narrow acute leaves, slender peduncles, smaller heads, and purple rays. Mr. Cox states that a white-rayed form, called *makora* by the natives, is occasionally found.

2. **O. Chathamica,** *T. Kirk in Trans. N.Z.I.* xxiii. (1890) 444. A branched shrub, 5ft.–6ft. high. Branches robust, tomentose. Leaves exceedingly coriaceous, 1in.–3in. long, ½in.–1½in. broad, lanceolate or broadly lanceolate or oblong-lanceolate, narrowed into a short broad petiole, acute, serrate or with teeth with obtuse callous tips, white with loosely-appressed tomentum beneath, midrib and lateral nerves prominent beneath. Heads 1½in. in diameter, on rather slender peduncles with few linear bracts; involucral bracts 2-seriate, the outer tomentose. Ray-florets white or purplish. Disk-florets violet. Achenes striated, pubescent.—*O. operina,* Hook. f., Handbk. 731. *O. angustifolia,* Hook f., var. Buch. in Trans. N.Z.I. vii. (1874) 336, t. 15.

CHATHAM Islands: *Travers! Enys! Cox!* In swampy places on the higher parts of the islands and on high cliffs. *Keketerehe.* Nov., Dec.

This fine plant is distinguished from *O. operina* and *O. angustifolia* by the broader leaves, slender peduncles, and few linear bracts. There is some reason to think that two species are included under this description.

3. **O. operina,** *Hook. f., Fl. N.Z.* i. 115. An erect sparingly-branched shrub, 6ft.–12ft. high. Branches stout, loosely tomentose. Leaves very coria-

ceous, spreading, 2in.–4in. long, ½in.–¾in. broad, white with appressed tomentum beneath, narrow-obovate-lanceolate, acute, gradually narrowed into the short winged petiole; teeth close, callous at the tips. Peduncles 6–15, stout, 1in.–3in. long, sheathed with short imbricating linear obtuse cottony bracts. Heads large; involucral leaves in 2–3 series, obtuse, tomentose. Rays white. Disk yellow. Achenes silky.—Handbk. 124. *Arnica operina*, G. Forst., Prod. n. 299.

Var. **robusta.** Leaves shorter, excessively coriaceous, more deeply toothed. Peduncles shorter. Heads smaller.

SOUTH Island: Martin's Bay to Preservation Inlet.

Forster's drawing of this fine plant is remarkably poor.

4. **O. angustifolia,** *Hook. f., Fl. N.Z.* i. 115. A shrub or small tree, 6ft.–20ft. high, with robust tomentose branches. Leaves 3in.–5in. long, ½in.–⅔in. broad, sessile, glossy above, white with appressed tomentum beneath, excessively rigid and coriaceous, narrow-lanceolate, acuminate, narrowed below, crenate or doubly crenate or serrate, the teeth usually rounded or callous; midrib and principal nerves obvious below. Heads 1½in.–2in. in diameter, on stout peduncles, shorter or longer than the leaves, clothed with short laxly-imbricating leafy bracts, white beneath; involucral bracts in 2 series, the outer densely tomentose. Ray-florets white, ligulate, each with a linear scale at its base. Disk-florets violet. Achenes grooved, silky. Pappus short, unequal.—Handbk. 124; T. Kirk, Forest Fl. N.Z. t. 138.

SOUTH Island: Puysegur Point (rare), *T. K.* Near the Bluff Hill, *Aston!* STEWART Island: in exposed places south of Paterson's Inlet, *T. K.* Sea-level to 100ft. *Tete-aweka.* Nov., Dec.

A noble plant, distinguished from all other species of this section by the foliaceous bracts. The peduncles are invariably terminal, but in this, as in other species, often appear lateral from the development of one or more strong usurping shoots after flowering. The heads are deliciously fragrant.

5. **O. Traillii,** *T. Kirk in Trans. N.Z.I.* xvi. (1883) 372. A shrub or small tree, 15ft. high or more, with robust tomentose branchlets. Leaves crowded near the tips of the branchlets, 4in.–6in. long, 1in.–1¾in. broad, lanceolate or narrow-obovate-lanceolate, narrowed into a broad winged petiole, glossy, very coriaceous, white beneath; margins doubly crenate, with narrow rounded callous points. Heads in erect terminal 3–8-flowered racemes with few deciduous leafy bracts. Rhachis, peduncles, and undersurface of bracts white with appressed tomentum. Peduncles 2in.–3in. long. Involucral leaves in 3 series, scarious, acute, the outer sparingly tomentose at the tips. Ray-florets shortly ligulate, white. Disk violet. Pappus reddish-brown, 1-seriate. Achenes grooved, silky.

STEWART Island: rare and local, *T. K.* Nov., Dec.

A noble species, which fitly commemorates a keen and enthusiastic naturalist, the late Charles Traill.

6. **O. Colensoi,** *Hook. f., Fl. N.Z.* i. 115, t. 29. A bushy shrub or tree, 4ft.–10ft. high, with trunk 2ft. in diameter. Branches stout. Leaves very

coriaceous, 1½in.–6in. long, 1in.–3in. broad, broadly oblong, oblong-lanceolate or obovate-lanceolate, acute or obtuse, acutely serrate or doubly serrate or crenate, narrowed into a short stout petiole, shining above, white with loosely appressed tomentum beneath. Heads in fascicled terminal racemes with a bract at the base of each peduncle, ⅓in.–1in. in diameter, discoid; involucral scales in 1 or 2 series, acute, villous at the tips. Outer florets female, rayless; inner hermaphrodite, campanulate at the mouth. Achenes silky. Pappus-hairs in several series, unequal.—Handbk. 124; T. Kirk, Forest Fl. N.Z. t. 102.

NORTH Island: Mount Hikurangi, Ruahine and Tararua Ranges, but local and never descending below 3,000ft. SOUTH Island: Mount Stokes and Mount Arthur to Foveaux Strait, but local except on the west coast of Otago; ascends to 5,000ft. STEWART Island: plentiful from sea-level to summit of Mount Anglem. *Tupari*. Nov. to Jan.

The leaves are clothed with white tomentum above as well as below when young, and vary greatly in size and shape.

7. **O. Lyallii**, *Hook. f., Fl. N.Z.* i. 116. A shrub or small tree, sometimes nearly 30ft. high, with trunk 2ft. in diameter. Branches open, robust, tomentose. Leaves 6in. long, 4in.–6in. broad, broadly ovate or orbicular-ovate, abruptly acuminate, rarely narrowed below, excessively rigid and coriaceous, doubly serrate or crenate, white with floccose tomentum above and closely appressed tomentum beneath; petiole short, stout, sheathing at the base. Racemes 4in.–7in. long. Rhachis, bracts, peduncles, and outer involucral bracts white with appressed tomentum. Involucral bracts in 5–8 series. Heads discoid as in *O. Colensoi*, but larger; florets darker. Achenes silky. Pappus unequal.—Handbk. 125. *Eurybia antarctica*, Hook. f., Fl. Antarc. ii. 543.

The SNARES; EWING Island and ROSS Island, Lord Auckland Group. Dec., Jan.

Very closely allied to *O. Colensoi*, of which it should perhaps be considered a variety. Its chief points of difference are the more open habit, stouter branches, orbicular-ovate leaves, and especially the many seriate involucre. The leaves on young plants often attain a large size, 9in.–12in. long (including the petiole) and 7in.–9in. broad.

8. **O. insignis**, *Hook. f., Fl. N.Z.* ii. 331. A low spreading shrub, 1ft.–3ft. high, rarely attaining 6ft.–8ft. Branches stout, densely pubescent. Leaves crowded at the tips of the branches, varying from oblong to ovate or narrow-obovate, 3in.–7in. long including the stout petiole, 1in.–4in. broad, excessively thick and coriaceous, shining above, quite entire, equal or unequal at the base, white with appressed tomentum beneath; petioles ¾in.–2in. long. Heads on terminal usually naked pubescent peduncles. Head hemispherical, with the expanded rays 3in. in diameter or more; involucral bracts imbricating, in 6–8 series, tomentose. Rays very narrow. Disk-florets gradually narrowed to the base. Achene narrow, silky. Pappus-hairs in 1 series, slightly thickened at the tips.—Handbk. 125. *O. marginata*, Col. in Trans. N.Z.I. xv. (1884) 32.

SOUTH Island: Marlborough and Nelson: Blind Bay to the Mason River. Sea-level to 4,000ft. Dec., Jan.

The most magnificent species of the genus. Sometimes 1 or several sessile or pedunculate bracts, sometimes of irregular shape, at others foliaceous, are produced on the peduncle or immediately below the involucre. This curious state has been described as a species by Colenso, but its inconstancy deprives it of all value even as a variety.

9. **O. Traversii,** *Hook. f.* ex *F. Muell., Handbk.* 731. An erect shrub or small tree, 15ft.–30ft. high; trunk 1ft.–2ft. in diameter. Bark furrowed. Branchlets tetragonous, and with the petioles and peduncles pubescent, silky, or hoary. Leaves opposite, shortly petioled, oblong-ovate or broadly ovate, acute or acuminate, entire, white and satiny beneath. Panicle cymose, much branched, axillary, exceeding the leaves. Heads small, numerous; involucral bracts short, silky. Outer florets oblique at the mouth. Rays 0. Disk-florets campanulate at the mouth; segments shortly recurved. Achene pubescent, faintly striated. Pappus 1-seriate.—T. Kirk, Forest Fl. N.Z. t. 34. *Eurybia Traversii*, F. Muell., Veg. Chath. Isds. 19, t. 2.

CHATHAM Islands: *Mair! H. Travers! Cox! Akeake.* Oct.

Distinguished by the opposite leaves and discoid heads. Originally discovered by Dieffenbach, who mistook it for *Avicennia officinalis*.

10. **O. Buchanani,** *n. s.* An erect shrub or small tree. Branchlets as thick as a goose-quill. Bark red, glabrous. Leaves all opposite, 2in.–3½in. long, ½in.–1in. broad, almost oblanceolate, sharply narrowed into the short slender petiole, entire, finely reticulated above, whitish with thin laxly-appressed tomentum beneath. Corymbs small, lax. Peduncles very slender, glabrous, equalling the leaves; branches opposite, capillary, the lower with pauperated leaves at the base, the upper pubescent. Heads mostly sessile in small terminal fascicles; involucral bracts about 10, lax, very unequal, pubescent or sparingly villous. Ray-florets 3–4. Disk-florets 3–4. Mouth narrow, campanulate, one-third as long as the tube. Achene short, ribbed, strigose.

NORTH Island: probably Taranaki or Auckland; the exact locality uncertain. *Buchanan!* 1870.

I have only a single specimen of this very distinct plant, so that the description must be considered imperfect. Except *O. Traversii*, it is the only New Zealand species with large opposite leaves. Its nearest ally appears to be *O. furfuracea*.

11. **O. furfuracea,** *Hook. f., Handbk.* 125. A much-branched shrub or small tree, 6ft.–16ft. high. Twigs softly pubescent, grooved. Leaves 2in.–3in. long, 1½in.–2½in. broad, ovate-oblong or elliptic-oblong, coriaceous, obtuse or rarely acute, margins undulate or flat, rarely sinuate-dentate, often unequal at the base, reticulated on both surfaces, silvery below with shining appressed tomentum; petioles ⅓in.–1in. long. Corymbs on long naked much-branched peduncles. Heads numerous, turbinate; involucral bracts broadly oblong, villous or fimbriate. Florets 8–12; 3–5 rays. Disk-florets with a funnel-shaped mouth abruptly contracted into a narrow tube; segments 5, recurved. Style-arms obtuse. Outer pappus-hairs short. Achenes faintly grooved, pubescent or silky.—*Eurybia furfuracea*, DC., Prod. v. 257; Hook. f., Fl. N.Z. i. 117. *Shawia furfuracea*, Raoul, Enum. 45. *Haxtonia furfuracea*, Caley *ex* Don in Edin. N. Phil. Journ. 1831; A. Cunn., Precurs. n. 440. *Aster furfuracea*, A. Rich., Fl. N.Z. 246. *A. elaeagnifolius*, A. Cunn., MSS.

Var. **rubicunda.** Branchlets grooved. Bark red, glabrous. Leaves ovate-lanceolate, narrowed at both ends, 1½in.-2in. long, less coriaceous than the type. Branches of panicle very slender, short. Heads turbinate, short; involucral bracts about 10, oblong-ovate. Florets about 5. Pappus-hairs slightly thickened at the tips.

Var. **angustata.** Leaves oblong or elliptical-oblong; margins often waved. Heads narrow, tubular; involucral bracts linear-oblong, slightly erose, puberulous or glabrous. Ray-florets 2; rays very broad. Disk-florets 3. Achene faintly grooved, silky.

Sub-var. **dubia.** Leaves broader. Heads shorter; involucral bracts broadly oblong, villous. Florets 5–8. Pappus hairs unequal, slightly thickened at the tips.

NORTH Island: Auckland; Hawke's Bay; Taranaki: from the North Cape to the White Cliffs. Dec., Jan. Var. *rubicunda*: mouth of the Mokau River. Var. *angustata*: northern portion of the Auckland District.

This species varies greatly in the shape of the flower-heads, in the number of florets in each head, in the form of the leaves, and in the colour of the branches. Var. *angustata* might be considered distinct were it not directly connected with the typical form by its sub-variety.

12. **O. nitida,** *Hook. f., Handbk.* 125. A shrub, 3ft.–12ft. high. Branchlets stout or slender, angular. Leaves petioled, $1\frac{1}{2}$in.–3in. long, $\frac{3}{4}$in.–$1\frac{1}{2}$in. broad, more or less coriaceous, silvery and shining beneath, ovate or broadly ovate-lanceolate, acute or acuminate, obscurely or acutely toothed or sinuate. Heads $\frac{1}{8}$in.–$\frac{1}{4}$in. long, shortly turbinate, very numerous, forming large rounded corymbs, with stout or slender silky or pubescent branches; the lower involucral bracts short, ovate, pubescent or villous; upper narrow-linear, ciliate or nearly fimbriate, glabrous, or sparingly pubescent at the tips. Florets 16–20; rays 8–10. Disk-florets abruptly contracted below the funnel-shaped mouth into a narrow silky tube; segments 5, recurved. Achene short, broad, silky. Pappus unequal.—*O. populifolia*, Col. in Trans. N.Z.I. xvii. (1885) 243. *O. suborbiculata*, Col., *l.c.* xviii. 263. *Eurybia nitida*, Hook. f., Fl. N.Z. i. 17. *E. alpina*, Lindl. & Paxt., Fl. Gard. ii. 84. *Shawia arborescens*, DC., Prod. v. 345. *Solidago arborescens*, G. Forst., Prod. n. 298; A. Rich., Fl. N.Z. 252 (not of Cunn.).

Var. **cordatifolia.** Leaves orbiculate, shortly acuminate, rounded or cordate at the base, very coriaceous. Heads broadly turbinate; involucral bracts about 20, the upper oblong or oblong-lanceolate, ciliate. Florets 20 or more; rays very long and narrow; disk-florets almost campanulate at the mouth.

Var. **capillaris.** Branchlets stout or slender. Bark pale, furrowed. Leaves $\frac{1}{4}$in.–1in. long, ovate or almost obovate, membranous, obtuse or subacute; petioles short and slender. Corymbs exceeding the leaves; branches capillary. Heads 3–6. Florets 8–10. Achenes rather longer than in the type.—*O. capillaris*, Buch. in Trans. N.Z.I. iii. (1870) 212.

NORTH and SOUTH Islands: from the East Cape and Urewera Country to Foveaux Strait. STEWART Island. Sea-level to nearly 4,000ft. Var *cordatifolia*: STEWART Island. Var. *capillaris*: Nelson mountains. *Travers! Dall!* Dec., Jan.

An extremely variable species, easily recognised by the satiny lustre of the lower surfaces of the leaves, the dense corymbs, and numerous florets.

13. **O. macrodonta,** *Baker in Gard. Chron.* (1884) i. 604. A shrub or small tree, 6ft.–20ft. high. Branchlets, panicles, and leaves below whitish with closely-appressed pubescence. Leaves 2in.–4in. long, 1in.–$1\frac{1}{2}$in. broad, oblong or ovate-oblong or ovate, acute or acuminate, pubescent on both surfaces when young, strongly toothed, truncate or slightly narrowed at the base, flat or margins slightly waved; veins divergent, not forming a right angle with the midrib. Corymbs large, rounded. Heads numerous, $\frac{1}{4}$in. long, obconic; involucral bracts few, linear, obtuse, pubescent or villous, ciliate at the tips, the

lower ovate, short. Florets 8–12; rays 3–5, very narrow. Achenes ribbed, pilose or pubescent.—*O. dentata*, Hook. f., Handbk. 126 (not of Moench.). *Eurybia dentata*, α, Hook. f., Fl. N.Z. i. 118.

NORTH and SOUTH Islands: from the East Cape and Urewera Country to Foveaux Strait; chiefly in mountain districts. From about 2,000ft. to nearly 4,000ft. Jan.

The branches of the corymb frequently develop depauperated leaves at the base, and the entire plant emits a musky fragrance.

14. **O. ilicifolia,** *Hook. f., Handbk.* 126. A shrub or small tree. Branchlets grooved, pubescent or nearly glabrous. Leaves rigid, glaucous when fresh, linear-oblong-lanceolate or linear-ovate-lanceolate, 3in.–4in. long, ½in.–¾in. broad, truncate or rarely narrowed at the base, acute or acuminate, margins with spinous teeth, waved, naked below, or sparsely clothed with thin appressed tomentum; veins at right angles to the midrib. Corymbs usually larger than those of the preceding. Branches with one or two depauperated leaves at the base. Heads rather smaller; involucral bracts lax, spreading, villous, ciliate at the tips. Rays 5–7. Disk-florets 5–6; mouth tubular, campanulate, as long as the tube. Style appendages subulate. Achenes ribbed, pubescent or pilose. Pappus equal.—*Eurybia dentata*, β, Hook. f., Fl. N.Z. i. 118. *O. multibracteolata*, Col. in Trans. N.Z.I. xvii. (1885) 242.

Var. **mollis.** Branchlets, petioles, and panicles white with fine pubescence. Leaves white beneath with laxly-appressed tomentum, the spinous teeth softer and smaller; veins prominent beneath. Panicles very slender, exceeding the leaves.

NORTH and SOUTH Islands: from the East Cape and Urewera Country to Foveaux Strait. STEWART Island. Var. *mollis*: Nelson, *Dall!* Westland and Broken River, *Cockayne!* Sea-level to fully 3,500ft. Jan.

Specimens of this plant assume a yellowish hue when drying. Although very different from the preceding in general appearance, the differences are exhibited by the vegetative organs alone; the reproductive organs show no really distinctive characters, and specimens with intermediate foliage are occasionally met with. This and the preceding species are valued by the Maoris on account of their strong musky fragrance. Captain Gilbert Mair informed me that the Urewera Natives when visiting the sea carried large bundles of this plant as tribute to the coastal Natives for permission to pass through their land.

15. **O. Cunninghamii,** *Hook. f., Handbk.* 126. A much-branched shrub or small tree, 6ft.–25ft. high; trunk sometimes 1ft. in diameter. Branchlets finely grooved, and with the petioles, panicles, and involucres more or less clothed with white or brown tomentum. Leaves 2in.–5in. long or more, ¾in.–2in. broad, oblong ovate-oblong linear-oblong or broadly elliptical, acute or obtuse, narrowed towards the base, margins irregularly cut into coarse or acute teeth, white beneath with appressed soft tomentum; petioles slender or robust. Panicles 3in.–7in. long; branches spreading. Heads very numerous, ¼in. in diameter, shortly pedicellate, broadly turbinate or campanulate; involucral bracts short, ovate, pilose. Florets 16–24; rays short. Disk-florets half-fewer than the rays; mouth short, campanulate; segments recurved. Achene glabrous.—T. Kirk, Forest Fl. N.Z. t. cxiv. *Eurybia Cunninghamii*, Hook. f., Fl. N.Z. i. 117, t. 30. *Brachyglottis Rani*, A. Cunn., Precurs. n. 465. *Solidago canescens*, Banks and Sol. MSS.

Var. **colorata.** Leaves broadly lanceolate. Involucral bracts almost glabrous, rather smaller than in the type and with fewer flowers. *O. colorata*, Col. in Trans. N.Z.I. xii. (1879) 362.

Var. **miniata.** A low shrub. Branchlets slender. Leaves 1in.–2in. long, ¾in.–1in. broad, elliptic-oblong; petioles slender. Panicles much branched and pedicels slender. Heads smaller; involucral bracts short, almost subulate, red, sparingly pubescent but never villous. Achene pubescent.

NORTH Island: from the North Cape to Cook Strait. SOUTH Island: Marlborough and Nelson. Sea-level to 2,800ft. Reported from the west coast of Otago, doubtless in error. *Heketara. Daisy-tree.* Nov.

16. **O. lacunosa,** *Hook. f., Handbk.* 732. A shrub or small tree. Branchlets stout, grooved, and with the petioles, leaves beneath, and panicles clothed with pale rusty tomentum. Leaves 3in.–6in. long, narrow-linear or linear-oblong, ½in.–¾in. broad, acuminate, cuneate at the base, entire or obscurely sinuate-toothed, glabrous and strongly reticulate above; lateral nerves stout, forming a right angle with the midrib and rendering the lower surface strongly rugose; petioles very short, broad. Peduncles 2–5, slender, laxly branched, spreading, forming a corymbose mass 3in.–6in. in diameter. Heads numerous, ⅓in. long, on long slender pedicels; involucre turbinate; bracts few, laxly imbricate, the lower very small, villous, the upper linear-oblong, ciliate and pubescent at the tips. Florets 8–12; rays 5 or 6, short, broad. Disk-florets with a tubular campanulate mouth, longer than the silky tube; segments recurved. Achene ribbed, silky. Pappus unequal.

SOUTH Island: Nelson and Canterbury: Rotoroa, *Travers!* Source of the Takaka River; Mount Arthur plateau; mountains between the Hope and Owen Rivers; *Cheeseman!* Canaan, *Dall!* Harper's Pass, *Haast!* 3,000ft. to 4,000ft. Dec., Jan.

Easily distinguished by the large linear acuminate leaves (rugose beneath), the large corymbs, and the small involucres.

17. **O. excorticata,** *Buch. in Trans. N.Z.I.* vi. (1873) 241. A shrub or small tree, 12ft.–15ft. high.; trunk 1ft. in diameter. Bark loose, papery, brown. Branchlets, petioles, and panicles clothed with whitish tomentum. Leaves linear-oblong, lanceolate, acuminate, 1in.–4in. long, ½in.–¾in. broad, coriaceous, obscurely sinuate-toothed, glabrous above, white with appressed tomentum beneath, shortly petiolate; lateral nerves forming rather less than a right angle with the midrib. Peduncles exceeding the leaves. Corymbs rounded. Heads numerous, small, ⅛in.–3/16in. long; pedicels short; involucral bracts few, lower minute, broadly ovate, convex, villous; upper linear-oblong, obtuse, ciliate and pubescent at the tips. Florets about 12; rays 5–7, very short. Disk-florets with a tubular mouth much longer than the very short silky tube; segments not recurved. Achene very short, broadly subulate, furrowed. Pappus unequal.

NORTH Island: Tararua Range, *Mitchell!* Mount Holdsworth, *Arnold!* Oct.

A very distinct species, of which I have seen but two poor specimens, and can only give an imperfect description.

18. **O. alpina,** *Buch. in Trans. N.Z.I.* xix. (1886) 215. A shrub or small tree, 8ft.–12ft. high; trunk 6in.–8in. in diameter. Branches and leaves beneath clothed with pale-buff tomentum. Leaves 5in.–6in. long, ¼in. broad, linear; margins entire; midrib very stout; veins close-set, forming a right angle with the midrib, and giving rise to a series of lacunae on each side,

Panicles large, much-branched. Heads numerous; involucres turbinate; flowers not seen. Pappus reddish. The branches of the panicle and the involucres are clothed with brownish tomentum.

NORTH Island: Tararua Range and mountains towards Whanganui.

Not having seen specimens of this remarkable plant, I have copied Buchanan's description. He remarks that the thin brownish papery bark peels off in large sheets.

19. **O. Allomii,** *T. Kirk in Trans. N.Z.I.* iii. (1870) 179. A low shrub, 1ft.–2ft. high, sparingly branched from the base. Branches stout. Leaves excessively thick and coriaceous, rather close-set, 1in.–2in. long, ¾in.–1½in. broad, broadly oblong-ovate or elliptical-ovate, rounded at the apex, truncate or slightly unequal at the base, shining above, silvery-white with smooth appressed tomentum beneath. Peduncles few, stout, exceeding the leaves. Corymbs spreading. Heads few; pedicels short, downy; involucres ½in. long or more, broadly campanulate; bracts laxly imbricate, pubescent or villous, broadly lanceolate. Ray-florets about 8, spreading. Disk-florets 6–8, the mouth narrow-campanulate, about as long as the tube, into which it is abruptly contracted; segments 5, spreading. Achenes deeply grooved, hispid. Pappus unequal.

GREAT BARRIER Island: not unfrequent from sea-level to 2,500ft. *Hutton* and *Kirk*. Nov., Dec.

A beautiful and distinct species. When fully expanded the rays are ¾in. in diameter.

20. **O. rigida,** *Col. in Trans. N.Z.I.* xx. (1887) 194. A diffuse shrub, 2ft. high. Branches as thick as a lead-pencil, rigid, clothed with appressed greyish tomentum. Leaves numerous, opposite, very thick and rigid, elliptic, usually about 1in. long, but sometimes reduced to ¼in. or ½in. on the same branchlet, tips recurved, glabrous or rugulose above, pilose beneath, entire; petioles ¼in.–½in. long, clasping. Heads in terminal subrotund corymbs, 1in.–1½in. in diameter, turbinate; pedicels very short, stout; involucral bracts in 2 series, oblong, obtuse, grooved, villous. Florets about 20. Achene glabrous, grooved. Pappus very short, unequal.

NORTH Island: high plains, Waimarino, west side of Tongariro, *Hill.*

Not having seen specimens, I have given, although in a condensed form, the original description. The inflorescence approaches *Cassinia* very closely, but the florets are not described.

21. **O. moschata,** *Hook. f., Handbk.* 127. A much-branched shrub or small tree, 5ft.–14ft. high, emitting a musky fragrance. Branches short and stout; branchlets, petioles, leaves below, and peduncles clothed with soft appressed whitish tomentum. Leaves ⅓in.–⅔in. long, obovate-oblong, obtuse, coriaceous, narrowed into a very short petiole, minutely reticulated above and sometimes pubescent near the tip; veins hidden below. Corymbs 2–4 times as long as the leaves. Peduncles very slender. Heads few, on long slender pedicels, campanulate or broadly turbinate; involucral bracts few, lax, villous, the upper linear. Florets 12–20; rays 7–12, long. Disk-florets with a cam-

panulate mouth and short recurved segments. Achenes ribbed, silky. Pappus unequal.

SOUTH Island: Canterbury, Westland, and Otago, but often local. 2,500ft. to 4,000ft. Jan., Feb.

22. **O. Haastii,** *Hook. f., Handbk.* 126. A much-branched shrub, 5ft.–8ft. high. Branches rather stout, hoary, as are the petioles and peduncles. Leaves $\frac{1}{2}$in.–1in. long, spreading, oblong-ovate or elliptical, obtuse, shortly petioled, reticulate on both surfaces, shining above, white with appressed pubescence below; veins not forming a right angle with the midrib, rather obscure. Peduncles slender, naked, exceeding the leaves. Corymbs lax or compact. Heads numerous, $\frac{1}{5}$in.–$\frac{1}{4}$in. long, sessile and fascicled or on short pedicels; involucre subcylindrical; upper involucral bracts linear-oblong, obtuse, ciliate, slightly pubescent or nearly glabrous. Rays 3–4, short. Disk-florets 4–5; mouth very short, abruptly contracted into the tube; segments spreading or recurved. Achene ribbed, pubescent.

SOUTH Island: near the glacier of Lake Ohau, 4,000ft. to 5,000ft., *Haast!*

The spreading habit of this shrub and its short broad obtuse leaves contrast strongly with the characters of the next species.

23. **O. oleifolia,** *T. Kirk in Trans. N.Z.I.* xi. (1878) 463. A much-branched shrub, 5ft.–8ft. high. Branchlets grooved, crowded, pubescent or hoary, erect, slender. Leaves 2in.–3in. long, $\frac{1}{3}$in.–$\frac{1}{2}$in. wide, shortly petioled, coriaceous, lanceolate or oblong-lanceolate, acute, erect, white with appressed tomentum beneath, forming a smooth surface; veins obscure. Peduncles twice as long as the leaves. Heads on slender pedicels, numerous; involucre narrow, cylindrical; bracts narrow, linear-oblong, pubescent at the tips, the inner almost membranous and nearly glabrous. Rays 2–3, obovate. Disk-florets 2–3. Achenes furrowed, pubescent.—*O. angustata,* J. B. Armst. in Trans. N.Z.I. xiii. 337.

SOUTH Island: Canterbury: Ashburton, *Potts!* Rangitata, *J. F. Armstrong!* Otago: West Coast sounds, *Hector* and *Buchanan!* Resolution Island, *Enys!* 1,500ft. to 3,000ft. Jan.

This plant bears much the same relationship to *O. Haastii* that *O. ilicifolia* bears to *O. macrodonta.* It differs greatly in general appearance, but the floral differences are slight: the heads are smaller and have fewer flowers; the involucral bracts are narrower and more membranous; while forms with intermediate leaves occur, although but rarely.

24. **O. suavis,** *Cheesem. in Trans. N.Z.I.* xxiv. (1891) 409. A shrub or small tree, 6ft.–18ft. high, rarely prostrate, excessively branched. Branches stout; branchlets slender, short, clothed with pale fulvous pubescence. Leaves shortly petioled, oblong or narrow-oblong or ovoid, $\frac{3}{4}$in.–1$\frac{1}{2}$in. long, $\frac{1}{2}$in.–$\frac{3}{4}$in. broad, entire, obtuse, rounded or almost cuneate at the base, subcoriaceous, rarely undulate, brownish-yellow with laxly-appressed tomentum beneath; veins slender. Peduncles exceeding the leaves, very slender, much branched, tomentose. Heads numerous, on rather long pedicels, $\frac{1}{6}$in.–$\frac{1}{5}$in. long; involucres turbinate, narrow; bracts few, lax, the lower short, pubescent or villous, the upper narrow-linear-lanceolate, acute, pubescent, and ciliate at the tips. Florets

6–10, small; rays 5–7. Disk-florets with a campanulate mouth equalling the tube; segments recurved. Achene faintly ribbed, pubescent or silky. Pappus 1-seriate.

SOUTH Island: Nelson: Mount Arthur Plateau, *Cheeseman!* Canaan, *Dall!* 4,000ft. Dec., Jan.

The small tassel-like corymbs are quite unlike those of any other species.

25. **O. nummularifolia,** *Hook. f., Handbk.* 127. An erect shrub, 2ft.–10ft. high. Branches stout, rigid, often viscid. Leaves ¼in.–½in. long, close-set, erect or spreading, coriaceous, suborbicular or orbicular-oblong; margins more or less recurved, shining above, white with appressed tomentum beneath, sessile or abruptly contracted into an extremely short broad petiole. Heads ⅓in.–½in. long, solitary, axillary, turbinate, on very slender peduncles ¼in.–2in. long, often with a pair of bractlets about the middle; involucral bracts few, imbricate, villous, pubescent or glabrate, upper linear. Florets 6–10; rays broad. Disk-florets with a tubular mouth. Achenes pubescent or silky.—*O. Hillii,* Col. in Trans. N.Z.I. xx. (1887) 194.

Var. **cymbifolia,** Hook. f., *l.c.* 732. Leaves ⅛in. long, spreading or deflexed, obtuse, convex above, smooth and nerveless; margins strongly revolute all round, cymbiform, with the concavity below. Heads few or 0.

NORTH and SOUTH Islands: in mountain districts from the East Cape to Foveaux Strait. Rare and local in Otago. Sea-level to 4,500ft. Var. *cymbifolia*, SOUTH Island.

This species varies greatly in the degree to which the margins of the leaves are revolute, and in the length of the petiole. Var. *cymbifolia* appears to be a depauperated condition, largely caused by the ravages of insects; it produces but few flower-heads, and the upper leaves are greatly reduced in size.

26. **O. angulata,** *T. Kirk in Trans. N.Z.I.* xiii. (1881) 384. A much-branched shrub, 8ft.–12ft. high. Branches short, spreading, grooved, almost hoary. Leaves 1½in.–2½in. long, 1in. broad, shortly petioled, oblong or broadly elliptic, rounded at the apex, truncate at the base, undulate at the margins, coriaceous, clothed with appressed white tomentum beneath; principal nerves prominent, forming a right angle with the midrib. Panicles exceeding the leaves, spreading. Heads about ⅙in. long; involucral bracts laxly imbricating, the lower farinose, the upper linear, obtuse, ciliate or pubescent. Florets 3–5. Achenes strigose. Pappus unequal.

NORTH Island: North Cape district and Spirits Bay, *T. K.*

Rather closely related to *O. albida*, but a more spreading plant, with shorter and more coriaceous waved leaves and smaller heads. The leaves closely approximate those of *O. Forsteri*. In the Handbook it appears to be confused with *O. albida*.

27. **O. albida,** *Hook. f., Handbk.* 128. A shrub or small tree, 10ft.–16ft. high. Branches erect; branchlets furrowed, downy pubescent or hoary. Leaves 3in.–4in. long, 1in.–1¼in. broad, petioled, oblong or ovate-oblong, obtuse, narrowed below, not reticulated above, white with soft appressed tomentum beneath. Panicles 2in.–4in. long. Heads numerous, on short pedicels, subcylindrical, ¼in. long; involucral bracts imbricating, often slightly farinose, the lower ovate, short, pubescent, the upper longer, linear-oblong, slightly ciliate. Florets 3–5; rays 1–2. Disk-florets with a very short campanulate mouth; segments 5,

recurved. Style elongate; style-arms very long. Achenes ribbed, downy or pubescent. Pappus with short hairs at base.—*Eurybia albida*, Hook. f., Fl. N.Z. i. 118.

NORTH Island: from the North Cape to the East Cape and the Mokau River, but remarkably local in many districts. April.

Most nearly related to *O. avicenniaefolia*, but the leaves are more narrowed upwards and the heads are larger.

28. **O. avicenniaefolia,** *Hook. f., Handbk.* 127. A shrub or small tree, 5ft.–20ft. high or more. Branches grooved, hoary. Leaves 1½in.–4in. long, ¾in.–1½in. broad, lanceolate-oblong or broadly lanceolate, narrowed at both ends, subacute, flat, reticulated on both surfaces, white with appressed tomentum beneath; petioles short, rather stout. Corymbs on long slender peduncles, dense; pedicels short, rarely 0. Heads small; upper involucral bracts linear, glabrous, entire, obtuse. Ray-florets 1, rarely 2; ray broad. Disk-florets usually 2; mouth shortly funnel-shaped, abruptly narrowed into a tube; segments 5, broad, spreading. Achenes silky. Pappus 1-seriate.—T. Kirk, Forest Fl. N.Z. t. 111. *Eurybia avicenniaefolia*, Hook. f., Fl. N.Z. i. 120. *Shawia avicenniaefolia*, Raoul, Enum. 45.

SOUTH Island: frequent from Nelson to the Bluff; most abundant below 1,600ft., but ascends to 3,000ft. STEWART Island. *Akeake*. Jan., Feb.

29. **O. fragrantissima,** *Petrie in Trans. N.Z.I.* xxiii. (1890) 398. An erect or much-branched deciduous (?) shrub, 6ft.–15ft. high. Branchlets erect, zig-zag but rigid, finely grooved. Bark black. Leaves distant, alternate or fascicled, ¾in.–1½in. long, elliptical or elliptical-lanceolate, acute, narrowed into a short petiole, glabrous above, silky beneath. Flower-heads fragrant, forming a globose 10–12-flowered capitulum, sessile in old axils from which the leaves have fallen, pedicellate or sessile or subracemose, fascicled or solitary with a woolly bract at the base of the lower flowers or fascicles. Involucral bracts in 2–3 series, woolly, the upper linear-oblong. Florets 4–8; rays 2–5, very short and broad, recurved at the tips. Disk-florets with a campanulate mouth and long recurved segments; both ligules and segments pubescent or ciliate at the tips. Achene grooved, silky. Pappus-hairs equal.

SOUTH Island: Canterbury: Lake Forsyth, *T. K.* Otago Heads, *Buchanan!* Dunedin, Saddle Hill, Tairoa Heads, Catlin's River, *Petrie!*

Originally discovered by Buchanan, but confused with *O. Hectori*, the flowers of which were at that time unknown. The only New Zealand species with globose fascicles of yellow flowers.

30. **O. Hectori,** *Hook. f., Handbk.* 128. An erect much-branched deciduous shrub or small tree, 6ft.–14ft. high. Branchlets slender, grooved, terete or angular at the tips, glabrous. Leaves in opposite fascicles, linear-obovate or linear-spathulate, 1in.–1½in. long, ¼in.–½in. broad, membranous, white with silvery tomentum beneath, glabrous above except when young; petioles very slender. Heads in opposite fascicles of 2–5. Peduncles silky, slender, ¼in.–½in. long; involucre saucer-shaped; bracts in 2 series, linear-oblong, lax, spreading,

woolly. Florets 20–26. Disk-florets about 8; mouth very short, funnel-shaped; rays very short. Achene silky or pubescent. Pappus white, unequal.

SOUTH Island: Marlborough: Pelorus Sound, *Rutland!* *T. K.* Canterbury: Banks Peninsula, *J. B. Armstrong.* Otago: Kaitangata, Catlin's River, Invercargill, Kawarau Gorge, Matukituki Valley, *Petrie!* Sea-level to 2,000ft. Oct., Nov.

31. **O. laxiflora,** *n. s.* Approaching *O. odorata*, but a larger plant, with slender divaricating almost pendulous branches. Leaves opposite or fascicled, narrow linear-spathulate or almost lanceolate, narrowed into very short petioles, obtuse but not rounded at the apex, subcoriaceous, whitish or ferruginous beneath with finely-appressed tomentum. Heads very numerous, in opposite lax fascicles; peduncles usually 5–10 or more, ¾in. long, glabrate or puberulous, very slender; involucre turbinate; bracts few, in 3 series, lax, villous at the tips, the upper linear-oblong, twice as long as the lower. Ray-florets 3–4, broad. Disk-florets 2–4; mouth campanulate; segments recurved; tube very long. Style-arms short, obtuse. Achene ribbed, glabrate.

SOUTH Island: Hokitika, *H. Tipler!*

Nearly allied to *O. odorata*, but differing in the hard tomentum on the under surface of the leaves, the larger open fascicles, long slender peduncles, lax involucres, and fewer florets. A very handsome species.

32. **O. odorata,** *Petrie in Trans. N.Z.I.* xxiii. (1890) 399. An erect spreading much-branched divaricating shrub, 8ft.–10ft. high. Branches stout, terete, rarely angular at the tips, grooved. Leaves opposite, usually fascicled, ½in.–1in. long, linear-spathulate, narrowed into very short petioles, rounded at the apex, coriaceous or subcoriaceous, glabrate or almost silky above, white with soft appressed tomentum beneath. Heads in opposite fascicles; peduncles 2–5, ⅛in.–⅜in. long, silky; involucres broadly campanulate; involucral bracts in 3–4 series, short, viscid, tips often spreading. Florets 15–30; rays 6–15, very short. Disk-florets with a short mouth. Style-arms more obtuse than in *O. laxiflora*. Achene silky or pubescent. Pappus white, unequal.

SOUTH Island: Westland and Otago: not unfrequent.

I have not seen North Island specimens. It may be that this plant is the typical *O. virgata*, but this could only be determined by an examination of the original specimens at Kew.

33. **O. virgata,** *Hook. f., Handbk.* 128. A much-branched spreading shrub, 4ft.–10ft. high, often forming thickets. Branchlets usually angled, grooved, stout or slender, puberulous, pubescent or glabrous. Leaves ¼in.–½in. long, linear-obovate or linear-spathulate, opposite or in opposite fascicles, sessile or narrowed into a short petiole, coriaceous, white beneath with appressed tomentum, glabrate or pubescent above. Heads solitary or fascicled, sessile or on very short peduncles, turbinate or almost campanulate, ¼in. long; bracts in 3 series, usually villous, upper oblong-lanceolate, subacute. Florets 7–12; rays very short; mouth of disk-florets shortly funnel-shaped. Style-arms acute. Achene glabrous or almost pubescent. The lips of the ligula and segments are usually pubescent on the outside, and rarely the rays are divided into 2 or 3 linear lobes.—*O. fasciculifolia,* Col. in Trans. N.Z.I. xxv. (1892) 330.

O. consimilis, O. quinquefida, and *O. parvifolia,* Col., *l.c.* xxviii. 596, 598. *Eurybia virgata,* Hook. f., Fl. N.Z. i. 119.

Var. **aggregata.** Branchlets slender or robust, terete or almost angular. Leaves in fascicles of 8–15, coriaceous, margins recurved, sessile or nearly so, white beneath. Heads numerous, forming large fascicles, sessile or on very short pedicels. Florets 15–17.—*O. aggregata,* Col. in Trans. N.Z.I. xxviii. 597.

Var. **ramuliflora.** Branches slender, angled. Leaves in fascicles of 2–6. Heads in fascicles of 2–6; involucres subcylindrical, $\frac{1}{4}$in. long; bracts in 3 series, pubescent or villous, imbricate, inner linear-oblong, villous at the tips. Rays 5. Disk-florets about 3; mouth funnel-shaped, shorter than the tube; segments acute, recurved. Achene glabrous.—*O. ramuliflora,* Col. in Trans. N.Z.I. xxii. 467.

Var. **lineata.** Branchlets very slender, often pendulous, angled, glabrous or pubescent. Leaves $\frac{1}{2}$in.–1$\frac{1}{4}$in. long, excessively narrow-linear; margins much recurved, glabrate or pubescent above, silky beneath. Heads fascicled; peduncles very slender; involucre broadly campanulate; bracts villous. Rays 8–13, very short. Disk-florets 6–10. Achene glabrous.

NORTH Island: from the East Cape and Hawke's Bay to Wellington, but local in many districts. SOUTH Island: Nelson to Foveaux Strait. Sea-level to 2,000ft. Dec., Jan.

An extremely variable plant, the numerous forms of which pass gradually into each other. Var. *ramuliflora* is a transition form between *O. virgata* and *O. odorata;* var. *lineata* may be worthy of specific honours, although the leaves are often as short as in the typical form.

34. **O. Solandri,** *Hook. f., Handbk.* 128. A much-branched shrub, 3ft.–12ft. high. Branchlets, leaves, and involucres often viscid, the former stout or slender, angled, puberulous. Leaves opposite, on very young plants $\frac{1}{2}$in. long, linear oblong-spathulate, flat, narrowed into short petioles, white beneath; on mature plants mostly in opposite fascicles, $\frac{1}{8}$in.–$\frac{1}{4}$in. long, linear-obovate or narrow-linear, obtuse, coriaceous, clothed with whitish-yellow tomentum beneath; margins recurved. Heads solitary, terminal, sessile or rarely pedicellate, $\frac{1}{4}$in. long; involucre subcylindrical, golden-yellow; involucral bracts in 4 series, imbricate, the upper lanceolate, subacute, puberulous or pubescent. Ray-florets 12–14; rays long, revolute. Disk-florets 6–8; mouth shortly funnel-shaped, with very long revolute segments. Achene ribbed, glabrate. Pappus white, equal.—*Eurybia Solandri,* Hook. f., Fl. N.Z. 128. *Calea axillaris,* Banks and Sol. MS.

NORTH Island: common near the coast. SOUTH Island: Queen Charlotte and Pelorus Sounds, *Rutland! MacMahon!* D'Urville Island, *Bryant!*

Usually a cupule formed of 4–5 diminutive leaves with short broad gold-coloured petioles is developed at the base of the involucre. The tube of the disk-florets is as long as in the species with large heads. On old specimens the involucres often become woody and persistent.

[**O. coriacea.** A sparingly-branched rigid shrub, 6ft.–8ft. high. Branches erect or ascending, rather stout, pubescent. Leaves very thick and coriaceous, $\frac{1}{2}$in.–$\frac{3}{4}$in. long on very short petioles, patent, shortly ovate or orbicular-ovate, obtuse, reticulate above, white with appressed tomentum beneath. Heads not seen. Dead panicles about twice as long as the leaves, apparently with 2–5 sessile or pedicellate heads.

SOUTH Island: Marlborough: Awatere and Mount Fyffe, *T. K.*

Notwithstanding the imperfect materials available, I have ventured to present a provisional description of what appears to be a very distinct species, quite unlike any other.]

5. SHAWIA, Forst.

A shrub or small tree, with alternate coriaceous leaves and much-branched axillary panicles. Heads cylindrical, narrow; involucral bracts imbricating, the upper narrow-linear. Floret 1, hermaphrodite; ray-florets 0. Achene pubescent. Pappus 1-seriate or with a few short hairs at the base.

This genus was suppressed owing to the erroneous idea that the characters given by Forster were inconstant. Thus, Hooker writes, "Florets variable in number; generally one is ligulate and female, with one to two discoid and hermaphrodite." I have examined numerous recent specimens collected over the entire area of the species, but have never seen a ray-floret or anything that could be mistaken for one, nor have I been able to find even a solitary involucre containing two florets, although experience has shown that a mistake on this point may easily be made if only dried specimens are examined. It therefore seems advisable to revive the genus, as its constancy is fully established.

1. **S. paniculata,** *Forst., Char. Gen.* 95, *t.* 48. A much-branched shrub or small tree, 5ft.-20ft. high. Leaves 1½in.–3in. long, 1in.–2in. wide, oblong or ovate-oblong, coriaceous, clothed beneath with short closely-appressed tomentum, obtuse; margins waved; petioles short. Corymbs longer or shorter than the leaves, spreading; heads often densely fascicled, glabrous or glabrate, sessile. Anthers sometimes wholly exserted.—A. Rich., Fl. N.Z. 243; A. Cunn., Precurs. n. 434; Raoul, Choix t. 13. *Eurybia Forsteri,* Hook. f., Fl. N.Z. i. 119. *Olearia Forsteri,* Hook. f., Handbk. 127; T. Kirk, Forest Fl. N.Z. t. 137.

Var. **obtusa.** Leaves subcoriaceous, broadly ovate or oblong, rounded at the tips. Heads small, very narrow.—*O. uniflora,* Col. in Trans. N.Z.I. xxii. 469!

Var. **elliptica.** Leaves narrow-oblong or elliptic-oblong, 2in.–2½in. long, ½in.–¾in. broad, almost flat. Heads rather short.

NORTH and SOUTH Islands: from the East Cape to Oamaru and Greymouth. Ascends to 1,600ft. *Akiraho.* April, May.

6. PLEUROPHYLLUM, Hook. f.

Involucre hemispherical, herbaceous; involucral bracts in 2–3 series. Receptacle flat, pitted, toothed. Florets very numerous. Ray-florets in 3 series, female, ligulate or shortly bilobate. Disk-florets campanulate at the mouth, abruptly narrowed into a slender filiform tube, 4–5-toothed or -lobed; segments recurved. Anthers with very short tails. Achenes striated, setose. Pappus-hairs in 3 series, scabrid, unequal. Tall robust succulent herbs. Root-fibres fleshy. Leaves chiefly radical, with stout parallel ribs. Heads racemose on erect scapes.

All the species are endemic on the Antarctic islands within the New Zealand area.

Heads with conspicuous ray-florets.

Leaves sessile on a broad base, longitudinally ribbed or furrowed .. .. 1. *P. speciosum.*

Heads with inconspicuous 2-3-lobed ray-florets.

Leaves radical and cauline on longer or shorter petioles; nerves few, slender .. 2. *P. criniferum.*

Leaves all radical, silvery, narrowed to a broad sheathing base, acute .. .. 3. *P. Hookeri.*

1. **P. speciosum,** *Hook. f., Fl. Antarc.* i. 31, *t.* 22, 23. Leaves all radical, 6in.–18in. long, 5in.–10in. broad, usually appressed to the ground, forming a huge rosette, broadly ovate or obovate or unequally rhomboid,

rounded at the apex or shortly acuminate, thick when fresh, with 15–20 longitudinal ridges, loosely tomentose below, villous or setose above, the bristles being intermixed with rather long moniliform hairs. Scapes erect, with several oblong leafy bracts below. Heads 8–20 or more, 1½in.–2½in. in diameter. Peduncles 1in.–6in. long. Ray-florets purplish-white, showy. Disk-florets purple. Achenes strigose. Pappus-hairs not thickened upwards.—Handbk. 129; T. Kirk in Trans. N.Z.I. xxiii. (1890) 433.

Var. **suberecta.** Leaves erect or suberect, narrower; moniliform hairs more numerous. Ray-florets violet or purple.

AUCKLAND and CAMPBELL Islands: sea-level to 800ft. Dec., Jan.

One of the most magnificent plants in the flora. The linear-acuminate involucral bracts are more or less tomentose.

2. **P. criniferum,** *Hook. f., Fl. Antarc.* 32, *t.* 24, 25 (except the leaf). Radical leaves 1ft.–4ft. long, 4in.–12in. broad, with long sheathing suberect petioles varying greatly in length; blade varying from oblong-lanceolate ovate-lanceolate or almost ovate to linear-lanceolate obovate-lanceolate or orbicular-ovate, usually acute, membranous but firm, white with thin tomentum beneath, slightly scabrid or setose above, with a few short subulate moniliform hairs; principal nerves 7–15, slender; margins with a few distant projecting teeth. Scapes 2ft.–6ft. high, stout, strongly grooved. Cauline leaves sessile, white above and beneath. Heads 15–30 or more; peduncles erect, 1in.–6in. long; involucral bracts ovate-acuminate or oblong, sparingly ciliate. Ray-florets very short and inconspicuous, 2–3-fid or -partite. Achenes strigose. Pappus-hairs slightly thickened upwards.—Handbk. 129; T. Kirk in Trans. N.Z.I. xxiii. (1890) 434. *P. Hombronii,* Decne., Bot. Voy. au Pole Sud, 36. *Albinea oresigenesa,* Homb., *l.c.* Dicot. Phan. t. 4.

ANTIPODES Islands, *T. K.* AUCKLAND and CAMPBELL Islands. Not found on Macquarie Island. Dec., Jan.

Distinguished by the large discoid heads and petiolate leaves.

3. **P. Hookeri,** *J. Buch. in Trans. N.Z.I.* xvi. (1883) 395 (except t. 37). Leaves all radical, 6in.–10in. long, 3in.–4in. broad, white on both surfaces with silky lax or close tomentum, flat, appressed to the ground, forming a rosette, obovate or oblong-obovate, abruptly acuminate, narrowed into a broad sheathing membranous base; principal ribs 9–13, very slender; marginal teeth reduced to minute points. Scapes 15in.–24in. high, strict, naked below except 3 or 4 linear bracts at the base, villous or silky. Heads hemispherical or almost globose, ¾in. in diameter; involucral bracts linear, acute or acuminate, the outer with a few scattered hairs. Ray-florets few or none; rays very short, bilobate, inconspicuous. Achenes silky. Pappus-hairs not thickened upwards. —*P. Hookerianum,* T. Kirk in Trans. N.Z.I. xxiii. (1890) 435, t. 40. *P. Gilliesianum,* T. Kirk MS.

AUCKLAND Islands, *T. K.* CAMPBELL Islands, *J. Buchanan! T. K.* MACQUARIE Island, *Scott! Hamilton!* Dec., Jan.

Distinguished by the small silvery leaves, rigid scapes, and small heads.

7. CELMISIA, Cass.

Involucre broadly hemispherical; involucral bracts in few or many series, imbricating, pubescent cottony silky or glandular; tips scarious, often recurved. Receptacle plane or convex, often deeply pitted. Ray-florets female, 1-seriate, spreading, flat or revolute, often very long, invariably white. Disk-florets hermaphrodite, tubular, 5-toothed. Anthers usually sagittate at the base, with short acute or rarely obtuse tails. Style-arms flattened, tipped with long or short cones or appendages, papillose on the back. Achenes linear, slightly compressed or angled, with 2 or 3 prominent ribs on each side. Pappus unequal, the hairs often shortly bifid. Tufted herbs, with fleshy root-fibres or short branched rhizomes, rarely suffruticose and erect. Leaves all radical or rarely cauline and densely imbricating, usually narrowed into sheathing bases, clothed beneath with appressed or lax tomentum. Heads large, solitary, terminating bracteate scapes, or rarely sessile or on axillary peduncles.

A beautiful genus, comprising the various "mountain daisies" of the settlers. Many of the species are extremely variable, and pass into each other by easy gradations. It differs from *Olearia* only in habit, such species as *C. Walkeri* and *C. ramulosa* having an equal claim to be placed in either genus. From *Pleurophyllum* it differs only in the solitary heads and white rays. The habit is exactly that of *Aster alpinus*, L., which, except for the anthers, might very well be referred to *Celmisia*. Under cultivation some of the stronger-growing species frequently develop branched scapes. All the species are endemic in New Zealand except *C. longifolia*, which extends to Australia, and the doubtful *C. Lechleri*, Sch. Bip., which is restricted to Peru.

I. PSEUDOCELMISIA.—SUFFRUTICOSE SPECIES WITH ELONGATING BRANCHES, THE LEAVES MORE OR LESS IMBRICATING (see also Nos. 13 and 36).

Leaves spreading, flat, toothed. Peduncles slender, axillary .. .. 1. *C. Walkeri.*
Leaves spreading, revolute. Peduncles slender, axillary .. .. .. 2. *C. rupestris.*
Leaves erect, imbricating. Peduncles strict, terminal .. .. .. 3. *C. ramulosa.*
Branches short. Leaves spreading or imbricate, glabrate or glandular. Peduncles axillary .. .. 4. *C. lateralis.*

II. EUCELMISIA.—BRANCHES SHORT. LEAVES MORE OR LESS ROSULATE. SCAPES USUALLY TERMINAL.

. *Leaves more or less toothed and clothed, white or buff-coloured tomentum beneath (rarely glabrous in 14). Disk-florets yellow or white.*

Leaves 6in.–10in. by 1½in.–2½in., acutely serrate; sheath very short .. 5. *C. holosericea.*
Leaves 3in.–6in. by ¾in.–1½in., narrow-oblong, crenate-dentate, white beneath 6. *C. densiflora.*
Leaves 5in.–8in. by 1in.–2in., obovate-oblong, acutely serrate, pale-buff-coloured beneath. Bracts leafy .. .. 7. *C. Dallii.*
Leaves 1in.–3in. by ½in.–1in., broadly oblong, serrulate, yellow or buff-coloured beneath .. .. 8. *C. hieracifolia.*
Leaves 1½in.–3in. by ½in.–1in., lanceolate, acute, rugose, furrowed above .. 9. *C. prorepens.*
Leaves ½in.–2½in. by ¼in.–⅓in., viscid, linear, coriaceous. Scapes slender .. 10. *C. discolor.*
Leaves 1in.–2in. by ½in.–1in., obovate-oblong, white beneath .. .. 11. *C. Haastii.*
Leaves 1in.–2½in. by ½in.–¾in., obovate, furrowed above, white beneath .. 12. *C. incana.*
Leaves 3in.–7in. by ½in.–1in., viscid, linear-oblong, white beneath. Scape often axillary .. .. 13. *C. Lindsayi.*
Leaves 1in.–3in. by ¼in.–⅔in., obovate-spathulate, white beneath or glabrous. Scapes slender .. .. 14. *C. Sinclairii.*

B. *Leaves 3in.–18in. long, entire or rarely denticulate, silvery silky or woolly below, rarely glabrate or glabrous.*

Leaves 8in.–18in. by 1in.–3in., oblong-spathulate, coriaceous, white or ferruginous beneath .. .. 15. *C. verbascifolia.*

Leaves 8in.–14in. by 1in.–2in., oblong-lanceolate, clothed with velvety ferruginous tomentum beneath .. .. 16. *C. Traversii.*

Leaves 7in.–9in. by 1¼in.–2in., broadly lanceolate or oblong, narrowed into the broad petiole, whitish beneath.. .. .. 17. *C. Brownii.*

Leaves 5in.–12in. by 1in.–2½in., ovate-lanceolate, silvery beneath .. .. 18. *C. Rutlandii.*

Leaves 3in.–9in. by ¾in.–2½in., oblong linear-oblong or oblong-lanceolate, glabrous or clothed with white or fulvous tomentum beneath .. 19. *C. petiolata.*

Leaves 6in.–14in. by 2in.–3in., acuminate, glabrous or glabrate .. .. 20. *C. Mackaui.*

Leaves 3in.–5in. by ½in.–¾in., rigid, clothed with appressed whitish tomentum beneath; sheath equalling the blade .. .. 21. *C. spectabilis.*

Leaves 8in.–20in. by ¾in.–3in., erect, clothed with a silvery pellicle above and appressed white tomentum beneath .. .. 22. *C. coriacea.*

Leaves 3in.–7in. by ⅓in.–⅔in., spreading, furrowed, linear-oblong or lanceolate, silvery above, clothed with appressed tomentum beneath .. .. 23. *C. Monroi.*

Leaves 6in.–12in. by ½in.–¾in., flat, membranous, denticulate, clothed with lax white tomentum beneath .. .. 24. *C. Adamsii.*

C. *Leaves narrow-linear, erect, rigid, flat except in* 25 *and* 27.

Leaves 1in.–16in. by 1/12in.–⅜in., often soft and spreading, longitudinally grooved above, margins strongly revolute, white with appressed tomentum beneath 25. *C. longifolia.*

Leaves ½in.–1in. by 1/20in., acerose, pungent, silky.. .. .. .. 26. *C. laricifolia.*

Leaves 5in. by ⅛in.–⅙in., rigid, margins recurved, shining above, white with lax appressed tomentum beneath .. .. 27. *C. linearis.*

Leaves 9in.–18in. by ¼in.–½in., ensiform, rigidly coriaceous, white beneath with thinly-appressed tomentum .. .. 28. *C. Lyallii.*

Leaves 6in.–18in. by ¼in.–¾in., ensiform, rigid, white with satiny tomentum beneath .. .. 29. *C. Armstrongii.*

Leaves 3in.–4in. by ¼in., linear, excessively viscid, grooved above, whitish beneath .. .. 30. *C. viscosa.*

D. *Leaves ½in.–1½in. by ⅛in. broad, silky or white below or on both surfaces.*

Leaves densely rosulate, 1in.–1½in. by ¼in.–⅜in., clothed with long silky hairs above and below .. .. 31. *C. MacMahoni.*

Leaves ¾in.–1½in. by ⅛in.–¼in., lanceolate, acute, white beneath .. .. 32. *C. parva.*

Leaves ½in.–¾in. by ⅛in.–¼in., linear-spathulate, silvery above, cottony or tomentose below .. .. 33. *C. Hectori.*

E. *Leaves ½in.–1in. long, very narrow-linear, densely imbricating all round the stem. Heads sessile.*

Leaves ½in.–1in. long, margins revolute, hoary or silky above and beneath. Heads ½in.–1in. in diameter .. .. 34. *C. sessiliflora.*

Leaves 1/16in.–1/10in. long, margins involute, acute, silvery. Heads ⅜in.–½in. in diameter .. .. 35. *C. argentea.*

F. *Leaves ½in.–1in. long, glabrate or minutely glandular-pubescent.*

Leaves ¼in.–½in. long, glabrous, narrowed into a short broad cottony petiole .. 36. *C. bellidioides.*

Leaves glabrate or glandular-pubescent, serrate; petiole slender .. .. 37. *C. glandulosa.*

G. *Disk-florets violet-purple.*

Leaves 1in.–4in. by ⅛in.–¼in., rigid, coriaceous, shining, glabrous .. .. 38. *C. vernicosa.*

Leaves 3in.–5in. by ½in.–¾in., lanceolate, grooved below and sparingly tomentose 39. *C. Chapmanii.*

1. **C. Walkeri,** *T. Kirk in Trans. N.Z.I.* ix. (1876) 549. Stem woody, procumbent or suberect, much or sparingly branched, 1ft.–3ft. long or more. Leaves slightly viscid, crowded, linear, 1in.–1½in. long, flat, acute, patent, with broad imbricating bases wider than the blade, slightly coriaceous, serrulate, glabrous above, clothed with snow-white tomentum beneath. Peduncles 5in.–8in.

long, axillary, solitary, near the tips of the branches, very slender; bracts short, subulate, acute. Heads 1½in. in diameter; involucral bracts unequal, linear-subulate, glandular-pubescent; tips recurved. Rays 30–40, narrow, spreading. Achenes not seen.

SOUTH Island: Canterbury: Arthur's Pass, *Cheeseman!* Mountains above Lake Harris, Otago, *Walker* and *Kirk*. Mount Alta, *Buchanan!* Mountains near Mount Aspiring, *Petrie*. 3,000ft. to 4,000ft. Jan.

A very distinct species, which cannot be mistaken for any other. This and the two following have perhaps equal claim to be included under *Olearia*, but the involucres and the sheathing leaf-bases are essentially those of *Celmisia*.

2. **C. rupestris,** *Cheesem. in Trans. N.Z.I.* xvi. (1883) 409. Habit and general aspect of *C. Walkeri*, but more branched. Leaves densely crowded, ½in.–1in. long, linear-spathulate, obtuse, narrowed towards the base, but there expanding into the imbricating sheath, suberect or spreading, silky above, clothed with lax tomentum beneath; margins strongly revolute. Peduncles as in *C. Walkeri*, one or two near the tips of the branches. Heads immature, apparently about 1in. in diameter when expanded; involucral bracts very numerous, linear-subulate, glandular-pubescent, unequal. Ray-florets rather numerous. Achenes not seen.

SOUTH Island: ravines on Mount Peel, Nelson, 5,000ft.

3. **C. ramulosa,** *Hook. f., Handbk.* 733. Stems woody, simple or much branched, 2in.–7in. long. Branches short, ascending. Leaves densely imbricating, ¼in.–⅓in. long, linear-oblong, obtuse, subcoriaceous, attached by broad membranous sheathing bases, glabrous above, clothed with soft white tomentum beneath; tips erect or spreading; margins strongly revolute. Peduncles solitary, terminal, strict, ½in.–2in. long, with 1 or 2 very short bracts. Heads ¾in.–1in. in diameter; involucral bracts unequal, linear-lanceolate, acute, glabrate or glandular-pubescent. Rays spreading, narrow. Achenes not seen.—*Olearia Buchanani*, Hook. f. MS.

SOUTH Island: Otago: mountains above Dusky Bay, *Hector* and *Buchanan!* Mountains above Lake Hauroto, *Thomson!* Hector Mountains and Mount Pisa, *Petrie!* Mount Cardrona, *Goyen*. 3,500ft. to 5,000ft. Jan.

Distinguished from the preceding by its smaller size, short usually erect leaves, and strict terminal peduncles.

4. **C. lateralis,** *Buch. in Trans. N.Z.I.* iv. (1871) 226, *t.* 15. A much-branched slightly-suffrutescent species. Root woody. Branches crowded, 1in.–5in. long, forming a compact mass. Leaves erect or spreading, ¼in.–½in. long, linear-subulate, acute, acuminate or apiculate, flat, attached by a short sheathing base, not narrowed below, glandular-pubescent at the margins or glabrous. Peduncles 2in.–3in. long, lateral or terminal, very slender; bracts short, subulate, flat. Heads ½in.–¾in. in diameter; involucral bracts crowded, linear-subulate, acute, glandular or the inner glandular-silky, scarious. Rays numerous, spreading, narrow, ¼in.–⅓in. long. Achenes silky. Pappus unequal.

SOUTH Island: Nelson: Mount Arthur, *Spencer! Cheeseman!* Mountains near Lake Guyon, *H. Travers!* Paparoa Range, Westland, *Helms!* 4,000ft. to 6,000ft. Dec., Jan.

A very distinct species, easily recognised by the short acute or apiculate leaves and glandular involucral bracts.

5. **C. holosericea,** *Hook. f., Fl. Antarc.* i. 36; *Fl. N.Z.* i. 121, *t.* 21. Leaves 6in.–10in. long, 1½in.–2½in. broad, coriaceous but not thick, flat, lanceolate, acute, acutely distantly serrate, glabrous above, white with appressed silvery tomentum beneath; midrib and veins obvious on both surfaces; petiole very short and broad, sheathing, glabrous. Scapes few, slender, 1ft.–1½ft. long or more, glabrous, with 3 or 4 acute bracts on the upper part. Heads 2in.–3in. or more in diameter; involucral bracts in 2 or 3 series, sometimes 1in. long or more, viscid, linear-lanceolate, with an obvious midrib, the outer ones almost leafy. Rays long, spreading. Achene pilose.—Handbk. 130. *Aster holosericeus,* G. Forst., Prod. n. 296; A. Rich., Fl. Nouv.-Zel. 248; A. Cunn., Precurs. n. 438; DC., Prod. v. 226.

SOUTH Island: Dusky Bay, *Forster, Wesley!* Port Preservation, *Lyall.* Mountains west of Te Anau Lake, 2,000ft. to 4,000ft.

A noble species, formerly cultivated by a Maori lady on Stewart Island.

6. **C. densiflora,** *Hook. f., Handbk.* 130. Leaves viscid above, narrow-oblong or lanceolate, 3in.–6in. long, ¾in.–1½in. broad, obtuse or acute, almost coriaceous, crenate-dentate or crenulate-serrate, glabrous above, white with appressed tomentum beneath; midrib prominent beneath; sheath broad, half the length of the blade, membranous; veins obscure. Scapes 6in.–16in. high, with few or many linear acute or acuminate bracts. Heads 1in.–2in. broad, viscid; involucral bracts crowded, linear-subulate; tips recurved, glabrous or pubescent. Rays long, flat, ultimately contorted. Achenes slightly curved, silky. Pappus almost equal.

NORTH Island: Wellington: Tararua Mountains, *Buchanan.* SOUTH Island: Canterbury: mountains near Lake Ohau, Hopkins and Dobson Rivers, *Haast!* Otago: lake district, *Haast* and *Buchanan!* Mihiwaka; Mount Ida Range; Mount St. Bathans, &c.; 800ft. to 3,000ft.: *Petrie!* Jan., Feb.

A handsome species, which, unhappily, is in danger of speedy extirpation. I have not seen North Island specimens.

7. **C. Dallii,** *Buch. in Trans. N.Z.I.* xiv. (1881) 357, *t.* 35. A stout species. Leaves rosulate, clothed with thin pale-buff tomentum beneath, glabrous above, 5in.–8in. long, 1in.–2in. broad, subcoriaceous, oblong or obovate-oblong, obtuse, rarely acute, apiculate, sharply serrate or entire, narrowed below and suddenly expanded into the broad membranous ribbed glabrous sheath a quarter to half as long as the blade; midrib and veins obvious on both surfaces. Scapes 1 or several, 8in.–18in. high, usually much compressed, glabrous; bracts few or many, 1in.–2in. long, leafy, often aggregated beneath the head and forming a kind of spurious involucre. Head 1½in.–2½in. in diameter; involucral bracts viscid, linear-acuminate, tomentose, inner cottony or ciliate. Rays broad, spreading. Achene hispid.

SOUTH Island: Nelson: Mount Arthur, *Spencer! Cheeseman!* Gouland Downs; head of the Aorere River; *Dall!* Mount Rochfort, *Spencer!*

A handsome species, best distinguished by the large leaves and leafy bracts clothed with pale-buff tomentum beneath, and by the broad strongly-ribbed glabrous sheaths.

8. **C. hieracifolia,** *Hook. f., Fl. N.Z.* i. 124, *t.* 34B. Branches very short. Leaves viscid, 1in.–3in. long, ½in.–1in. broad, broadly oblong, narrowed at both ends, subacute or acute or obovate-oblong or linear-oblong, rounded at the apex and narrowed to the short sheath, coriaceous, serrulate, pubescent, glabrous or puberulous above, clothed beneath with appressed gold-coloured or buff tomentum; midrib and veins prominent beneath. Scapes erect, rather stout, viscid, 2in.–10in. high, pubescent or glandular-pubescent; bracts 2–10, pubescent. Heads ½in.–1½in. in diameter; involucral bracts linear-acuminate, the outer recurved, glandular-pubescent, viscid, rarely cottony. Rays numerous. Achene furrowed, very silky, larger than the pappus.—Handbk. 131.

Var. **oblonga**, T. Kirk in Trans. N.Z.I. xxvii. 328. Leaves linear-oblong, broadly rounded above. Heads usually smaller.

SOUTH Island: Marlborough: Mount Stokes, *MacMahon! T. K.* Nelson: Dun Mountain, &c., *Bidwill, Monro, Sinclair!* 4,000ft. Dec., Jan.

This and *C. Dallii* are the only species with yellow or buff tomentum.

9. **C. prorepens,** *Petrie in Trans. N.Z.I.* xix. (1885) 326. Stem much branched at the base, stout, prostrate, densely clothed with the remains of old leaf-sheaths below. Leaves viscid, crowded, 1½in.–3in. long, ½in.–1in. broad, subcoriaceous, lanceolate or oblong-lanceolate, acute, unequally coarsely serrate, longitudinally furrowed and rugose, glabrous on both surfaces or pubescent beneath; margins slightly recurved; midrib and veins obvious below; sheath very short, narrower than the leaves, viscid. Scapes few, 3in.–7in. long, slender, glabrate, viscid; bracts linear-obtuse. Head ¾in.–1in. in diameter; involucral bracts linear-lanceolate, puberulous, viscid. Rays spreading. Achenes slender, silky.

SOUTH Island: Otago: Upper Waipori; Rock and Pillar Range; Old Man Range; Mount St. Bathans. 3,000ft. to 5,000ft. *Petrie.* Nov., Dec.

A local species, easily recognised by the rugose leaves.

10. **C. discolor,** *Hook. f., Fl. N.Z.* i. 123. Stems branched below. Branches prostrate, stout or slender, usually clothed with old leaves. Leaves viscid, spreading or erect, 1in.–3in. long, ¼in.–½in. broad, linear-oblong to oblong-spathulate, coriaceous, entire or serrulate, subacute or obtuse, white with silvery appressed tomentum beneath; midrib and veins obvious on both surfaces or only beneath; sheaths one-third to one-half the length of the leaves, imbricating, glabrous. Scapes 1 or several, terminal, very slender, glandular-pubescent; bracts few or many, linear-subulate, acute. Head ¾in.–1in. in diameter; involucral bracts linear-subulate, the outer with recurved tips, glabrate, pubescent or pilose. Rays spreading, rather narrow. Anthers with very short basal appendages. Achene silky.—Handbk. N.Z. Fl. 131. *Erigeron Novae-Zelandiae,* Buch. in Trans. N.Z.I. xvii. (1884) 287, t. 15.

SOUTH Island: not uncommon in mountain districts from Cook Strait to Foveaux Strait. 2,500ft. to 5,000ft. Dec., Jan.

This species occasionally produces strong autumnal branches with strict lateral peduncles having only 1 or 2 bracts.

11. **C. Haastii,** *Hook. f., Handbk.* 131. A small tufted species. Leaves oblong broadly oblong or oblong-spathulate, 1in.–2in. long, ½in.–1in. broad, subcoriaceous; petioles very short, glabrous above and often longitudinally furrowed, whitish with appressed tomentum beneath; margins slightly recurved. Scape 3in.–8in. high, glabrate or laxly tomentose; bracts numerous, linear, acute, cottony. Heads ¾in.–1½in. in diameter; involucral bracts linear, acute, glandular-pubescent or almost villous. Rays spreading. Achene glabrous.

SOUTH Island: Canterbury, Westland, and Otago; most plentiful in the south and west· 3,000ft. to 5,000ft. Jan., Feb.

In the Handbook the involucral bracts are described as obtuse. I do not find them so.

12. **C. incana,** *Hook. f., Fl. N.Z.* i. 123, *t.* 34A. Usually stout. Stems branched. Branches short, prostrate, rather woody, densely clothed with dead leaves below. Leaves numerous, 1in.–2½in. long, ½in.–¾in. broad, oblong-spathulate or obovate-spathulate, sharply narrowed towards the base and suddenly expanded into a sheathing petiole which is about half the length of the blade, longitudinally furrowed above, margins serrulate or entire, both surfaces clothed with lax snow-white tomentum or the upper pubescent only. Scapes 1–3, 8in.–10in. high, with numerous acute or obtuse linear bracts. Heads 1in.–1½in. in diameter, viscid; involucral bracts subulate-lanceolate, often acuminate, pubescent, the outer often recurved. Rays numerous, spreading. Appendages of the anthers very short. Achene silky.—Handbk. 131. *C. robusta,* Buch. in Trans. N.Z.I. xix. (1886) 215, t. 18.

NORTH Island: Te Moehau, Cape Colville, *Adams!* Hikurangi, *Colenso, Lee!* Ruahine and Tararua Ranges. SOUTH Island: common in the mountains of Marlborough, Nelson, Canterbury, and Westland. Western mountains of Otágo, *Buchanan!* 2,000ft. to 5,000ft. Dec., Jan.

Var. **petiolata.** Leaves oblong-lanceolate, narrowed into a distinct petiole above the expanded sheath, glabrous above, almost silvery beneath. Heads small.

Kelly's Hill, Westland, *Petrie!* Approaches states of *C. discolor.*

13. **C. Lindsayi,** *Hook. f., Handbk.* 132. A robust much-branched plant, often forming large clumps. Stems very stout, woody, prostrate. Leaves numerous, 3in.–7in. long, ½in.–1in. broad, coriaceous but not thick, linear-oblong, subacute or obtuse, gradually narrowed below and suddenly expanding into a grooved imbricating petiole 1in. long, white beneath with appressed tomentum, often longitudinally furrowed above; midrib and veins obvious beneath. Peduncles axillary, numerous, 2in.–8in. long, slender, flexuous, glabrate or pubescent above; bracts linear-acuminate. Heads 1in.–2in. in diameter; involucral bracts linear-subulate, acuminate, unequal. Rays spreading, rather distant; tube thickened. Anthers obtuse at the base. Achene silky.—Lindsay, Contrib. to N.Z. Bot. t. 3. *Erigeron Bonplandii,* J. Buch. in Trans. N.Z.I. xix. (1886) 213.

SOUTH Island: Otago: sea-cliffs, mouth of the Clutha and Catlins Rivers; probably not uncommon westward to Waikawa and Fortrose. Jan., Feb.

A beautiful littoral species, flowering profusely. On the face of cliffs, where it was collected by Lindsay, the leaves rarely exceed 2in.–3in. in length, but when growing under sheltering scrub on the tops of the cliffs it often forms masses 6ft. in diameter or more. The reported habitats at Mount Bonpland and Lake Harris are erroneous.

14. **C. Sinclairii,** *Hook. f., Handbk.* 132. Stems branched below. Branches slender or stout, prostrate. Leaves 1in.–3in. long or more, ¼in.–⅔in. broad, linear-oblong or linear-obovate-spathulate, obtuse or subacute, serrulate, membranous or rarely subcoriaceous, glabrous above, white beneath with closely-appressed tomentum, or rarely glabrous on both surfaces; midrib obvious beneath; sheath membranous, shining. Scape slender, 3in.–8in. high; bracts few, linear-acuminate. Heads ¾in.–1½in. in diameter; involucral bracts subulate-acuminate, tips usually recurved, pilose. Rays ⅓in.–1in. long, spreading. Achene silky.

SOUTH Island: not unfrequent in the mountains from Cook Strait to Otago, 2,000ft. to 4,000ft. STEWART Island: summit of Mount Anglem, 3,300ft., *T. K.* Dec., Jan.

Perhaps best distinguished by the membranous dull-green leaves with thin appressed tomentum beneath. I find the sheaths invariably glabrous. Mr. Cockayne sends a form from Mount Fyffe with serrated very coriaceous oblong bracts.

15. **C. verbascifolia,** *Hook. f., Fl. N.Z.* i. 121. Leaves oblong-lanceolate or oblong-spathulate, 8in.–18in. long including the broad sheathing petiole, which is about the length of the blade, 1in.–3in. broad, coriaceous, acute or subacute, sometimes crenulate, thickly clothed with soft white or ferruginous tomentum beneath; veins prominent on both surfaces. Scapes tomentose, exceeding the leaves; bracts numerous, linear-spathulate, tomentose, 2in.–4in. long. Head 2in. in diameter; involucral bracts woolly or cottony, narrow-linear-subulate. Rays slender. Tube of corolla thickened below. Achene glabrous.—Handbk. 132.

SOUTH Island: Canterbury, *J. B. Armstrong.* Otago: Horse Ranges, *T. K.*, 1871; Flag Swamp, Macrae's, and other places in the north-eastern portion of the district, 80ft. to 2,000ft. *Petrie!* CAMPBELL Island, *Lt. Rathouis!* Jan., Feb.

The Canterbury habitat requires confirmation.

16. **C. Traversii,** *Hook. f., Handbk.* 134. A stout tufted species. Leaves erect, 8in.–14in. long including the sheath, which is about one-third of the entire leaf, oblong or oblong-lanceolate, brownish-green above, densely clothed beneath with rich velvety ferruginous tomentum except the deep purple midrib. Sheaths and midrib above clothed with snow-white tomentum. Scapes 12in.–18in. long, densely clothed with ferruginous tomentum; bracts few, obtuse. Heads 1in.–2in. in diameter; involucral bracts numerous, linear-lanceolate, acute, or the outer obtuse, thickly covered with soft velvety tomentum. Rays narrow. Achene glabrous.

SOUTH Island: Nelson: Mount Arthur; Mount Peel, &c.; Raglan Mountains; *Cheeseman.* Discovery Peaks, *Travers!* Valley of Waiau-ua and its tributaries; Mount Captain Range, Amuri; *T. K.* Canterbury (?), *J. B. Armstrong.* 3,800ft. to 4,800ft. *Brown mountain daisy.* Jan.

A beautiful species, the rich brown velvety tomentum of the leaves and scapes forming a pleasing contrast with the silky snow-white villous tomentum of the sheath and the rich purple of the midrib.

17. **C. Brownii,** *F. R. Chapm. in Trans. N.Z.I.* xxii. (1889) 444. Leaves 7in.–9in. long including the petiole, 1½in.–2in. broad, broadly lanceolate, acute, entire, narrowed into the broad sheathing petiole, sparingly clothed with brown pubescence above, whitish with rather loose tomentum beneath. Scapes equalling the leaves, villous; bracts linear, obtuse, villous. Involucral bracts numerous, linear-acuminate, villous or cottony. Rays very numerous, narrow. Achenes pilose.

SOUTH Island: Otago: between Lake Manapouri and Smith Sound, *Chapman, Matthews!* Hector Mountains; Clinton Pass; Te Anau; *Petrie*. Jan.

I have only seen a single specimen of this handsome plant, and that in a very imperfect state. Considerable allowance must be made for the brief diagnosis. It seems nearly related to *C. verbascifolia*, but the leaves and involucral bracts are very different. Unfortunately, not even a leaf of the type specimen was preserved by Mr. Chapman.

18. **C. Rutlandii,** *T. Kirk in Trans. N.Z.I.* xxvii. (1894) 329. Leaves with the grooved petioles 5in.–12in. long, 1in.–2½in. broad; petiole shorter than the blade, broad, clothed with snow-white tomentum; blade broadly ovate-lanceolate, acute or slightly acuminate, silvery beneath; margins entire or denticulate, often revolute; midrib and lateral nerves obvious beneath. Scape erect, 6in.–10in. high, hoary; bracts narrow-linear, purplish. Heads 1in.–1½in. in diameter; involucral bracts narrow linear-acuminate, the outer cottony. Rays numerous. Achenes silky.

SOUTH Island: Mount Stokes, *J. MacMahon! T. K.* Jan.

A handsome species, allied to *C. petiolata*, from which it differs in the white tomentum of the petiole, the acute broad leaves, satiny beneath, and the very numerous chaffy involucral bracts, which are not recurved.

19. **C. petiolata,** *Hook. f., Handbk.* 134. A slender or stout tufted species. Leaves 3in.–12in. long including the petiole; blade 2in.–8in. long, ¾in.–2½in. broad, oblong linear-oblong or oblong-lanceolate, membranous or coriaceous, entire, obtuse or acute, glabrate or villous above, whitish with appressed tomentum beneath, or rarely glabrous on both surfaces; petiole longer or shorter than the blade, grooved, usually purple, expanded into a short broad sheath below, laxly tomentose on the margins. Scapes 1 or several, 4in.–16in. long, villous; bracts few, linear, slender. Heads 1½in.–2½in. in diameter; involucral bracts linear-subulate, densely tomentose or rarely glabrate. Rays ¾in. long. Achene glabrous or silky.

Var. **rigida.** Leaves oblong-lanceolate or ovate-lanceolate, rigid and coriaceous; tomentum white or ferruginous beneath; midrib obvious on both surfaces, purple. Head larger. Rays sometimes 1in. long. Achene perfectly glabrous.

Var. **membranacea.** Leaves narrowed at both ends, acute, membranous, glabrate or glabrous. Scape pubescent; veins obvious below. Head small; involucral bracts glabrate or glabrous.

Var. **cordatifolia.** Leaves oblong or ovate-oblong, usually cordate at the base, green and opaque above, densely clothed with ferruginous tomentum beneath. Scape slender. Achene glabrous.—*C. cordatifolia*, Buch. in Trans. N.Z.I. xi. 427, t. 18.

SOUTH Island: Nelson: Amuri, *T. K.* Not unfrequent in mountain valleys, Canterbury and Westland, *Travers!* *Sinclair* and *Haast!* Otago: Clinton Valley, 1,800ft. to 4,000ft, *Petrie.* Dec., Jan. Var. *rigida*: STEWART Island, *T. K.*; descends to sea-level. Var. *cordatifolia*: Mount Starveall, Nelson, *Mackay! Bryant!*

Nearly related to *C. Rutlandii*, which is distinguished by the broader leaves with their satiny undersurfaces.

20. **C. Mackaui,** *Raoul, Choix.* xix. *t.* 18. Leaves 6in.–14in. long, 2in.–3in. broad, broadly lanceolate-acuminate, entire, membranous, narrowed into broad sheaths 1in.–2in. long, cottony on the inner surface; midrib pubescent or slightly tomentose beneath; veins obvious on both surfaces. Scape 12in.–18in. high, stout, sparingly clothed with cottony tomentum; bracts few, distant or numerous, glabrate, acuminate, each with a broad base and obvious midrib. Head 2in. in diameter; involucral bracts herbaceous, lanceolate-acuminate, glabrate, 1in.–1½in. long. Rays narrow, acuminate, spreading. Disk-florets very numerous. Corolla-tube thickened at the base. Anther-cells obtuse at base. Achene glabrous.—Hook. f., Fl. N.Z. i. 122; Handbk. 133.

SOUTH Island: Mount Fyffe, Kaikoura, *H. B. Kirk!* (identified from leaves only). Mount Herbert, Banks Peninsula, *Raoul, W. Gray!* Jan., Feb.

A handsome species, distinguished from all others by the broad acuminate glabrate leaves and large heads with acuminate rays. Raoul's plate is excellent, but the bracts are much shorter than in my specimens, in which they are from 3in.–8in. long, leafy and imbricating.

21. **C. spectabilis,** *Hook. f., Fl. N.Z.* i. 122, *t.* 33. A rather small but very stout species, forming broad patches, the stems with the leaf-sheaths 1in.–2in. in diameter. Leaves 3in.–5in. long, ½in.–¾in. broad, linear-oblong, acute or obtuse, slightly narrowed at the base, very thick and coriaceous, entire or serrulate, glabrous and longitudinally furrowed above, clothed beneath with closely-appressed white or primrose-coloured tomentum; sheath equalling the blade or nearly so, shaggy on both surfaces with silky white tomentum. Scape exceeding the leaves, rather stout, with few linear bracts, hoary or villous. Head 1in.–1½in. broad; involucral bracts subulate-acuminate, outer recurved, glabrate, cottony or woolly. Rays numerous, very short, narrow, scarcely spreading. Corolla-tube thickened at base. Achene glabrous.—Handbk. 134. *C. Ruahinensis*, Col. in Trans. N.Z.I. xxvii. (1894) 388.

NORTH Island: Hikurangi, East Cape, *Colenso.* Tongariro, Ruapehu, *T. K.* Mount Egmont, *Buchanan!* Castlepoint, &c., *T. K.* SOUTH Island: common in all mountain districts, 300ft. to 4,000ft. *Leather-plant. Puhaeretaiko.* Jan., Feb.

Best distinguished by the glabrous achenes and very coriaceous linear leaves, which are scarcely longer than the snow-white sheath. The compact tomentum of the undersurface of the leaves is used for lamp-wicks.

22. **C. coriacea,** *Raoul in Ann. Sc. Nat. ser. III.* ii. (1844) 119. Leaves 8in.–20in. long, ¾in.–3in. broad, lanceolate, acute, coriaceous, longitudinally furrowed, narrowed into rather short broad woolly or cottony sheaths, clothed with a fine pellicle of matted hairs above and white with appressed silvery tomentum beneath. Scapes 6in.–30in. high, erect, hoary or cottony; bracts numerous, linear. Head 1½in.–3in. in diameter or more; involucral bracts numerous, subulate-acuminate, glabrate or cottony. Rays very numerous,

spreading, 1½in. long. Achene linear, compressed, pubescent.—Hook. f., Fl. Antarc. i. 36; Fl. N.Z. i. 121, t. 32; Handbk. 132. *C. Martinii*, Buch. in Trans. N.Z.I. xix. (1886) 213. *Aster coriaceus*, G. Forst., Prod. n. 297; A. Rich., Fl. Nouv.-Zel. 250; A. Cunn., Precurs. n. 439.

Under cultivation this plant varies to a remarkable extent. Two forms may be specially mentioned :—

(*a.*) **foliacea.** Scape with crowded elongated foliaceous bracts often 6in. long.

(*b.*) **corymbifera.** Scape much branched above. Heads forming a loose corymb.

Var. **ensata.** Leaves 8in.–12in. long, ½in. broad, ensiform from the tip of the sheath or sometimes lanceolate, acute, the pellicle on the upper surface sometimes 0. Scapes very slender, irregularly branched; bracts numerous, linear. Heads smaller. Approaching *C. longifolia.*

I have only seen cultivated specimens of this form, and doubt its occurrence in the wild state.

NORTH Island : Tararua Range and East Cape coast, *Buchanan.* SOUTH Island : common in mountain-valleys, 1,000ft. to 4,000ft. *White mountain daisy.* Dec., Jan. Var. *ensata :* Lake Harris, *H. Matthews !*

A magnificent plant. The appressed tomentum can be entirely detached from the under-surface of the leaf; it is used by the shepherds for lamp-wicks.

23. **C. Monroi,** *Hook. f., Handbk.* 133. A tufted species, with spreading leaves 3in.–7in. long, ⅓in.–¾in. broad, linear-oblong or lanceolate, usually acute, longitudinally wrinkled above and clothed with a silvery pellicle of fine matted hairs, white with appressed tomentum beneath except the midrib; sheath about half as long as the blade, more or less clothed with snow-white cottony wool. Scapes numerous, 8in.–12in. high, rather stout, white with cottony wool; bracts numerous, linear. Head 1in.–1½in. in diameter; involucral bracts numerous, linear-subulate, the outer recurved, glabrate or tomentose. Rays numerous, spreading. Achenes glabrous or rarely hispidulous.

NORTH Island : Bay of Islands, *Colenso.* Mount Manaia and Whangarei Harbour, *T. K.* SOUTH Island : Marlborough : Upper Awatere, *Monro, T. K.* Mountains of Nelson, *Cheeseman.* Canterbury : Hopkins River, Mount Cook, &c. Sea-level to 4,500ft. Dec., Jan.

A handsome species, in some respects intermediate between *C. coriacea* and large forms of *C. longifolia.* I have not seen ripe achenes of the North Island plant.

24. **C. Adamsii,** *T. Kirk in Trans. N.Z.I.* xxvii. (1894) 329. Leaves 6in.–12in. long or more including the petiole, ½in.–¾in. broad, membranous but firm, narrow, oblong-lanceolate, sparingly denticulate, nearly glabrous above, white with loose tomentum beneath; midrib prominent beneath; petiole one-fourth to one-third the length of the blade, cottony, grooved. Scape 6in.–15in. long, slender, often flexuous, more or less clothed with loose tomentum; bracts few, short, acute. Heads 1in.–1½in. in diameter; involucral bracts lanceolate-acuminate, the outer cottony, the inner glabrate or glabrous. Rays few, spreading. Achene glabrous.

NORTH Island: Auckland: Cape Colville Peninsula; Castle Rock, Coromandel; *Cheeseman !* Kauoranga Creek, Whakairi, *Adams ! Cheeseman !* 2,000ft.

The flat membranous leaves distinguish this plant from *C. longifolia* and some forms of *C. discolor.*

25. **C. longifolia,** *Cass. in Dict. Sc. Nat.* 37, 256. Solitary or densely tufted. Leaves numerous or few, never rigid, 1in.–16in. long, $\frac{1}{12}$in.–⅜in. broad,

narrow-linear, acute or very acuminate, silvery and furrowed or wrinkled above, clothed with soft appressed tomentum beneath; midrib distinct beneath; margins usually strongly revolute; sheaths long or short, usually broader than the blade, membranous, cottony or silky. Scapes equalling or exceeding the leaves, slender or rather stout, hoary or cottony; bracts few or many, linear. Head ½in.–1¼in. broad; involucral bracts lanceolate-subulate, glabrate or the outer cottony, often black at the tips. Rays scarcely spreading, usually narrowed towards the apex. Achene glabrous, rarely puberulous or silky, furrowed.—DC., Prod. v. 209; Hook. f., Handbk. 134; Benth., Fl. Austr. iii. 489. *C. gracilenta*, Hook. f., Fl. Antarc. i. 35 (note); Fl. N.Z. i. 122; Fl. Tasm. i. 181; Gaud. in Freyc. Voy. 470, t. 91. *C. asteliaefolia*, Hook. f., Fl. Antarc. *l.c.* *C. setacea*, Col. in Trans. N.Z.I. xxi. (1888) 88. *Aster celmisia*, F. Muell., Fragm. v. 84. *Aster gracilenta*, Banks and Sol. MS. and Icon.

Var. **alpina.** Rootstock woody, much branched. Branches very short. Leaves ½in.–1½in. long, 1/20in.–1/16in. wide; margins slightly recurved; sheaths with a few scattered hairs. Scape strict, 2in. long. Involucral bracts glabrate. Achenes glabrous. In alpine bogs.

The following forms depend largely upon situation and may be easily recognised, although from their not being permanent they are not eligible for even varietal rank:—

(1.) *gracilenta.* Usually solitary. Leaves slender, erect, soft.—*Aster gracilenta*, Banks and Sol. MS. and Icon.

(2.) *asteliaefolia.* Leaves recurved, stouter, rigid; margins very revolute.

(3.) *major.* Leaves very long, ¼in.–⅜in. broad, very acuminate; margins much or slightly recurved.

(4.) *graminifolia* (sp.), Hook. f., Fl. Antarc. i. 35 (in note). Leaves very slender, narrow, flat, acute. Scapes very slender.—*C. perpusilla*, Col. in Trans. N.Z.I. xxii. 470.

NORTH Island: rare and local in the north; Great Barrier Island, Bay of Islands, Manukau Harbour. Common from the head of the Hauraki Gulf to STEWART Island. Sea-level to 5,500ft. Dec. to Feb. Also in Australia.

I am indebted to the Director of the Royal Gardens, Kew, for one of Cunningham's specimens of *C. graminifolia* from the Bay of Islands.

The most abundant species in the colony, and the most widely distributed, exhibiting a wonderful range of variation; *gracilenta*, the form figured by Banks and Solander, being the most common. The broad-leaved forms approach *C. Monroi*, but the leaves differ from that species in having longer and more acuminate leaves with more or less revolute margins. Var. *alpina* approaches large forms of *C. laricifolia*.

26. **C. laricifolia,** *Hook. f., Fl. N.Z.* ii. 331. Much branched at base. Branches 1in.–2in. long, densely leafy. Leaves crowded, linear, sessile, ⅓in.–1in. long, 1/20in. broad, spreading or recurved, pungent, glabrate or glabrous and shining above or rarely cottony, clothed with silvery or shining tomentum beneath. Scape filiform, very slender, 2in.–4in. long, glabrate or cottony; bracts 1 or 2, very short. Head ½in. in diameter; involucral bracts few, linear, subulate, cottony, purple. Rays few, short. Achene hispid.—Handbk. 135.

SOUTH Island: Marlborough: Awatere and Mount Stokes, *T. K.* More frequent in the mountains of Nelson, Westland, Canterbury, and Otago. 3,000ft. to 6,000ft. Dec., Jan.

Cottony specimens from the Heaphy River have the leaves longitudinally furrowed above, approaching *C. longifolia*.

27. **C. linearis,** *J. B. Armst. in Trans. N.Z.I.* xiii. (1880) 337. Stem sparingly branched. Branches short, rather stout. Leaves densely crowded,

radiating, 1in.–5in. long, ⅛in.–⅙in. broad, narrow-linear, acute, not pungent, margins strongly recurved, longitudinally furrowed, and clothed with a shining pellicle of fine matted hairs above, beneath with lax short whitish tomentum except the stout midrib; sheaths one-third to half the length of the leaves and rather broader, flat, membranous, grooved, clothed with shining silky tomentum on the outer surface. Scapes 1 or several, 1in.–10in. high, silky or tomentose, glistening; bracts few or many, subulate-acuminate, springing from a dilated sheathing-base. Heads 1in. in diameter; involucral bracts lanceolate-acuminate, with a distinct midrib, the outer cottony and recurved, the inner glabrate or glabrous. Rays numerous, short. Achene pubescent.

SOUTH Island: Canterbury, *J. B. Armstrong.* Otago: Maungatua, *Petrie!* STEWART Island: summit of Mount Anglem; Rakiahua; Taylor's Lookout; *T. K.* Fraser Peaks, *Petrie* and *Thomson!* 600ft. to 3,5000ft. Dec., Jan.

Allied to *C. longifolia*, but distinguished from all forms of that species by the acute shining rigid leaves, the recurved margins of which usually hide the stout midrib.

28. **C. Lyallii,** *Hook. f., Handbk.* 133. A strong tufted plant, with radiating slightly-curved ensiform leaves, 9in.–18in. long, ¼in.–½in. broad at the base, tapering to the acute but not pungent apex, rigidly coriaceous, finely grooved above or on both surfaces, white beneath with thinly-appressed tomentum; sheath a quarter the length of the blade and broader, clothed with snow-white wool. Scape 10in.–20in. high, rather slender, white with dense lax tomentum; bracts numerous on the upper part of the scape, linear, tomentose. Head 1¼in.–2in. in diameter; involucral bracts very numerous, subulate-acuminate, the outer recurved, glabrate or more or less cottony. Rays rather short. Achene silky.

SOUTH Island: from Cook Strait to Foveaux Strait. Common in mountain districts. 1,000ft. to 4,500ft. *Blunt-leaved Spaniard.* Dec., Jan.

29. **C. Armstrongii,** *Petrie in Trans. N.Z.I.* xxvi. (1893) 269. A strong tufted species, with radiating straight ensiform leaves 6in.–18in. long, ¼in.–⅓in. broad or more, rigidly coriaceous, acute, tapering towards the apex, slightly narrowed just above the sheath, margins not recurved when fresh, serrulate above, upper surface clothed with a silvery pellicle, lower with satiny appressed tomentum; midrib very stout, with several parallel longitudinal veins on each side; sheaths rather broad, the length of the blade, clothed with snow-white cottony tomentum on both surfaces. Scapes shorter or longer than the leaves, tomentose; bracts linear. Head 1in.–1½in. in diameter; involucral bracts subulate-acuminate, the outer recurved, glabrous or cottony at the margins. Rays very narrow. Achenes glabrous.—*C. gladiata,* T. Kirk MS.

SOUTH Island: Nelson: Heaphy River, *Dall!* Canterbury: Arthur's Pass, *Enys* and *Kirk* (1876). Westland: Kelly's Hill, *Petrie* and *Cockayne* (1893)! 3,000ft. to 4,000ft. Jan.

Nearly related to *C. Lyallii*, but easily distinguished by the perfectly straight leaves with stout midribs and silvery undersurface. The stems with the short sheaths are sometimes 3in. in diameter.

30. **C. viscosa,** *Hook. f., Handbk.* 133. Viscid in all its parts. Stem sparingly branched. Branches very short, stout, with leaf-sheaths 1in.–1½in. in

diameter. Leaves densely crowded, erect, rigid, 3in.–4in. long, linear, ¼in. broad, coriaceous, acute or subacute, glabrous and longitudinally grooved above, white with closely appressed tomentum beneath, viscid; sheaths about 1in. long, rather broader than the blade, glabrous. Scapes exceeding the leaves, stout, pubescent or tomentose; bracts numerous, short, acute. Head 1½in. in diameter, viscid; involucral bracts narrow-subulate, crowded, short, pilose. Rays short, spreading. Achene silky.

SOUTH Island: Nelson: Mount Starveall, *W. H. Bryant!* Canterbury: Mount Torlesse, *Haast!* Mount Enys, &c., *Enys!* Otago: Lake district, *Hector* and *Buchanan!* Central and western Otago. 4,000ft. to 6,000ft. Jan.

Easily recognised by the erect grooved viscid leaves and numerous short crowded bracts.

31. **C. MacMahoni,** *T. Kirk in Trans. N.Z.I.* xxvii. (1894) 327. A tufted species forming small patches; the whole plant densely clothed with long white silky hairs. Leaves numerous, densely rosulate, 1in.–1½in. long including the short petiole, ¼in.–⅜in. broad, lanceolate or oblong-lanceolate, acute or subacute, thick, 5–7-nerved beneath. Scape 3in.–5in. long; bracts very numerous, narrow-linear, obtuse. Heads ¾in.–1in. in diameter; involucral bracts numerous, the outer acute, villous, inner acuminate, more or less clothed with short hispid brown hairs. Rays broad. Achene hispid.

SOUTH Island: Mount Stokes, Marlborough, *MacMahon!* 3,800ft. Rare. Jan.

A charming little species, most nearly related to *C. incana*, but easily recognised by the long silky tomentum, the crowded linear leaves with short petiolar sheaths and very numerous bracts.

32. **C. parva,** *T. Kirk in Trans. N.Z.I.* xxvii. (1894) 328. A small densely-tufted species. Leaves, including the petiole, ¾in.–1½in. long, ⅛in.–¼in. broad, membranous, tomentose; blade linear-lanceolate or lanceolate, acute, scarcely denticulate, rigid or subcoriaceous, white with appressed tomentum beneath; margins sometimes revolute; midrib obvious beneath. Scape 1½in.–3in. high, extremely slender, naked or with 2–4 filiform bracts on dilated bases. Head small; involucral bracts numerous, narrow-linear, acute, the outermost chaffy, sparingly pilose with distinct midrib. Rays few, narrow. Achenes hispid.

SOUTH Island: Nelson: source of the Heaphy River, *J. Dall!* Dec., Jan.

A curious little plant, scarcely larger than the English daisy. Most nearly related to *C. spectabilis*.

33. **C. Hectori,** *Hook. f., Handbk.* 135. "A densely-tufted species forming extensive patches." Leaves ½in.–¾in. long, ⅛in.–¼in. broad, densely crowded, imbricate, coriaceous, linear-spathulate spathulate-oblong or obovate, obtuse or rarely subacute, slightly recurved, clothed with a silvery pellicle of matted hairs above, cottony or sparingly tomentose beneath; sheath slightly expanded, very short. Scape 4in.–5in. high, villous with few or many linear silky bracts. Head 1in. in diameter; involucral bracts few, linear-lanceolate, subacute, pubescent. Rays short. Achene silky.

NORTH Island: Tararua Range, *Budden!* SOUTH Island: Canterbury: Mount Cook, *Dixon!* Mount Brewster, *Haast!* Otago: Mount Alta, *Hector* and *Buchanan!* The Remarkables, Humboldt Mountains, Mount Tyndall, &c., *Petrie!* 4,500ft. to 6,000ft.

A very distinct species, of which I have only seen poor specimens.

34. **C. sessiliflora,** *Hook. f., Handbk.* 135. Much branched below. Branches short, stout, with the leaves 1in.–1$\frac{1}{2}$in. in diameter, forming broad masses 2in.–3in. high. Leaves excessively crowded, yellowish-grey, hoary or silky above and beneath, $\frac{1}{4}$in.–1in. long, usually shorter and narrower than the scarious silky or villous sheath, $\frac{1}{16}$in.–$\frac{1}{10}$in. wide at base, strict, coriaceous, subacute or obtuse, flat above, concave beneath; margins revolute. Head terminal, sessile, usually sunk among the apical leaves, $\frac{1}{2}$in.–1in. in diameter; involucral bracts few, subulate-acuminate, scarious, the outer cottony and revolute, the inner glabrate. Rays few, spreading. Achene silky. Pappus scarcely barbed.

Var. **pedunculata.** Heads on terminal silky or villous ebracteate peduncles $\frac{3}{4}$in.–1$\frac{1}{2}$in. long.

SOUTH Island: frequent in mountain districts, 2,500ft. to 6,000ft. STEWART Island: summit of Mount Anglem, *T. K.* Dec., Jan.

In all probability the pedunculate form is not uncommon, although I have only once met with it. Possibly the peduncle may not be developed until after the florets have withered. The Stewart Island plant is identified in the absence of flowers, and may be different.

35. **C. argentea,** *n. s.* Habit of *C. sessiliflora*, but more slender, branches $\frac{1}{4}$in.–$\frac{1}{2}$in. in diameter, erect, forming small patches 2in.–5in. high. Leaves densely imbricating all round the stem, erect, $\frac{3}{16}$in.–$\frac{4}{16}$in. long, shorter and narrower than the silky scarious sheath, subulate, subacute or acute, margins involute, glabrate or glabrous, silvery or shining on both surfaces. Head sessile, $\frac{3}{8}$in.–$\frac{1}{2}$in. in diameter, deeply sunk among the apical leaves; involucral bracts few, linear-acuminate, glabrate or ciliate. Rays few, short. Achenes pubescent.—*C. sessiliflora*, var. *minor*, Petrie in Trans. N.Z.I. xv. (1882) 359.

SOUTH Island: Maungatua, Southland, *Petrie!* STEWART Island: Mount Anglem; Rakiahua, Smith's Lookout, *T. K.* 500ft. to 3,300ft. Dec., Jan.

I formerly agreed with Mr. Petrie in considering this to be a variety of *C. sessiliflora*, but an examination of recent specimens grown under varying conditions has convinced me of its distinctness. It differs essentially in the very short silvery involute leaves, fewer bracts, much shorter rays, and especially in the shorter achenes.

36. **C. bellidioides,** *Hook. f., Handbk.* 135. Rootstock much brauched. Branches usually slender, creeping. Leaves spreading, usually crowded, linear-oblong or linear-spathulate, rounded at the apex, $\frac{1}{4}$in.–$\frac{1}{2}$in. long, $\frac{1}{8}$in.–$\frac{1}{6}$in. broad, narrowed into a broad petiole, cottony on the inner surface, entire or obscurely toothed, green, subcoriaceous, glabrous. Scapes slender, terminal, 2in.–5in. high, clothed with numerous leafy bracts. Head $\frac{3}{4}$in. in diameter; involucral bracts green or purple, acute or subacute, linear-lanceolate. Rays numerous, short. Corolla-tube pilose. Achene silky.

SOUTH Island: Nelson: Tarndale, *Sinclair!* Mount Peel, *Cheeseman.* Amuri, *T. K.* Canterbury: Mount Torlesse, *Haast!* Bealey, Arthur's Pass, &c., *Enys* and *Kirk.* Macaulay and Hopkins Rivers, &c., *Haast!* Ashburton Mountain, *Potts!* Westland, *Cockayne!* Otago: Mount Ida, above Arrowtown, Mount Tyndall, *Petrie!* 2,500ft. to 5,500ft. Dec., Jan.

Usually on watery shingle. In sheltered places luxuriant shoots are developed with leaves 1in. long, the elongated petioles being clothed with long tomentum.

37. **C. glandulosa,** *Hook. f., Fl. N.Z.* i. 124. Solitary or tufted. Leaves few, membranous, ½in.–1in. long, ovate or oblong-spathulate, narrowed into a short winged petiole with a slightly expanded base, acute or apiculate, with 2–4 acute teeth on each side, glabrate or clothed with minute glandular pubescence. Scape slender, glandular-pubescent, 1in.–5in. high, with few linear-acuminate bracts. Head ½in.–¾in. in diameter; involucral bracts in 2 series, linear-subulate, glabrate and ciliate or cottony. Rays few or many, broad. Achene pubescent or silky.—Handbk. 136. *C. membranacea,* Col. in Trans. N.Z.I. xxii. (1889) 470.

NORTH Island: Tongariro, *Colenso!* Ngauruhoe and Ruapehu, *T. K.* Taupo, *G. Mair! Hill!* SOUTH Island: Mount Arthur, *Spencer!* Canterbury and Westland: Arthur's Pass, &c., *Enys* and *Kirk.* Mount Cook, &c., *Haast!* Otago: near Mount Aspiring, &c., *Petrie!* 1,500ft. to 4,000ft. Dec., Jan.

38. **C. vernicosa,** *Hook. f., Fl. Antarc.* i. 34, *t.* 26, 27. Densely tufted, glabrous in all its parts, shining. Leaves very numerous, rosulate, sessile, 1in.–4in. long, linear, obtuse, ⅛in.–¼in. broad, rigid, thick and coriaceous, flat or convex above, slightly expanded at the base, sometimes serrulate towards the apex; midrib stout and obvious beneath. Scapes usually numerous, flexuous, 1in.–8in. high, green or purple, clothed with rather broad subimbricating bracts. Heads 1in.–1½in. in diameter; involucral bracts lanceolate, acute, often ciliate. Rays white, spreading. Disk-florets purple. Corolla-tube pilose. Achene hispid.—Handbk. 136.

AUCKLAND and CAMPBELL Islands: sea-level to 1,200ft.; abundant. Nov., Dec.

The expanded base of the leaf is usually 3-ribbed, but never forms a membranous sheath.

39. **C. Chapmani,** *T. Kirk in Gard. Chron. ser. III.* ix. (1891) 731. Solitary or tufted. Leaves sessile, rosulate, 3in.–5in. long, ½in.–¾in. broad, broadly lanceolate, narrowed into a short broad sheathing-base, obtuse or subacute, serrate, scarcely coriaceous, glabrous above and longitudinally furrowed, pubescent or sparsely tomentose beneath, longitudinal veins prominent below. Scapes 6in.–10in. high, sparingly tomentose; bracts rather numerous, linear, the lower obtuse, the upper acuminate. Head 1½in.–2in. in diameter or more; involucral bracts with a distinct midrib, linear-lanceolate, acute, glabrate or sparsely cottony. Rays spreading. Disk-florets purple. Corolla-tube pilose. Achenes densely hispid.—*C. Campbellensis,* Chapman in Trans. N.Z.I. xxiii. 407.

CAMPBELL Island: Perseverance Harbour, *Chapman* and *Kirk.*

The leaves are sessile by a broad clasping base and destitute of a membranous sheath.

* ERIGERON, Linn.

Receptacle flat or convex. Involucral bracts numerous, narrow, nearly equal, 2–3-seriate. Ray-florets numerous, female, in 2 or more rows, all ligulate, narrow, or the inner short and filiform. Disk-florets hermaphrodite, tubular. Anthers obtuse below. Style-arms narrow, tips lanceolate, papillose outside. Achenes flattened, usually 2-nerved, pubescent. Pappus equal or unequal. Herbs, with radical and alternate usually sessile leaves. Heads solitary or paniculate or in elongated corymbs. Ray-florets white or purple.

1ft.–5ft. high. Female ray-florets all ligulate .. .. * *E. Canadensis.*
1ft.–3ft. high, hirsute. Female ray-florets mostly filiform and tubular. Ligule very short or 0 .. .. * *E. linifolius.*

* **E. Canadensis,** *L., Sp. Pl.* 863. Annual. Stem 1ft.–5ft. high, simple or rarely branched, glabrous or sparingly hispid. Leaves 1in.–3in. long, the radical often toothed, the upper entire. Panicle terminal, oblong. Heads on slender peduncles, numerous, very small; involucral bracts narrow, acute, nearly glabrous. Ray-florets short, scarcely exceeding the involucre. Ligule shorter than the tube, white. Disk-florets fewer.

NORTH and SOUTH Islands: abundantly naturalised, especially in the North. *Canadian flea-bane.* Feb. to April. North America.

* **E. linifolius,** *Willd., Sp. Pl.* iii. 1955. Annual, erect, 1ft.–3ft. high. A coarser plant than the preceding, with longer usually softer hairs, or sometimes scabrid, pubescent. Radical leaves oblong, toothed or pinnatifid; petioles short; cauline leaves entire or remotely toothed, acute, sessile or petioled. Heads on slender peduncles, paniculate; involucre broadly campanulate, pubescent or almost villous; involucral bracts few, mostly 2-seriate, acute. Female florets shorter than the pappus, mostly tubular, the outer minutely ligulate. Disk-florets few. Achenes small, pubescent.

NORTH Island: naturalised from Waitemata to the North Cape; Wellington. Rare. Feb. to April. Tropics.

8. VITTADINIA, A. Rich.

Involucre obconic or campanulate; involucral bracts in about 3 series, narrow, acute, margins scarious. Receptacle pitted. Ray-florets numerous, short, ligulate, spreading. Disk-florets usually fewer, hermaphrodite, 5-lobed or -toothed. Anthers obtuse below. Style-arms with subulate tips, flattened. Achenes narrow, compressed, sometimes ribbed. Pappus unequal. Suffrutescent plants, with alternate leaves and solitary terminal flower-heads.

A small genus of about 13 species, most plentiful in the Sandwich Islands. Four or five species are found in Australia, one of which is indigenous in New Zealand.

1. **V. australis,** *A. Rich., Fl. Nouv.-Zel.* 251. Stems 3in.–10in. high, subcrect or ascending from a woody base. Branches numerous or few, usually slender. The whole plant glabrate, pubescent, hispid or glandular. Leaves $\frac{1}{3}$in.–$\frac{1}{2}$in. long, obovate-spathulate or linear-cuneate, entire or 3-toothed or -lobed at the apex; margins lobed or entire; petioles very short. Head terminal, solitary on slender peduncle; involucre broadly obconic; involucral bracts in 2 or 3 series, linear-subulate, acute, glabrate or pubescent. Rays white, usually exceeding the pappus, narrow, spreading. Disk-florets longer than the involucre. Achene subulate, puberulous, pubescent or almost hispid, smooth or with 5–8 fine ribs on each face. Pappus equal or unequal, longer than the achene.—A. Cunn., Precurs. n. 441; DC., Prod. v. 280; Hook. f., Handbk. 136; Benth., Fl. Austr. iii. 490. *V. triloba* and *V. cuneata*, DC., Prod. v. 281. *V. scabra* and *V. cuneata*, Hook. f., Fl. Tasm. i. 181, 182.

Eurybiopsis australis, Hook. f., Fl. N.Z. i. 125. *E. Hookeri,* F. Muell. in Linnaea, xxv. 453. *Aster humilis,* Banks and Sol. MSS. and Icon.

NORTH and SOUTH Islands: from the Great Barrier Island and Auckland Isthmus to Southland.

Var. **dissecta.** Stems numerous, erect, more woody, often fastigiate. Leaves usually pinnatifid; segments 3-lobed. Heads with purple rays.—Benth., Fl. Austr. iii. 491. Nelson: naturalised; *Cheeseman.*

Var. **linearis.** Stems numerous, slender, hispid, 6in.-8in. high. Leaves 1in.-1½in. long or more, $\frac{1}{20}$in. broad, acute, irregularly toothed or lobed or pinnatifid; segments acute, hispid. Heads larger. Rays purple. Otago: Taieri Plains, naturalised; *Petrie!*

Var. **erecta.** Stems erect, 1½ft.-2ft. high. Branches slender, numerous, dense. Leaves linear-spathulate, entire or with 3 small teeth, pubescent or hoary. Rays purple. Wellington: Orongorongo to White Rock. Marlborough: near Blenheim, &c. Nelson: abundant. Otago: Taieri Plains, naturalised; *T. K.*

9. HAASTIA, Hook. f.

Flower-heads large, solitary, deeply sunk amongst the apical leaves. Involucre broadly campanulate; involucral bracts in 2 or more series, imbricate, very narrow, free or connate at the base, the outer densely villous, the inner shorter, membranous at the tips, glabrous. Receptacle flat, papillose. Ray-florets female, in 2 or more series. Corolla tubular, very short, mouth minutely crenulate or dentate; style-arms very long, exserted, tips papillose. Disk-florets numerous, hermaphrodite, tubular, mouth expanded, minutely 5-toothed; style-arms shorter. Anthers obtuse at the base. Achene glabrous, subterete, narrow, even or obtusely 4–5-ribbed. Pappus 1-seriate; bristles exceeding the achene, thickened upwards. Perennial pulvinate or caespitose herbs. Rootstock woody, sometimes much branched. Leaves alternate, densely woolly, patent or recurved or densely imbricating.

A singular genus, sometimes forming large dense woolly shapeless masses, often several feet in diameter; the "vegetable sheep" of the shepherds. Although closely related to *Gnaphalium* and *Helichrysum,* it is referred to *Asteroidea* on account of the tailless anthers. Named in honour of the late Sir Julius von Haast, F.R.S., &c. The genus is endemic.

Fulvous. Leaves most densely imbricate. Pappus-hairs free	1. *H. pulvinaris*
Fulvous or rufous, woolly. Leaves laxly imbricate, recurved. Pappus-hairs united below	2. *H. recurva.*
White, cottony. Leaves laxly imbricate, erect or suberect. Pappus-hairs free ..	3. *H. Sinclairii.*
Fulvous, woolly, small. Leaves densely imbricate	4. *H. Greenii.*

1. **H. pulvinaris,** *Hook. f., Handbk.* 156. Rootstock stout, woody. Plant forming dense balls or pulvinate masses, 3in.–6ft. across. Branches with the leaves ¾in.–1in. in diameter. Leaves thickly clothed with fulvous or buff-coloured wool, patent, densely imbricate, broadly obcuneate, with rounded lobed crenulate tips, membranous, 3-nerved. Heads ⅓in.–½in. broad; involucre hemispherical; involucral bracts 1–2-seriate, acuminate. Ray-florets very short, tubular, minutely 5-toothed; style long, exserted. Disk-florets hermaphrodite; mouth expanded; teeth larger; style scarcely exserted. Achene glabrous, not ribbed. Pappus-hairs free.—Hook. f. in Hook. Ic. Pl. t. 100, 3.

SOUTH Island: Marlborough: Mount Mouatt, Awatere, *Sinclair! T. K.* Kaikoura Mountains, *Buchanan!* Nelson: Mount Captain, Amuri, *T. K.* Discovery Peaks, *Travers!* Mountains above the Wairau Gorge, *Sinclair! Cheeseman.*

One of the most remarkable plants in the world; it is impossible to thrust one's finger between the branches of even a young plant. *Vegetable sheep.* Jan., Feb.

2. **H. recurva,** *Hook. f., Handbk.* 156. Branches laxly tufted, 3in.–8in. long, forming loose rounded masses or spreading in a fan-shaped manner, with the leaves ½in.–¾in. in diameter; the whole plant densely clothed with long fulvous or rufous wool. Leaves ½in.–¾in. long, loosely imbricating, obovate-spathulate, recurved from the middle, woolly on both surfaces. Head ⅓in.–¾in. in diameter; involucre turbinate; involucral bracts few, recurved at the tips. Style-arms of female florets often exceeding the pappus. Pappus-hairs coherent at the base.

SOUTH Island: chiefly on loose shingle. Marlborough: Kaikoura Mountains, *Buchanan!* Wairau Gorge, Mount Peel, *Cheeseman*. Tarndale, *Sinclair!* Discovery Peaks, *Travers!* Canterbury: Mount Torlesse, *Haast!* Mount Enys and other mountains above the Broken River Basin, *Enys!* Mount White, *J. B. Armstrong.* 4,000ft. to 6,500ft. Jan., Feb.

Usually a deep-rooted species with slender branches, woody at the base. When growing it may easily be mistaken for a dead or withered plant.

3. **H. Sinclairii,** *Hook. f., Handbk.* 156. Rootstock slender. Branches lax, ascending or erect, as thick as a crow-quill, 3in.–10in. long. Leaves erect or rarely patent, laxly imbricating, ½in.–¾in. long, ⅓in.–½in. broad, oblong-obovate or broadly oblong-obovate, obtuse or rarely subacute, 5–7-nerved, clothed with white cottony wool on both surfaces. Heads ¾in.–1½in. in diameter; involucral bracts more acuminate, the outer lanuginous below, the inner longer, ciliate. Achene obscurely grooved, glabrous. Pappus-hairs free. —Hook. f. in Hook. Ic. Pl. t. 1003.

SOUTH Island: Marlborough: Upper Awatere, *Sinclair!* Nelson: St. Arnaud Mountains, *Cheeseman.* Mountains above the Wairau Gorge, *Sinclair!* Mountains above the Rainbow, *Bryant!* Canterbury: Mount Torlesse, Mount Enys, and other mountains above the Broken River basin, *Enys!* Mount Darwin, Mount Cook, and Mount Brewster, *Haast!* Otago: Mount Alta, *Hector* and *Buchanan!* Hector Mountains, Mount Bonpland, Mount Arnauld, &c., *Petrie!* 4,000ft. to 6,500ft. Jan., Feb.

Easily distinguished by the white cottony leaves, furrowed on the upper surface, and the large heads.

4. **H. Greenii,** *Hook. f. MS.* Densely tufted, about 2in. high. Stems with leaves ⅓in. in diameter. Leaves densely imbricating all round the stem, ⅕in. long, obovate-cuneate, rounded at the tip, thickly clothed on both surfaces with long straight hairs which meet beyond the margin and completely hide the leaves. Flowers unknown.

SOUTH Island: Mount Cook, 6,500ft., *Rev. W. S. Green.*

In the absence of flowers it is impossible to determine the genus with certainty. I have only seen a solitary stem.

10. GNAPHALIUM, Linn.

Involucre turbinate ovoid or campanulate; involucral bracts in several rows, narrow, scarious, imbricate, sometimes with short spreading apices. Receptacle naked, flat or conical. Florets all fertile, the outer in 2 or more rows, filiform, numerous. Disk-florets fewer, hermaphrodite, funnel-shaped, 5-toothed. Anthers with slender tails. Achenes oblong or obovate, pubescent.

Pappus-hairs in 1 series, often slightly united at the base. Annual or perennial herbs, more or less cottony or tomentose. Leaves alternate. Flower-heads small, solitary fascicled or corymbose.

A large genus, comprising many species of wide distribution.

I. *Heads corymbose. Inner involucral bracts radiating, white.*

Stems robust, erect or suberect. Leaves 3-nerved below .. .. .. 1. *G. Lyallii.*
Stems slender, usually prostrate. Leaves short, obscurely 3-nerved above .. 2. *G. trinerve.*
Stems slender, suberect. Leaves with midrib prominent beneath .. .. 3. *G. Keriense.*
Leaves less than $\frac{1}{20}$in. broad .. .. var. *linifolia.*

II. *Heads fascicled, corymbose, or spicate.*

Leaves woolly or cottony. Heads fascicled and corymbose .. .. .. 4. *G. luteo-album.*
Leaves woolly beneath. Heads fascicled below, spicate above .. .. * *G. purpureum.*

III. *Heads terminal, solitary or very rarely in fascicles of 2–4.*

Leaves radical, cottony or silky. Head large .. .. 5. *G. Traversii.*
Leaves radical, spathulate, acute. 1in.–2in. high .. .. 6. *G. paludosum.*
Leaves filiform; margins recurved. 1in.–2in. high .. .. 7. *G. minutulum.*
Tufted. Stems short. Leaves silky on both surfaces .. .. .. 8. *G. nitidulum.*

IV. *Heads aggregated, forming a globose capitulum, usually subtended by linear bracts.*

Annual, much branched, slender .. .. 9. *G. japonicum.*
Perennial, emitting short scions .. .. 10. *G. collinum.*

1. **G. Lyalli,** *Hook. f., Fl. N.Z.* i. 137. Stems 1ft.–2ft. high, suffrutescent or suffruticose, rather stout, branched from the base, cottony above. Leaves close-set below, upper distant, spreading, sessile, 2in.–4in. long, ¼in.–⅔in. broad, oblong-lanceolate or narrow oblong-lanceolate, glabrous above, white with thinly-appressed tomentum beneath, narrowed below, but scarcely petiolate. Heads forming dense terminal corymbs 2in.–4in. across; pedicel tomentose; involucral bracts numerous, scarious, white, with short claws, radiating. Outer florets in 2 or 3 series, female. Disk-florets few. Achene glabrous. Pappus-hairs few, equal.—Handbook 152. *G. adhaerens*, Col. in Trans. N.Z.I. xvii. 244.

NORTH Island: Ruahine Range, *Andrew!* Maohonga, *Petrie!* Rimutaka Range, *T. K.* Opunake, *T. K.* SOUTH Island: Nelson: Golden Bay, *Lyall.* West Coast, *T. K.* Canterbury: "alpine," *J. B. Armstrong.* Westland: Otira Gorge, *T. K.* Westport: *Gaze!* Sea-level to 2,500ft. Nov., Dec.

A beautiful plant. The lower leaves are obviously 3-nerved beneath. I have not seen Canterbury specimens, and doubt the occurrence both of this and *G. trinerve* in alpine localities, as stated by J. B. Armstrong.

2. **G. trinerve,** *G. Forst., Prod. n.* 289. Stems 1ft.–2ft. long, suffrutescent, branched, prostrate or suberect, glabrate or more or less cottony. Leaves distant, flat, obovate or spathulate-lanceolate, rounded at the apex, acute or apiculate, glabrous and obscurely 3-nerved above, white beneath. Heads 5–10 on terminal or axillary bracteate tomentose peduncles; involucral bracts numerous, radiating, shortly clawed, the outer densely clothed at the base with woolly tomentum. Achenes small, glabrous. Pappus-hairs few.—Willd., Sp. Pl. i. 1871; A. Rich., Fl. Nouv.-Zel. 239; A. Cunn., Precurs. n. 455; DC., Prod. vi. 236; Hook. f., Fl. N.Z. i. 138; Handbk. 153.

NORTH Island: Rimutaka Range, *T. K.* SOUTH Island: west coast of Nelson, *T. K.* Westland, *T. K.* Canterbury: "alpine," *J. B. Armstrong.* Otago and Southland: not unfrequent. STEWART Island, *T. K.* Sea-level to 2,000ft. Nov., Dec.

Best distinguished from the preceding by the bracteate peduncles, tomentose outer bracts of the involucre, and the shorter leaves, which are never 3-nerved beneath. The corymbs are sometimes reduced to a single head.

3. **G. Keriense,** *A. Cunn., Precurs. n.* 454. Suffrutescent stems prostrate or decumbent, 3in.–9in. long or high. Leaves mostly close-set, spreading, ½in.–2½in. long, $\frac{1}{20}$in.–⅓in. broad, narrow-linear, oblong-lanceolate or oblong-spathulate, acute or almost acuminate, glabrate or glabrous above, white beneath with the midrib prominent, sessile. Heads ⅓in. in diameter, in many-flowered terminal corymbs. Peduncles bracteate or leafy; pedicels slender, cottony. Involucral bracts numerous, clawed, radiating, outer cobwebby at base. Achenes minute, glabrous, shorter than the pappus-hairs.—Hook. f., Fl. N.Z. i. 138; Handbk. 153. *G. Novae-Zelandiae,* Sch. Bipont in Bot. Zeit. iii. (1845) 176. *Helichrysum micranthum,* A. Cunn. in DC. Prod. vi. 189.

Var. **linifolia,** Hook. f., Handbk. 153. Stems decumbent or erect. Leaves ½in.–1in. long, $\frac{1}{20}$in.–$\frac{1}{10}$in. broad, acute, sometimes cottony above; margins slightly revolute. Peduncles usually ebracteate; pedicels capillary.—*G. subrigidum,* Col. in Trans. N.Z.I. xvii. 244.

NORTH Island: frequent by the margins of streams and on wet cliffs from Mangonui to Cook Strait. SOUTH Island: Nelson, *Travers.* Dusky Bay, *Lyall.* Var. *linifolia:* NORTH Island: usually on dry banks or cliffs; local. Sea-level to 1,800ft.

The leaves of this species are extremely variable in size, although fairly constant in shape.

4. **G. luteo-album,** *L., Sp. Pl.* 851. Annual or biennial. Stems simple or branched from the base, and with the leaves densely clothed with white cottony tomentum, 8in.–18in. high. Lower leaves petiolate, narrow-linear or oblong-spathulate, the upper linear or lanceolate, sessile. Heads ⅙in.–¼in. long, fascicled or corymbose, dusky-brown or yellowish; involucre nearly globose; bracts numerous, scarious, shining; tips incurved, obtuse. Female florets very numerous. Disk-florets few. Achene finely punctulate.—A. Cunn., Precurs. n. 431; Hook. f., Fl. N.Z. i. 139; Handbk. 154.

Var. **compactum.** Branched from the base. Branches filiform, cottony, prostrate or sub-erect, 1in.–3in. long. Leaves ¼in.–½in. long, oblong or oblong-spathulate, cottony on both surfaces. Heads 1–4 in terminal fascicles; involucral bracts fewer. Lake Lyndon, *Enys* and *Kirk.*

Common from the KERMADEC Islands and the North Cape to STEWART Island and the AUCKLAND Islands. Sea-level to 3,000ft. Dec. to March.

In all temperate and warm countries.

* **G. purpureum,** *L., Sp. Pl.* 584. Annual or biennial. Stems simple or branched from the base, 6in.–9in. high, white with silvery wool. Leaves oblong-spathulate, sessile or narrowed into a broad petiole, obtuse or the upper rarely linear, clothed on both surfaces with white tomentum or green above. Heads in dense sessile clusters in the axils of the upper leaves, and forming a leafy spike at the apex of the stem; involucral bracts linear-oblong, woolly but scarious, pale-brown, the inner often purplish. Female florets very numerous. Disk-florets 2 or 3. Pappus-hairs united in a ring at the base.

NORTH Island: naturalised in several localities from the North Cape to the Auckland Isthmus. Wellington: on ballast; rare. *Purple cudweed.* Jan. to March. North America.

Doubtless introduced from Australia, where it is abundantly naturalised.

5. **G. Traversii,** *Hook. f., Handbk.* 154. Perennial, tufted, 1in.–3in. high. Leaves radical, linear-spathulate or linear-obovate, ½in.–2in. long, clothed on both surfaces with white cottony wool; petiole long or short. Scapes erect, simple, with 1 or more linear bracts. Head solitary, terminal; involucre broadly ovoid; involucral bracts in 3 series, linear, hyaline, the outer cottony at the base, brown or brownish-red, mostly obtuse, the inner pale. Female florets numerous; hermaphrodite few. Achene glabrous or puberulous. Pappus-hairs cohering at the base, very fine.—Benth., Fl. Austr. iii. 655; F. Muell., 2 Syst. Cens. Austr. Pl. 134.

Var. **McKayi.** Tufted, branched from the base, ½in.–1in. high. Leaves laxly imbricate, spreading, obovate or linear-obovate, white with laxly-appressed tomentum on both surfaces, minutely apiculate. Heads small, solitary, sessile or on very short scapes; involucral bracts few, the outer obtuse, the inner acute.—*Raoulia McKayi*, Buch. in Trans. N.Z.I. xiv. 354, t. 34.

SOUTH Island: not unfrequent in the mountains from Marlborough and Nelson to Southland; 1,000ft. to 5,000ft. Jan., Feb. Also in Australia.

Best distinguished by the rather large solitary head and almost silvery tomentum.

6. **G. paludosum,** *Petrie in Trans. N.Z.I.* xxii. (1889) 441. Solitary or tufted, 1in.–2in. high. Leaves all radical, ½in.–1½in. long, narrow linear-spathulate; blade rather shorter than the petiole, subacute or acute, glabrous or glabrate above, white with appressed tomentum beneath; midrib prominent; margins flat or slightly recurved. Scapes few, cottony, capillary, equalling the leaves during anthesis, but elongating to two or three times their length in fruit; bracts few, small. Head terminal, solitary; involucre ⅛in.–$\frac{3}{16}$in. in diameter; involucral bracts few, the outer cottony, the inner glabrous, linear-subulate, dark at the tips, scarious. Female florets very numerous. Achene clavate, compressed, pilose. Pappus-hairs very fine, coherent at the base.

NORTH Island: chiefly in swamps. Rangipo Plain and Ruahine Mountains, *Petrie!* SOUTH Island: Arthur's Pass, Canterbury, *T. K.* Otago: Maniototo Plain, Dunstan Mountains, Rock and Pillar Range, Hector Mountains, *Petrie!* 1,000ft. to 4,000ft. Jan.

Probably common, but easily overlooked, from its small size and inconspicuous appearance.

7. **G. minutulum,** *Col. in. Trans. N.Z.I.* xxii. (1889) 473. Rootlets long, wiry. Leaves 6–10, all radical, linear-filiform, ½in.–¾in. long, $\frac{1}{10}$in. broad, obtuse, glabrous above, white and cottony beneath; margins incurved. Scape solitary, ¾in.–1in. long, slender, almost setaceous, finely cottony with 1–2 acuminate bracts. Head solitary, terminal, 2½ lines in diameter; involucral bracts about 10, ovate-lanceolate, acuminate, scarious, glabrous. Achene $\frac{1}{24}$in. long, terete, slightly papillose. Pappus-hairs scabrid below, obtuse.

NORTH Island: boggy ground near eastern base of Tongariro, *Hill.*

This plant may be a state of *G. paludosum*, Petrie, described in the same volume of Trans. N.Z.I., but not having seen specimens I am unable to determine the point.

8. **G. nitidulum,** *Hook. f., Handbk.* 154. "A small densely-tufted species, covered with appressed silky shining yellowish tomentum. Leaves closely imbricated at their bases, above spreading, flat, ⅓in. long, linear, obtuse; lower one-third membranous, glabrous, upper two-thirds densely silky. Heads terminal, solitary, large, ½in. broad, on very short slender peduncles; involucral scales in

2 series, erect, linear, hyaline, shining, with pale erect tips. Florets not seen." —Doubtfully referred to *Raoulia Planchoni*, Hook. f. in Gen. Pl. ii. 307.

SOUTH Island: "Nelson Mountains, *Sinclair*. Wairau and Clarence Valleys, 3,500ft., *Travers*."

I have copied the above from the Handbook, not having seen New Zealand specimens. The plant has not been collected since its original discovery, nearly fifty years ago.

9. **G. japonicum,** *Thunb., Fl. Jap.* 311. Annual, erect, 1ft.–1½ft. high. Stems stiff, rather slender, usually branched, more or less tomentose. Leaves linear-lanceolate or oblong-spathulate, sessile or the lower petiolate, glabrous or glabrate above, white with cottony tomentum beneath. Heads small, ⅙in.–¼in. long, aggregated in dense globose fascicles, usually subtended by 3–6 linear spreading foliaceous bracts, axillary or terminal; the terminal heads sometimes ¾in. in diameter on long naked peduncles, the axillary smaller on short peduncles; involucres oblong, densely clothed with white wool at the base; involucral bracts erect, hyaline, linear, the outer obtuse, the inner acute. Outer florets numerous. Disk-florets 1–3. Achene slightly compressed, glabrous. Pappus-hairs slender, almost free at the base.—Miq., Prolus. Fl. Jap. 109; F. Muell., Fragm. v. 150; Benth., Fl. Austr. iii. 653. *G. involucratum*, G. Forst., Prod. n. 291; A. Rich., Fl. Nouv.-Zel. 241; A. Cunn., Precurs. n. 453; Hook. f., Fl. N.Z. i. 139; Handbk. 155. *G. lanatum*, G. Forst., Prod. n. 290; A. Cunn., Precurs. n. 452. *G. virgatum*, Banks and Sol. MSS.

KERMADEC Islands and the North Cape to Southland; STEWART Island; CHATHAM Islands. Sea-level to 2,500ft. Nov. to Jan. Also in Norfolk and Lord Howe's Island, Australia, the Eastern Archipelago, and Japan.

A common weed.

10. **G. collinum,** *Lab., Nov. Holl. Pl.* ii. 44, *t.* 189. Perennial, emitting scions; 2in.–7in. high. Leaves lanceolate-spathulate, acuminate, the lower petiolate, cottony on both surfaces or glabrate or glabrous above. Heads smaller and less compact than in *G. japonicum*, with fewer subtending bracts or 0; involucres broader. Florets and achenes similar to the preceding, or the achenes puberulous.—DC., Prod. vi. 235; Hook. f., Fl. N.Z. i. 139; Handbk. 155; Benth., Fl. Austr. iii. 654. *G. simplex*, A. Rich., Fl. Nouv. Zel. 237; A. Cunn., Precurs. n. 451.

Var. **obscurum.** Forming matted grey patches ½in.–1in. high. Leaves radical or cauline, ¼in.–¾in. long, linear, glabrate above, grey beneath. Scape when present leafy. Head solitary or 2–4, in terminal fascicles; involucral bracts 3–10, linear-lanceolate, obtuse, scarious. Achene linear, not compressed. Pappus-hairs cohering at the base.

Var. **monocephalum.** Leaves all radical, ¼in. long, obtuse or acute. Head solitary, sessile or sometimes on a slender filiform scape ¾in.–1in. high in fruit.

KERMADEC Islands, and North Cape to STEWART Island; CHATHAM Islands. Sea-level to 4,500ft. Jan. to March. Also in Australia. Var. *obscurum* and var. *monocephalum*, Broken River Basin, 2,000ft. to 3,000ft., *T. K.*

In luxuriant forms the heads are sometimes crowded on the upper part of the scapes. Resembles *G. purpureum* in habit, but may be distinguished by the broader involucres and narrower and more acute leaves. This species is united with *G. japonicum* by F. von Mueller.

11. RAOULIA, Hook. f.

Involucre oblong, ovoid or hemispherical; bracts imbricating in 2 or more series, the inner often with white radiating tips. Receptacle very narrow, flat or convex, naked. Outer florets in 1 or 2 series, female, filiform. Disk-florets numerous, hermaphrodite; mouth funnel-shaped, 5-toothed. Anthers with capillary tails. Style-branches exserted, usually truncate. Achenes small, oblong or ovate. Pappus-hairs in 1 series, sometimes thickened at the tips. Very short densely-tufted perennial herbs, sometimes with a woody rootstock. Leaves minute, densely imbricating. Heads solitary, terminal, sessile or shortly pedunculate.

A small genus, confined to New Zealand, with the exception of 1 or possibly 2 species found in Australia and Tasmania. The species are distributed from the Thames Goldfield to Stewart Island, but are most numerous in the South Island. I fully agree with Sir Joseph Hooker that it is impossible to maintain *Raoulia* as a separate genus, there being no character to distinguish it from *Gnaphalium* and *Helichrysum* except the peculiar habit. The species in Section I. should be referred to *Gnaphalium*, those in Section II. to *Helichrysum*. For the present they are retained under *Raoulia* simply as a matter of convenience.

Name, in honour of M. M. E. Raoul, surgeon to "L'Aube" during her expedition to New Zealand in 1840–41.

I. Leptopappus. Pappus-hairs copious, slender, not thickened upwards.

* *Involucral bracts without white or radiating tips.*

Leaves spathulate, erect or recurved, clothed above with snow-white wool. Florets about 12 .. .. 1. *R. australis.*

Leaves mostly acute, linear, recurved, glabrous or glabrate or densely tomentose 2. *R. tenuicaulis.*

Leaves broadly ovate-subulate, glabrate. Florets 6–8 .. .. .. 3. *R. Haastii.*

Leaves patent, linear, silky, involute .. .. 4. *R. Monroi.*

** *Involucre with the inner bracts white-tipped and radiating.*

Stems slender. Leaves loosely imbricate, glabrous or glabrate .. .. 5. *R. glabra.*

Stems short. Leaves densely imbricate, silky or glabrous .. .. .. 6. *R. subsericea.*

II. Imbricaria. Pappus-hairs few in 1 series, thickened upwards.

* *Involucral bracts without white or radiating tips.*

Leaves subulate, spreading, rigid, glabrous .. .. 7. *R. subulata.*

Leaves densely imbricate, linear, obovate-oblong, rounded, hidden by the dense tomentum .. .. 8. *R. eximia.*

Leaves ovate, $\frac{1}{12}$in. long, imbricate, silvery .. .. 9. *R. Hectori.*

** *Involucres with white or radiating tips.*

Leaves closely imbricate, $\frac{1}{8}$in.–$\frac{1}{4}$in. long, linear-oblong .. .. .. 10. *R. grandiflora.*

Stems irregularly branched, stout. Leaves recurved at the tips. Florets about 40 .. .. 11. *R. Petriensis.*

Forming dense patches. Leaves densely imbricate, hairy on both surfaces. "Florets crimson" .. .. 12. *R. rubra.*

Forming large dense woolly masses. Leaves densely imbricate, with a tuft of hairs at the tip, cottony beneath .. .. 13. *R. mammillaris.*

Forming large dense masses. Leaves densely imbricate, linear-oblong, truncate at the apex, hairy above, glabrous beneath .. .. .. 14. *R. Goyeni.*

Leaves most densely imbricate; tips silky with appressed silvery tomentum .. 15. *R. bryoides.*

Forming dense woolly masses. Leaves densely imbricate, subacute, hidden by the dense tomentum .. .. *R. Brownii.*

Leaves broadly cuneate or flabellate, truncate at the apex, corrugated beneath, with a broad margin; hairy above .. .. *R. Buchanani.*

1. **R. australis,** *Hook. f., Fl. N.Z.* i. 135. Densely tufted. Stems prostrate, 1in.–4in. long, giving off numerous short erect branches 1in. high or less. Leaves densely or laxly imbricate, spathulate, erect or recurved, $\frac{1}{10}$in.–$\frac{1}{6}$in. long, rounded at the apex, usually clothed on both surfaces with silky appressed tomentum. Heads $\frac{1}{10}$in.–$\frac{1}{5}$in. long; outer involucral bracts spathulate, tomentose; inner linear, scarious with membranous margins, rarely with dark tips. Florets about 12–18, the hermaphrodite disk-florets fewer than the outer female or very rarely equalling them. Achene angled, glabrate or glabrous. Pappus-hairs numerous, slightly pilose, not thickened at the tips.—Handbk. 148; Raoul, Choix. t. 15. *R. McKayi*, Buch. in Trans. N.Z.I. xiv. 354, t. 34.

Var. **albo-sericea** (*sp.*), Col. in Trans. N.Z.I. xx. (1887) 195. Leaves linear-spathulate or rotund. Florets few, 6–8.

Var. **apice-nigra** (*sp.*), T. Kirk in Trans. N.Z.I. xi. 164. Leaves excessively woolly. Outer involucral bracts black at the tips.

Var. **lutescens.** Stems short, often forming compact masses. Leaves $\frac{1}{18}$in.–$\frac{1}{12}$in. long, densely imbricating. Heads small; involucral rays yellow. Florets very short.

NORTH Island: from the East Cape to Cook Strait, but often local. SOUTH Island: from Nelson to Foveaux Strait. STEWART Island: Mason Bay, very rare, *T. K.* Sea-level to fully 5,000ft. Dec., Jan.

Extremely variable in habit, but very constant in its floral characters. The preponderance of female florets shows a close relationship to *Gnaphalium*. The gradual transition of the apical leaves into involucral bracts is very instructive. Some forms of var. *apice-nigra* approach *R. Monroi*, but the leaves are never linear.

2. **R. tenuicaulis,** *Hook. f., Fl. N.Z.* i. 135, t. 36A. Stems laxly tufted, slender, 1in.–10in. long, prostrate, creeping, much branched, branches ascending. Leaves spreading or loosely imbricating, sometimes recurved, $\frac{1}{12}$in. long, concave above, linear-oblong, acuminate or in the autumn often obovate-spathulate, apiculate, glabrate, pubescent or densely clothed on both surfaces with cinereous tomentum. Heads oblong, $\frac{1}{10}$in. long, sessile, tubular; involucral bracts in 3 series, the inner linear-oblong, with obtuse brown tips. Florets 12–18; outer, female, predominating. Disk-florets tubular, 5-toothed. Achenes glabrate or puberulous.—Handbk. 148.

NORTH and SOUTH Islands: from the Thames Goldfield to Southland; chiefly in river-beds. Sea-level to 2,000ft. Nov. to Jan.

Var. **pusilla.** Glabrous in all its parts; much branched. Branches $\frac{1}{8}$in.–$\frac{1}{4}$in. high, the latter including the head. Leaves on sterile shoots $\frac{1}{10}$in.–$\frac{1}{12}$in. long, linear, concave, acute or subacute, patent; on flowering shoots shorter, with broad membranous bases. Female florets filiform, twice as many as those of the disk. NORTH Island: forming broad mossy patches on the Rimutaka Range. Nov.

The only species with strongly-marked dimorphic leaves.

3. **R. Haastii,** *Hook. f., Handbk.* 148. Glabrous or nearly so, densely tufted. Stems much branched, rather stout, prostrate. Branches $\frac{1}{2}$in.–1in. high. Leaves coriaceous, densely imbricate, $\frac{1}{16}$in. long, erect patent, broadly ovate-subulate, sheathing at the base, obtuse, rarely with a few loose hairs but never cottony or silky. Head sessile amongst the apical leaves, $\frac{1}{8}$in.–$\frac{1}{6}$in. long; involucral bracts few, in 2 series; the inner linear-oblong, scarious, not radi-

ating. Florets 6–8; those of the disk 2–3, narrow, tubular, minutely 5-toothed. Achene puberulous.

SOUTH Island: Nelson: Waiau-ua River, *Sinclair! Travers.* Amuri, *T. K.* Canterbury: Kowhai River, *Haast!* Beds of Thomas, Porter, Cass, and Bealey Rivers and Upper Waimakariri, *T. K.* Otago, *Buchanan!* Kyeburn Crossing, *Petrie!* 1,100ft. to 3,000ft. Nov., Dec.

Distinguished from all forms of *R. tenuicaulis* by the short obtuse coriaceous leaves and few florets.

4. **R. Monroi,** *Hook. f., Handbk.* 148. Stems wiry, creeping, with long filiform rootlets. Branches ascending, silky, 1in.–3in. high, slender. Leaves $\frac{1}{8}$in.–$\frac{1}{4}$in. long, linear or linear-spathulate, obtuse or rounded at the apex, laxly or densely imbricate, rarely distant, tips patent or recurved, clothed with white or greyish appressed tomentum on both surfaces, margins mostly involute. Heads oblong, narrow, $\frac{1}{6}$in. long; involucral bracts in 3–4 series, the outer short, ovate-oblong, tomentose; the inner linear-oblong, scarious, with deep brown obtuse tips. Florets 15–20, female predominating. Disk-florets funnel-shaped, 5-toothed. Achenes oblong, sparingly papillose.

SOUTH Island: Nelson: Cheviot, *Haast!* Waihopai Valley, *Monro.* Canterbury Plains, &c., *T. K.* Otago: more frequent; attaining its southern limit on Riverton beach, *T. K.* Sea-level to 3,500ft. Nov. to Jan.

A very constant and distinct species, frequent on sand-dunes.

5. **R. glabra,** *Hook. f., Fl. N.Z.* i. 135. Stems prostrate, elongated, slender, branching, forming lax patches 2in.–2ft. in diameter. Branches ascending, 2in.–3in. high. Leaves laxly imbricate, erect patent or sometimes distant and spreading, $\frac{1}{6}$in. long, linear or linear-oblong, acute or subacute or obtuse, 1-nerved, glabrous or glabrate. Heads $\frac{1}{4}$in.–$\frac{1}{3}$in. in diameter; involucral bracts in 3–4 series, the outer short, broadly-ovate, acuminate, the inner linear-oblong, scarious, the innermost with white radiating tips. Receptacle conical, hispid. Florets 30–50; female in 2 series, 12–21. Disk-florets tubular; mouth 5-toothed. Achene puberulous. Pappus-hairs numerous, soft, not thickened upwards.—Handbk. 149.

NORTH Island: Tararua and Rimutaka Ranges; descending to within 100ft. above sea-level near the White Rock. SOUTH Island: mountain districts from Nelson to Southland. Ascends to 5,000ft. Dec., Jan.

In late shoots the apex is elongated, bearing smaller distant leaves resembling a bracteate peduncle. The leaves are sometimes tipped with a minute pencil of white hairs.

6. **R. subsericea,** *Hook. f.,* Fl. N.Z. i. 136. Resembling *R. glabra,* but more robust, densely tufted, forming smaller patches with shorter branches. Leaves rather longer and broader, closely imbricate, usually clothed on both surfaces with loose silvery tomentum, obtuse, rarely linear-spathulate, sometimes glabrate. Heads similar to *R. glabra,* but rather larger; involucral bracts broader. Receptacle conical, glabrate or glabrous. Florets fewer. Achene glabrous. Pappus-hairs slightly thickened upwards.—Handbk. 150.

SOUTH Island: common in mountain districts. 1,500ft. to 5,000ft. Dec., Jan.

Perhaps best distinguished from *R. glabra* by the silky leaves and pappus-hairs more or less thickened upwards.

7. **R. subulata,** *Hook. f., Handbk.* 149. Densely tufted, rigid, glabrous in all its parts. Stems $\frac{1}{2}$in.–1in. high. Leaves densely imbricate, $\frac{1}{4}$in. long or more, rigid, subulate, acuminate, 1-nerved. Heads $\frac{1}{6}$in.–$\frac{1}{5}$in. in diameter; involucral bracts in 2–3 series, scarious, linear-lanceolate, acuminate. Receptacle convex, hispid. Florets 20–25, those of the disk predominating; mouth almost campanulate, 5-toothed. Achene puberulous or silky. Pappus-hairs slightly thickened at the tips.

SOUTH Island: Nelson: mountains above Wairau Gorge, *Sinclair, Cheeseman!* Alps of Canterbury, *J. B. Armstrong.* Otago: Lake district, *Hector* and *Buchanan!* Old Man Range, Hector Mountains, Mount Pisa, Mount Tyndall, &c., *Petrie!* 4,000ft. to 6,000ft. Jan.

A very distinct species, not easily mistaken for any other. I have not seen Canterbury specimens.

8. **R. eximia,** *Hook. f., Handbk.* 149. Forming small or large densely-compacted grey tomentose hummocky masses on the mountains, from a few inches to several feet long, 1ft. high or more. Rootstock woody. Branches with the leaves fully $\frac{1}{4}$in. in diameter. Leaves in many series, densely imbricated all round the stem, $\frac{1}{10}$in.–$\frac{1}{8}$in. long, membranous, narrow obovate-oblong, bearing near the tips on both surfaces a dense tuft of long silky hairs meeting beyond the tip, which is completely hidden; lower three-quarters glabrous above, cottony or woolly beneath. Heads very numerous, sunk amongst the apical leaves; involucral bracts in 2 or 3 series; outer linear-spathulate, with a long scarious claw and a tuft of hairs on both surfaces at the very tip; inner linear-oblong, tips not radiating. Receptacle very small, narrow, convex. Female florets 6, hermaphrodite 4, tubular; teeth extremely minute. Achene with long erect silky hairs. Pappus-hairs very few, thickened above.

SOUTH Island: Canterbury: Mount Torlesse, *Haast! Enys* and *Kirk.* Mount Dobson, *Haast!* Otago: *Buchanan!* Mount Ida Range, *Petrie!* 4,000ft. to 5,000ft. *Vegetable sheep.* Jan.

Closely related to *R. mammillaris*, but distinguished by longer branches, more woolly leaves, and by the absence of white radiating involucral bracts, although in old specimens the scarious bracts sometimes show a tendency to curve outwards. The leaves are excessively woolly, giving a most deceptive appearance of softness to the mass, which in dry weather is as hard and unyielding as the rock on which it grows. The Mount Torlesse plant is rightly referred here, and I have no certain knowledge of any other stations for *R. eximia* than those already mentioned. It is desirable that some Canterbury botanist should visit the Ribbonwood Range, Mount Dobson, and Mount Arrowsmith, to determine the species reported from those localities.

Var. **lata.** Leaves less densely imbricate than in the typical form, $\frac{1}{15}$in. long, broadly oblong, concave above, rounded at the tip. Hairs long but less uneven than in the type. Heads not seen. This is only known to me from a specimen collected by Enys and forwarded by the Director of the Royal Gardens, Kew, without mention of locality, as *R. eximia*, with the following note by Mr. N. E. Brown: "This may be a variety of *R. eximia*, but I doubt it; the heads seem smaller, the hairs shorter, and the phyllotaxis different." In the absence of flower-heads I am unable to separate it from *R. eximia*.

9. **R. Hectori,** *Hook. f., Handbk.* 149. Stems prostrate, 1in.–3in. long, densely tufted, much branched. Branches erect or ascending, $\frac{3}{4}$in.–2in. high. Leaves densely imbricate, broadly ovate, obtuse, coriaceous, silvery on both surfaces except the membranous base, longitudinally grooved at the back. Heads sunk amongst the apical leaves, $\frac{1}{12}$in.–$\frac{1}{8}$in. in diameter; involucral bracts few, in 2 series, the outer cottony, the inner linear-oblong, scarious,

subacute or obtuse, glabrous. Receptacle conical, pubescent, scarcely pilose. Florets 10–12; 2–4 female. Disk-florets with a shortly campanulate mouth, 5-toothed. Achene silky. Pappus-hairs thickened upwards, barbellate.

SOUTH Island: Otago: Lake district, *Hector* and *Buchanan!* Mount St. Bathan's, Mount Pisa, Old Man Range, Ben Lomond, &c., *Petrie!* 4,500ft. to 6,000ft. Jan.

Var. **mollis**, Buch. MS. Leaves laxly imbricate, cottony at the base, broader, softer, more obtuse and spreading. Heads smaller. Florets 6–8. Achenes faintly ribbed, glabrous. Mount St. Bathan's, *Petrie!*

This pretty species has the habit of *R. Haastii*, but with more robust branches and silvery leaves. It is included in J. B. Armstrong's list of Canterbury plants, but his specimens belong to *R. australis*, var. *lutescens*.

10. **R. grandiflora,** *Hook. f., Fl. N.Z.* i. 136, *t.* 36B. Stems mostly simple, 1in. high or more, densely tufted, erect, forming small patches with strong wiry roots. Stem with the leaves $\frac{1}{3}$in.–$\frac{1}{2}$in. in diameter. Leaves densely imbricating all round the stem, $\frac{1}{6}$in.–$\frac{1}{4}$in. long, clothed with shining silvery tomentum, ovate-subulate, cottony at the base, rigid, striate or grooved beneath. Heads $\frac{1}{3}$in.–$\frac{2}{3}$in. in diameter, sunk amongst the apical leaves; receptacle convex, hispid; involucral bracts in 2 series, linear-oblong, the outer few, scarious, short, the inner white, spreading, subacute or acute, $\frac{1}{4}$in. long. Florets 25–30, 5–10 female, the remainder hermaphrodite, tubular; mouth campanulate. Anthers with rather long tails. Achene silky. Pappus-hairs thickened towards the tip.—Handbk. 150.

NORTH and SOUTH Islands: not unfrequent in mountain districts from the East Cape to Southland. 3,000ft. to 6,300ft. Jan.

It is impossible to separate this and other fine species from *Helichrysum* except by habit.

11. **R. Petriensis,** *T. Kirk in Trans. N.Z.I.* ix. (1876) 549. Laxly or densely tufted. Stems 2in.–5in. long, prostrate or suberect. Branches 1in.–1$\frac{1}{2}$in. long, ascending or erect. Leaves laxly imbricate, about $\frac{1}{10}$in. broad, coriaceous, broadly spathulate, erect patent, tips rounded, recurved, nerveless, clothed above with appressed white tomentum, greenish below. Heads $\frac{1}{4}$in.–$\frac{1}{3}$in. in diameter, sessile amongst the apical leaves; involucral bracts in 2 series, the inner scarious, linear, obtuse, a few with very short white radiating tips. Receptacle conical. Florets about 40; female filiform, 8–12. Disk-florets very narrow, tubular; mouth shortly campanulate, 5-toothed. Achene faintly striate, glabrous or puberulous. Pappus-hairs thickened upwards.

SOUTH Island: Otago: Mount Ida and Mount St. Bathan's, *Petrie!* 4,000ft. to 5,000ft. Jan.

A beautiful species, best distinguished by the recurved leaves, conical receptacle, and numerous florets.

12. **R. rubra,** *Buch. in Trans. N.Z.I.* xiv. (1881) 349, *t.* 30, *f.* 2. Densely tufted, forming hemispherical cushions or patches 6in.–12in. in diameter and 4in.–8in. high. Branches with the leaves $\frac{1}{6}$in.–$\frac{1}{5}$in. in diameter. Leaves in many series, densely imbricating, $\frac{1}{8}$in. long, broadly obovate-spathulate, rounded at the tips, rather membranous, the upper portion thickly clothed on both surfaces with long hairs slightly projecting beyond the margins, naked at

the base. Heads sessile, deeply sunk amongst the apical leaves, $\frac{1}{10}$in. in diameter; involucral bracts numerous, glabrous, narrow-linear, obtuse, the inner linear-spathulate with short radiating tips. Disk-florets broadly tubular, 5-toothed. Florets 10–14. Corolla dark-crimson. Achene pilose. Pappus-hairs thickened upwards, barbellate.

NORTH Island: Mount Holdsworth, Tararua Range, 4,500ft., *Arnold* and *Beck! Buchanan* and *Logan!* Jan.

I have not seen flowering specimens of this curious plant, and have therefore copied portions of Mr. Buchanan's description.

13. **R. mammillaris,** *Hook. f., Handbk.* Forming large densely-compacted rounded or amorphous knobby grey masses or patches, from a few inches to several feet in length and 1ft.–2ft. high. Branches very short, thick, with the leaves fully $\frac{1}{4}$in. in diameter. Leaves imbricated all round the stem in many series, densely compacted, spreading, $\frac{1}{14}$in.–$\frac{1}{10}$in. long, narrow obovate-spathulate or cuneate, rounded at the apex, membranous, cottony or woolly beneath, both surfaces bearing a dense tuft of short hairs just below the tip, produced beyond the margins, which are completely hidden. Heads deeply sunk amongst the apical leaves, $\frac{1}{6}$in. long. Florets about 10. Inner involucral bracts linear-oblong, scarious, with very short white acute radiating tips. Receptacle convex, naked. Achenes with a thickened areole at the base, compressed, broad, copiously clothed with long erect or spreading white silky hairs. Pappus-hairs few, thickened at the tips.

SOUTH Island: Nelson: Mount Starveall, *Bryant!* Canterbury: the plant found on the Ribbonwood Range, Mount Arrowsmith, and Mount Dobson by *Sinclair* and *Haast* probably belongs to this, but I have not seen specimens. 5,000ft. to 6,000ft. *Vegetable sheep.*

Perhaps best distinguished from *R. eximia* by the "knobby" habit, short branches, small leaves, and short radiating involucral bracts. Hooker describes the hairs of this species as not exceeding the tip of the leaf, but all the specimens seen by me have much longer hairs; those so named at Kew agree in this particular. Much confusion exists with regard to the habitats of this species and *R. eximia*. Haast's Mount Torlesse habitat certainly belongs to the latter, although it is the only station given in the Handbook for *R. mammillaris*; and it may well be that Sinclair and Haast's stations cited above are really identical with Mount Starveall; at any rate, until observed by Mr. Bryant on that mountain it had not been seen by any New Zealand botanist since its original discovery. A specimen named *R. mammillaris*, kindly given me by the Director of the Royal Gardens, is certainly *R. eximia*. I have not seen florets in a good condition.

14. **R. Goyeni,** *T. Kirk in Trans. N.Z.I.* xvi. (1883) 373. Forming hard dense compact rounded or amorphous much-branched greenish masses from a few inches to 2ft.–3ft. long, but rarely exceeding 6in. high. Rootstock stout, woody. Branches short, with the leaves fully $\frac{1}{4}$in. in diameter. Leaves densely imbricated in many series, extremely membranous, sessile by a broad base, narrow linear-oblong, $\frac{1}{7}$in.–$\frac{1}{5}$in. long, rounded, truncate at the apex with a narrow margin, slightly contracted at the middle; the apical portion of the upper surface clothed with a dense brush of short white uneven hairs slightly exceeding the margin, glabrous at the base. Heads deeply sunk amongst the apical leaves, short; involucral bracts in 1 or 2 series, the outer narrow linear-spathulate, with a scarious margin and tufts of hairs at the tip, the inner narrow-linear, scarious, mostly obtuse, a few with short scarcely radiating tips.

Receptacle convex. Florets few, female almost filiform, hermaphrodite with large funnel-shaped mouth and 5 strong teeth. Achene with a thickened areole, ribbed, hispid. Pappus-hairs few, thickened upwards, barbellate, stout.

STEWART Island: Mount Anglem, *T. K.* Rakiahua, *P. Goyen! T. K.* Smith's Lookout, *T. K.* 800ft. to 3,300ft. Jan.

Distinguished by the linear-oblong almost truncate leaves, with a narrow margin and tuft of hairs above. My flower-heads are not in good condition.

15. **R. bryoides,** *Hook. f., Fl. N.Z.* ii. 332. Rootstock stout and woody, with long wiry roots. Stems much branched below, ¾in.–3in. long, forming dense hard knobby hoary patches 2in.–6in. in diameter. Branches ½in.–1in. long or more, cylindric, with the leaves $\frac{1}{10}$in.–⅛in. in diameter. Leaves densely imbricate all round the stem, erect patent or rarely with the minute tips slightly recurved, narrow obovate-spathulate, $\frac{1}{12}$in.–$\frac{1}{10}$in. long; upper third coriaceous, clothed on both surfaces with appressed silvery wool, shining when young, cottony at the margin; lower two-thirds membranous, 1–3-nerved. Heads ¼in.–⅓in. in diameter, deeply sunk amongst the apical leaves; involucral bracts in 2 or 3 series, linear-oblong, membranous, with white radiating tips. Receptacle conical. Florets 9–13, female 2–5, hermaphrodite 5–9; mouth campanulate, 5-toothed. Achene clothed with long erect silky hairs, and a thickened areole at the base. Pappus-hairs few, with slightly thickened tips. —Handbk. 150.

SOUTH Island: Marlborough: Mount Mouatt, *T. K.* Nelson: Ben Nevis, *Gibbs!* Mount Starveall, *Bryant!* Gordon's Nob, *Monro.* St. Arnaud and Raglan Mountains, *Cheeseman.* Wairau Gorge and Tarndale, *Sinclair!* Clarence Valley, &c., *Travers!* Mount Captain Range, *T. K.* Canterbury: Mount Torlesse and mountains above Broken River, *Enys!* Otago: Mount Pisa and Hector Mountains, *Petrie!* 3,000ft. to 6,000ft. Jan., Feb.

[Pending fuller information the following are provisionally described:—

R. Brownii. Habit and appearance of *R. eximia* so far as known, differing chiefly in the linear-oblong leaves, which are abruptly narrowed just below the apex and subacute. The hairs are long and even, as if cut with a pair of scissors, when viewed from beneath. Heads not seen.

SOUTH Island: Canterbury Alps, *Enys!* in Herb. Kew., but the locality not stated.

The only species with sublanceolate obtuse leaves. Sent to me as *R. mammillaris* by the Director of Kew, with the following note by Mr. N. E. Brown: "This is certainly not *R. mammillaris*, Hook. f.; it appears to be a new species." I have not met with it. In all probability it was collected by Enys on Mount Torlesse or on one of the lofty peaks above the Broken River.

R. Buchanani. Stems apparently much branched, probably forming a compact mass similar to *R. eximia*, but not woolly. Leaves membranous, densely imbricate in several series, broadly cuneate, truncate at the apex, ⅛in.–⅙in. long and almost as broad, with a broad patch of straight hairs extending the whole breadth of the upper surface and projecting beyond the margin, forming a kind of fringe, the apical portion glabrous or glabrate with a broad margin beneath, corrugated cottony or hairy at the base. Heads not seen.

SOUTH Island: Mount Alta, *Buchanan!*

The broad truncate cuneate leaves are quite unlike those of any other species.]

12. HELICHRYSUM, Vaill.

Involucre cylindrical ovoid or hemispherical, the bracts imbricating in several series, the tips often radiating and petal-like. Receptacle without scales, flat convex or conical. Outer florets few, mostly filiform, female. Disk-florets hermaphrodite, tubular, 4–5 toothed. Anthers sagittate at the base. Style-branches nearly terete, usually truncate. Pappus-hairs simple or barbellate. Herbs or small shrubs, with alternate leaves, tomentose below or on both surfaces. Heads solitary or corymbose.

A large genus, comprising about 280 species, distributed through most temperate and tropical regions, but the species are very local. All the New Zealand species are endemic.

NAME, from the Greek, supposed to have been applied to a European species.

I. XEROCHLAENA. Involucre broad, hemispherical, the outer broad bracts sessile, gradually passing into the inner, with scarious linear claws and radiating tips. Achenes glabrous or pubescent. Perennial herbs, sometimes slightly woody at the base.

* *Stems prostrate, very slender.*

Heads solitary on bracteate peduncles or sessile .. .. .. 1. *H. bellidioides.*
Heads 3–7, in small cymes or corymbs .. .. 2. *H. Purdiei.*

** *Stem very slender, erect, 4in.–8in. high.*

Leaves small. Heads terminal, solitary .. .. 3. *H. filicaule.*

*** *Stem tufted, 2in.–4in. high. Leaves woolly.*

Flowers in small terminal corymbs .. .. 4. *H. Sinclairii.*

**** *Stems 1in.–4in. high, densely tufted. Leaves imbricating all round the stem.*

Leaves silvery, densely imbricating, erect. Heads 1–4 towards the apex of the stem .. .. 5. *H. fasciculatum.*
Leaves woolly. Heads solitary, terminal, sessile .. .. .. 6. *H. Youngii.*
Leaves densely tufted, silky. Head deeply sunk amongst the apical leaves .. 7. *H. Loganii.*

II. OZOTHAMNUS. Small shrubs, with terminal cymose or solitary heads.

* *Leaves flat. Heads cymose.*

Leaves suborbicular or ovate .. .. 8. *H. glomeratum.*
Leaves lanceolate .. .. 9. *H. lanceolatum.*

** *Leaves imbricate. Heads solitary, terminal.*

Heads free, sessile.

Leaves in about 6 series, linear, hoary or silky .. .. 10. *H. depressum.*
Leaves in about 4 series, keeled and polished at the back .. .. .. 11. *H. microphyllum.*
Leaves in 6–8 series, keeled and shining at the back .. .. .. 12. *H. Selago.*

Heads deeply sunk amongst the apical leaves.

Leaves in many series, broadly obtuse and convex at the back .. .. 13. *H. coralloides.*
Leaves in many series, silvery; margins recurved .. .. .. 14. *H. pauciflorum.*

III. LEONTOPODIOIDES. Leaves silvery, densely imbricating. Heads small, in dense terminal cymes, subtended by woolly foliaceous bracts.

Leaves linear-oblong; tips erect .. .. 15. *H. Leontopodium.*
Leaves oblong-spathulate; tips recurved .. .. 16. *H. grandiceps.*

1. **H. bellidioides,** *Willd., Sp. Pl.* iii. 1911 (*Elichrysum*). Stems slender, prostrate, slightly woody at the base, 6in.–20in. long, much-branched. Leaves loosely imbricating or rather distant, $\frac{1}{4}$in.–$\frac{1}{3}$in. long, spreading or recurved, flat or concave, ovate-spathulate or obovate-spathulate, subacute or apiculate, 1-nerved, glabrous above, white beneath or on both surfaces with

appressed cottony tomentum. Heads ½in. in diameter, on terminal bracteate cottony peduncles 1in.–5in. long; receptacle convex; involucral bracts in many series, white, radiating, with scarious tomentose claws. Female florets few. Achene glabrous. Pappus-hairs slender.—*Gnaphalium bellidioides*, Hook. f., Fl. N.Z. i. 137; Handbk. 152. *Xeranthemum bellidioides*, G. Forst., Prod. n. 293.

NORTH and SOUTH Islands: chiefly in mountain districts from the East Cape to Foveaux Strait. CHATHAM Islands; STEWART Island; AUCKLAND and CAMPBELL Islands; ANTIPODES Island. *T. K.* Sea-level to 5,000ft. Nov. to Feb.

Var. **erectum**. Stem erect or suberect, rather stout. Leaves linear-lanceolate, spathulate, acute, coriaceous. Peduncles very short and stout. SOUTH Island: Dun Mountain, *P. Lawson!*

Var. **prostratum** Stems shorter than in the type, spreading, irregularly branched. Leaves similar. Heads sessile at the tips of the branches. Receptacle usually conical.—*Helichrysum prostratum*, Hook. f., Fl. Antarc. 30, t. 21. *Gnaphalium prostratum*, Hook. f., Fl. N.Z. i. 137; Handbk. 152. NORTH Island: local. Titiokura, *Colenso*. Mount Egmont, *Dieffenbach*. SOUTH Island: Broken River basin, *T. K.* Dunedin, *Petrie!* AUCKLAND and CAMPBELL Islands, *Hooker* and others.

A gradual transition may be traced from forms with sessile heads to those on slender elongate peduncles, but the conical receptacle appears to be peculiar to var. *prostratum*.

2. **H. Purdiei,** *Petrie in Trans. N.Z.I.* xxii. (1889) 440. Stems 1ft.–2ft. long, very slender, wiry, prostrate, woody, spreading, much branched. Leaves rather distant, ⅓in.–½in. long, broadly obovate-spathulate, mostly rounded at the apex and minutely apiculate, pubescent or glabrate above, clothed with laxly-appressed whitish tomentum beneath. Heads in 3–7-flowered terminal cymes or corymbs. Peduncles and pedicels very short, cottony. Involucral bracts numerous, scarious, the outer very short, obtuse, tomentose; inner with white radiating tips, much shorter than the scarious claw. Female florets few. Disk-florets numerous. Achene glabrous. Pappus-hairs barbellate, cohering at the base.

SOUTH Island: Dunedin Harbour, *Petrie! Aston!*

Easily recognised by the ovoid heads and short radiating tips of the involucral bracts.

3. **H. filicaule,** *Hook. f., Fl. N.Z.* i. 140, *t.* 36B. Stems simple, rarely branched, very slender, filiform, cottony, 3in.–9in. high. Leaves ¼in.–⅓in. long, distant, obovate-oblong or lanceolate, obtuse or apiculate, glabrous glabrate or loosely cottony above, white with cottony tomentum beneath. Heads terminal, solitary, ⅓in.–½in. in diameter; involucral bracts in 4 series, scarious, linear-lanceolate, acute, the outer tomentose or cottony at the base. Female florets few. Receptacle narrow, convex. Achene puberulous or slightly papillose. Pappus-hairs very fine and delicate.—*Gnaphalium filiforme*, Hook. f., Handbk. 153. *Conyza uniflora*, Banks and Sol. MSS. and Icon.

NORTH and SOUTH Islands: from Rotorua to Foveaux Strait. CHATHAM Islands; STEWART Island. Sea-level to fully 4,000ft., chiefly in dry grassy places. Dec. to March.

4. **H. Sinclairii,** *Hook. f., Handbk.* 153. "A small subalpine species. Stems and branches ascending, leafy, 2in.–4in. high. Leaves close-set, spreading, ¼in.–⅓in. long, ⅛in. broad, linear-oblong or obovate-spathulate, obtuse,

densely covered with pale cottony tomentum on both surfaces. Heads ¼in. in diameter, in numerous rounded terminal dense corymbs ½in.–1in. across. Peduncles and pedicels short, densely cottony. Outer scales of involucre cottony, inner shortly radiating. Female florets in 1 series. Pappus of few stout hairs, thickened towards the tip. Achene glabrous."

SOUTH Island: Upper Awatore, *Sinclair*.

Not having seen specimens, I have copied the description from the Handbook. Hooker remarks, "Very closely allied to the *Raoulia catipes* of Tasmania; but the leaves are much smaller, the heads not half the size and much more numerous."

5. **H. fasciculatum,** *Buch. in Trans. N.Z.I.* ix. (1876) 529, *t.* 19. Densely tufted, forming small patches 1in.–1½in. high. Stems with the leaves ½in. in diameter. Leaves densely imbricating, erect, ¼in. long, narrow-oblong or lanceolate, acute, sessile by a broad base, concave, furrowed beneath, covered on both surfaces with appressed silvery-white tomentum, the base clothed with loose silky hairs. Heads 1–4, sessile in the axils of the upper leaves, ⅓in.–½in. in diameter; involucre turbinate; involucral bracts in 3 series, scarious, linear-oblong, obtuse, with loose silky tomentum at the base. Receptacle narrow, conical. Florets 12–18; female few in 1 series. Disk-florets trumpet-shaped, acutely 5-toothed. Achene silky, pubescent. Pappus-hairs few, not scabrid.

NORTH Island: Tararua Range, *H. Travers!* SOUTH Island: Nelson: Mount Starveall, *Bryant!* 4,000ft. to 5,000ft. Dec., Jan.

A singular plant, uniting the habit of *Raoulia* with the inflorescence of *Gnaphalium* and the few female florets of *Helichrysum*.

6. **H. Youngii,** *Hook. f., Handbk.* 152. Densely tufted, forming depressed patches 1in.–2in. high. Stems rather stout. Branches very short, erect, with the leaves ⅜in. in diameter. Leaves densely imbricating, broadly oblong or obovate, rounded at the apex, abruptly contracted just below the middle and expanded into a broad membranous base, densely tomentose on both surfaces; tips erect patent. Heads ⅓in.–½in. in diameter, sessile amongst the apical leaves; receptacle very narrow; involucral bracts in 3 series, the outer very short and tomentose, inner oblong, scarious, with white membranous margins, innermost with white radiating obtuse tips shorter than the scarious claw. Florets 12–18 or more, the outer in 1 series, female, few, the hermaphrodite disk-florets always predominating, funnel-shaped, 5-toothed. Achene pubescent or glabrate. Pappus-hairs slightly thickened upwards.

SOUTH Island: Canterbury: mountains above Lake Hawea, Mount Torlesse, Mount Cook, *Haast!* Otago: Lake district, *Hector* and *Buchanan!* Mount Pisa, Mount Cardrona, mountains of the west coast of Otago ("covering acres of the bare tops of the Old Man Range"), *Petrie!* 1,200ft. to 7,000ft.

A beautiful plant, with the habit of *Raoulia*, but separated by the paucity of female florets.

7. **H. Loganii,** *T. Kirk.* Forming pulvinate masses 6in.–12in. in diameter. Branches slender, woody at the base, with the leaves ⅓in.–⅜in. in diameter, the whole plant clothed with soft white or greenish-white wool. Leaves densely imbricating, ¼in. long, obovate or obovate-oblong, rounded at the tip or subacute, membranous, 3-nerved, tips recurved, clothed with

long soft hairs, which are restricted to a dense tuft above, projecting beyond the margin. Heads $\frac{1}{2}$in.–$\frac{3}{4}$in. in diameter; involucral bracts in 3 series, oblong, mostly obtuse, the outer villous, the inner broader, glabrate, scarious, pale, not radiating. Achene compressed, "covered with long silky hairs." Pappus-hairs barbellate at base, irregularly thickened towards the apex.—*Haastia Loganii*, Buch. in Trans. N.Z.I. xiv. (1881) 350, t. xxx. f. 3.

NORTH Island: Mount Holdsworth, Tararua Range, 4,500ft., *Arnold* and *Beck! Buchanan* and *Logan*.

My specimens, for which I am indebted to T. Arnold, are in poor condition, and do not contain florets or achenes; but the involucral bracts, the included style as shown in Buchanan's drawing, and the compressed silky achene show that this plant belongs either to *Helichrysum* or *Raoulia*.

8. **H. glomeratum,** *Benth. and Hook. f., Gen. Pl.* ii. 311. A much-branched shrub, 2ft.–8ft. high. Branches spreading, slender, grooved, cottony or tomentose. Leaves $\frac{1}{4}$in.–1in. long or more, $\frac{1}{8}$in.–$\frac{3}{4}$in. broad, flat, orbicular to broadly ovate-spathulate, abruptly narrowed into short slender petioles, rounded at the tip or minutely apiculate, white with appressed tomentum beneath. Heads in lateral or terminal subglobose corymbs, sessile or shortly pedunculate; involucres sessile or pedicellate, $\frac{1}{10}$in. long; involucral bracts few, linear, obtuse or subacute, scarious, tomentose at the base. Female flowers few, very slender. Disk-florets tubular, acutely 5-toothed. Achene glabrate or puberulous. Pappus-hairs slightly thickened at the tips.—*Ozothamnus glomeratus*, Hook. f., Fl. N.Z. i. 133; Handbk. 146. *Swammerdamia glomerata*, Raoul, Choix de Pl. 20, t. 16.

NORTH and SOUTH Islands: from Mount Manaia, Whangarei, and Great Barrier Island to Banks Peninsula. Sea-level to 2,000ft. Dec., Jan.

9. **H. lanceolatum,** *T. Kirk.* A shrub with slender grooved branches, 2ft.–4ft. high. Leaves narrow-lanceolate, $1\frac{1}{2}$in. long, acute, spathulate or abruptly contracted into a short winged petiole, white with appressed cottony tomentum beneath; margins slightly waved, entire or obscurely crenate. Heads in small lateral peduncled corymbs; involucres and fruits as in the preceding. Achene glabrous.—*Ozothamnus lanceolatus*, Buch. in Trans. N.Z.I. ii. (1869) 88.

NORTH Island: Maungataniwha Hills, 2,000ft., *Buchanan!*

This plant is probably a variety of the preceding, differing only in the lanceolate leaves and glabrous achenes. My specimens being very poor, I have copied the author's description.

10. **H. depressum,** *Benth. and Hook. f., l.c.* Suberect or rarely prostrate; the entire plant hoary with silver-grey tomentum. Branches distant, rigid, divaricating, irregular; branchlets usually numerous. Leaves $\frac{1}{12}$in.–$\frac{1}{10}$in. long, laxly imbricating, linear, obtuse, silky or woolly, concave and woolly above, closely appressed to the branches. Heads solitary, terminal, sessile, $\frac{1}{4}$in. long; involucral bracts linear-oblong, obtuse, glabrate or cottony at the base. Female florets few, long and slender. Disk-florets trumpet-shaped. Achene puberulous or glabrous. Pappus-hairs in several series.—*Ozothamnus depressus*, Hook. f., Fl. N.Z. i. 134, t. 35B; Handbk. 146.

NORTH Island: Tukituki River, Hawke's Bay, *Petrie!* SOUTH Island: mountain districts of Marlborough; Nelson; Canterbury; North Otago, but often local. Most frequent in shingly river-beds. 1,000ft. to 5,000ft. Jan., Feb.

In the Handbook the involucral bracts are described as "acuminate." I do not find them so.

11. **H. microphyllum,** *Benth. and Hook. f., l.c.* A small excessively-branched shrub, 6in.–15in. high. Branchlets slender, short, irregularly tetragonous, tomentose. Leaves $\frac{1}{16}$in.–$\frac{1}{12}$in. long, almost quadrifariously imbricate, triangular, ovate, appressed to the branchlet, concave and woolly on the inner face, keeled and polished at back. Heads turbinate, solitary, terminal, sessile, $\frac{1}{6}$in.–$\frac{1}{4}$in. in diameter; involucral bracts in 3 series, linear-oblong, obtuse, scarious. Female florets in 1 series, few. Disk-florets trumpet-shaped, 5-toothed. Achene pubescent. Pappus-hairs not thickened at the tips.—*Ozothamnus microphyllus*, Hook. f., Fl. N.Z. i. 134, t. 35A; Handbk. 146.

SOUTH Island: in mountain districts from Marlborough and Nelson to Southland. 2,000ft. to 4,000ft. Jan.

Best distinguished by the slender crowded branchlets, minute leaves, and pubescent achenes.

12. **H. Selago,** *Benth. and Hook. f., l.c.* A small much-branched shrub, 4in.–9in. high. Branches sometimes flabellate, with the leaves $\frac{1}{8}$in.–$\frac{1}{4}$in. in diameter. Leaves densely imbricating in 5–6 series, rhomboid-ovate, coriaceous, trigonous and keeled in the upper part, shining, subacute or obtuse, membranous at the very base, concave and woolly on the surface, appressed except at the apex. Head solitary, terminal, sessile, not sunk amongst the leaves, hemispherical; involucral bracts linear-oblong, the outer obtuse, slightly cottony at the base; the inner subacute, cartilaginous below, with short recurved scarious tips. Florets numerous; female few. Disk-florets almost filiform, with a narrow 5-toothed mouth. Achene puberulous. Pappus-hairs not thickened upwards. —*Ozothamnus Selago,* Hook. f., Fl. N.Z. ii. 332; Handbk. 147.

SOUTH Island: Marlborough: Awatere, *T. K.* Kaikoura Range, *Buchanan!* Nelson: Wairau Gorge, *Cheeseman.* Amuri, *T. K.* Canterbury: Mount Torlesse, *Carrington!* Rangitata, *J. B. Armstrong!* Otago: Mount Kurow, *Petrie!* 3,000ft. to 4,500ft. Jan.

Best distinguished from *H. microphyllum*, to which it is closely allied, by the stouter branchlets, 6-ranked leaves, and larger heads with broader involucral bracts.

13. **H. coralloides,** *Benth. and Hook. f., l.c.* A much-branched shrub, 3in.–10in. high. Branches cylindrical, $\frac{1}{4}$in.–$\frac{1}{3}$in. in diameter, densely tomentose beneath the leaves, which resemble minute knobs or tubercles imbedded in the tomentum. Leaves $\frac{1}{6}$in.–$\frac{1}{4}$in. long, densely imbricating in many series all round the stem, oblong or slightly obovate, $\frac{1}{6}$in.–$\frac{1}{4}$in. long, shining, convex, rounded at the apex and coriaceous but membranous below, 3-nerved, inner surface closely appressed to the stem and densely tomentose. Heads small, solitary, deeply sunk amongst the apical leaves; involucre hemispherical; involucral bracts cartilaginous, linear-oblong, acuminate, with narrow membranous margins and short white recurved tips; a few of the outer short, tomentose at the base and obtuse. Florets numerous; female few, filiform. Disk-florets slender; mouth narrow, tubular, 5-toothed. Achene short, puberulous.—*Ozothamnus coralloides,* Hook. f., Fl. N.Z. ii. 332; Handbk. 147.

SOUTH Island: Marlborough: Kaikoura Mountains, *McDonald*. Upper Awatere, *Sinclair! D. Rough!* Medway Creek, *T. K.* 3,500ft.

One of the most remarkable plants in the flora, and one of the rarest. I made careful search for the plant in various parts of the district, but only found it in a single station, which is doubtless the place where it was first discovered.

14. **H. pauciflorum,** *T. Kirk in Trans. N.Z.I.* xxvii. (1894) 351. Rootstock woody. Stems 4in.–7in. high, excessively branched and woody at the base. Branches with the leaves about ½in. in diameter. Leaves closely imbricating, oblong or oblong-spathulate with rather broad bases, clothed with silvery-white tomentum on both surfaces; margins slightly recurved. Heads solitary, almost hidden amongst the terminal leaves; involucres turbinate; involucral bracts lanceolate, acute, scarious, sparingly silky at the base, scarcely exceeding the florets. Female florets 3–4. Disk-florets 8–9, tubular, 5-toothed. Achene pubescent or hispid, with a thickened ring at the base. Pappus-hairs free, slightly thickened upwards.

SOUTH Island: Canterbury: Craigieburn Mountains, 3,000ft., *L. Cockayne!*

A remarkable plant, bearing the closest external resemblance to *H. grandiceps*, but differing in the total absence of the conspicuous woolly bracts so characteristic of that species, and in the sessile solitary heads deeply sunk amongst the apical leaves.

15. **H. Leontopodium,** *Hook. f., Fl. N.Z.* i. 141, *t.* 37B. Slightly woody at the base. Stems much branched, decumbent or ascending, 2in.–6in. high. Leaves ⅓in.–¾in. long, flat, densely imbricated, erect, rarely patent or reflexed, striate, linear-oblong or lanceolate, acute or subacute, evenly clothed with silvery appressed tomentum on both surfaces. Peduncles terminal, clothed with silvery foliaceous imbricating bracts. Heads numerous, forming a dense capitulum subtended by 10–15 ovate or ovate-oblong bracts, densely woolly, obtuse, ⅓in.–½in. long; involucres ⅛in. long; involucral bracts erect, linear-lanceolate, acute, scarious, woolly at the back, shining on the inner surface. Female florets few. Disk-florets tubular, 5-toothed. Achene silky. Pappus-hairs scabrid and slightly thickened upwards.—*Gnaphalium Colensoi*, Hook. f., Handbk. 154.

NORTH Island: Mount Hikurangi, East Cape, and Ruahine Range, *Colenso*. Tararua Range, *Budden! H. Travers!* SOUTH Island: Nelson: Mount Arthur, *Bryant!* Raglan Mountains and mountains above the Wairau Gorge, *Cheeseman!* Tarndale, *Sinclair!* Mountains above the Rainbow, *Bryant!* 3,000ft. to 6,000ft. Jan., Feb.

A beautiful plant, distinguished from the next species by the lanceolate leaves and larger heads.

16. **H. grandiceps,** *Hook. f., Handbk.* 154. Tufted, woody at the base. Stems mostly slender, 1in.–7in. high, decumbent and ascending. Leaves clothed on both surfaces with white silvery tomentum, densely imbricating, ¼in.–⅓in. long, obovate-spathulate, flat or the tips recurved. Peduncles terminal, solitary, usually leafy to the tips, but the leaves shorter than those below. Heads and bracts similar to *H. Leontopodium*, but the female florets are more numerous and the teeth of the disk-florets are more acute. Achene puberulous.

32

Pappus short, the hairs slightly thickened upwards.—*Gnaphalium grandiceps*, Hook. f., *l.c.*

SOUTH Island: not uncommon in mountain districts from 2,800ft. to 5,000ft. Jan., Feb.

13. CASSINIA, R. Br.

Involucre oblong, cylindric, or campanulate; involucral bracts imbricate, of few or many short obtuse oblong scarious or coloured scales, the innermost with short radiating tips. Receptacle narrow, with scarious chaffy scales amongst the florets. Florets tubular, the outer when present filiform, female; the inner hermaphrodite. Anthers short, tailed. Style long, terete, truncate, glandular. Achenes usually papillose, short, angular or terete. Pappus-hairs in 1–4 rows, slightly thickened at the tips, simple or barbellate. Shrubs, with entire alternate leaves. Heads in terminal panicles or corymbs.

A small genus, comprising about 20 species, of which about 14 are restricted to Australia, 1 or 2 to South Africa, and 5 to New Zealand. Some states of the local species are not easily distinguished.

Name, in honour of M. Henri Cassini, a French botanist.

* *Receptacle with numerous scales amongst the florets.*

Leaves whitish below, ½in. long, linear-obovate .. .. 1. *C. retorta.*
Leaves whitish below, $\frac{1}{15}$in.–$\frac{1}{10}$in. long, linear .. .. 2. *C. leptophylla.*
Leaves fulvous below, ¾in. long, flat. Involucral bracts red .. .. .. 3. *C. rubra.*

** *Scales amongst the florets few or 0.*

Leaves fulvous below, glutinous above, obovate .. .. 4. *C. Vauvilliersii.*
Leaves whitish below, ¼in.–¾in. long, linear-spathulate .. .. .. 5. *C. amoena.*
Leaves fulvous below, glutinous above, linear .. .. 6. *C. fulvida.*

1. **C. retorta,** *A. Cunn. ex DC., Prod.* vi. 154. Sparingly or densely branched, 5ft.–12ft. high. Branches clothed with whitish tomentum. Leaves close-set, ⅛in.–⅓in. long, linear-obovate or linear-oblong, obtuse, margins recurved, coriaceous, never glutinous, narrowed into a short petiole which is closely appressed to the branch; midrib obvious beneath. Heads 3–8, in terminal corymbs; pedicels stout; involucres turbinate; involucral bracts ovate or ovate-oblong, the outer pubescent or cottony. Receptacle paleaceous. Florets about 8. Achene faintly striate, glabrous.—Hook. f., Fl. N.Z. i. 132; Handbk. 145. *C. leptophylla*, A. Cunn., Prod. n. 447 (not *Calea leptophylla* of Forst.).

NORTH Island: from the North Cape to the East Cape, but chiefly littoral; not unfrequent on blown sand. Nov. to Jan.

A single unnamed specimen of this plant is in the Banksian collection.

2. **C. leptophylla,** *R. Br. in Trans. Linn. Soc.* xii. (1817) 126. Similar to *C. retorta*, but smaller in all its parts, with more slender branches. Leaves erect, spreading or recurved, $\frac{1}{15}$in.–$\frac{1}{10}$in. long, narrow-linear or narrow linear-spathulate, obtuse, margins recurved, clothed with appressed white tomentum beneath. Heads numerous, ⅙in.–¼in. long, in small terminal corymbs; pedicels very short; involucres turbinate; involucral bracts few, ovate or broadly oblong,

glabrate or rarely pubescent. Florets 6–10.—DC., Prod. vi. 155; Hook. f., Fl. N.Z. i. 133; Handbk. 145. *Calea leptophylla*, G. Forst., Prod. n. 287. *Calea axillaris*, Banks and Sol. MSS.

Var. **spathulata.** Leaves linear-spathulate, nearly flat, clothed below with yellowish tomentum. Involucres more tubular. Florets 9–10.—*C. spathulata*, Col. in Trans. N.Z.I. xxii. 472.

NORTH and SOUTH Islands: from the East Cape to Marlborough and Nelson. *Tauhinu. Cottonwood.*

3. **C. rubra,** *Buch. in Trans. N.Z.I.* xix. (1886) 216. A slender shrub, 2ft.–4ft. high. Branches, pedicels, and leaves below clothed with appressed yellow or fulvous tomentum. Leayes oblong-spathulate, erect or spreading, $\frac{1}{4}$in.–$\frac{5}{16}$in. loug, rounded at the tips, flat, or margins slightly recurved. Heads $\frac{3}{16}$in. long, very numerous, in terminal slender-branched corymbs; pedicels short, slender; involucres cylindric or narrow, turbinate; involucral bracts red, glabrous, elliptic-oblong, narrow, the inner with very short white tips. Receptacle with numerous scales amongst the florets.

NORTH Island: Wanganui River. Collector's name not stated, but I am indebted to Mr. Buchanan for one of his specimens.

A charming plant, resembling *C. fulvida* in aspect, but distinguished by the broad leaves, smaller heads, and numerous scales among the florets. From all other New Zealand species it is separated by the red involucral bracts, with extremely short white tips.

4. **C. Vauvilliersii,** *Hook. f., Fl. N.Z.* i. 133. Erect, 2ft.–8ft. high, much branched. Branches fastigiate or spreading, stout, clothed with viscid fulvous or yellowish tomentum, grooved. Leaves spreading or erect, $\frac{1}{4}$in.–$\frac{1}{3}$in. long, linear-obovate or oblong-spathulate, coriaceous, obtuse, narrowed into a short broad petiole or sessile, glabrous and glutinous above, clothed below with fulvous tomentum; margins flat or recurved. Heads numerous, in terminal globose corymbs; pedicels short; involucre $\frac{1}{5}$in.–$\frac{1}{4}$in. long, turbinate, scarious, woolly or glabrate; outer involucral bracts ovate-lanceolate, subacute, the inner oblong, obtuse. Florets about 10; scales amongst the florets numerous.—Handbk. 146. *Ozothamnus Vauvilliersii*, Homb. and Jacq. in Voy. au Pole Sud, Bot. Dicot. t. 5; Hook. f., Fl. Antarc. 29. *Olearia xanthophylla*, Col. in Trans. N.Z.I. xx. (1887) 193? (from description only).

NORTH and SOUTH Islands: from Taupo to Foveaux Strait; STEWART Island; AUCKLAND Islands.

Var. **albida.** Branchlets and leaves below clothed with whitish tomentum. Leaves linear-spathulate, strongly costate beneath. Kaikoura Mountains: not unfrequent, *T. K.* West Cape, *Buchanan!*

A beautiful species, distinguished from all states of *C. fulvida* by the broader leaves and numerous scales among the florets.

5. **C. amoena,** *Cheesem. in Trans. N.Z.I.* xxix. (1896) 391. Densely branched, 1ft.–2ft. high, the younger branches clothed with greyish tomentum. Leaves close-set, spreading or ascending, narrow linear-obovate or linear-spathulate, obtuse, $\frac{1}{4}$in.–$\frac{2}{3}$in. long, clothed with white tomentum beneath, glabrous above; margins recurved, narrowed into a very short petiole. Heads numerous, in crowded terminal hemispherical corymbs; pedicels short; involucres

narrow, turbinate; involucral bracts narrow, the outer ovate-oblong, pubescent, the inner oblong, glabrous, membranous, with short radiating tips. Florets 5. Scales of receptacle few or 0.

NORTH Island: cliffs near the North Cape; abundant; *Cheeseman!*

Near to *C. Vauvilliersii*, but distinguished by the smaller leaves, the narrow heads, and fewer florets.

6. **C. fulvida,** *Hook. f., Handbk.* 145. Erect, 2ft.–5ft. high, much branched, rather slender, glutinous. Branches clothed with subviscid tomentum. Leaves ⅙in.–⅓in. long, spreading or ascending, sessile, linear or narrow linear-spathulate or linear-obovate, obtuse, margins slightly recurved, clothed with fulvous tomentum, glutinous above, midrib obvious below. Heads very numerous, ⅙in.–¼in. long, in terminal simple or compound corymbs, cylindric or oblong; involucres cylindric, pubescent or glabrate. Florets 6–10. Scales among the florets few or 0.—*C. leptophylla* γ, Hook. f., Fl. N.Z. i. 133.

NORTH and SOUTH Islands: from the East Cape to Foveaux Strait. STEWART Island. Sea-level to 3,500ft. Dec. to Feb.

Var. **linearis.** Leaves ¼in.–½in. long, distant, very narrow, linear-lanceolate or spathulate-lanceolate, obtuse, clothed with white tomentum beneath, flat. Florets 4–6. Dunedin, *Aston!*

Distinguished from *C. retorta* and *C. leptophylla* by the fulvous or yellowish tomentum and the paucity or absence of scales amongst the florets.

14. CRASPEDIA, R. Br.

Flower-heads in clusters of 3–8, very numerous, forming a dense globose or ovoid compound head, subtended by a common involucre consisting of several bracts. Heads narrow; involucral bracts linear, hyaline, membranous. Receptacle very narrow. Florets 5–8, hermaphrodite, tubular, 5-toothed, with scales intermixed. Anthers tailed. Style-branches terete, truncate, included. Achene narrow, silky. Pappus-hairs in 1 row, plumose. Perennial erect simple leafy herbs, glabrous silky or woolly.

A small genus of about 5 species, restricted to Australia and New Zealand.

NAME, from the Greek, signifying *a fringe*, in reference to the white margins of the leaves.

1. **C. uniflora,** *G. Forst., Prod. n.* 306. A perennial herb, 4in.–20in. high, glabrate cottony or excessively woolly in all its parts. Rootstock simple or tufted. Leaves mostly radical, 1in.–8in. long, orbicular-obovate or obovate, abruptly narrowed into a short broad petiole, usually fringed with white tomentum; cauline leaves narrow, the upper reduced to short distant bracts, ovate to linear-oblong. The compound head ½in.–2in. in diameter, globose or often disciform; outer bracts herbaceous, ovate, with a broad scarious margin. Receptacle globular, rarely ovoid. Partial heads 3–8-flowered; involucral bracts in 1 series, free, hyaline, linear-oblong or broadly oblong, shorter than the florets. Achene compressed, silky. Pappus-hairs as long as the florets, plumose.—A. Cunn., Precurs. n. 446; Willd., Sp. Pl. iii. 2392; A. Rich., Fl. Nouv.-Zel. 245. *C. Richea*, Cass. in Dict. Sc. Nat. xi. 353; Benth., Fl.

Austr. iii. 579. *C. fimbriata*, DC., Prod. vi. 152; Hook. f., Fl. N.Z. i. 131; Handbk. 144. *Richea glauca*, Labill, Voy. i. 186, t. 16. *Cartodium apricum*, Banks and Sol. MSS.

NORTH and SOUTH Islands: from the East Cape to Southland. STEWART Island: rare, *T. K.* Sea-level to 5,000ft. Dec. to Feb.

Var. **pedicellata.** Glabrous or minutely scaberulous. Leaves never margined, broadly obovate or narrow oblong-obovate, subacute. Scape leafy. Heads sometimes very large; secondary heads usually on longer pedicels. Florets more numerous, with a longer tube; mouth broadly campanulate. Pappus longer. STEWART Island, *T. K.* Perhaps identical with *C. macrocephala*, Hook. f., Fl. Tasm. i. 197, but I have not seen authenticated specimens.

Var. **lanata.** The entire plant densely clothed with snow-white wool.—Hook. f., Fl. N.Z. i. 132. *C. alpina*, Backh. in Hook., Lond. Journ. Bot. vi. 119; Hook. f., Handbk. 144. In alpine districts.

Var. **viscosa.** Leaves with minute raised viscid points. Compound head broadly subconical; secondary heads usually with 3 florets; involucral bracts ovate, acute. Achene linear-ovate, strigillose.—*C. viscosa*, Col. in Trans. N.Z.I. xvi. (1883) 333. Matamau, Hawke's Bay. I have not seen specimens.

An extremely variable plant; found also in Australia.

* XANTHIUM, Linn.

Flower-heads in terminal racemes or clusters, monoecious. Male: globose; involucral bracts small, 1-seriate; anthers free; receptacle cylindrical, paleaceous; florets tubular. Female: heads ovoid; involucral bracts 2- or 3-seriate, the outer small, the 2 innermost large, coherent, forming a hard ovoid mass, 2-celled, prickly outside, terminating in 2 conical tubercles; florets 2; corollas 0. Style-arms filiform, protruding. Achene obovoid, enclosed in the indurated prickly involucres. Pappus 0. Annuals, with coarse alternate leaves.

* **X. spinosum,** *Sp. Pl.* 987. A coarse spreading herb, 1ft.–2ft. high. Branches rigid. Leaves alternate, with a strong 3-fid spine below each, lanceolate, acute or acuminate, cuneate at base, 3-fid, the lateral lobes very short, white with appressed tomentum beneath. Flower-heads in axillary fascicles or solitary, sessile, the upper male; the lower female, forming in fruit an oblong burr, ½in. long, clothed with hooked prickles, the terminal conical beaks reduced to minute tubercles.

NORTH Island: naturalised in many places from Auckland to Wellington, but sporadic. SOUTH Island: Marlborough: rare. Feb. to April.

15. SIEGESBECKIA, Linn.

Involucre broadly campanulate, glandular-hispid; the outer bracts linear-spathulate, spreading; the inner shorter, ovate or oblong-ovate. Receptacle paleaceous, the scales half enclosing the achene. Outer florets in 1 row, female. Disk-florets hermaphrodite, tubular, 5-toothed. Anthers without tails. Style-branches short, usually obtuse. Achenes somewhat turgid, usually curved. Pappus 0. Herbs, with opposite leaves and paniculate flowers.

A small genus, chiefly found in warm regions.

1. **S. orientalis,** *L., Sp. Pl.* 900. A sparingly branched pubescent annual, 1ft.–2ft. high. Leaves 1in.–2in. long, membranous, lanceolate or broadly ovate, triangular, petiolate, entire or irregularly sinuate-toothed or

lobed. Flower-heads ¼in.–½in. broad in a dichotomous leafy panicle; outer involucral bracts usually longer than the inner, spreading. Outer florets with very short irregularly toothed rays.—Benth., Fl. Austr. iii. 535.

KERMADEC Islands. NORTH Island: from Mangonui to the East Cape district, *Banks* and *Solander!* Great Barrier Island, *T. K.* *Punawaru*. Jan., Feb.

16. BIDENS, Tourn.

Involucral bracts in 2–3 series, few, erect, narrow, slightly connate at the base. Receptacle paleaceous. Ray-florets in 1 series, sterile; rays short or 0. Disk-florets hermaphrodite, tubular, 5-toothed. Style-arms subulate. Achene 4-angled, or rarely compressed, narrow. Pappus of 2–4 rigid retrorsely hispid persistent bristles. Erect herbs, with opposite leaves. Flowers on terminal peduncles.

Species, about 125, distributed through most warm and temperate countries.

Name, from the two rigid pappus-bristles of some species.

1. **B. pilosa**, *L., Sp. Pl.* 832. An erect annual or rarely perennial herb, 1ft.–2ft. high, with angular slightly hairy branches. Leaves membranous, simple or pinnate; segments mostly 3, rarely 5, petiolulate, ovate or ovate-lanceolate, serrate, ¾in.–2in. long. Heads ½in. in diameter, on slender terminal peduncles; involucral bracts ¼in. long. Ray-florets few, short or 0. Achenes slender, 4-angled, the inner longer than the outer, 2–4-awned, striate.—G. Forst., Prod. n. 283; A. Cunn., Precurs. n. 442; Hook. f., Handbk. 138; *B. aurantiacus*, Col. in Trans. N.Z.I. xxvii. (1894) 388.

KERMADEC Islands. NORTH Island: north of the East Cape.

A common weed in all warm and temperate countries.

* MADIA, Linn.

Involucral bracts in 1 or 2 series. Receptacle convex, paleaceous. Ray-florets in 1 or 2 series, fertile; rays short. Disk-florets hermaphrodite, fertile or sterile, tubular, 5-toothed. Style-arms appendiculate. Achene compressed, usually enclosed in one of the receptacular scales. Pappus 0 or consisting of short lacerate scales. Annual or perennial herbs, with entire or rarely pinnatifid leaves, glandular or villous. Heads on slender peduncles, globular.

* **M. sativa**, *Molina, Sagg. Chil. ed.* i. 136. An erect annual, 1ft.–3ft. high. Stem simple or branched, glandular, viscid, hirsute. Leaves alternate, sessile, narrow-oblong or oblong-lanceolate, 1in.–3in. long, villous or glandular. Heads on solitary peduncles or racemose; involucres globose, viscid; outer involucral bracts glandular-hispid. Achenes curved, angled.

SOUTH Island: naturalised; Renwicktown, Marlborough, *Reader!* Otago: between Balclutha and Catlins, *T. K.* Bannockburn, *Petrie!* Chili. Feb., March.

* ACHILLEA, Linn.

Involucral bracts in 3 series, oblong, usually with scarious margins, imbricated. Receptacle paleaceous, narrow. Ray-florets few, fertile; ligule short and broad. Disk-florets hermaphrodite, tubular, compressed, 5-toothed. Anthers obtuse at the base. Achenes oblong, flattened, margined. Pappus 0. Perennial herbs, with rather small corymbose flowers.

Leaves 2–3-pinnate; segments narrow. Rays white .. .. * *A. Millefolium.*
Leaves pinnate or 1–2-pinnatifid. Rays red .. .. * *A. tanacetifolia.*

* **A. Millefolium,** *L., Sp. Pl.* 899. Rootstock creeping. Stems simple, erect, leafy, pubescent or villous, 1ft.–2ft. high. Leaves alternate, 2in.–6in. long, linear-oblong in outline, 2–3-pinnate; segments linear-acute, very close. Heads numerous, ovoid; pedicel short. Rays white or pink, shorter than the involucre. Achene glabrous.

NORTH and SOUTH Islands: naturalised; common. *Yarrow.* Feb., March. Europe, &c.

* **A. tanacetifolia,** *All., Fl. Pedem.* i. 183. Erect, 1½ft.–2½ft. high. Stem and leaves glabrate or pilose. Leaves 2in.–6in. long; radical pinnate, leaflets 1–2-pinnatifid, segments narrow, acute; cauline pinnatifid, rhachis broadly winged, segments pinnatifid lobed or toothed. Involucres broadly oblong, pubescent. Rays very broad, red.

SOUTH Island: Canterbury: naturalised near Green Park, Lincoln. Feb. to June. Southern Europe.

* ANTHEMIS, Mich.

Involucre hemispherical; involucral bracts numerous, imbricating in several series, shorter than the disk. Receptacle conical, paleaceous. Ray-florets in 1 series, female or neuter, ligulate. Disk-florets hermaphrodite, tubular. Anthers without tails. Style-arms short, obtuse. Achene terete, striated or ribbed. Herbs, or rarely suffruticose. Leaves alternate, much divided. Heads solitary.

Leaves 2-pinnate; segments pinnatifid. Bracteoles of receptacle mucronate .. * *A. arvensis.*
Leaves 3-pinnate, dissected. Bracteoles of receptacle setaceous .. .. .. * *A. Cotula.*
Leaves 2-pinnate. Bracteoles of receptacle obtuse .. .. * *A. nobilis.*

* **A. arvensis,** *L., Sp. Pl. ed.* i. 894. Annual or biennial, 1ft.–2ft. high, spreading. Leaves pubescent, 2-pinnate or pinnatifid; segments linear-acute. Heads on stout terminal peduncles, 1in.–1½in. in diameter; bracteoles of the receptacle mucronate. Ray-florets female. Achenes broadly truncate, unequally ribbed, glabrous. Pappus a minute border.

NORTH and SOUTH Islands: naturalised on waste and cultivated ground. *Corn chamomile.* Feb., March. Europe, &c.

* **A. Cotula,** *L., Sp. Pl.* 894. Annual, erect, 1ft.–1½ft. high, corymbosely branched, foetid. Leaves 3-pinnate; dissected segments almost capillary, gland-dotted. Heads on slender peduncles; involucral bracts narrowed towards the tip; bracteoles of the receptacle setaceous. Ray-florets mostly neuter. Achene faintly ribbed. Pappus 0.

NORTH and SOUTH Islands: naturalised on waste and cultivated ground. *Stinking chamomile.* Feb., March. Europe.

* **A. nobilis,** *L., Sp. Pl.* 894. Perennial, aromatic. Stem branched, more or less pubescent, 1ft. long. Leaves bipinnate, downy or pilose; leaflets linear-subulate, acute, rather fleshy. Peduncles very slender. Ray-florets female. Disk-florets cylindric. Bracteoles of the receptacle lanceolate, obtuse. Pappus 0.

SOUTH Island: the Bluff, Southland; apparently well established (1887), *T. K.* *Chamomile.* Jan. to March. Southern Europe.

* MATRICARIA, Tourn.

Involucral bracts in few series; margins scarious. Receptacle at first flat, conical in fruit, naked. Ray-florets 1-seriate, female, ligulate, sometimes 0. Disk-florets hermaphrodite, tubular, 4–5-toothed. Anthers without tails. Achenes 3–5-costate. Pappus a membranous epigynous disk or 0. Annual or rarely perennial herbs, with dissected leaves.

Branches short. Heads rayless. Receptacle conical .. .. .. * *M. discoidea.*
Branches spreading. Rays spreading. Receptacle conical .. .. .. * *M. Chamomilla.*
Branches spreading. Rays spreading. Receptacle broadly ovate-conical .. * *M. inodora.*

* **M. discoidea,** *DC., Prod.* vi. 51. Annual, erect, glabrous, rather stout. Stem furrowed, 6in.–9in. high, sparingly branched. Leaves 2–3-pinnatifid; segments usually short, linear, acute. Heads on short pedicels, rayless; involucral bracts shorter than the receptacle, oblong, obtuse, with scarious margins. Receptacle elongate, conical in fruit. Achene almost terete. Pappus 0.

NORTH Island: plentifully naturalised about Auckland. Feb., March. Oregon.

* **M. Chamomilla,** *L., Sp. Pl.* 891. Annual, faintly aromatic, erect, 1ft. high, rather slender, branched. Leaves glabrous, 2-pinnate; segments capillary. Heads terminal or subcorymbose, ½in.–¾in. in diameter. Receptacle elongating, conical, hollow. Ray-florets sometimes 0. Achenes 5-ribbed on the inner face.

NORTH Island: Auckland, *T. K., Cheeseman.* Sparingly naturalised in cultivated land and waste places, but often confused with *M. inodora.* Europe.

* **M. inodora,** *L., Fl. Suec.* ii. 765. Annual, much branched, slender. Leaves 2-pinnatifid; segments capillary. Heads 2in. in diameter, on long naked peduncles; involucral bracts with brown margins. Receptacle broadly conical or ovate in fruit. Rays numerous. Achenes with 3 strong ribs on the inner face and 2 apical pits on the outer face. Pappus a short cup or crown.

NORTH and SOUTH Islands: commonly naturalised. Feb., March. Europe.

* CHRYSANTHEMUM, Tourn.

Involucre campanulate; bracts imbricating in several series, small, with scarious margins. Receptacle broad, flat or convex, naked. Ray-florets in 1 series, female, ligulate, large, white or yellow. Disk-florets hermaphrodite, tubular, 4–5-toothed. Achenes of the ray-florets ribbed or winged, of the disk-florets compressed. Pappus 0. Herbs, with alternate or radical toothed lobed or pinnatifid leaves.

Stems much branched. Heads corymbose .. .. * *C. Parthenium.*
Stems sparingly branched. Heads solitary, large, golden-yellow .. .. * *C. segetum.*
Stems usually simple. Heads solitary. Rays white .. .. .. * *C. Leucanthemum.*

* **C. Parthenium,** *Bernh., Syst. Verz. Erf.* 145. Perennial, erect, branched, 1ft.–2ft. high. Leaves petioled, pinnate; leaflets pinnatifid; segments lobulate or toothed. Heads numerous, corymbose, ½in.–¾in. in diameter, on slender terminal peduncles; involucral bracts ribbed; margins scarious. Disk-florets yellow; ray white. Achene ribbed.

NORTH Island: chiefly in waste places; naturalised. *Fever-few.* Feb. Europe.

* **C. segetum,** *L., Sp. Pl.* 889. Annual, erect, glabrous, 1ft.–1½ft. high, sparingly branched. Leaves obovate, the lower petioled and pinnatifid, the upper clasping, lobulate and toothed. Heads terminal, solitary, 2in. in diameter; involu-

cral bracts very broad, obtuse, with scarious margins. Rays golden-yellow. Achenes of the ray with 2 narrow wings, ribbed; of the disk, wingless.

NORTH and SOUTH Islands: in cultivated land, but not common. *Corn marigold.* Feb., March. Europe.

* **C. Leucanthemum,** *L.*, *Sp. Pl.* 888. Perennial. Stems 1ft.–2ft. high, erect, pubescent or glabrous, usually simple, bearing a single head 1in.–2in. in diameter. Lower leaves petioled, spathulate or linear-obovate; cauline leaves sessile, clasping, linear-oblong, lobed or pinnatifid, obtuse. Involucral bracts small, numerous, with brown or purple margins. Rays white. Disk-florets yellow. Achenes terete, ribbed.

NORTH and SOUTH Islands: abundantly naturalised. A troublesome weed in pastures. *Ox-eye.* Feb., March. Europe.

17. COTULA, Tourn.

Involucre hemispherical or campanulate, with few nearly equal bracts; margins scarious or coloured. Receptacle flat, convex, or conical, papillose. Outer florets in 1 or more rows, female; corolla short, broad, conical, or 0. Disk-florets numerous, hermaphrodite, 4–5-toothed, sometimes sterile. Anthers without tails. Style-arms obtuse or truncate or style undivided. Achenes compressed or unequally plano-convex, sometimes winged. Pappus 0. Small perennial flaccid or succulent decumbent herbs, usually with numerous oil-glands, often aromatic. Leaves alternate or radical. Heads small, on slender scapes or axillary peduncles, rarely dioecious.

SPECIES, about 50, chiefly distributed through warm and temperate countries, but extending to the Antarctic islands; about 8 species are found in Australia. One of the New Zealand species is common to Europe, the cooler parts of South America and South Africa, and Australia; another is found in Australia and Tristan d'Acunha; the others are endemic.

NAME, from the Greek, signifying *a cup*, in reference to the form of the involucre.

The student should be careful to obtain fully matured achenes for examination.

I. COTULA. Receptacle flat or convex. Achenes of the female florets in a single row, stipitate. Corolla 0 in the female florets.

Stems stout, creeping. Leaves entire or lobed, distant, forming a membranous sheath round the stem. Flower-heads yellow .. .. .. 1. *C. coronopifolia.*

II. STRONGYLOSPERMA. Receptacle flat or convex. Female florets numerous, in several rows. Corolla 0.

Stems slender. Leaves pinnate or bipinnate .. .. 2. *C. australis.*

III. LEPTINELLA. Receptacle convex or conical. Female florets in 1 or several series. Corolla always present.

* *Flower-heads bisexual.*

Scapes fleshy, leafy. Leaves much divided. Heads large. Florets dark .. 3. *C. atrata.*

Stem and leaves softly woolly. Leaves 2in.–6in. long, dissected. Florets glandular .. .. 4. *C. plumosa.*

Stem stout. Leaves rather fleshy. Peduncles short, stout, terminal .. 5. *C. lanata.*

Stems prostrate, forming hoary depressed silky patches. Leaves ¼in.–½in. long. Heads terminal .. .. 6. *C. Maniototo.*

Stems prostrate. Leaves tufted, 2in.–3in. long. Heads hemispherical. Florets numerous, eglandular .. .. 7. *C. Muelleri.*

Stems prostrate. Leaves tufted. Heads hemispherical. Female florets few, glandular .. .. 8. *C. Traillii.*

Stems creeping. Leaves solitary or tufted. Female florets glandular. Corolla almost equalling the achene .. .. 9. *C. minor.*

Stems wiry. Leaves $\frac{1}{4}$in. long. Head $\frac{1}{10}$in. in diameter .. .. .. 10. *C. filiformis.*

Leaves tufted, pubescent, glandular, incised. Heads $\frac{1}{8}$in.–$\frac{1}{4}$in. in diameter. Achene unequally plano-convex .. .. 11. *C. Haastii.*

Leaves tufted; segments subulate. Achenes cuneate .. .. .. 12. *C. pectinata.*

Branches numerous, prostrate. Leaves oblong-cuneate, sessile, 3-toothed at the apex, downy .. .. 13. *C. Featherstonii.*

Stems prostrate, rather stout. Leaves laxly imbricating, vertically pectinate. Heads terminal, $\frac{1}{8}$in. in diameter .. .. 14. *C. Goyeni.*

** *Heads unisexual.*

Leaves linear-spathulate, entire. Scapes with 4–8 short bracts, strict, erect .. 15. *C. linearifolia.*

Stout, glabrous. Leaves pinnatifid, fleshy or coriaceous. Scapes $\frac{1}{2}$in.–3in. long with linear bracts .. .. 16. *C. pyrethrifolia.*

Rhizomes wiry. Leaves tufted, $\frac{1}{4}$in.–1in. long, clothed with shining silky hairs. Heads $\frac{1}{8}$in.–$\frac{1}{4}$in. in diameter .. .. 17. *C. perpusilla.*

Leaves $\frac{1}{3}$in.–$\frac{1}{2}$in. long, shortly lobed or pinnatifid; lobes entire .. .. 18. *C. obscura.*

Leaves $\frac{1}{3}$in.–2$\frac{1}{2}$in. long, linear oblong, deeply pinnatifid; lobes toothed .. 19. *C. pulchella.*

Stems rather stout. Leaves usually pinnatifid, lobed or crenate. Heads $\frac{1}{8}$in.–$\frac{1}{3}$in. in diameter .. .. 20. *C. dioica.*

Slender. Leaves much cut and divided, silky, pubescent. Heads $\frac{1}{8}$in.–$\frac{1}{4}$in. in diameter.. .. .. 21. *C. squalida.*

1. **C. coronopifolia,** *L., Sp. Pl.* 892. Glabrous. Stems succulent, creeping, rooting at the nodes. Branches ascending, 3in.–10in. high. Leaves alternate, often distant, sheathing at the base, linear-lanceolate to oblong, entire toothed lobed or pinnatifid. Peduncles slender, axillary, exceeding the leaves. Heads $\frac{1}{8}$in.–$\frac{1}{2}$in. in diameter, yellow; involucral bracts narrow-oblong, obtuse. Receptacle flat or convex. Ray-florets on slender flattened pedicels, 1-seriate; corolla 0; achene flat, broadly winged, lobed at both ends, the upper lobe equalling the bifid style, glandular on the inner face. Disk-florets on shorter pedicels; corolla cylindric or almost tetragonous, 4-toothed; wing of disk-achenes narrower.— G. Forst., Prod. 300; A. Cunn., Precurs. n. 443; Hook. f., Fl. N.Z. i. 127; Handbk. 141; Benth., Fl. Austr. iii. 549.

Var. **integrifolia.** Stems usually simple, 1in.–2in. high. Leaves linear, entire, obtuse. Peduncles terminal, solitary. Heads $\frac{1}{3}$in. in diameter.—*C. integrifolia*, Hook. f., Fl. Tasm. i. 192, t. 50B.

NORTH and SOUTH Islands: from the North Cape to Foveaux Strait; chiefly in lowland situations. STEWART Island, *T. K.* CHATHAM Islands, *Cox!* *Yellow-button.* Sept. to March. Also in Australia, South America, South Africa, and Europe.

2. **C. australis,** *Hook. f., Fl. N.Z.* i. 128; *Fl. Tasm.* i. 191, *t.* 50A. Stems slender, much branched, diffuse, flaccid, glabrate, pubescent or woolly at the nodes, 2in.–5in. high. Leaves pinnate or bipinnatifid, $\frac{1}{2}$in.–1in. long; pinnae entire, 3-lobed or pinnatifid; segments linear, acute or mucronate. Heads $\frac{1}{10}$in.–$\frac{1}{8}$in. broad; peduncles slender, larger or shorter than the leaves; involucral bracts in 2 series, linear-oblong. Female florets in 3 rows, pedicellate; corolla 0; achenes winged, wing retuse, glandular on the inner face. Disk-florets on shorter pedicels, tubular, 4-toothed; achenes wingless.—Handbk. 141. *C. venosa*, Col. in Trans. N.Z.I. xxiii. (1890) 388. *Strongylosperma australis*, Less., Syn. Comp. 261.

KERMADEC Islands. From the North Cape to Otago; CHATHAM Islands. Chiefly in lowland districts. Oct. to March. Also in Australia and Tristan d'Acunha.

3. **C. atrata,** *Hook. f., Handbk.* 142. Rootstock rather stout, woody, shortly creeping. . Stem 1in.–5in. high, succulent, ascending or erect, leafy, pubescent. Leaves pubescent, fleshy, $\frac{1}{2}$in.–1in. long, linear-oblong or obovate, pinnatifid ; segments lobed, toothed, or crenate ; cauline leaves shorter, pinnatifid or lobed. Heads $\frac{1}{3}$in.–$\frac{3}{4}$in. in diameter, subglobose or discoid ; involucral bracts in several series, pubescent, entire, toothed or pinnatifid. Florets very numerous, rugose ; female in 3–4 series, cylindric, 3–4-toothed ; achene linear-oblong, rugose. Disk-florets tubular, 4-toothed.

SOUTH Island : chiefly on alpine shingle-slips. Nelson : Tarndale, *Sinclair !* Wairau Gorge, *Travers !* Mount Captain Range and Spencer Mountains, *T. K.* Canterbury : Mount Torlesse, Ashburton Glacier, and Macaulay River, *Haast !* Valley of the Mathias and mountains above the Broken River, *Enys !* Otago : Mount St. Bathan's and Mount Kyeburn, *Petrie !* 2,500ft. to 6,500ft. Jan., Feb.

4. **C. plumosa,** *Hook. f., Handbk.* 141. A handsome tufted feathery soft villous species. Stems creeping, rather stout. Leaves villous or almost woolly, petioled, oblong, membranous, 2in.–6in. long, pinnate ; leaflets recurved, the upper side 2-pinnatifid at the base, ultimate division toothed on one side, $\frac{1}{8}$in. long, acute. Scapes pilose or woolly, shorter than the leaves, filiform with a pinnatifid bract about the middle. Heads $\frac{1}{2}$in. in diameter ; involucral bracts in 2–3 series, broadly oblong, with black or purple margins. Receptacle conical. Female florets in 2–3 series on short pedicels ; corolla dilated laterally, contracted at the mouth, unequally 4-toothed ; style-arms short ; achene obovoid. Disk-florets tubular ; mouth campanulate, 5-toothed. — *Leptinella plumosa*, Hook. f., Fl. Antarc. i. 26.

AUCKLAND and CAMPBELL Islands, *Hooker.* ANTIPODES Island, *T. K.* MACQUARIE Island, *Scott ! Hamilton !* Dec., Jan. Also in Kerguelen Land.

A beautiful species, forming dense matted patches with the heads hidden amongst the leaves.

5. **C. lanata,** *Hook. f., Handbk.* 141. Stems 3in.–12in. long, robust, prostrate or creeping, glabrate or woolly. Leaves 1in.–2in. long, rather fleshy ; blade obovate, pinnate, pinnatifid or shortly lobed ; segments entire or 3–5-lobed or -toothed along the upper margins, glandular ; petiole dilated at the base, woolly on the inner surface. Scapes terminal, stout, shorter than the leaves, woolly ; involucral bracts broadly oblong, green. Florets clothed with pellucid conglobate glands ; female very narrow ; style-arms very short ; achene narrow, obovate, papillose.—*Leptinella lanata*, Hook. f., Fl. Antarc. i. 25, t. 19. *L. propinqua*, Hook. f. *l.c.* i. 27.

AUCKLAND and CAMPBELL Islands. Dec., Jan.

6. **C. Maniototo,** *Petrie in Trans. N.Z.I.* xiv. (1882) 362. Rhizome creeping, forming dense matted depressed grey patches. Branches short, leafy, and the whole plant thickly covered with long white hairs. Leaves very numerous, silky on both surfaces, $\frac{1}{8}$in.–$\frac{1}{3}$in. long, narrow linear-oblong, pinnatifid to the expanded sheathing base ; segments linear-subulate, acute or subacute. Heads terminal, $\frac{1}{4}$in. in diameter ; peduncle $\frac{1}{12}$in.–$\frac{1}{10}$in. long, stout ; involucral bracts in 2 series, shortly obovate, the outer silky with erose scarious

margins. Receptacle conical. Female florets in 2–3 series; corolla filiform, scarcely expanded at the 2-lipped mouth; achenes oblong, with a narrow border, slightly turgid. Hermaphrodite florets longer, narrow, trumpet-shaped, 4-lobed.

SOUTH Island: Canterbury: Lake Lyndon, *Enys* and *Kirk* (1876). Central and south Otago, Kakanui Mouth, Maniototo Plain, Mossburn, Te Anau, &c., *Petrie!* Sea-level to 2,800ft. Jan., Feb.

Petrie describes the female florets as 1-seriate, but in my specimens they are invariably 2–3-seriate.

7. **C. Muelleri,** *T. Kirk.* Habit of *C. Traillii,* but stems and leaves more robust. Leaves shorter, gland-dotted, glabrate; petiole stout; segments subovate, crenate-toothed or lobed, never apiculate. Scapes axillary, downy or pubescent. Heads hemispherical; involucral bracts about 30, in 3 series, the outer short, ovate, acute; the inner oblong-ovate or cuneate-ovate, membranous, glabrate or glabrous. Receptacle convex. Female florets in about 3 rows, eglandular; corolla shortly ovate, one-third to one-half the length of the achene; style-arms very short; achenes oblong, clavate, narrowly winged. Hermaphrodite florets less numerous; corolla slightly tetragonous, 4-lobed; stigma undivided.—*Leptinella potentillina,* F. Muell., Veg. Chath. Isds. 28, t. 6.

CHATHAM Islands, *H. Travers! Cox!*

This bears a close resemblance to *C. Traillii,* but the leaves are larger and more fleshy, the involucral bracts narrower, while the female florets are very numerous, usually exceeding the male, and the achenes are much narrower. *C. lanata,* to which Sir Joseph Hooker seems inclined to refer it (Handbook 733), differs in the robust woolly branches, very short stout terminal peduncle and glandular florets, with the corolla of the female equalling the achene.

8. **C. Traillii,** *n. s.* Stems slender, creeping, 3in.–12in. long. Leaves tufted, membranous, linear-obovate, with a few scattered hairs on one or both surfaces, deeply pinnatifid; segments in 7–9 pairs, acutely toothed or apiculate, especially above; terminal lobes very large and broad; petiole slender, winged at the base. Peduncles axillary, equalling the leaves or shorter, pubescent or hairy. Heads hemispherical, ⅓in. in diameter; involucral bracts few, in 2–3 rows, orbicular or orbicular-ovate, with erose scarious margins. Receptacle flat or slightly convex. Female florets few, sparingly glandular, shortly stipitate; corolla ovate, compressed; achene broadly ovate or suborbicular, scarcely twice as long as the corolla and much broader, obviously 3-winged, slightly turgid and convex at the back. Disk-florets very numerous, funnel-shaped, 4-lobed; stigma discoid.

STEWART Island: often on blown sand, *T. K.* Dec. to Feb.

This species is closely related to *C. lanata* and *C. Muelleri,* from both of which it is distinguished by the more slender habit, acutely toothed leaves, and the short broad achenes, which are very few in number and invariably confined to the outside row.

9. **C. minor,** *Hook. f., Handbk.* 142. Stems very slender, creeping, glabrate or pubescent, 1in.–12in. long. Leaves radical, or alternate on slender creeping runners, linear-oblong or obovate in outline, ½in.–2in. long, pinnatifid nearly to the base; the lower segments rather distant, and deeply lobed or toothed on the upper margins; the upper segments close-set and recurved,

deeply toothed on both margins; teeth acute. Scapes very slender, equalling or exceeding the leaves. Heads ⅙in.–¼in. in diameter; involucral bracts about 12, broadly oblong or suborbicular, usually with deep purple margins. Female florets in 3–4 series, glandular; corolla scarcely shorter than the achenes, inflated, ovoid; mouth narrow, denticulate or entire; achene obcuneate, glandular. Disk-florets funnel-shaped; mouth 4-toothed.—*Leptinella minor*, Hook. f., Fl. N.Z. i. 129. *Soliva tenella*, A. Cunn., Precurs. n. 445.

NORTH and SOUTH Islands: from Mangonui to Marlborough and Nelson, but often local. Sea-level to 2,000ft. Nov. to Jan.

10. **C. filiformis,** *Hook. f., Handbk.* 142. A very slender rigid creeping plant, glabrous or pilose. Leaves minute, ¼in. long, oblong, pinnatifid; segments subulate. Scapes filiform, 1in. long, naked. Heads $\frac{1}{10}$in. in diameter; involucral bracts 6–8, orbicular, with purple edges. Receptacle conical. Ray-florets about 20; corolla short, compressed, inflated, very broad, oblong, 2-lobed above; achene obconic, glandular. Disk-florets funnel-shaped, 4-lobed; lobes glandular.

SOUTH Island: amongst grass, Canterbury Plains, *Haast.*

Not having seen specimens of this species I have copied the original description. A scrap given me by Haast without name or locality may possibly be the same, but the involucral bracts and florets are more numerous.

11. **C. Haastii,** *n. s.* Tufted, emitting wiry runners. Leaves pubescent, glandular, ¾in.–1in. long, linear oblong-obovate, pinnate or pinnatifid to the base; leaflets or segments oblong or oblong-flabellate, cuneate at base, shortly lobed toothed or incised; lobes acute. Scapes slender, pubescent or pilose. Heads ⅕in.–¼in. in diameter; involucral bracts in 2–3 series, broadly oblong, obtuse, with purple tips. Receptacle broadly convex. Female florets in several rows; corolla short, ovate, compressed; mouth greatly contracted, 4-toothed; achene narrow oblong-cuneate, ultimately tumid and convex at the back when fully ripe. Disk-florets short, broadly funnel-shaped, with 4 large lobes.

SOUTH Island: Banks Peninsula and Canterbury Plains, *Haast! T. K.* Otago: Ahuriri and Kurow, *Petrie!* Sea-level to 2,000ft. Dec., Jan.

This appears to have been confused with *C. pectinata* in the Handbook, but differs in the incised leaves, convex receptacle, shorter and broader disk-florets, and especially in the convex achenes. Mr. Petrie's *C. pectinata* from the Maniototo Plain may belong to this, but I have not seen specimens.

12. **C. pectinata,** *Hook. f., Handbk.* 142. Stems tufted, short, often emitting wiry runners, glabrous, silky or pilose. Leaves few, somewhat rigid, 1in.–1½in. long, linear-oblong, pectinately pinnatifid; segments subulate, entire. Scapes slender, 1in.–3in. long, pubescent or glabrous, naked or with a minute bract about the middle. Heads ¼in.–⅓in. in diameter; involucral bracts in 2–3 series, usually purple, pubescent, oblong or broadly oblong. Receptacle conical. Female florets in several series; corolla ovate-oblong, much compressed, minutely 2–4-toothed; achene narrow, cuneate, glandular or eglandular, compressed and slightly winged. Disk-florets trumpet-shaped, 4-toothed, glandular.

SOUTH Island: Nelson: Mount Captain Range, Amuri, *T. K.* Canterbury: Mount Torlesse, *Haast!* Limestone gravel by the Thomas River, *T. K.* Mountains above Broken River, *Enys!* Otago: Lake district, *Hector* and *Buchanan!* Common in the mountains, *Petrie!* Waitaki Valley and Maniototo Plains, *Petrie*. 1,200ft. to 5,000ft. Dec., Jan.

Var. **sericea.** Leaves ⅓in.–⅔in. long, hoary, with long silky hairs; segments close-set, flat, acute. Peduncle ⅓in.–⅔in. long, hoary. Heads ⅓in. in diameter; involucral bracts silky, tomentose. Receptacle flat. Florets as in the type, but more glandular. Mount Cardrona and Old Man Range, 4,000ft. to 5,000ft., *Petrie!*

13. **C. Featherstonii,** *F. Muell.* ex *Hook. f., Handbk.* 733. Stem prostrate, rather stout, branched, the entire plant downy. Leaves obovate or oblong-cuneate, narrowed to the sessile base, fleshy; apex crenately 3-toothed. Peduncles ½in.–2in. long. Heads ¼in.–⅓in. in diameter; involucral bracts 8–14, downy, unequally 2-seriate, the outer ovate-lanceolate, the inner obtuse. Receptacle hemispherical, depressed. Female florets in several rows, glandular, stipitate; corolla minute, conical, half as long as the achene; style-arms very short; achene obovate-cylindric, streaked, glandular. Disk-florets tubular, 4-toothed, glandular; stigma minutely toothed.—*Leptinella Featherstonii*, F. Muell., Veg. Chath. Isds. 27, t. 5.

CHATHAM Islands, *H. Travers! Cox!*

This fine species resembles *Centipeda* in habit.

14. **C. Goyeni,** *Petrie in Trans. N.Z.I.* xviii. (1885) 295. Stem with leaves ¼in. in diameter, creeping and rooting, much branched. Branches short, ascending at the tips. Leaves imbricating, glabrous glabrate or almost woolly, broadly oblong, ⅛in.–⅕in. long, sessile by a broad sheathing membranous base, the blade above pectinately divided into 5–7 very narrow linear acute segments. Heads about ⅙in. in diameter, on short terminal woolly peduncles as long as the head; involucral bracts in 2 series, oblong-ovate, membranous, with scarious purple margins. Receptacle conical. Female florets 4–6; corolla broadly oblong or nearly ovate, compressed; mouth minutely 4-toothed; stigma minute, rounded or shortly bifid, with flattened arms; achene (immature) apparently oblong. Male florets tubular below, funnel-shaped above, 4-lobed.

Var. **pinnatisecta.** Leaves silky, tomentose, pinnatisect.

SOUTH Island: Otago: Mount Pisa, Hector Mountains, and Old Man Range, *Petrie!* 5,000ft. to 6,000ft. Feb., March.

A remarkable species, with leaves resembling those of *Azorella Selago*, but more deeply pectinate.

15. **C. linearifolia,** *Cheesem. in Trans. N.Z.I.* xv. (1883) 299. Stem branched, prostrate, ascending at the tip. Leaves gland-dotted, fleshy, aromatic, narrow-linear or linear-spathulate, ½in.–1½in. long, obtuse, entire, narrowed into the short sheathing petioles. Scape slender, erect, minutely pilose, 2in.–4in. long, with 4–8 minute bracts. Heads unisexual, ¼in.–⅓in. in diameter; involucral bracts in 3 rows, linear-oblong, obtuse, herbaceous, with scarious margins. Receptacle convex. Florets with rounded transparent glands. Female corolla slightly tetragonous, swollen at the base, thick and fleshy, narrowed

above; mouth 4-lobed; lobes erect; achene linear-obovate, compressed. Male florets smaller and more slender.

SOUTH Island: mountains flanking the Wairau Valley, Nelson, 3,000ft. to 4,500ft.

Closely related to *C. pyrethrifolia* in the structure of the florets, bracteate scape, and texture, but separated by the entire leaves and strict habit. I have only seen 2 small specimens, which might readily be mistaken for *Abrotanella linearis* by a casual observer.

16. **C. pyrethrifolia,** *Hook. f., Handbk.* 142. Rhizome short, stout, often woody, glabrous or sparingly pubescent. Leaves $\frac{1}{2}$in.–1in. long, fleshy or coriaceous, the upper half pinnatifid; lobes or segments 3–5, alternate, linear, obtuse, $\frac{1}{12}$in.–$\frac{1}{4}$in. long, rarely bifid at the tip. Scapes $\frac{1}{2}$in.–3in. long, with 1 or several linear bracts. Heads unisexual, large, $\frac{1}{3}$in.–$\frac{3}{4}$in. in diameter; involucral bracts in 2–3 series, linear-oblong, obtuse, with red or purple margins. Florets numerous, glandular; female tubular, inflated, base truncate; mouth contracted, 4-toothed; achene oblong-cuneate. Disk-florets narrow, tubular below; mouth broadly funnel-shaped, 4-lobed.

SOUTH Island: Marlborough: Kaikoura Mountains, *Buchanan!* Nelson: Mount Arthur, Mount Peel, and Raglan Mountains, *Cheeseman.* Fowler's Pass, Mount Captain, &c., *T. K.* Franklin Mountains, *Gibbs* and *Bryant!* Wairau Gorge, *Travers!* Tarndale, *Sinclair!* Canterbury: Kowhai River and Mount Torlesse, *Haast!* Mount Enys and mountains above Broken River, *Enys!* Mount Peel, *W. Barker!* Otago: *Hector* and *Buchanan.* 2,500ft. to 6,500ft. Jan., Feb.

17. **C. perpusilla,** *Hook. f., Handbk.* 143. Rhizomes rigid, wiry, extensively creeping, often forming large patches. Leaves tufted, $\frac{1}{4}$in.–1in. long, $\frac{1}{6}$in. broad, white with rather long silky hairs, linear-oblong, obovate in outline, pinnatifid to the base or shortly petiolate; segments slightly recurved, close-set, the upper margin deeply serrate or entire. Scapes short, rigid, silky, $\frac{1}{4}$in.–$\frac{3}{4}$in. long. Heads unisexual: male $\frac{1}{8}$in.–$\frac{1}{7}$in. in diameter; involucral bracts in 2 rows, broadly obovate, with purple margins, pubescent; receptacle conical; florets funnel-shaped; mouth broad, 5-lobed: female $\frac{1}{6}$in.–$\frac{1}{5}$in. in diameter; involucral bracts broader than in the male; receptacle small, convex, slightly incurved; corolla narrow obovoid, excessively contracted at the erose-denticulate mouth, compressed; achene curved, shortly clavate, obscurely trigonous, with a narrow obtuse marginal ridge, convex at the back.—*Leptinella pusilla,* Hook. f., Fl. N.Z. i. 129.

NORTH Island: East Cape, *Colenso.* SOUTH Island: Nelson: Mount Arthur, *Cheeseman.* Tarndale, *Sinclair!* Canterbury: Aylesbury, *Enys!* Burnham Plains, &c., *T. K.* Westland: Okarito, *A. Hamilton!* Otago: not uncommon except on the western side; Maniototo, Cromwell, Old Man Range, &c., *Petrie!* Riverton, on sand-dunes, *T. K.* Sea-level to 4,000ft. Oct. to Feb.

18. **C. obscura,** *n. s.* Stems very short, slightly creeping, $\frac{3}{4}$in.–1in. long, rather stout for the size of the plant. Leaves few, tufted, glabrous, fleshy, $\frac{1}{3}$in.–$\frac{1}{2}$in. long including the petiole, oblong-lanceolate, obtuse, shortly pinnatifid or with 3 obtuse lobes on each side. Scapes longer or shorter than the leaves, glabrous. Heads unisexual, about $\frac{1}{10}$in.–$\frac{1}{8}$in. in diameter; involucral bracts 5–6, oblong or oblong-ovate, fleshy, glabrous or glabrate. Receptacle conical. Female florets few; corolla tubular, as long as the ovary; mouth minutely

3–4-toothed, eglandular; achenes (immature) linear-oblong, cuneate. Male florets trumpet-shaped; mouth broad, 4-lobed.

SOUTH Island: in swamps at Woodend, Southland, *T. K.*, Nov. (1887).

Better specimens of this curious little plant are much wanted, the material collected not being sufficiently advanced. The female scapes seen are very slender.

19. **C. pulchella,** *n. s.* Stems slender, rather wiry, 1in.–10in. long or more, hairy. Leaves ⅓in.–2½in. long, glabrate or pilose, membranous, linear-oblong, obovate, obtuse, pinnatifid or pinnate at the base; segments in 6–9 pairs, narrow, deeply 2–4-toothed at the tips: teeth sometimes piliferous; petiole long, slender. Heads unisexual; peduncles axillary, usually shorter than the leaves. Female: ⅛in.–½in. in diameter; involucre hemispherical; involucral bracts in 3 rows, outer orbicular-ovate with erose purple or green margins, glabrate or pubescent, inner linear-oblong; corolla eglandular, ovoid-conical, about one-third as long as the ovary; mouth denticulate; achene stipitate, slightly curved, turgid, plano-convex. Male: rather smaller; corolla funnel-shaped, 5-lobed.

SOUTH Island: in boggy ground near Lincoln, Canterbury. Invercargill: mouth of the Oreti River; the Bluff Hill. STEWART Island: rare and local, *T. K.*

Nearly related to *C. dioica*, but the leaves are much longer and more deeply divided, never flaccid; the female corolla is longer and narrower, and the stipitate achene is turgid and convex at the back. Occasionally the heads are slightly heterogamous, one or two female florets being found in the outer row of male florets, and more rarely a male floret in the centre of the female head. In minute specimens linear-oblong bracts are not developed in the involucre and the leaf-segments are often entire.

20. **C. dioica,** *Hook. f., Handbk.* 143. Stems creeping, usually rather robust, 3in.–12in. long. Leaves rather flaccid, tufted or solitary, glabrous; petiole ½in.–2in. long, obtuse, crenate-serrate or lobed or semipinnatifid; lobes serrate on the upper margin or entire; teeth acute. Scape naked, longer or shorter than the leaves, pubescent. Heads unisexual, ⅛in.–⅓in. broad. Female: involucral bracts in 3 rows, pubescent, broadly ovate to oblong-orbicular; margins erose, purple or green, the innermost linear-oblong; receptacle slightly convex (not conical); corolla broadly ovoid or almost orbicular, eglandular, slightly inflated; mouth minutely denticulate; style-arms short. Male: smaller; receptacle usually conical; corolla funnel-shaped; mouth broad; stigma discoid; achene broadly obconic.—Banks and Sol. MSS. and Icon. *Leptinella dioica*, Hook. f., Fl. N.Z. i. 129.

Var. **crenatifolia.** Leaves sometimes 3in. long on longer petioles, crenate-lobed or pectinately pinnatifid.

NORTH and SOUTH Islands: from the Great Barrier Island to Southland. Not unfrequent in littoral situations. Var. *crenatifolia* less frequent in inland swamps. Dec. to Feb.

21. **C. squalida,** *Hook. f., Handbk.* 143. Stems slender, creeping, woolly or silky, 6in.–15in. long or more. Leaves 1in.–2in. long, linear-oblong, obovate, flaccid, glabrate or silky, pinnatifid, petiolate; segments usually lax, soft, incised along the upper margin or both. Scapes 1in.–3in. long, weak, silky. Heads unisexual. Male: ⅛in. in diameter; involucral bracts few, mostly in 2 series,

broadly obovate, with purple margins, pubescent; receptacle convex; florets numerous, trumpet-shaped. Female: ¼in. in diameter; involucral bracts in 3–4 series, mostly incurved, with purple margins, the outer suborbicular, the inner linear-obtuse; receptacle narrow, conical; corolla short, ovate, inflated, greatly contracted at the mouth, minutely 4-toothed. Achene curved, almost clavate, convex at the back.—*Leptinella squalida,* Hook. f., Fl. N.Z. i. 129.

NORTH Island: East Cape and Hawke's Bay, *Colenso.* Hikurangi, *J. B. Lee!* SOUTH Island: Nelson: Wairau Valley, *Cheeseman!* Clarence and Waiau-ua Valleys; Spencer Mountains; *T. K.* Canterbury: Akaroa, *Raoul!* Upper Canterbury Plain, *Travers!* Upper Waimakariri, *Berggren! Enys!* Westland: Jackson's Bay, *T. K.* Otago: Lake District, St. Bathan's, Te Anau, &c., *Petrie!* Martin's Bay, *T. K.* Sea-level to 2,500ft. Jan., Feb.

Easily distinguished by the soft pale incised leaves and the incurved bracts of the pistillate heads, completely hiding the florets.

18. CENTIPEDA, Lour.

Involucre hemispherical; bracts in about 2 rows, membranous, with scarious margins. Receptacle naked, depressed. Female florets very numerous, in many series; corolla minute, tubular; style-arms short. Male florets few, central; corolla short, broadly campanulate; style-arms very short. Achenes thick, tetragonous or nearly so. Pappus 0. Small annual or perennial herbs, with erect or prostrate stems, alternate leaves, and axillary or rarely terminal heads.

A genus comprising about 4 species, 1 of which is found in Australia, China, India, extra-tropical South America, and New Zealand. The others are Australian.

1. **C. orbicularis,** *Lour., Fl. Cochinch.* ii. 493. A strong-smelling much branched prostrate or rarely suberect annual. Stems 3in.–6in. long. Leaves ¼in.–⅜in. long, oblong or oblong-lanceolate, cuneate below, sparingly irregularly toothed or almost lobed, glabrous or glabrate, rarely hairy. Heads axillary, ⅛in.–¼in. in diameter, numerous, sessile or rarely on very short peduncles; involucral bracts in 2 rows, broadly oblong or cuneate-oblong, membranous. Receptacle hemispherical, much depressed. Female florets in several rows, excessively numerous; corolla minute, 4-lobed; style lobes short, obtuse. Disk-florets few; style-arms short, truncate. Achenes obtusely tetragonous, streaked, hairy.—F. Muell., Key to Vict. Pl. 312. *Myriogyne minuta,* Less. in Linnaea vi. (1831) 219; Hook. f., Handbk. 144.

NORTH and SOUTH Islands: from Awanui and Mangonui to the Lake district and central Otago. Sea-level to 2,000ft. Dec. to Feb.

* SOLIVA, Ruiz and Pav.

Involucral bracts in about 2 rows; margins scarious. Receptacle flat, naked. Outer florets female, in several rows; corolla 0. Disk-florets tubular, narrowed at the base, hermaphrodite but usually sterile; style-arms short, truncate, or stigmas entire. Achenes flattened, bordered by a thick or rigid wing, which invests the base of the persistent style, or is produced into divergent points. Pappus 0. Small diffuse herbs, with alternate finely-dissected leaves and sessile flower-heads.

Heads crowded. Achenes with a broad thick wing, truncate above .. .. * *S. anthemifolia.*
Heads rather distant. Achenes with a broad rigid wing, ovate .. .. * *S. sessilis.*

* **S. anthemifolia,** *R. Br. in Trans. Linn. Soc.* xii. (1817) 102. Densely tufted. Stems prostrate, 1in.–3in. long. Leaves petiolate, 2in.–4in. long, 2–3-pinnate; segments linear, entire or 3-fid, glabrate or clothed with long soft hairs. Heads sessile, crowded, ¼in.–½in. in diameter, nearly globular in fruit; involucral bracts oblong or lanceolate. Achenes bordered by a thick transversely rugose wing, truncate and villous above, as long as the naked persistent style.

NORTH Island: naturalised. Near Dargaville and Mangawhare, *Cheeseman!* Jan. Australia, Brazil.

* **S. sessilis,** *Ruiz and Pav., Prod.* 113, *t.* 24. Densely tufted. Stems suberect, much branched. Leaves 1in.–3in. long, pilose or villous, petiolate, 2-pinnate or pinnatipartite; segments 3–5-lobed; lobes linear, acute. Heads sessile, rather distant; involucral bracts ovate or ovate-lanceolate, strigose. Achenes broadly ovate; wing broad, flat, membranous, with the upper margin produced into short horn-like processes. Style as long as the achene itself.

NORTH Island: naturalised. Lower and middle Waikato, *Cheeseman! T. K.* Jan. to March. Chili.

19. ABROTANELLA, Cass. in Dict. Sc. Nat. xxxvi. 27 (1825).

Involucral bracts few, in about 2 rows, erect, coriaceous, nearly equal or the outer shorter. Receptacle narrow, flat, naked. Florets all tubular; mouth unequally 3- or 4-toothed. Marginal florets female, inflated; style-arms short, truncate. Disk-florets male or hermaphrodite; anthers nearly free, obtuse or shortly tailed at the base; style truncate, minutely denticulate. Achene angled or compressed, ribbed or horned, glabrous or rarely setose. Small moss-like herbs, with minute glabrous imbricating leaves and sessile or subsessile inconspicuous heads, rarely with lax leaves and an elongated leafy scape.

A small genus of about 14 species, of which 7 are endemic in New Zealand, the Auckland and Campbell Islands, 3 are restricted to Australia, and about 4 others are found in Fuegia, the Falkland Islands, &c.

* *Heads numerous.*

Scape erect, 1in.–3in. high. Heads 2–8 on bracteate peduncles .. .. 1. *A. spathulata.*
Stems prostrate or suberect. Heads 3–6, partially hidden amongst the apical leaves .. .. 2. *A. rosulata.*

** *Heads solitary.*

Scapes bracteate, erect, ½in.–1½in. high. Leaves ½in.–2in. long. Achenes clavate 3. *A. linearis.*
Scapes ⅜in.–½in. high. Leaves minute .. .. 4. *A. caespitosa.*
Stems prostrate, ½in.–2in. long. Achene linear-clavate, with 4 prominent ribs .. 5. *A. inconspicua.*
Stems 1in. long. Achene linear-clavate, 4-angled .. .. 6. *A. pusilla.*
Minute. Stems ½in.–¾in. long. Achene tetragonous, setose .. .. .. 7. *A. muscosa.*

1. **A. spathulata,** *Hook. f., Handbk.* 139. Tufted, 1in.–3in. high. Leaves narrow linear-spathulate, ½in.–1in. long, obtuse or acute, rather close-set, spreading, coriaceous. Scapes sparingly leafy or with 1 or 2 or more leafy bracts near the apex. Heads on short bracteate peduncles crowded near the top of the scape; involucral bracts 8–10, oblong, with 3 translucent nerves. Florets 8–12: female with short spreading teeth; achene obovoid, compressed, with 3 cellular ribs: male with erect teeth; achene tetragonous, compressed.—*Trineuron spathulatum*, Hook. f., Fl. Antarc. i. 23, t. 17.

AUCKLAND Islands: 1,000ft. to 2,000ft. CAMPBELL Islands: 500ft. to 800ft. Jan., Feb.

The largest species. The cellular structure of the veins of the involucral bracts and of the ribs of the achenes becomes obscured after maturation.

2. **A. rosulata,** *Hook. f., Handbk.* 139. Tufted. Stems prostrate or suberect, ½in.–1½in. high. Leaves ¼in.–⅓in. long, narrow-ovate or lanceolate, imbricating, acute or subacute, patent or recurved, concave above, rigid, very coriaceous, 5-nerved beneath. Heads 3–6, terminal, partly hidden by the apical leaves; involucral bracts 8–10, linear-oblong, 3-nerved. Female florets tubular, swollen at the base, 4-toothed; achene tetragonous, 4-ribbed, narrowed below, each rib produced upwards into a short horn. Male florets 4-angled; teeth erect.—*Ceratella rosulata,* Hook. f., Fl. Antarc. i. 25, t. 18.

CAMPBELL Island: rare and local; 800ft. to 1,000ft. Jan., Feb.

3. **A. linearis,** *Bergg. in Minneskr. Fisiog. Sälisk, Lund.* viii. (1877) 14, *t.* 3, *f.* 28–38. Tufted, scapigerous, rarely exceeding 2in. in height. Leaves equalling or slightly exceeding the scape, ½in.–2in. long, linear, obtuse or apiculate, with sheathing slightly hairy bases. Scapes ½in.–1½in. high, with 1 or several linear obtuse bracts. Head solitary; involucral bracts subacute, obscurely 3-nerved. Female florets about 4, deeply 4-toothed; achenes clavate, rounded above, 4-ribbed. Male: corolla longer; achene tetragonous, abruptly truncate above.—T. Kirk in Trans. N.Z.I. xxiv. (1891) 420.

SOUTH Island: mountains of Nelson, Westland, Canterbury, and Otago, 2,000ft. to 3,500ft. STEWART Island, sea-level to 2,000ft. Dec., Jan.

Diminutive specimens are sometimes less than ⅜in. high; on the other hand, the leaves occasionally exceed 2in. in length, and the rootstock is sometimes largely developed, the decaying bases of the old leaves being more or less persistent.

4. **A. caespitosa,** *Petrie MS.* ex *T. Kirk in Trans. N.Z.I.* xxiv. (1891) 420. Tufted. Stems ⅜in.–½in. high. Radical leaves recurved, ⅜in. long, linear obtuse, concave above, with a scarious margin when young, coriaceous. Scapes naked or with 1 or 2 short bracts. Head solitary; involucral bracts 7 or 8, broadly rounded at the apex, 3-nerved; margins broad, scarious. Florets 6–12. Female achenes clavate, rounded above, obscurely 4-ribbed. Disk-achenes tetragonous, abruptly truncate.

SOUTH Island: mountains above the Broken River basin; Craigieburn Mountains, Canterbury; Clarke's Diggings, Mount Ida and Mount Kyeburn, Otago; *Petrie!* 3,500ft. to 5,000ft. Nov. to Jan.

A curious little plant, combining the leaves of *A. spathulata* with the solitary head of *A. linearis,* but smaller in all its parts than either. In the original description only immature achenes were available; better specimens are still wanted. Perhaps only an aberrant form of *A. linearis,* although it looks very different.

5. **A. inconspicua,** *Hook. f., Handbk.* 140. Stems prostrate, ½in.–2in. high. Leaves ¼in.–⅓in. long, densely crowded, spreading or ascending, subulate or linear-oblong, concave at the base, subacute, rigid and rather flat when dry. Head solitary, nearly hidden amongst the upper leaves; involucral bracts linear-oblong, 3-nerved, obtuse. Florets 15–20: female slightly swollen at the base;

mouth with 4 narrow spreading lobes; achene linear-clavate, with 4 prominent ribs: male with larger corolla.—Buch. in Trans. N.Z.I. xiv. (1881) 354, t. 34, f. i.

SOUTH Island: Otago: Mount Alta and Black Peak, *Hector* and *Buchanan!* Common on all high mountains, 4,000ft. to 6,000ft., *Petrie!* Dec., Jan.

When dry the leaves appear to be partially flattened, with a strong marginal nerve, which is scarcely observable in fresh specimens. Closely related to *A. pusilla*, of which it may be a variety.

6. **A. pusilla,** *Hook. f., Handbk.* 139. Stems slender, prostrate, emitting wiry roots. Leaves ½in.–⅓in. long, narrow-linear, acute, flat above, with a prominent midrib below. Head solitary, shortly pedunculate, sunk amongst the upper leaves; involucral bracts linear, obtuse, ribbed. Achenes linear-clavate, 4-angled.

NORTH Island: snowy places amongst the Ruahine Mountains, *Colenso!*

I have copied the description from Hooker, as the species has not been found since its discovery nearly fifty years ago. My thanks are due to the Director of the Kew Herbarium for a precious fragment of Colenso's specimen.

7. **A. muscosa,** *T. Kirk in Trans. N.Z.I.* xxiv. (1891) 422, *t.* 36. Minute. Stems ⅛in.–¼in. high, or forming depressed patches ½in.–1in. in greatest diameter. Leaves densely imbricating, ⅛in.–⅐in. long, erect, linear, concave above, excessively coriaceous, truncate or retuse, with a strong marginal nerve, rarely rounded at the tip. Heads solitary, sunk amongst the apical leaves, very shortly pedunculate; involucral bracts 5, oblong, acuminate, obtuse or acute, nerves indistinct. Florets 4 or 5. Female corolla narrow, tubular, indistinctly 4-toothed, with a bristle springing from each angle of the ovary; achene truncate above, tetragonous, setose, with an erect bristle at each corner as long as the achene itself. Male achenes smaller, abortive.

STEWART Island: summit of Rakiahua, 2,300ft., *T. K.* Jan.

A singular little plant, closely resembling a sterile *Tortula* or *Bryum*. It closely approaches the original *A. emarginata*, Cass., from the Falkland Islands, and, with the exception of *Lemna*, is the smallest flowering-plant in the colony. It is the only species with setose achenes.

* TANACETUM, Tourn.

Head discoid or nearly so; involucral bracts imbricating in several series. Receptacle convex, naked. Florets all tubular, fertile; female marginal, 1-seriate, 3–4-toothed. Disk-florets 4–5-toothed; anthers obtuse at the base. Achenes angled or ribbed, with an epigynous disk, crowned with a membranous border. Strong-scented herbs, with erect stems, sometimes suffruticose. Leaves alternate. Heads solitary or corymbose.

* **T. vulgare,** *L., Sp. Pl.* 844. Stems erect, 2ft.–3ft. high, grooved, glabrous. Leaves 2in.–4in. long, oblong, the lower petioled, the upper half-amplexicaul, pinnate; leaflets linear, pinnatifid or 2-pinnatifid or inciso-serrate, glandular-dotted. Heads numerous, corymbose; involucral bracts coriaceous, the outer acute, the inner longer, obtuse. Female florets terete, obliquely truncate, 3-toothed, rarely 0. Achene obovoid, 5-ribbed. Disk membranous, 5-lobed.

NORTH and SOUTH Islands: established in several localities from Auckland to Otago, but nowhere abundant. *Tansy*. March. Europe.

*ARTEMISIA, Linn.

Heads small, discoid; involucral bracts in several series, imbricating, margins scarious. Receptacle narrow, flat or convex, naked, pilose, or fimbriate. Florets tubular, few; marginal female, 1-seriate; corolla 3-toothed, rarely 0. Disk-florets hermaphrodite; corolla 5-toothed; anthers obtuse at the base. Achenes obovoid. Disk minute. Pappus 0. Herbs, sometimes suffruticose, often bitter and aromatic. Leaves alternate, much divided. Heads racemose or paniculate.

*A. **Absinthium**, *L.*, *Sp. Pl.* 848. Silky-pubescent in all its parts. Stems numerous, erect, 1ft.–2ft. high, often suffruticose, grooved. Leaves 1in.–3in. long, 2–3-pinnatifid or pinnatipartite; segments numerous, gland-dotted, oblong or lanceolate. Heads nodding or drooping, hemispheric in erect leafy panicles; involucres broadly campanulate, silky. Receptacle pilose. Female corollas dilated below.

NORTH and SOUTH Islands: plentifully naturalised in waste places and on sheep-runs. Sea-level to fully 3,000ft. *Wormwood*. Feb. to April. Europe.

20. ERECHTITES, Rafin.

Involucre of 1 series of nearly equal linear bracts, usually with a few smaller at the base, herbaceous, appressed. Receptacle naked. Rays 0. Florets all narrow, tubular; female always filiform, extremely slender, in 2–3 marginal rows, 3–4-toothed. Disk-florets about half as many as the female, 4–5-toothed, hermaphrodite; anthers obtuse at the base; style-arms truncate. Achenes striate or angular, glabrous or pubescent, sometimes contracted immediately beneath the disk-like apex. Erect annual or perennial glabrous or cottony herbs, with alternate simple or pinnatifid leaves and corymbose cylindric heads.

SPECIES, about 17, of which 4 are endemic in Australia and 3 in New Zealand, 4 others are common to both countries, the remainder being distributed through North and South America and Java. The ancient name of a species of groundsel.

Leaves glabrous, membranous, toothed or pinnatifid. Involucral bracts 8–10 . . 1. *E. prenanthoides.*

Involucral bracts 10–14.

Leaves lobed toothed or pinnatifid, cottony. Achenes short, hispidulous .. 2. *E. arguta.*
Scabrid. Leaves pinnatifid. Achene linear, slender, pubescent .. .. 3. *E. scaberula.*
Leaves mostly linear, entire, cottony. Achene hispid .. .. .. 4. *E. quadridentata.*
Glabrous or glabrate. Leaves erect; lower on long petioles, obtuse; upper sessile, acute. Achenes pubescent.. .. .. 5. *E. diversifolia.*
Glabrous. Leaves spreading, unequally pinnatifid. Achenes glabrous .. 6. *E. glabrescens.*

1. **E. prenanthoides**, *DC.*, *Prod.* vi. 296. Annual or rarely biennial, erect, 1ft.–4ft. high, simple or branched, glabrous or slightly hairy. Leaves 2in.–6in. long, linear-oblong or lanceolate, lower petioled, upper sessile, with toothed auricles, regularly or irregularly denticulate or toothed. Heads numerous, in lax terminal corymbs, glabrous; pedicels slender, with minute subulate bracts; involucres narrow-cylindric; bracts 8–10, linear-lanceolate, with scarious margins. Florets about 18–20; female in 2 rows, filiform, 4-toothed. Disk-florets 6–7, 4-lobed. Achenes angular, truncate above, glabrate or pubescent. Pappus-hairs rigid.—Hook. f., Fl. N.Z. i. 141; Handbk. 156; Benth., Fl. Austr. iii. 658. *E. sonchoides*, DC., Prod. vi. 296.

E. picridioides, Turc. in Bull. Soc. Nat. Hist. Mosc. xxiv. (1851) I. 200. *Senecio prenanthoides*, A. Rich., Sert. Astrol. 96. *S. heterophylla*, Col. in Trans. N.Z.I. xxvii. (1894) 389. *S. tabidus*, Banks and Sol. MSS. and Icon.

NORTH and SOUTH Islands: from Mangonui to Southland. STEWART Island. *T. K.* CHATHAM Islands. Sea-level to 2,000ft. Oct. to Jan. Also in Australia.

2. **E. arguta,** *DC., Prod.* vi. 296. Annual or biennial, erect. Stem furrowed, stout, 2ft.–4ft. high, more or less scabrid, cottony or woolly, especially on the undersurface of the leaves, rarely glabrate. Leaves 2in.–4in. long, linear-oblong or lanceolate, acute or obtuse, the lower petioled, the upper with toothed auricles, irregularly pinnatifid, lobed or toothed, the teeth obtuse or acute. Heads smaller than in *E. prenanthoides;* corymb dense. Involucral bracts about 12, often with a few outer at the base, shorter than the florets, linear-lanceolate, acute; tips sometimes squarrose. Achene short, angular, hispidulous or glabrate.—Hook. f., Fl. N.Z. i. 142; Handbk. 157; Benth., Fl. Austr. iii. 659. *E. Bathurstiana*, DC., Prod. vi. 297. *Senecio arguta*, A. Rich., Fl. N.Z. 258; Endlich., Prod. Fl. Norf. n. 101; A. Cunn., Precurs. n. 466. *S. Lessoni*, F. Muell., Cat. Hort. Melb. 1858. *S. plebeius*, Banks and Sol. MSS. and Icon.

NORTH and SOUTH Islands: from the Three Kings Islands and North Cape to Foveaux Strait. STEWART Island. Nov. to Feb. Also in Australia.

The involucres are often purple.

3. **E. scaberula,** *Hook. f., Handbk.* 157. Annual, slender, erect, 1ft.–2ft. high. Stem grooved, hispid. Leaves linear-oblong or linear-lanceolate, acute, 2in.–4in. long, hispid on both surfaces, irregularly pinnatifid, lobed or toothed, teeth acute; lower petioled, upper sessile with small auricles. Heads in a lax spreading corymb; pedicels and involucres glabrous; involucral bracts 12–14, subulate, acuminate. Receptacle convex. Achene linear-oblong, grooved, pubescent.—*E. hispidula*, Hook. f., Fl. N.Z. i. 142 (not of DC.). *E. pumila*, J. B. Armst. in Trans. N.Z.I. xiii. 338. *Senecio hispidulus*, A. Cunn., Precurs. n. 462 (not of A. Rich.). *S. incomptus*, Banks and Sol. MSS. and Icon.

NORTH and SOUTH Islands: from the North Cape to Southland. STEWART Island; CHATHAM Islands. Nov. to Feb.

Armstrong's *E. pumila* is merely a starved state in which the inflorescence is reduced to a single head.

4. **E. quadridentata,** *DC., Prod.* vi. 295. Usually much branched from the base and suffrutescent, 1ft.–3ft. high, the entire plant more or less cottony or hoary. Leaves linear, linear-lanceolate, rarely oblong-lanceolate, the lower sometimes petioled, with or without a few distant teeth, the upper sessile, rarely auricled; margins revolute. Heads very slender, in a corymbose panicle; involucral bracts 12–14, glabrous or cottony, outer bracts very few or 0. Florets about 20–30. Disk-florets about 8–9. Achene linear, striate, abruptly contracted at the neck, glabrous or pubescent, slender.—Hook. f., Fl. N.Z. i. 142; Handbk. 157; Benth., Fl. Austr. iii. 660; *E. glandulosa* and *E. glabrescens*, DC., Prod. vi. 295. *E. tenuifolia*, DC., *l.c.* 296. *E. Gunnii*,

Hook. f. in Lond. Journ. Bot. vi. (1847) 122. *E. incana,* Turc. in Bull. Soc. Nat. Mosc. xxiv. (1851) II. 85. *Senecio quadridentatus,* Labill, Pl. Nov. Holl. ii. 48, t. 194; A. Cunn., Precurs. n. 461. *S. angustifolius,* Banks and Sol. MSS. and Icon.

NORTH and SOUTH Islands: from the Three Kings Islands and North Cape to Southland. CHATHAM Islands. Nov. to Jan.

Var. **lanceola.** Radical leaves 2in.-5in. long, on long petioles, ½in.-¾in. broad, lanceolate, irregularly toothed or serrate. Heads immature. SOUTH Island: Lynton Downs, Kaikoura, *T. K.* Near Lincoln, *T. K.* Possibly a new species. Also in Australia.

5. **E. diversifolia,** *Petrie in Trans. N.Z.I.* xix. (1887) 324. Slender, erect, 1ft.–2ft. high. Stems grooved, simple or sparingly branched, leafy below. Leaves 2in.–4in. long, erect, oblong or linear-oblong, the lower on long petioles, irregularly crenate-toothed or sinuate-toothed or the teeth rarely acute, obtuse, puberulous beneath. The cauline leaves are narrow, sessile or petioled, acute or acuminate, glabrate or slightly cottony. Heads small, forming a lax irregular corymb; pedicels slender, minutely bracteolate; involucral bracts 10–14, linear-subulate. Receptacle convex. Florets 20–40. Achenes linear-oblong, grooved, pubescent or hispid, sometimes contracted at the neck.

SOUTH Island: Central Otago, but rather local, *Petrie!* The Bluff Hill, *Enys!* STEWART Island, *T. K.* Nov. to Jan.

6. **E. glabrescens,** *T. Kirk in Trans. N.Z.I.* ix. (1877) 550 (not of DC., Prod. vi. 295). Annual, erect, glabrous or glabrate, 1ft.–3ft. high. Stem grooved, simple or sparingly branched. Leaves membranous, 3in.–6in. long, 1in.–1½in. broad, oblong or lanceolate-oblong, more or less pinnatifid or rarely pinnate below with a large terminal segment, lobed, toothed, or the margins sinuate-dentate, the lower petioled, the upper sessile with large toothed auricles. Heads laxly corymbose; involucral bracts about 12, linear-acute, with scarious margins. Florets about 20, all minutely toothed. Achenes linear, faintly grooved, glabrous.

SOUTH Island: not unfrequent in mountain woods from Marlborough to Southland. STEWART Island. Jan., Feb.

The achenes are longer than those of any other New Zealand species, and the leaves are often purple beneath.

21. BRACHYGLOTTIS, Forst.

Heads heterogamous; involucral bracts 1-seriate, shining, scarious, linear, obtuse, the outer minute. Receptacle narrow, alveolate. Marginal florets female; corolla irregularly lobed or 2-lipped, outer lip short, broad, inner narrow, revolute. Disk-florets hermaphrodite, tubular; mouth campanulate; anthers with short tails; style-arms truncate, papillose. Achenes very short, terete, papillose. Pappus-hairs 1-seriate. Shrubs or small trees, with tomentose branches and large leaves clothed with milk-white tomentum beneath. Heads very numerous, in large crowded much-branched panicles.

The genus is endemic in the northern part of the colony.

Leaves dull. Involucres white, shining .. .. 1. *B. repanda.*

Leaves larger, glossy. Involucres purplish, opaque .. .. .. 2. *B. Rangiora.*

1. **B. repanda,** *Forst., Char. Gen.* 46, *t.* 40. A shrub or small tree, 8ft.–20ft. high. Branches and petioles covered with soft white downy tomentum. Leaves 1½in.–6in. long, broadly ovate-oblong, sometimes cordate at the base; margins sinuate or irregularly lobed, clothed with milk-white tomentum beneath; petioles 1in.–3in. long. Panicles usually terminal, exceeding the leaves, often drooping. Branches slender, with 1 or 2 angular leaves at the base. Heads excessively numerous, $\frac{1}{7}$in.–$\frac{1}{8}$in. long, whitish; involucral bracts oblong, shining. Florets 10–12. Female florets with the outer lip lobed or entire.—A. Cunn., Precurs. n. 463; Hook. f., Handbk. 163. *Senecio Forsteri*, Hook. f., Fl. N.Z. i. 148, t. 41. *Cineraria repanda*, G. Forst., Prod. n. 295; Willd., Sp. Pl. iii. 2076; A. Rich., Fl. N.Z. 250. *C. dealbata*, Banks and Sol. MSS. and Icon.

NORTH and SOUTH Islands: from the North Cape to the Kaikoura Mountains on the east coast, and Greymouth on the west. Sea-level to 2,300ft. *Pukapuka. Wharangi-tawhito* (Buchanan). Aug. to Oct.

2. **B. Rangiora,** *Buch. in Trans. N.Z.I.* xiv. (1880) 357. A shrub, 8ft.–14ft. high, resembling *B. repanda*, but the branches are stouter and the leaves larger. Leaves 6in.–9in. long, more coriaceous and glossy, often unequal at the base; petiole stouter, 3in.–5in. long. Panicles axillary and terminal, usually with an entire ovate or oblong-ovate leaf at the base of the principal branches. Involucres purple; bracts opaque. Female florets with the outer lobes always (?) entire.

NORTH Island: Mokoia Island, Rotorua, *H. B. Kirk!* Silver Creek, Wellington, *H. B. Kirk!* SOUTH Island: Westland, *Hon. W. B. D. Mantell! Rangiora.* July to Sept.

A much handsomer plant than the preceding, but scarcely satisfactory as a species. The late Mr. Mantell, who had it under cultivation for many years, strenuously urged its claims to specific rank, and was supported by Mr. Buchanan. I have seen leaves 12in. broad and 15in. long without the petiole.

22. SENECIO, Linn.

Heads heterogamous and radiate or homogamous and discoid; involucral bracts nearly equal, in 1 or 2 rows, linear or ovate, herbaceous or coriaceous, usually with a few small outer bracts at the base. Receptacle naked or pitted. Outer florets female and ligulate, rarely sterile or 0. Disk-florets tubular, hermaphrodite, 5-toothed. Anthers obtuse at the base or with very short tails. Style-arms truncate, often shortly penicillate. Achenes terete, angular or striate. Pappus-hairs in 1 or several series, smooth or scabrid. Herbs, shrubs, or trees. Leaves alternate, entire toothed or pinnatifid. Heads terminal, solitary corymbose or paniculate. Florets yellow, rarely white or purple.

A large genus, comprising nearly 1,000 species, distributed through all countries, although many of the species are remarkably local. The indigenous forms comprise some of the finest members of the flora; one extends to Australia, the others are endemic. Most of the herbaceous kinds are extremely variable, and sometimes difficult to determine, but the ligneous species are easily recognised.

NAME, from *senex*, "an old man," probably in allusion to the copious white hairs of the pappus.

I. HERBACEOUS, MOSTLY PERENNIAL. HEADS YELLOW.

1. *Herbs with broad radical leaves and naked simple or branched scapes. Involucral bracts in 2 series.*

Leaves rugose and hispid above, cottony or woolly, 1in.–6in. by 1in.–3in. Scape glandular	1. *S. lagopus.*
Leaves ½in.–4in. by ⅓in.–1½in., rugose and hispid above, glabrate or cottony beneath. Scape cottony	2. *S. bellidioides.*
Leaves 2in.–5in. by 1in.–3in. Entire plant snow-white	3. *S. Haastii.*
Leaves 3in.–6in. by 1in.–4in., woolly on both surfaces. Scape glandular ..	4. *S. saxifragoides.*

2. *Scape erect, simple, leafy. Heads corymbose.*

Heads large. Achenes silky or hispid	5. *S. Lyallii.*

3. *Branched, leafy, glabrous cottony or woolly. Involucral bracts in 1 series.*

Annual, soft, ½ft.–1ft. high. Heads corymbose. Rays 0	* *S. vulgaris.*
Annual, 1ft.–3ft. high, much branched above. Rays very short	* *S. sylvaticus.*
Scandent. Leaves palmate, 3–7-lobed. Rays 0	* *S. mikanioides.*
Erect. Leaves pinnatifid or 2-pinnatifid. Flowers in large corymbs ..	* *S. Jacobaea.*
Lower leaves entire or lobed; upper lyrately pinnatifid. Heads in a lax spreading corymb	* *S. aquaticus.*
Annual or biennial. Branches stout, fistulose. Heads corymbose. Rays 0	6. *S. antipodus.*
Annual or biennial. Leaves pinnatifid. Rays 10–15	7. *S. lautus.*
Perennial, erect, 2ft.–3ft. high, much branched, glabrous. Leaves toothed, lobulate or pinnatifid. Rays 12–20	8. *S. latifolius.*
Leaves entire, oblong-ovate to oblong-lanceolate or variously lobed or pinnatifid, glabrate or tomentose on one or both surfaces. Rays revolute ..	9. *S. Colensoi.*
Stem zigzag. Leaves glaucous, oblong-ovate, amplexicaul. Heads ½in. broad. Rays about 10	10. *S. Banksii.*
Stem erect, 2ft.–5ft. high. Leaves 2in.–4in. long, 1½in. broad, amplexicaul, broadly oblong; veins prominent beneath. Heads ½in. broad ..	11. *S. pumiceus.*
Leaves lanceolate, acute or acuminate	var. *angustatus.*
Stems numerous, 1ft.–3ft. high, everywhere glaucous. Leaves 2in.–4in. long, oblanceolate to obovate-spathulate, sinuate dentate or serrate ..	12. *S. glaucophyllus.*
Stems slender, decumbent, 3in.–6in. long, white. Leaves ovate, white beneath	13. *S. Pottsii.*

II. SHRUBS OR SMALL TREES. BRANCHES AND INVOLUCRES GLABROUS, TOMENTOSE, WOOLLY, OR COTTONY. HEADS PANICLED, CORYMBOSE OR RARELY SOLITARY.

4. *Leaves linear, revolute to the midrib.*

Head solitary on a leafy peduncle	14. *S. bifistulosus.*

5. *Glabrous or nearly so. Leaves membranous.*

a. Ray-florets always present.

Leaves 6in.–12in. long, broadly lanceolate, pinnate or pinnatifid at base, scaberulous above. Heads 1in.–2½in. in diameter. Rays white ..	15. *S. Hectori.*
Leaves 2in.–4in. long, oblong to cuneate-rhomboid. Heads 1½in.–2in. in diameter. Rays distant, white	16. *S. Kirkii.*
Branches flexuose, climbing. Leaves ⅓in.–1in. long, orbicular crenate. Rays 4–6, yellow	17. *S. sciadophilus.*
Erect, much branched. Leaves ⅓in.–2in. long, ovate or elliptic-oblong, toothed. Ray-florets 2, yellow	18. *S. perdicioides.*

6. *Leaves usually white beneath, very coriaceous (membranous in 19). Heads paniculate or corymbose.*

Leaves membranous, lanceolate-oblong or elliptic-lanceolate. Panicle 12in.–18in. long, narrow. Heads small, numerous. Rays white	19. *S. myrianthos.*
Leaves lanceolate or elliptic-oblong. Panicles subcorymbose, terminal, glandular. Rays few	20. *S. Huntii.*

Leaves few, lanceolate-acuminate, 3in.–7in. long. Panicles terminal. Ray-florets about 12, narrow, contorted	21. *S. Muelleri*.
Leaves narrowed at both ends. Corymb lax. Heads few	22. *S. laxifolius*.
Leaves rounded at both ends. Corymb dense, glandular-pubescent. Heads numerous	23. *S. Greyii*.
Leaves viscid, oblong-lanceolate, flat. Corymbs terminal on a leafy peduncle	24. *S. revolutus*.
Leaves excessively viscid and coriaceous; margins recurved. Corymb lax. Peduncle naked. Heads few	25. *S. Adamsii*.
Leaves ¾in.–1½in. long, narrow oblong-obovate, crenate. Corymbs slender, lax. Heads numerous; involucres glandular	26. *S. Monroi*.
Leaves ⅔in.–1½in. long, obovate or obovate-oblong. Heads few, in leafy racemes; involucres tomentose	27. *S. compactus*.
b. Ray-florets 0.	
Leaves 2in.–6in. by 1in.–2½in., broadly lanceolate or oblong, narrowed below. Panicle slender. Heads chiefly racemose	28. *S. elaeagnifolius*.
Leaves orbicular or suborbicular, 2in.–5in. in diameter. Panicle stout. Branches numerous. Heads corymbose	29. *S. rotundifolius*.
Leaves broadly oblong, excessively coriaceous, ½in.–1in. long. Corymbs short. Peduncles short	30. *S. Bidwillii*.
Leaves 1½in.–3in. long, oblong-obovate, narrowed towards the base. Corymbs 3in.–6in. long, naked below	31. *S. viridis*.
Leaves obovate-lanceolate or spathulate. Heads on slender glabrous pedicels; involucral bracts 8, rigid	32. *S. geminatus*.
7. *Leaves crowded, ¼in. long, white beneath.*	
Heads numerous, solitary, terminal	33. *S. cassinioides*.

1. **S. lagopus,** *Raoul in Ann. Sc. Nat. Ser. III.* ii. (1844) 119, *t.* 18, *and Choix* 21, *t.* 17. Rootstock stout, thickly clothed above with long matted hairs, as is the base of the petiole. Leaves all radical, 1in.–8in. long, usually coriaceous, broadly cordate-oblong, rounded at the tip, entire or crenate or crenulate, rugose and bristly above, densely covered with white wool beneath; petioles ½in.–4in. long, stout or slender, villous. Scapes 1in.–12in. high, simple or much branched, glandular-pilose; bracts few, small, obtuse. Heads solitary or numerous. Pedicels rather slender. Involucre broadly obconic; involucral bracts oblong, tomentose. Rays ¼in.–½in. long, spreading. Achenes narrow-linear, glabrous.

NORTH and SOUTH Islands: from the Ruahine Mountains to South Canterbury, but often local. Sea-level to 4,000ft. Nov. to Jan.

Alpine specimens, with almost sessile leaves and solitary heads, are sometimes difficult to distinguish from small states of *S. bellidioides*.

2. **S. bellidioides,** *Hook. f., Fl. N.Z.* i. 144. Habit of *S. lagopus*, but smaller and more slender. Leaves ½in.–4in. long, membranous or subcoriaceous, broadly oblong or linear-oblong, entire or crenulate, usually rounded at the apex, narrowed into the slender petiole, glabrate or clothed with white or brownish tomentum beneath, more or less setose above; petioles usually woolly, rarely glabrous. Scapes very slender, 1in.–12in. high, glabrate cottony woolly or glandular-pubescent; bracts few, linear-acute. Heads solitary or numerous; involucral bracts broadly oblong, glabrate or tomentose, acute or obtuse.

Achenes linear, subulate, less slender than in *S. lagopus*.—Handbk. 159. *S. Traversii*, F. Muell. in Trans. Bot. Soc. Edinb. vii. (1861) 154.

Var. **glabratus.** Leaves broadly oblong, glabrate or glabrous below, sparingly setose above.

Var. **angustatus.** Leaves narrow linear-oblong, apex rounded or subacute.

SOUTH Island: in subalpine and alpine districts from Nelson and Marlborough to Southland. STEWART Island, *T.K.* 2,000ft. to 5,000ft. Jan., Feb.

Hooker describes the leaves as sometimes cordate, but I have not found them so, and suspect that states of *S. lagopus* have been mistaken for this.

3. **S. Haastii,** *Hook. f., Handbk.* 159. Rootstock woody, rather stout, the crown usually covered with whitish wool. Leaves orbicular-oblong, truncate or slightly cordate at the base, 2in.–5in. long, 1in.–3in. broad, obtuse or rounded at the apex, obscurely crenulate, covered with white appressed or lax cottony wool below, downy or pubescent rarely tomentose above; petioles slender, 2in.–6in. long, tomentose or woolly at the base. Scapes slender or stout, 2in.–12in. high, simple or sparingly branched, downy cottony or slightly glandular, with few linear bracts. Peduncles long. Heads 1–3, ¾in.–1in. in diameter, broadly campanulate; involucral bracts linear-oblong, obtuse or subacute, glandular tomentose or pubescent. Rays linear, spreading. Disk-florets numerous, tubular; mouth narrow. Achene linear, subulate, obscurely grooved, glabrous.

SOUTH Island: Canterbury: Castle Hill basin, *Enys* and *Kirk*. Shores of Lake Ohau; mountains near the source of the Ahuriri; Mount Cook; *Haast!* Otago: Lake Hawea, *Haast!* Maniototo, Manuherikia, *Petrie*. 1,500ft. to 4,000ft. Jan.

4. **S. saxifragoides,** *Hook. f., Fl. N.Z.* i. 144. Rootstock very stout, ¾in.–1½in. in diameter, densely clothed with long shaggy wool. Leaves all radical, 3in.–6in. long, orbicular or broadly oblong, slightly cordate, crenate or crenulate, thick, silky or villous above, densely clothed with white woolly tomentum beneath; petioles 1in.–4in. long, stout, woolly or villous. Scape 2in.–12in. high, stout, erect, bracteate, cottony or tomentose below, densely clothed with glandular white or reddish-purple tomentum above. Heads few or many, broadly campanulate; involucral bracts linear, acute, densely tomentose. Rays 25–30, broad, spreading. Disk-florets narrow. Achenes glabrous.—Handbk. 159.

SOUTH Island: Port Lyttelton; Banks Peninsula. Jan. to March.

A short obtuse tooth is sometimes developed between two crenatures. The leaves are often glabrous or glabrate on the upper surface, but never bristly as in *S. lagopus*.

5. **S. Lyallii,** *Hook f., Fl. N.Z.* i. 146. Rootstock stout, crowned with matted silky hairs. Radical leaves 1–5-nerved, linear or very narrow-linear, $\frac{1}{14}$in.–¼in. broad, acute or subacute, glabrous or pubescent or glandular-pubescent, narrowed towards the base, which is densely clothed with matted silky hairs. Stems simple, erect, 3in.–18in. high, leafy or clothed with linear bracts. Heads 1in.–2in. in diameter, on slender bracteate peduncles 2in.–5in. long, forming a simple corymb, or rarely a spreading umbel; involucral bracts in 1–2 series, linear-acuminate, usually acute, glabrate or pubescent. Ray-florets ½in.–1in. long, yellow, spreading; tube very short. Disk-florets numerous,

with a long cylindric mouth and extremely short tube. Achenes slender, slightly contracted at the neck, deeply ribbed, silky or hispid. Pappus-hairs rigid, unequal.—Handbk. 160.

Var. **scorzoneroides.** Glandular or glandular-pubescent. Stems more robust and leafy than the typical form. Radical and cauline leaves shorter and broader, fewer, 2in.–9in. long, ½in.–¾in. broad, petiolate, linear-lanceolate, acute, 5-nerved. Heads larger. Rays broader, salmon-coloured. *S. ? scorzoneroides*, Hook. f., Fl. N.Z. i. 146.

SOUTH Island: mountain-valleys from Nelson and Marlborough to Southland. Descends to sea-level on Stewart Island. Ascends to 5,000ft. above the Wairau Gorge. Jan., Feb.

* **S. vulgaris,** *L., Sp. Pl.* 867. Annual, 3in.–12in. high, branched from the base, succulent, eglandular. Leaves usually glabrous, oblong, obtuse, half amplexicaul, pinnatifid, acutely and unequally toothed. Heads in clustered cymes, small, pedicellate, drooping; involucre oblong-conical, conical in fruit; outer involucral bracts ovate-subulate; tips black. Rays 0. Achene ribbed, silky.

Naturalised throughout the colony. Sea-level to 3,000ft. *Groundsel.* Europe.

* **S. sylvaticus,** *L., Sp. Pl.* 868. Annual, erect, 1ft.–3ft. high, much branched above, slightly glandular-pubescent. Leaves 1in.–3in. long, downy, deeply and irregularly pinnatifid. Heads numerous, erect or spreading. Pedicels slender. Involucre cylindric; outer involucral bracts very few or 0. Rays very short, revolute. Achene faintly ribbed, silky.—*S. areolatus*, Col. in Trans. N.Z.I. xxvi. (1893) 317.

Naturalised from the North Cape to Southland. *Hogweed.* Jan. to April. Europe.

* **S. mikanioides,** *Otto* ex *Walp. in Otto and Dietrich Allg. Gartenz* xiii. (1845) 42. Glabrous in all its parts, rarely suffruticose. Stems 5ft.–15ft. long, flexuous, scandent. Leaves 2in.–3in. broad, palmate, truncate or deeply cordate at the base, 3–7-lobed; lobes acute; petiole 1in.–3in. long, sheathing at the base. Heads in crowded terminal corymbs or panicles. Pedicels slender, minutely bracteolate. Involucre turbinate; bracts oblong-lanceolate, acute; margins scarious. Receptacle narrow, alveolate. Florets 7–9. Rays 0. Achenes oblong, faintly ribbed, glabrous.

NORTH Island: naturalised in many places from Mangonui to Wellington. *Climbing groundsel.* June to Oct. South Africa.

* **S. Jacobaea,** *L., Sp. Pl.* 870. Perennial, erect, glabrous glabrate or rarely cottony. Stems 2ft.–4ft. high, furrowed, leafy. Lower leaves petioled, 2in.–6in. long; upper stem leaves sessile, clasping, pinnatifid or irregularly 2-pinnatifid, lobed or toothed. Heads ⅔in. in diameter, numerous, forming a dense or rarely spreading corymb; involucre campanulate; bracts oblong, acuminate, outer few, small, subulate. Rays spreading. Achenes of the ray angled, glabrous; achenes of the disk subterete, hairy.—*S. dimorphocarpos*, Col. in Trans. N.Z.I. xxvi. (1893) 316!

NORTH Island: naturalised. East Cape, *Bishop Williams!* Kaweka Range, *Sturm* (*Colenso!*). SOUTH Island: between Forest Gap and Winton, extending for miles, *T. K.* Tapanui district, *Petrie! Ragwort.* Feb. to April. Europe.

* **S. aquaticus,** *Hill, Veg. Syst.* ii. 120. Biennial, erect, glabrous or glabrate, rarely cottony, 1ft.–2ft. high. Radical leaves petioled, entire, crenate or variously lobed or divided, oblong-ovate, narrowed towards the base; cauline leaves lyrately pinnatifid, segments oblong or linear, entire or toothed. Heads in a lax spreading corymb; involucre campanulate; bracts ovate-lanceolate. Rays long, spreading. Achenes all ribbed, glabrous.

SOUTH Island: naturalised in the valley of the Buller, *Rev. F. H. Spencer!* Feb. to March. Europe.

6. **S. antipodus,** *n. s.* Annual or biennial, erect, 1ft.–2ft. high. Branches spreading from the base, very stout, grooved, ⅓in.–½in. in diameter, fistulose, glabrous except the peduncles and young leaves, mealy-tomentose beneath. Radical leaves apparently narrowed into a petiole; upper sessile by a broad auriculate base, membranous, sparingly and irregularly pinnatifid or partite, 2in.–5in. long, 1in.–3in. broad; segments toothed lobed or almost pinnatifid, acute. Heads numerous, in terminal corymbs, discoid. Peduncles bracteolate, slender. Involucres broadly campanulate, ⅓in.–½in. in diameter; involucral bracts about 20, oblong or oblong-lanceolate, acute, 2-ribbed. Receptacle flat. Florets all hermaphrodite, very numerous, trumpet-shaped, 5-lobed; anthers nearly free. Achenes short, linear-oblong, striate, glabrous or minutely puberulous.

ANTIPODES Islands, *T. K.* Jan.

A striking plant, not easily mistaken for any other New Zealand species, and nearly allied to the Fuegian *S. candidans.* In some specimens the sessile leaf-bases are from 1in.–1½in. broad.

7. **S. lautus,** *Soland.* ex *Hook. f., Fl. N.Z.* i. 145. A prostrate decumbent or erect glabrous or pubescent annual or biennial, 3in.–24in. high. Stem and branches stout or slender, grooved, flexuose. Leaves succulent or very fleshy, 1in.–2in. long, linear linear-lanceolate or broadly lanceolate, narrowed into a petiole or dilated and auriculate at the base, distantly toothed lobed or pinnatifid; segments broad or narrow. Heads ¼in.–¾in. in diameter, in few- or many-flowered corymbs; involucre campanulate; involucral bracts green, herbaceous, 2-ribbed, linear, acute, pubescent at the tips; outer bracts few, minute. Rays 10–15, spreading, sometimes revolute, rarely 0. Disk-florets numerous, equalling the involucre. Achene deeply grooved, glabrous, puberulous or pubescent.—G. Forst., Prod. n. 538 (name only); A. Cunn., Precurs. n. 457; A. Rich., Fl. Nouv.-Zel. 257; Hook. f., Handbk. 160; Benth., Fl. Austr. iii. 667. *S. neglectus,* A. Rich., Fl. Nouv.-Zel. 258. *S. rupicola,* A. Rich., Sert. Astrol. t. 37. *S. tripartitus, l.c.* 114. *S. crithmifolius,* A. Rich., *l.c.* 116. *S. pinnatifolius,* A. Rich., *l.c.* 117.

From the KERMADEC Islands, THREE KINGS Islands, and the North Cape to Southland. STEWART Island; CHATHAM Islands. Sea-level to 3,000ft., but comparatively rare inland. Oct. to March.

An extremely variable plant, many forms of which have been described as species, but pass into each other so gradually that they can hardly be distinguished even as varieties.

Var. **carnosulus.** Stems robust, prostrate. Leaves very fleshy, sessile by a broad base; lobes broad, obtuse. Heads obconic; bracts fleshy, thick. Rays very short. Achene silky. Not uncommon on maritime rocks.

Var. **radiolatus.** Lower leaves membranous, broadly cuneate-ovate, narrowed into slender petioles, toothed lobed or pinnatifid; lobes obtuse, downy beneath. Upper leaves sessile, auriculate, lobed toothed or pinnatifid above; teeth acute. Stems and branches grooved, glabrescent or almost hispid in old specimens. Rays short, broad. Achenes slender, grey, excessively mucilaginous when placed in warm water.—*S. radiolatus,* F. Muell., Veg. Chath. Isds. 25, t. 4.—*H. Travers! Cox!*

8. **S. latifolius,** *Banks and Sol.* ex *Hook. f., Fl. N.Z.* i. 145. Erect, much branched, glabrous, 2ft.–3ft. high. Stems flexuose, grooved or furrowed. Leaves membranous, polymorphic, 2in.–8in. long, ½in.–2½in. broad; radical

usually on long winged petioles with small toothed auricles, oblong, toothed, lobulate-pinnatifid or lyrate-pinnatifid; upper sessile, ovate-oblong or contracted below the middle and again expanding into broad usually toothed auricles, acute; uppermost leaves linear-lanceolate, toothed, in young specimens lanceolate, toothed, scaberulous. Heads ½in.–¾in. in diameter, in lax much branched corymbs; branches slender. Receptacle broad, flat. Involucral bracts 1-seriate, linear, acute, glabrous or pubescent, with 2–3 or more minute bracteoles at the base. Rays 12–20, slender, spreading. Disk-florets about 40; mouth funnel-shaped, 5-lobed, much shorter than the tube. Achene faintly or strongly ribbed, puberulous, hispidulous or hispid. Pappus-hairs soft, white.—Handbk. 159.

NORTH and SOUTH Islands: from the Wairoa Falls, Hunua, to Ross, Westland. Otago, *Thomson!* Apparently absent from the east coast of the South Island. Sea-level to nearly 4,000ft. Dec., Jan.

Var. **rufiglandulosus** (*sp.*), Col. in Trans. N.Z.I. xxviii. (1895) 599! Leaves about 4in. long, irregularly doubly dentate; blade ovate, deltoid, abruptly narrowed into a broadly winged and toothed auriculate petiole, glandular-pubescent or glabrate. Corymb very large; lower branches 12in.–15in. long. Achene densely pubescent.

Var. **sinuatifolius.** Stems very slender, flexuose, sparingly glandular-pubescent. Leaves distant; blade broadly oblong-ovate, acute or obtuse, entire, margins sinuate but not toothed, truncate, slightly cuneate at the base, narrowed into a long slender scarcely winged petiole with small entire auricles; midrib and veins glandular-pubescent. Heads few, small, in a short terminal corymb. Branches 1in. long. Peduncles and pedicels glandular.

NORTH Island: Patangata and other places in Hawke's Bay; local. *Tryon! Hamilton!*

S. latifolius, DC., Prod. vi. 387, has priority, but is considered a synonym of the South African *S. barbellatus*, DC.

9. **S. Colensoi,** *Hook. f., Fl. N.Z.* i. 147. Stems erect, simple or much branched, usually woody at base, flexuose, grooved, the whole plant covered with cobwebby tomentum, or the leaves white beneath or on both surfaces. Leaves polymorphic, 1in.–4in. long, ½in.–1½in. broad, entire, broadly obovate to lanceolate narrowed into short petioles with broad wings, or oblong-ovate to oblong-lanceolate sessile by a broad auriculate base, rounded at the apex or acute with the margins sinuate, crenate, serrate, or dentate, or irregularly lobed or lyrate-pinnatifid, sometimes hoary on both surfaces. Corymbs usually lax. Heads few or many, ¼in.–⅓in. long, campanulate. Pedicels with numerous bracteoles. Involucral bracts linear or oblong, acute or abruptly acuminate. Receptacle convex. Rays usually revolute, exceeding the tube. Disk-florets numerous; mouth funnel-shaped, longer than the narrow tube. Achenes slender, grooved, hispidulous or silky.—Handbk. 160.

NORTH Island: sea cliffs, Bay of Islands and East Cape, *Colenso*. Napier, *Bishop Williams!* Cape Kidnappers and other places on the coast as far south as Pourere, *T. K.*

The leaves vary greatly in form and tomentum, some specimens being hoary with white tomentum on both surfaces, but the florets and achenes are uniform in all. When growing in places close to the sea the leaves are glabrate and very fleshy.

10. **S. Banksii,** *Hook. f., Fl. N.Z.* 143. Glabrous in all its parts. Stem erect, sparingly branched, flexuose, grooved, 2ft.–4ft. high. Leaves glaucous, membranous but firm, 2in.–5in. long, 1in.–2in. broad, broadly ovate-oblong, subacute, sessile, amplexicaul; auricle broad, rounded; margins sinuate or irre-

gularly toothed; upper narrow, elliptic-oblong, acute. Corymbs lax or somewhat compact. Branches at first forming a right angle with the axis, erect. Pedicels very slender. Heads few or many, ¼in. broad, campanulate; involucral bracts oblong-acuminate, acute or obtuse, pubescent at the tips. Rays about 10, narrowed at both ends, revolute. Disk-florets numerous; mouth shorter than the tube, 5-toothed. Achene slender, deeply grooved, slightly contracted at the neck, pubescent or almost silky.—*S. odoratus*, Hook. f., Handbk. 160 (not of Hornem. Hort. Hafn. ii. 809.)

NORTH Island: on sea-cliffs, *Colenso*, *Bishop Williams!*

Distinguished from *S. pumiceus* at sight by the zigzag stem and branches, the glaucous leaves, and the peculiar branching of the corymb. I have not seen an authenticated specimen, and the material at my command is but scanty.

11. **S. pumiceus,** *Col. in Trans. N.Z.I.* xxi. (1888) 89. A rather stout erect glabrous or glabrate herb, 2ft.–5ft. high. Stems striate, slightly flexuous, erect or suberect, branched above. Leaves coriaceous or subcoriaceous, shining, 2in.–4in. long, 1in.–1½in. broad, broadly oblong, amplexicaul sessile, acute or obtuse, broadly auriculate; margins irregularly and distantly serrate or sinuate-serrate; veins prominent beneath; upper leaves smaller, oblong to lanceolate. Branches of corymb stout, erect. Heads few or many, about ½in. broad; involucral bracts and florets as in *S. Banksii*. Achene shorter, rather stout, grooved, not narrowed above, sparingly pubescent.—*S. Banksii*, *β. velleia*, Hook. f., Fl. N.Z. i. 146. *S. scabrosus*, Banks and Sol. MSS.

Var. **angustatus.** Stem less robust, sparingly clothed with scattered hairs. Leaves 1½in.–3in. long, ¼in.–½in. broad, subcoriaceous; lower lanceolate, acute or acuminate, irregularly toothed or sinuate, narrowed into slightly winged petioles, auricles small; upper sessile, auricle larger; veins less prominent than in the typical form.—*S. Banksii*, var. *γ scabrosus*, Hook. f., *l.c.*

NORTH Island: Makohinui Islands, *Col. Mus.!* Mercury Bay, Anaura, and Tolaga Bay, *Banks* and *Sol.!* East Cape Island, *Ross!* Between Tolaga and Poverty Bay, *Colenso*. Near Table Cape, *A. Hamilton!*

The Banksian specimens belong to the typical form; they are less coriaceous than those collected by Hamilton, and have a few scabrid hairs on the superior surface of the midrib.

12. **S. glaucophyllus,** *Cheesem. in Trans. N.Z.I.* xxviii. (1895) 536. Rootstock stout, woody. Stems herbaceous, 1ft.–3ft. high, glaucous, much branched, glabrous, strongly grooved, naked at the base or furnished only with minute scale-like leaves. Leaves numerous, 2in.–4in. long, ½in.–1in. wide, oblanceolate oblong-obovate or obovate-spathulate, obtuse or subacute, sinuate-dentate or serrate, gradually narrowed into broad flat petioles, not dilated at the base, very glaucous, somewhat thickened at the margins; upper leaves narrower, lanceolate or linear-lanceolate, serrate, passing into narrow-linear entire bracts. Heads several in a loose terminal corymb, broadly campanulate; involucral bracts linear, acuminate, 2-ribbed, glabrous, pilose at the tips. Ray-florets about 15. Disk-florets numerous. Achenes not seen.

SOUTH Island: on limestone rocks, Mount Arthur, Nelson, 4,000ft.

My specimens, for which I am indebted to Mr. Cheeseman, are very immature.

13. **S. Pottsii,** *J. F. Armst. in Trans. N.Z.I.* iv. (1871) 290. A small slender suffruticose species, with decumbent flexuose branches, 3in.–6in. long. Branches, petioles, and leaves beneath clothed with white cottony tomentum. Leaves ½in.–1in. long, ovate or spathulate, crenate, glabrate or glabrous above; petioles short. Heads solitary, on slender bracteate peduncles ⅓in. long, turbinate; involucral bracts 15–20, linear, spreading, obtuse, tomentose.

SOUTH Island: Mount Jollie, Rangitata district, 4,500ft., *W. Gray* and *J. F. Armstrong.*

I have not seen specimens.

14. **S. bifistulosus,** *Hook. f., Fl. N.Z.* i. 144. An erect much branched shrub, 1ft.–2ft. high. Bark pale, closely marked with the scars of fallen leaves. Leaves densely crowded near the tips of the branchlets, spreading, 1in. long, $\frac{1}{12}$in. broad, narrow-linear, sessile, obtuse or subacute, woolly beneath and revolute to the glabrous midrib except a woolly line on each side, the folded margin slightly and irregularly constricted so as to appear crenate. Peduncles 2in.–4in. long, terminal, on short branchlets with numerous imbricating leaflike bracts. Heads solitary, 1¼in. broad; involucral bracts few, ½in. long, broad, herbaceous, slightly woolly at the back. Achene short, obscurely ribbed, glabrous. Pappus very soft.—Handbk. 161.

SOUTH Island: Dusky Bay, *Lyall, Hector* and *Buchanan!* Chalky Bay (identified in the absence of flowers). 1,500ft. to 2,500ft.

I have copied Hooker's description of the flower-heads, my specimens being much too young.

15. **S. Hectori,** *Buch. in Trans. N.Z.I.* v. (1872) 348, vi. *t.* 23. An erect sparingly branched shrub, 6ft.–12ft. high. Branches stout, ascending. Leaves mostly near the tips of the branches, membranous, pubescent or thinly tomentose beneath, 6in.–12in. long or more, 2in.–4in. broad, broadly lanceolate or ovate-lanceolate, narrowed at both ends, acute, scabrid or scaberulous above, shortly dentate, pinnate or pinnatifid at the very base; segments ¼in.–⅓in. long, linear, acutely toothed; petiole short, stout, slightly sheathing. Panicle terminal, corymbose, 6in.–12in. long or more, slender, clothed with short dense glandular pubescence. Heads 1in.–2½in. in diameter, broadly campanulate; involucral bracts 2-seriate, glandular-pubescent, the outer lanceolate, acute, the inner oblong-lanceolate, with broad naked margins. Receptacle alveolate. Ray-florets 8–12; rays broad, spreading, white. Disk-florets numerous; limb campanulate, 5-toothed, shorter than the tube. Achene linear, grooved, glabrous. Pappus-hairs long and scabrid.

SOUTH Island: Nelson: mountains between Riwaka and Takaka, *T. K.* Collingwood and Wangapeka, *Travers.* Source of the Takaka River, Upper Motueka and its tributaries, *Cheeseman!* Valley of the Buller from the Mangles Gorge to the Inangahua Junction, *T. K.* Westland: left bank of the Grey, *T. K.* 200ft. to 3,000ft. Jan., Feb.

A magnificent species, distinguished from all others by the membranous leaves, pinnate or pinnatifid at their base.

16. **S. Kirkii,** *Hook. f. MS.* An erect glabrous sparingly branched shrub, 6ft.–12ft. high. Branches rather stout, brittle. Leaves 2in.–4in. long, ½in.–1½in. broad, lanceolate or oblong to ovate-oblong or ovate-rhomboid, acute

or obtuse, cuneate at base, entire or sinuate-toothed; petiole short, slender. Corymbs terminal, 5in.–8in. in diameter or more; lower bracts foliaceous; pedicels slender, 1in.–3in. long, bracteolate. Heads numerous, broadly campanulate, 1½in.–2in. in diameter; involucral bracts 2-seriate, broad, linear, membranous, acute, the inner linear-oblong, with scarious margins. Receptacle flat, alveolate. Rays few, white, distant, spreading, nearly 1in. long; tube very short. Disk-florets with a campanulate 5-cleft limb, much shorter than the slender tube. Achene linear, slender, grooved, glabrous, slightly thickened upwards. Pappus-hairs long and rigid, scarcely scabrid.—*S. glastifolius*, Hook. f., Fl. N.Z. i. 147, t. 39; Handbk. 161 (not of L. f., Supp. 372). *Solidago arborescens*, A. Cunn., Precurs. n. 435. *Cineraria glastifolius*, Banks and Sol. MSS. and Icon.

NORTH Island: in woods from the North Cape to Cook Strait; often epiphytic. Sea-level to 2,400ft. Oct. to June.

A noble plant. The corymbs are sometimes highly compound, 3ft. in diameter. Epiphytic specimens sometimes form a dome-shaped head, 12ft.–20ft. in diameter, the leaves being completely hidden by the snow-white flowers. A few reflexed bracteoles are sometimes developed at the base of the involucre.

17. **S. sciadophilus,** *Raoul in Ann. Sc. Nat. Ser. III.* ii. (1844) 119. A climbing shrub with slender flexuose branches, 2ft.–20ft. high, glabrous or pubescent. Leaves ¾in.–1in. long, membranous, orbicular or orbicular-ovate, coarsely dentate, rarely sparingly pubescent beneath; petiole slender, longer or shorter than the blade. Heads broadly campanulate, ½in. in diameter, in few-flowered corymbs forming an elongated terminal panicle. Peduncles and pedicels slender, puberulous or pubescent. Involucral bracts 6–8, lanceolate, subacute; margins broad, scarious. Receptacle convex. Rays 4–6, yellow, revolute, ¼in. long. Disk-florets 8–10; mouth deeply 5-lobed. Achene grooved, glabrate or hispid. Pappus-hairs in several series, denticulate.—Choix 21, t. 18; Hook. f., Fl. N.Z. i. 150; Handbk. 161.

SOUTH Island: local. Nelson: Wairoa Gorge, *Bryant!* Riwaka, *Rev. F. H. Spencer!* Canterbury: Banks Peninsula, *Raoul.* Alford Forest, *Enys!* Peel Forest, *W. Barker!* Otago: Dunedin, *Petrie!* Otago Heads and Saddle Hill, *Thomson!* Sea-level to 2,300ft. Jan. to April.

18. **S. perdicioides,** *Hook. f., Fl. N.Z.* i. 149. An erect bushy shrub, 2ft.–6ft. high. Branchlets pubescent, very numerous, slender, striated, scarred with the marks of the fallen leaves. Leaves glabrous, alternate, ¾in.–2in. long or more, on slender petioles, membranous, ovate-oblong or elliptic-oblong, serrate or crenate, obtuse or rounded at the apex. Heads in terminal leafy corymbs. Peduncles and pedicels slender, glabrate or pubescent. Involucres cylindric, turbinate in fruit; involucral bracts herbaceous, oblong, obtuse, with rather broad scarious margins. Receptacle very narrow, pitted. Ray-florets 2, yellow, ligule as long as the tube, broad, 3-toothed. Disk-florets 7–8; limb funnel-shaped, deeply 5-lobed. Achene oblong-subulate, grooved. Pappus-hairs 2-seriate, scabrid.—Handbk. 161. *S. multinerve*, Col. in. Trans. N.Z.I. xxv. (1892) 330. *Perdiceum senecioides*, Banks and Sol. MSS.

Var. **distinctus** (*sp.*), Col. in Trans. N.Z.I. xxvii. 390! Leaves oblong-lanceolate, 1½in.–2in. long, ⅝in. broad, narrowed at both ends, deeply crenate-dentate, sometimes waved.

NORTH Island: from Hicks Bay and East Cape to Mahia Peninsula. Nov. to Jan.

19. **S. myrianthos,** *Cheesem. in Trans. N.Z.I.* vii. (1874) 348. A shrub or small tree, 6ft.–15ft. high, with dark bark. Branchlets glabrate or young leaves thinly clothed with buff tomentum. Leaves alternate, 3in.–6in. long, 1in.–3in. broad, membranous, lanceolate oblong or elliptic lanceolate, acute or acuminate, narrowed towards the unequal or slightly cordate base, irregularly dentate, white with thinly appressed tomentum beneath; petioles 1in.–2in. long. Panicles slender, terminal, lax, sometimes leafy, glandular-pilose, 12in.–18in. long. Pedicels slender, bracteolate. Heads numerous, ⅓in. long, obconic; involucral bracts 8, glabrous or glabrate, rather shorter than the florets, narrow-linear, with shining margins, obtuse or acute, with few bractlets at the base. Receptacle narrow, shortly alveolate. Rays about 6, shorter than the tube, broad, white. Disk-florets 6–7; tube short; mouth long, campanulate, deeply 5-lobed. Anthers very shortly tailed. Pappus-hairs 1-seriate, denticulate. Achenes oblong, grooved, shortly hispid.—*S. Cheesemanii*, Hook. f. in Hook. Ic. Pl. 1201 (1877).

NORTH Island: in river gorges between the Puru and Kaueranga Creeks, Thames Goldfield, *Cheeseman!* Nov., Dec.

A beautiful species, easily recognised by its membranous leaves, narrow elongated panicles, and white florets. The leaves are sometimes lobulate.

20. **S. Huntii,** *F. Muell., Veg. Chath. Isds.* 23, *t.* 3. A shrub or small round-headed tree, 6ft.–20ft. high, more or less viscid and glandular-pubescent in all its parts. Branchlets marked with the scars of fallen leaves. Leaves close-set, 2in.–3in. long or more, ½in.–1in. broad, lanceolate or elliptic-oblong, acute or obtuse, sessile, glabrous or clothed beneath with thin fulvous tomentum, often downy above; margins subrevolute, entire; midrib prominent beneath. Panicle terminal, subcorymbose, 3in.–5in. broad. Heads numerous, on long slender hirsute glandular pedicels, broadly campanulate; involucral bracts 12–13, linear-oblong, usually obtuse, villous at the tips. Ray-florets few; ligula broad, about as long as its tube. Disk-florets numerous; limb campanulate, 5-lobed. Achene grooved, glabrous.—Hook. f., Handbk. 734.

Var. **oblongifolius.** Leaves elliptic-oblong, 3½in. long, 1¼in. broad, rounded at the tips, sessile by a broad base, flat. Panicle broader. Rays rather narrow.

CHATHAM Islands, *Mair! H. Travers! Cox! Rautini.* Dec. to Feb.

I am greatly indebted to my friend Mr. Cox for valuable specimens of this and other Chatham Island plants.

21. **S. Muelleri,** *T. Kirk in Trans. N.Z.I.* xv. (1882) 359 [*not of Regel, Ind. Sem. Hort. Petrop.* (1863) 31]. A shrub or small tree, 6ft.–26ft. high; trunk 8in.–24in. in diameter, with spreading cicatricose branches. Leaves 3in.–7in. long, 1¼in.–2½in. wide, gland-dotted when fresh, crowded near the tips of the branches, slightly coriaceous when fresh, broadly lanceolate or lanceolate-acuminate, narrowed into a broad base, sessile, clothed with thin appressed

white tomentum beneath. Heads numerous, in erect terminal panicles 4in.–9in. long, sparingly leafy; rhachis, pedicels, and involucres glandular or glandular-pubescent; pedicels short; involucral bracts about 12, linear, obtuse. Female florets about 12; rays narrow, contorted. Disk-florets 20–30; limb campanulate, 5-lobed. Anthers with short tails. Achenes subulate, grooved, glabrous. Pappus short, dirty white, scabrid.

HEREKOPERE Island, Foveaux Strait, and THE SNARES. Dec., Jan.

The finest species of the genus, and well worthy of the distinguished botanist whose name it bears. It is one of the rarest plants in the world, being restricted to the two rocky islands named above. *S. Stewartiae*, J. B. Armst. in Trans. N.Z.I. xiii. 339, is probably this species badly described; the flowers are unknown, and the leaves are said to be "linear-lanceolate."

22. **S. laxifolius**, *Buch. in Trans. N.Z.I.* ii. (1869) 89. A spreading shrub, 1ft.–2ft. high. Branchlets, petioles, leaves beneath, and inflorescence clothed with downy or short cottony tomentum. Leaves close-set, lanceolate or elliptic-lanceolate, narrowed at both ends, subacute, 1in.–2½in. long, ¾in. broad; petioles very slender, shorter than the blade, entire; midrib slightly pubescent above, prominent on both surfaces. Panicle terminal, very open, usually leafy below. Branches and peduncles very slender, wiry, downy-pubescent, bracteolate. Heads few, cylindric-campanulate, ½in.–¾in. in diameter; involucral bracts about 13–15, close-set, downy pubescent or tomentose. Receptacle alveolate. Rays few, narrow, spreading. Disk-florets as in *S. Greyii*, but the anthers are tailless. Achenes shortly stipitate, slightly compressed, 4-angled and grooved, glabrous.

SOUTH Island: Nelson: Mount Rintoul, above the Rainbow, *Bryant!* Wairau Gorge, *Cheeseman*. Spencer Mountains, *Gibbs!* Discovery Peaks, *Travers!* Fowler's Pass, *T. K.* 3,000ft. to 5,000ft. Jan., Feb.

Closely allied to *S. Greyii*, but distinguished by the small thin close-set leaves, slender panicle, eglandular involucres, and glabrous slightly compressed achenes. The name is singularly inappropriate; it was originally written "*laxiflorus*," but assumed its present form through a typographic error.

23. **S. Greyii**, *Hook. f., Fl. N.Z.* i. 148, *t.* 38. A spreading shrub, 2ft.–8ft. high. Branchlets, petioles, and leaves beneath densely clothed with appressed white cottony tomentum. Leaves 1½in.–3in. long, ½in.–1½in. broad, oblong or ovate-oblong, rounded at both ends, coriaceous, entire, sometimes unequal at the base; petioles slender, ½in.–1½in. long; midrib prominent below, glabrate above. Corymbs terminal, 2in.–5in. broad; bracts foliaceous. Pedicels slender, with one or two linear bracteoles, glandular-pubescent. Heads numerous, broadly campanulate, ¾in. in diameter; involucral bracts about 15, membranous, glandular-pubescent, acute. Receptacle corrugated. Outer florets with broad spreading rays, ⅓in.–½in. long. Disk-florets with a shortly 4-toothed campanulate limb, equalling the tube. Anthers shortly tailed. Achenes narrow-linear, silky. Pappus-hairs rigid, in several series.—Handbk. 164.

NORTH Island: from Cape Teneriffe, Pahou River, to Maungatiri Creek, Cape Palliser, *Colenso*, *T. K.* Sea-level to 1,500ft.

A rare and beautiful species, distinguished from *S. laxifolius* by the larger leaves, dense corymbs, glandular-pubescent heads, and silky achenes.

24. **S. revolutus,** *T. Kirk.* A suberect shrub, 8in.–15in. high. Stems decumbent. Bark purple with white blotches, striate. Leaves viscid, 1in.–2½in. long, ⅓in.–¾in. broad, oblong-lanceolate or elliptic-lanceolate, obtuse, narrowed at both ends, coriaceous, glabrous and reticulated above, white with appressed viscid tomentum beneath; petiole ¼in.–¾in. long. Corymb compact, terminal, simple, 5in.–8in. high. Peduncle thickly clothed with ascending linear-oblong foliaceous bracts. Pedicels slender, tomentose. Heads 5–15, campanulate, ½in.–¾in. in diameter; involucral bracts in 2 series, viscid, linear, acute, pubescent. Rays 12–15, broad, strongly revolute. Limb of disk-florets very shortly 5-toothed. Anthers tailless. Achene shortly stipitate, truncate, ribbed, glabrous.—*S. robustus*, Buch. in Trans. N.Z.I. vi. (1873) 240, t. 23, f. 1 [not of Sch. Bip. in Flora xxviii. (1845) 50].

SOUTH Island: Mount Eglinton, *J. Morton!* Mountains above Lake Harris, *T. K.* Ben Lomond; Mount Bonpland; not rare on the western mountains; *Petrie!* 3,000ft. to 4,500ft.

25. **S. Adamsii,** *Cheesem. in Trans. N.Z.I.* xxviii. (1895) 536. A much-branched shrub, 3ft.–5ft. high. Branches stout, and with the leaves and inflorescence excessively viscid. Leaves petiolate, 1in.–2in. long, oblong or oblong-ovate, broadly rounded at the tip or narrowed at both ends; margins revolute, extremely thick and coriaceous, clothed beneath except the midrib with dense white tomentum. Heads 6–20, in terminal lax bracteate corymbs or racemes, broadly campanulate; pedicels 1in.–2in. long, slender and extremely viscid; involucral bracts in 2 rows, linear, obtuse, pilose at the tips. Rays short, spreading. Anthers shortly tailed. Achenes glabrate or pubescent.—*S. pachyphyllus*, Cheesem. in Trans. N.Z.I. xvi. 410 (not of Remy in C. Gay., Fl. Chil. iv. 147).

SOUTH Island: Nelson: Mount Arthur, *MacKay!* Mount Peel, *Cheeseman!* Mount Rintoul and Ben Nevis, *Gibbs! Bryant!* 3,000ft. to 5,500ft.

Closely related to *S. robustus*: the leaves, however, are more coriaceous, the midrib is prominent beneath, the flower-heads are less numerous, while the peduncles, pedicels, and involucres are nearly glabrous.

26. **S. Monroi,** *Hook. f., Fl. N.Z.* ii. 333. A much-branched slender shrub, 2ft.–6ft. high. Branches, petioles, and leaves beneath clothed with appressed whitish tomentum. Leaves ½in.–1½in. long, coriaceous, narrow oblong-obovate, exceeding the slender petiole, obtuse, gradually narrowed below; margins slightly wrinkled, crenate. Corymbs slightly viscid, terminal, very slender, lax. Peduncles slender, bracteate, tomentose. Pedicels very slender, glandular-pubescent. Heads numerous, broadly turbinate, ½in.–¾in. in diameter; involucral bracts 14–15, acute, densely glandular-pubescent. Receptacle flat. Rays about 12, broad, revolute. Limb of disk-florets shortly funnel-shaped; teeth 5, very small. Anthers shortly tailed. Achenes grooved, pubescent.—Handbk. 163.

SOUTH Island: mountains of Marlborough, from the Awatere to the Conway Rivers. Nelson: Wangapeka, *Kingsley!* Ascends from 1,000ft. to 4,500ft. Dec., Jan.

Distinguished from *S. compactus* by the narrower leaves, broader corymbs, glandular involucres, and tailed anthers.

27. **S. compactus,** *T. Kirk in Trans. N.Z.I.* xii. (1879) 395. A much-branched compact shrub, 2ft.–3ft. high. Branches erect, stout, and with the petioles, undersurface of the leaves, pedicels, and involucres densely clothed with appressed snow-white tomentum, forming a smooth surface. Leaves $\frac{3}{4}$in.–$1\frac{1}{2}$in. long, $\frac{1}{3}$in.–$\frac{1}{2}$in. broad, obovate or obovate-oblong, rounded at the tip, narrowed into the slender petiole, coriaceous; margins tomentose, slightly waved and obscurely crenulate. Heads in 4–8-flowered terminal leafy racemes, rarely solitary, broadly campanulate, $\frac{3}{4}$in.–1in. in diameter; involucral bracts about 12, 2-seriate, acute, tomentose. Receptacle flat. Rays about 12, broad, spreading. Disk-florets very numerous; limb elongate, 5-toothed, as long as the tube. Anthers not tailed. Achenes terete, ribbed, silky.

NORTH Island: on limestone cliffs, Castlepoint, East Coast, *T. K.* Reported from Cape Turnagain, but this requires confirmation. Jan., Feb.

28. **S. elaeagnifolius,** *Hook. f., Fl. N.Z.* i. 150, *t.* 41. A slender or robust shrub, 5ft.–10ft. high. Branches grooved, and with the petioles, inflorescence, and leaves beneath clothed with dull-whitish appressed tomentum. Leaves 2in.–6in. long, 1in.–$2\frac{1}{2}$in. broad, broadly lanceolate lanceolate-oblong or obovate, rounded or subacute at the tip, narrowed below, coriaceous but usually thin, glabrous and shining above; margins rarely sinuate; petiole $\frac{1}{2}$in.–$1\frac{1}{2}$in. long, grooved; midrib and veins more or less obvious. Panicle terminal, stout or slender. Heads numerous; pedicels slender, $\frac{1}{3}$in. long, campanulate; involucral bracts 9–10, thick and coriaceous, woolly. Receptacle convex. Female florets 1–3 or 0; corolla-tube truncate, shortly toothed. Limb of disk-florets elongated, campanulate, 5-lobed for half its length; tube very short. Anthers shortly tailed. Achenes furrowed, pubescent. Pappus-hairs scabrid, rigid.—Handbk. 162.

Var. **Buchanani** (sp.), J. B. Armst. in N.Z. Country Journ. III. 56. More robust, 2ft.–3ft. high. Branches grooved. Leaves broadly oblong; veins reticulated. Panicle reduced to a short raceme.

NORTH and SOUTH Islands: from the East Cape mountains to Southland. STEWART Island. Ascends to 4,000ft. Descends to sea-level in Otago and Stewart Island. Jan.

The heads are mostly racemose.

29. **S. rotundifolius,** *Hook. f., Fl. N.Z.* i. 149. A shrub or small tree, 6ft.–30ft. high, with thin smooth bark and robust grooved tomentose branches. Leaves orbicular or orbicular-oblong, 2in.–5in. in diameter, very thick and coriaceous, unequal or slightly cordate at the base, glabrous and shining above; petioles stout, 1in.–3in. long, grooved, and with the leaves beneath densely clothed with white or buff tomentum. Panicle terminal, corymbose; branches numerous, very stout, and with the pedicels densely tomentose. Heads numerous, broadly campanulate; involucral bracts 9–12, linear, woolly. Receptacle deeply alveolate. Florets as in *S. elaeagnifolius* except those of the disk, which are shorter, with a funnel-shaped limb, 5-cleft almost to the base. Achenes grooved, shortly stipitate, glabrous.—Handbk. 162; T. Kirk, Forest

Fl. N.Z. t. 162. *Brachyglottis rotundifolia*, Forst., Char. Gen. 46; A. Cunn., Precurs. n. 464. DC., Prod. v. 210. *Cineraria rotundifolia*, G. Forst., Prod. 294; A. Rich., Fl. Nouv.-Zel. 254.

SOUTH Island: Westland: Westport, *W. Townson!* Jackson's Bay to Foveaux Strait. STEWART Island and islands in Foveaux Strait; abundant. Sea-level to almost 3,500ft. *Puheritaiko.* Dec., Jan.

30. S. **Bidwillii,** *Hook. f., Fl. N.Z.* i. 150. A small robust shrub, 1ft.–2ft. high. Branches, petioles, peduncles, and leaves beneath clothed with thickly appressed white tomentum. Leaves ½in.–1in. long, ⅓in.–⅔in. broad, broadly oblong or elliptical-oblong, rounded at both extremities, usually patent, excessively thick and coriaceous, brownish, glossy, glabrous above and usually reticulate; petiole ⅛in.–½in. long, stout, articulated at the base. Corymb ¾in.–2in. long. Heads few or many; pedicels short; involucral bracts about 8, spreading, thick and coriaceous, densely woolly. Receptacle convex. Florets about 20: female 4–5, rayless; corolla-tube very short, truncate, minutely 4-toothed; style greatly exserted. Disk-florets with a long tube; limb divided nearly to the base into 4 linear spreading lobes. Anthers very shortly tailed. Achenes deeply grooved, glabrous.—Handbk. 162.

NORTH Island: Hikurangi, East Cape, Ruahine Range, Kaimanawa, Tongariro, Ngauruhoe, Ruapehu, and Tararua Range, as far south as Mitre Peak and Mount Holdsworth, 2,500ft. to 5,500ft., *Colenso! Tryon! Lee! Hill!* Jan.

31. **S. viridis,** *n. s.* An erect sparingly-branched shrub, 2ft.–4ft. high. Young shoots slender, thinly tomentose. Leaves 1½in.–3in. long, ¾in.–1in. broad, coriaceous but thin, green, glossy above, 3–5-nerved, oblong-obovate, rounded at the tip, narrowed towards the base, white with appressed tomentum beneath; petiole ½in.–¾in. long. Corymb 3in.–6in. long. Peduncles naked, tomentose. Pedicels slender. Heads numerous, ⅓in.–½in. long, campanulate; involucral bracts 10–12, close-set, woolly. Receptacle flat. Female florets about 6; corolla-tube truncate, 4-toothed, or with a minute ray. Disk-florets less deeply divided than in *S. Bidwillii.* Achene grooved, glabrous.

SOUTH Island: local; mountains of Nelson and Marlborough, *Bidwill, Rough!* Discovery Peaks, *Travers!* Amuri, *T. K.* Canterbury: Bealey River, Arthur's Pass, and other localities in North Canterbury, *T. K.* 2,300ft. to 5,000ft. Jan.

Perhaps too close to *S. Bidwillii* to be altogether satisfactory, although the habit and foliage differ widely, but the characters drawn from the reproductive organs, although slight, are remarkably constant. This plant was included under *S. Bidwillii* by Hooker, but his description appears to have been drawn from Colenso's Ruahine specimens.

32. **S. geminatus,** *T. Kirk.* A small erect glabrous bushy species, 1ft.–2ft. high, woody below, the tips of flowering-shoots apparently dying back after flowering, much branched from the base. Branches brittle, slender, with raised lines from the base of each leaf. Leaves 1½in.–3in. long, ½in.–1¼in. broad, sessile, obovate-lanceolate or obovate-spathulate, acute or subacute, serrate, subcoriaceous. Heads few, ⅓in.–½in. in diameter, homogamous, on slender geminate peduncles from the axils of the upper leaves or forming a lax

terminal corymb; involucral bracts 8, in 1 series, shorter than the florets, coriaceous, oblong, with broad membranous margins, obtuse, rigid. Receptacle alveolate. Florets all hermaphrodite, tubular; limb campanulate, deeply divided into 5 narrow revolute lobes. Anthers tailless. Achene shortly stipitate, clavate, narrowed at the apex, ribbed, glabrous. Pappus-hairs in 2 series, rigid, spreading, scaberulous.—*Traversia baccharoides,* Hook. f., Handbk. 164; Hook., Ic. Pl. t. 1002.

SOUTH Island: mountains of Marlborough and Nelson; Upper Awatere; Kaikoura Range; valley of the Clarence; Rotoiti, &c.; Wairau Gorge; Hurunui Mountains; Waiau-ua Valley; Spencer Mountains; Lake Guyon; Discovery Peaks, &c. Canterbury: Upper Waimakariri, &c.

The leaves and involucres are more or less glutinous. *Traversia* is rightly referred to *Senecio* by the authors of the "Genera Plantarum." It is to be regretted that the old specific name has been applied to another species.

33. **S. cassinioides,** *Hook. f., Handbk.* 163. An erect shrub, usually branched from the base, 2ft.–10ft. high. Branches stout, with brown deciduous papery bark hanging in long flakes, excessively numerous, short, tomentose. Leaves ⅙in.–⅓in. long, laxly imbricating, linear-oblong, subacute or obtuse, clothed beneath with appressed whitish-yellow tomentum. Heads solitary, terminal, sessile, ⅓in. long, ovoid, with a few bracteoles at the base; involucral bracts about 8, 2-seriate, coriaceous, tomentose, the inner broadly oblong, with wide scarious margins. Receptacle alveolate, but not toothed. Florets about 10. Rays 4, short, broad, revolute. Disk-florets with a shortly campanulate 5-cleft limb. Anthers with minute tails. Achene shortly stipitate, expanded upwards, subulate, grooved, glabrous.

SOUTH Island: in mountain districts from Marlborough and Nelson to Mount Earnslaw, Otago. 2,000ft. to 4,500ft. Jan., Feb.

This singular plant has the habit of *Cassinia Vauvilliersii*, but with crowded branchlets and softer tomentum.

* CALENDULA, Linn.

Heads broad; involucre disk-shaped; involucral bracts 1–2-seriate, narrow, imbricate. Outer florets 2–3-seriate, female, fertile, ligulate; rays erect or spreading, entire or 3-toothed. Disk-florets sterile, tubular; mouth 5-toothed; style with an epigynous disk at base, conical; apex papillose, conical or shortly bidentate. Anthers shortly tailed. Achenes unequally incurved, more or less muricate or tuberculate, especially at the back, or winged or compressed. Pappus 0. Herbs, often glandular-pubescent. Leaves alternate, entire or sinuate. Heads solitary on terminal peduncles.

* **C. officinalis,** *L., Sp. Pl.* 921. Erect or spreading, 1ft. high, more or less branched from the base, pubescent or glandular-pubescent. Lower leaves oblong-spathulate, upper oblong or lanceolate, acute or obtuse, somewhat amplexicaul. Heads on terminal peduncles with acute bracts; involucral bracts in 2 series, linear, acuminate, glandular-pubescent. Rays 2–3-seriate, greatly exceeding the involucre, orange-coloured. Achenes all curved, boat-shaped, incurved, smooth on the inner surface, muricate at the back.

Not unfrequent by roadsides, waste places, sites of old gardens, &c., but rarely permanent. *Garden marigold.* Feb. to April. South Europe.

* OSTEOSPERMUM, Linn.

Involucre campanulate or hemispherical; involucral bracts in 2–3 series. Receptacle naked or setiferous. Outer florets female, 1-seriate, ligulate; rays spreading. Disk-florets hermaphrodite, sterile, tubular; style entire or 2-dentate. Anthers shortly tailed. Achenes straight, incurved or globose, drupaceous; putamen hard or bony. Pappus 0. Herbs or shrubs, with alternate or opposite entire toothed or sinuate leaves and racemose heads.

* **O. moniliferum,** *L., Sp. Pl* 923. A much-branched shrub, 3ft.–5ft. high, glabrous when mature, the young leaves sometimes pubescent. Branches rather stout, striated. Leaves alternate, 2in.–2½in. long, ¾in. broad, obovate, subacute, margins irregularly coarsely toothed, narrowed into a short winged petiole. Heads few, ¾in. in diameter, racemose on leafy peduncles in the upper axils; involucral bracts 2–3-seriate, lanceolate or oblong-acuminate, entire or serrate. Rays spreading, yellow. Disk-florets numerous. Achenes few, globose or ovate, bony, shining when fresh.

NORTH Island: a garden escape in several localities from Titirangi to the East Cape, but permanent only near the Auckland cemetery, where it has been observed for thirty-five years. Oct. to Feb. South Africa.

* CRYPTOSTEMMA, R. Br.

Involucre broadly hemispherical; bracts numerous, in several series, imbricating, the outer very short, obtuse, the innermost lanceolate, glabrate. Receptacle naked. Florets of the ray ligulate, sterile. Disk-florets numerous, fertile, shortly tubular, 5-toothed, hermaphrodite. Anthers shortly mucronate at the base. Style-arms very short. Achenes oblong, densely woolly. Pappus of 6–8 short scales, hidden by the wool. Perennial herbs, with radical and alternate pinnatifid or pinnately divided leaves and large solitary scapigerous or pedunculate heads.

* **C. calendulacea,** *R. Br. in Ait. Hort. Kew. ed. II.* 141. Stems very short or rarely elongate, prostrate, cottony. Leaves 3in.–6in. long, pinnatifid or unequally pinnately divided segments oblong, ovate or lanceolate, acutely toothed or lobed, white with appressed cottony tomentum beneath, scaberulous above. Scapes or peduncles cottony, naked, shorter than the radical leaves. Involucre ⅜in.–¾in. in diameter, glabrate. Rays ½in.–1in. long, narrow, distant. Disk-florets purple.

NORTH Island: naturalised in pastures and waste places from the Auckland Isthmus to Paikakariki. *Cape-weed.* Nov. to March. South Africa.

* ARCTIUM, Linn.

Heads homogamous; involucres glandular; bracts very numerous, imbricate, coriaceous, appressed at the base, narrowed into long stiff radiating spines with hooked tips. Receptacle flat, with rigid subulate scales. Florets hermaphrodite, tubular; limb campanulate, 5-lobed; filaments papillose. Anthers with a long terminal appendage. Style-arms connate, obtuse, pubescent below Achenes oblong, flattened, transversely wrinkled. Pappus deciduous, of numerous short rough subulate free bristles. Coarse branching unarmed biennial leafy herbs, with large petioled leaves and corymbose racemose or solitary heads.

Erect, much branched, 2ft.–4ft. high. Heads ¾in.–1½in. in diameter, corymbose .. * *A. majus.*
Erect, sparingly branched, 1ft.–2ft. high. Heads ½in.–¾in. in diameter, racemose .. * *A. minus.*

* **A. majus,** *Bernh., Syst. Verz.* 154. An erect branched biennial, 2ft.–4ft. high. Stem and leaves glabrous or cottony. Leaves 1ft.–2½ft. long, sometimes 1ft. broad, broadly cordate-ovate, obtuse, sinuate-denticulate or entire; petioles solid,

angular, furrowed. Heads ¾in.–1¼in. in diameter, mostly corymbose, glabrate or webbed; corolla-tube exceeding the limb. Achenes yellowish or dark-brown, irregularly rugose.

NORTH and SOUTH Islands: naturalised in Hawke's Bay, Taranaki, Wellington, and Nelson. *Greater burdock.* Feb. to April. Europe.

* **A. minus,** *Bernh., l.c.* Smaller than the preceding in all its parts, 1ft.–2ft. high. Radical leaves cordate-ovate, coarsely toothed or entire; petioles hollow, slightly furrowed. Stem-leaves mostly cuneate at the base. Heads ½in.–¾in. in diameter, racemose or subracemose, the terminal head solitary, slightly webbed. Florets longer than the spines. Corolla-tube equalling the limb. Achenes fuscous, with black blotches.

SOUTH Island: naturalised in waste places, Nelson. *Lesser burdock.* Feb. to April.

* CARDUUS, Tourn.

Involucres ovoid or globose; bracts imbricating in many rows, rigid, acuminate-spinous-tipped. Receptacle bristly or hairy, deeply pitted. Florets all tubular; tube short, ventricose above; lobes 5, narrow. Style-arms connate into a cylindrical bifid pubescent column. Achenes oblong, flattish, rarely terete, glabrous. Pappus, hairs in many series, naked, scaberulous, united into a ring at the base. Erect herbs, mostly with spinous-toothed leaves. Stems often winged, spinous.

Stems with interrupted spinous wings. Heads solitary, large, drooping .. * *C. nutans.*
Stems with continuous spinous wings. Heads numerous, sessile, small .. * *C. pycnocephalus.*

* **C. nutans,** *L., Sp. Pl.* 821. Erect, 1½ft.–2ft. high, biennial. Leaves 6in.–12in. long, hairy on both surfaces, veins woolly beneath, lanceolate, waved, entire, sinuate or pinnatifid, spinous. Stems cottony, furrowed; wings interrupted, spinous. Heads 1in.–2in. in diameter or more, solitary, hemispherical, crimson, drooping; involucral bracts with long spinous tips, cottony. Achenes glabrous, granulated.

SOUTH Island: Pomahaka; naturalised. *Musky thistle.* Dec., Jan. Europe.

* **C. pycnocephalus,** *L., Sp. Pl. ed. II.* 1151. An erect branched annual, 2ft.–4ft. high. Branches rather slender. Leaves broadly lanceolate, pinnatifid; lobes broad, sinuate-toothed or dentate, cottony beneath. Stems with uninterrupted broad deeply-lobed spinous wings. Heads fascicled, sessile, ¾in.–1in. in diameter; involucres glabrous or glabrate, narrow; bracts few, subulate-lanceolate, spinous, tips recurved. Achenes shining, minutely pitted.

NORTH Island: naturalised, Auckland, Napier, Whanganui, &c. SOUTH Island: Marlborough, Nelson, Canterbury, and Otago. Jan., Feb. Europe.

* CNICUS, Linn.

Characters of *Carduus*, but the bristles of the pappus plumose to the middle or below.

Stem winged. Leaves cottony beneath; lobes large and broad .. .. * *C. lanceolatus.*
Stem wingless. Leaves white beneath. Heads very large, clothed with dense white web .. .. * *C. eriophorus.*
Stem 3ft.–5ft. high, wingless. Leaves narrow-oblong, waved, spinous .. .. * *C. arvensis.*

* **C. lanceolatus,** *Willd., Prod. Fl. Berl.* 259. Erect, 2ft.–4ft. high. Stems winged, furrowed. Leaves obovate-lanceolate, pinnatifid, bristly above, cottony below; lobes bifid, each segment terminating in a long stout spine. Heads few,

erect, solitary or fascicled, 1in.–2in. in diameter, shaggy; involucral bracts numerous, ciliated, subulate; spine stout, recurved. Achenes shining, striped.

Abundantly naturalised throughout the colony. *Spear thistle. Plumed thistle.* Jan. to March. Europe.

* **C. eriophorus,** *Roth., Tent. Fl. Germ.* i. 345. Biennial. A handsome woolly much-branched plant, 3ft.–4ft. high. Stem wingless. Leaves 1ft.–2ft. long, half-clasping, deeply pinnatifid, white and cottony beneath, bristly above; lobes few, distant, bifid, the entire lanceolate segments alternately pointing upwards and downwards, each terminated by a strong spine. Heads 2in.–3in. in diameter, globose; involucres clothed with a dense white web; bracts lanceolate, with a long spinous reflexed point. Achenes shining.

NORTH Island: naturalised on limestone in the Upper Wairarapa. Jan., Feb.

* **C. arvensis,** *Hoffm., Deutschl. Fl. ed. II.* 130. Erect, dioecious. Rhizome extensively creeping. Stem 3ft.–5ft. high, angled and grooved, more or less cottony. Leaves shortly petioled, upper subsessile, oblong-lanceolate, slightly decurrent, waved, sinuate-lobed or pinnatifid, spinous. Heads many, in terminal panicles or corymbs, $\frac{3}{4}$in.–1in. in diameter; male subglobose, female ovoid. Involucres glabrate; bracts ciliate, broadly lanceolate, appressed; tips rigid, spinous. Achenes smooth, shining.

Naturalised in numerous localities from Auckland to Southland. *Creeping thistle. Californian thistle. Canadian thistle.* Feb. to April. Europe.

A dangerous pest.

* CYNARA, Vaill.

Involucre large, globose; bracts ovate, in many series, imbricate, coriaceous; apex produced into a spine. Receptacle flat, fleshy, setose. Florets all tubular; limb 5-lobed; lobes narrow; filaments papillose. Style-arms concrete. Achenes obovate, compressed or tetragonous. Pappus-hairs plumose, rigid. Perennial herbs, with simple or branching stems, spinous pinnatifid or pinnatipartite leaves, and large homogamous terminal solitary heads. Corollas violet or purple.

* **C. Cardunculus,** *L., Sp. Pl.* 827. A coarse herb with branched leafy stems, 2ft.–4ft. high, furrowed. Radical leaves 1ft.–2ft. long, bipinnatifid; segments deeply lobed; lobes lanceolate, acute, downy above, white with rather loose tomentum beneath; midrib and principal veins armed with slender rigid trifid spines below, rarely spineless. Heads globose, 3in.–5in. in diameter; bracts numerous, large, coriaceous, rigid, ovate, acuminate, terminating in a rigid spine; inner bracts linear-oblong, with an appendage at the apex. Receptacle flat, fleshy, with many long rigid bristles. Florets very numerous. Achene large, obovate, slightly compressed, glabrous.

NORTH Island: naturalised. Mangonui, *Cheeseman.* The Bluff, Napier, *T. K.* Southern Europe.

* SILYBUM, Vaill.

Heads large; involucres globose; bracts foliaceous, appressed, imbricating in many series, the outer broad, coriaceous, with a stout terminal subulate recurved spine and several smaller spines at the base. Receptacle flat, densely setose. Florets all hermaphrodite, filiform, with limb deeply 5-lobed. Filaments glabrous. Anthers sagittate. Achenes oblong, transversely wrinkled, black with grey streaks, glabrous. Pappus-hairs silky, connate at the base. An erect biennial, with large sinuate-dentate waved spinous leaves marked with milk-white veins.

* **S. Marianus,** *Gaertn., Fruct.* ii. 378, *t.* 168. Erect. Stems 2ft.–4ft. high, furrowed, glabrous. Leaves with milk-white veins.

Naturalised in many parts of the North Island; less frequent in the South. *Milk thistle.* Jan. to March. Europe.

The only species.

*CENTAUREA, Linn.

Involucre ovoid. Receptacle nearly flat, bristly. Involucral bracts appressed, imbricate, entire or spinous, margined or toothed. Florets all tubular; the outer usually larger; if radiating, neuter. Disk-florets hermaphrodite, slender; limb 5-lobed. Anthers with a long terminal coriaceous appendage. Achenes obovoid or oblong, compressed. Pappus-hairs in 2 or several series, usually scabrid or short or 0. Annual or perennial herbs, with branched or simple stems and entire divided rarely spinous leaves.

Involucral bracts with a black pectinate fringe .. .. * *C. nigra.*
Involucral bracts terminating in a long straight spine. Pappus 0 .. .. * *C. Calcitrapa.*
Involucral bracts terminating in a slender spine. Pappus white, soft. Annual .. * *C. solstitialis.*

* **C. nigra,** *L., Sp. Pl.* 911. Perennial. Stems 1ft.–3ft. high. Leaves lanceolate, entire or lobed, hispidulous. Heads globose, 1in.–1½in. in diameter; peduncles leafy; involucral bracts with a pectinate fringe. Achenes compressed, grey. Pappus reduced to scales or 0.

NORTH Island: sparingly naturalised in various localities. SOUTH Island: less frequent. *Knapweed.* Feb., March. Europe.

* **C. Calcitrapa,** *L., Sp. Pl.* 917. Biennial, much branched, 1ft.–2ft. high. Branches rigid, spreading, glabrate or cottony. Radical leaves 3in.–4in. long, deeply pinnatifid; segments lanceolate, toothed, of the cauline leaves narrow-linear, aristate. Heads ovoid, solitary, lateral, sessile or on short leafy peduncles, ½in. in diameter; outer bracts terminating in a stout subulate channelled spine, with 1–3 short lateral basal spines on each side; inner with a scarious obtuse appendage. Achenes compressed. Pappus 0.

NORTH and SOUTH Islands: more or less naturalised in all districts. *Star thistle.* Dec. to Feb. Europe.

* **C. solstitialis,** *L., Sp. Pl.* 917. Annual, 1ft.–2ft. high, erect. Stems slender but rigid, branched, leafy, cottony or hoary. Radical leaves lyrate-pinnatifid; segments rounded at the tips; upper entire, decurrent. Heads usually terminal, globose, cottony, woolly, ½in. in diameter; involucral bracts palmately spinous, the central spines very long, needle-shaped, not channelled, inner with an obtuse scarious appendage. Florets yellow. Achene white. Pappus as long as the achene.

NORTH and SOUTH Islands: sparingly naturalised in all the districts. *Yellow star thistle.* Feb. to April. Europe.

*CICHORIUM, Tourn.

Involucre cylindric; bracts in 2 series, herbaceous, outer about 5, short, appressed, inner 8–10, longer, erect, connate at the base. Receptacle rather flat, naked, setose or pilose. Florets numerous, all ligulate. Achenes crowded on the indurated receptacle, glabrous, obovoid or turbinate. Pappus of minute chaffy erect scales, 2-seriate. Branching perennial deep-rooted herbs, with milky juice and axillary crowded sessile or pedunculate heads.

* **C. Intybus,** *L., Sp. Pl.* 813. Erect, 9in.–18in. high. Branches straight, rigid, mostly at right angles to the stem. Lower leaves 3in.–10in. long, runcinate,

midrib hispid; upper oblong-lanceolate, clasping. Heads many, 1in.–1½in. in diameter, usually in pairs, sessile or shortly peduncled. Florets bright blue; ligule truncate, 5-toothed. Achenes angled.

Frequent in waste places and by roadsides from Auckland to Otago, but often sporadic. *Chicory*. Jan. to March. Europe.

23. MICROSERIS, D. Don.

Involucre herbaceous, campanulate; bracts few, nearly equal, 2-seriate, with a few shorter imbricating at the base, often with black tips. Receptacle naked. Florets all ligulate. Achene linear, cylindric, ribbed. Pappus of linear flat scales, bearing simple or plumose bristles, 2-seriate. Scapigerous herbs, with milky juice and usually radical pinnatifid leaves.

A small genus of about 20 species, 1 restricted to Australia and New Zealand, 1 from extra-tropical South America, the others from North America.

NAME, from the Greek, signifying *small* and *a lettuce*.

1. **M. Forsteri,** *Hook. f., Fl. N.Z.* i. 151. A glabrous scapigerous perennial, with fleshy or almost tuberous roots. Leaves radical, 2in.–10in. long, narrow-lanceolate, toothed or irregularly pinnatifid. Scape usually exceeding the leaves. Heads ½in.–⅔in. long; involucral bracts linear, acute, with membranous borders, shorter than the florets. Achenes ribbed. Pappus-bristles shortly plumose or sometimes only serrulate.—Fl. Tasm. i. 226, t. 66; Handbk. 164; Benth., Fl. Austr. vi. 676. *M. pygmaea*, Raoul, Enum. 45. *Monermios Lawrencii*, Hook. f. in Lond. Journ. Bot. vi. 124. *Scorzonera scapigera*, Banks and Sol. MSS. and Icon.; G. Forst., Prod. n. 534; A. Cunn., Precurs. n. 430.

NORTH and SOUTH Islands: from Cambridge, Waikato, and the East Cape southwards to Dunedin; common. Sea level to 2,800ft. Jan., Feb. Stated to have been found at the Bay of Islands by R. Cunningham, but apparently in error. Also in Australia.

* TOLPIS, Adans.

Involucres campanulate, pubescent or almost tomentose; outer bracts short, spreading, narrow-linear; inner 1–2-seriate, subulate-acuminate, erect. Receptacle flat, pitted. Florets all ligulate, exceeding the bracts, yellow. Achenes subterete, 5–8-ribbed, truncate at the apex, the outer embraced by the involucral bracts, puberulous. Pappus in 1 series of few short rigid bristles or with few longer bristles or scales intermixed. Annual or perennial herbs, rarely suffruticose. Stems slender, dichotomous. Leaves lanceolate, incised dentate or pinnatifid. Heads laxly corymbose.

* **T. umbellata,** *Bertol. in Mem. Soc. Emul. Gen.* ii. (1803) 133. Annual. Stems straggling, slender, somewhat dichotomously branched, grooved, glabrate. Leaves chiefly radical, 2in.–3in. long, oblong or oblong-lanceolate, acute, dentate, puberulous or pubescent. Corymb spreading, very lax. Peduncles long, filiform, almost capillary, pubescent, thickened upwards, proliferous. Heads ⅓in. in diameter; outer bracts spreading, rigid. Achenes cylindric, truncate above, 5–8-ribbed; the outer embraced by the involucral bracts, fertile. Pappus-bristles 1-seriate, very short and rigid, forming a crown or fringe; the inner mostly sterile, free, with 4 long scabrid bristles amongst the shorter.

NORTH Island: Remuera; plentiful; *Cheeseman!*

* LAPSANA, Linn.

Heads cylindric; involucral bracts 8 in one series, erect, with a few minute bracteoles at base. Receptacle flat, naked. Florets 8–12, all ligulate. Style-arms obtuse, pilose. Achenes not beaked, slightly compressed, striate, glabrous, not enclosed in the bracts. Pappus 0. Erect branched annuals, with small heads in lax panicles.

* **L. communis**, *L.*, *Sp. Pl.* 811. Erect, 1ft.–2ft. high, branched above, glabrate or nearly hispid. Leaves membranous, on slender petioles; lower lyrate, dentate or lobed, the upper entire or sinuate-toothed, narrowed into the petiole. Panicle spreading. Peduncles short, slender. Involucral bracts linear-acute, glabrous or sometimes glandular. Florets 8–12. Achenes slightly curved, pale.

Naturalised in cultivated lands and waste places from the North Cape to STEWART Island. *Nipplewort.* Jan. to April. Europe.

24. PICRIS, Linn.

Involucre hispid; outer bracts in 2–3 series, short, spreading; inner bracts larger, erect, linear, acute. Receptacle naked, flat. Florets all ligulate. Style-arms pilose. Achenes narrowed above, constricted or shortly beaked, transversely striate, muricate or rugose. Pappus of 2 or many rows of soft plumose bristles. Hispid branched annuals, with milky juice and alternate paniculate leaves.

The genus comprises about 24 species, chiefly natives of the temperate regions of the Northern Hemisphere. One species extends to Australia and New Zealand; another is naturalised in both countries.

NAME, from the Greek, in reference to the bitter taste of some species.

Erect. Leaves linear. Achenes not beaked, or very shortly .. .. .. 1. *P. hieracioides.*
Erect. Bristles glochidiate. Achenes with a distinct beak .. .. .. * *P. echioides.*

1. **P. hieracioides**, *L.*, *Sp. Pl.* 792. An erect hispid biennial, 1ft.–3ft. high, corymbosely branched, the bristles mostly barbed at the tip. Leaves 3in.–6in. long, linear-oblong or lanceolate, the lower petioled, the upper shorter, sessile, clasping. Peduncles slender. Involucres ½in.–¾in. long; bracts hispid and pubescent. Achenes contracted below the apex, sometimes forming a very short beak, transversely striate or muricate. Pappus-bristles in 1 series, plumose.—E. B. t. 196; A. Cunn., Precurs. n. 432; Hook. f., Fl. N.Z. i. 151; Handbk. 165; Benth., Fl. Austr. iii. 677. *P angustifolia*, DC., Prod. vii. 130. *P. attenuata*, A. Cunn., Precurs. n. 433.

NORTH Island: from the North Cape to the Auckland Isthmus. July, Aug.

* **P. echioides**, *L.*, *Sp. Pl.* 792. A coarse hispid or setose annual or biennial. Stem stout, 2ft.–3ft. high, paniculately branched, grooved. Radical leaves oblong-lanceolate, sinuate-toothed, petioled; cauline amplexicaul, hispid with rigid simple or trifid hairs hooked at the tip and springing from tubercular bases. Peduncles rather stout, diverging. Heads 1in. in diameter; involucre hemispheric, prickly and setose; outer bracts 3–5, foliaceous, lax, broadly cordate; inner narrow-lanceolate, acuminate, in 1 series, with a few smaller appressed at the base. Ligulae short. Achenes curved, compressed, transversely rugose, with a long erect slender beak. Pappus in several series, plumose.—*Helminthia echioides*, Gaertn.

NORTH and SOUTH Islands: naturalised from the Bay of Islands to Otago. *Ox-tongue.* Jan. to March. Europe.

25. CREPIS, Vaill.

Heads campanulate; involucral bracts in 2 series, with a few smaller at the base, herbaceous. Receptacle naked. Florets numerous, ligulate; tube short. Achenes linear or oblong, terete or scarcely compressed, striate, narrowed at the apex, sometimes beaked. Pappus in several series of very fine white or brown soft hairs. Annual or perennial branched rarely scapigerous herbs, with radical or alternate toothed or lobed leaves and corymbose or paniculate heads.

A rather large genus, widely distributed through the temperate and subtropical regions of the Northern Hemisphere; extremely rare in the Southern Hemisphere.

NAME, from the Greek, signifying *a sandal*.

Branched.

1ft.–3ft. high. Leaves glabrous. Achenes not beaked .. .. * *C. virens.*

Leaves hispid.

½ft.–1ft. high. Buds drooping. Outer achenes enclosed in the rigid involucral bracts; inner with long slender beaks .. .. .. * *C. foetida.*

Involucres herbaceous, tomentose. Achenes all with long slender beaks .. * *C. taraxacifolia.*

Peduncles and involucres clothed with simple rigid bristles .. .. * *C. setosa.*

Scapigerous .. .. 1. *C. Novae-Zelandiae.*

* **C. virens,** *L., Sp. Pl. ed. II.* 1134. A variable annual. Stem 1ft.–3ft. high, furrowed, branched. Leaves glabrous, 2in.–6in. long, oblong or oblong-lanceolate; radical petioled, toothed or runcinate, pinnatifid or pinnatisect; cauline narrow, sessile, half-clasping. Heads laxly corymbose, campanulate, ½in.–¾in. in diameter; involucres glandular, hairy; outer bracts few, short; inner linear-acuminate, erect, equalling the pappus. Achene terete, slightly narrowed upward, but not beaked; ribs about 10, smooth.

Naturalised throughout the colony. CHATHAM Islands. Jan. to March. Europe.

* **C. foetida,** *L., Sp. Pl.* 807. Branched from the base. Branches spreading, 6in.–12in. high, hispid. Leaves few, runcinate-pinnatifid, 1in.–2in. long. Heads ¾in. in diameter, solitary; peduncles bracteate or naked, strict, thickened upwards; buds drooping; involucral bracts ultimately rigid, erect, lanceolate-acute, downy hairy or setose, with a few minute bracteoles at the base. Outer achenes slightly beaked, embraced by the bracts; inner with long slender beaks equalling the involucres, ribbed, glabrous.

NORTH Island: sparingly naturalised near One-Tree Hill, Auckland, *T. K.* Remuera, *Cheeseman!* Dec., Jan. Europe.

* **C. taraxacifolia,** *Thuill, Fl. Par.* 409. Hispid. Stem 1ft.–2ft. high, grooved or furrowed, branched above. Leaves runcinate-pinnatifid, radical, 3in.–6in. long. Peduncles slender. Heads ¾in.–1in. in diameter, erect in bud; involucral bracts herbaceous, linear-acute, tomentose or glandular-hairy, with a few shorter spreading at the base. Achenes all equally beaked; ribs rough; beak very slender, setaceous, equalling the achene. Bracts never enclosing the achene.

NORTH Island: naturalised; Remuera, *T. K.* Lower Waikato, *Cheeseman.* Jan., Feb. Europe.

* **C. setosa,** *Haller f. in Roem. Arch.* i. II. 1. Hispid. Stem angled and furrowed, 1ft.–2ft. high, branched from the base, leafy. Lower leaves obovate-oblong, sinuate-dentate or runcinate-pinnatifid; upper amplexicaul, sagittate, toothed at the

base or entire. Peduncles grooved, bristly; buds erect. Heads campanulate, ½in. in diameter; involucral bracts erect, clothed with simple rigid bristles, narrow-linear, acute, with a few smaller spreading at the base, never enclosing the achene. Achenes all with slender beaks. Pappus partly enclosed by the involucre.

NORTH Island: Remuera, *T. K.* (1865), *Cheeseman.* Jan., Feb. Europe.

1. **C. Novae-Zelandiae,** *Hook. f., Handbk.* 161. A glabrous or tomentose scapigerous herb, 2in.–10in. high. Root stout and fleshy. Leaves usually forming a dense rosette, 2in.–6in. long, linear oblong-obovate, narrowed at the base or shortly petiolate, unequally pinnatifid or lobed; lobes usually toothed, the terminal segment very large and usually rounded. Scape slender, exceeding the leaves, simple or with a solitary branch, glabrous or sparingly clothed with black glandular hairs. Involucral bracts in 2–3 series, obtuse, often with black tips. Achene glabrous, short, compressed, ribbed.—Lindsay, Contrib. N.Z. Bot. 54, t. 3. *Hieracium fragile,* Banks and Sol. MSS. and Icon.

SOUTH Island: Marlborough: Queen Charlotte Sound, *Banks* and *Sol.!* The Brothers rocks, *Robson!* Nelson: Wairau Valley, *Cheeseman.* Clarence Valley, *T. K.* More frequent in the Canterbury and Otago Districts. Descends nearly to sea-level; ascends to fully 3,000ft. Jan., Feb.

The scape is frequently clothed with black hairs similar to those on the involucre.

* HYPOCHAERIS, Linn.

Involucral bracts herbaceous, in several series, imbricating, often elongating after flowering. Receptacle with a few linear scales amongst the florets. Florets numerous, all ligulate. Achenes usually striate, scabrous, all beaked, or sometimes the outer not beaked. Pappus-hairs in 2 rows, the inner long and plumose, the outer of short stiff bristles, sometimes 0. Annual or perennial herbs, with radical leaves and simple or branched scapes.

Leaves glabrous. Only the inner achenes beaked .. .. .. * *H. glabra.*
Leaves hispid. All the achenes beaked .. .. * *H. radicata.*

* **H. glabra,** *L., Sp. Pl.* 810. Annual. Leaves forming a rosette, glabrous or glabrate, 1in.–4in. long, narrow oblong-obovate, sinuate-toothed or pinnatifid, obtuse. Scapes several, 2in.–12in. long, the primary simple, the lateral branched, naked or with 1 or 2 bracteoles. Heads ¼in.–⅜in. long; involucres equalling the flowers; bracts few, with a strong midrib, very unequal, acute. Achenes subulate, terete, grooved, scabrous, the inner with a slender beak, which is absent in the outer.

NORTH and SOUTH Islands: naturalised throughout. Jan., Feb. Europe.

* **H. radicata,** *L., Sp. Pl.* 810. Perennial. Leaves hispid on both surfaces, narrow oblong-obovate, 5in.–10in. long, sinuate- or runcinate-pinnatifid, usually obtuse. Scapes 1ft.–2ft. high, much branched, naked or nearly so. Peduncles erect, slightly thickened above. Heads 1in.–1½in. in diameter, yellow; involucre subcylindric, shorter than the florets; bracts unequal, glabrous or ciliate. Achenes all beaked, muricate.

NORTH and SOUTH Islands: naturalised throughout the colony. STEWART Island; CHATHAM Islands. *Cat's-ear.* Dec. to March. Europe.

* LEONTODON, Linn.

Involucral bracts in several series, subimbricate; outer smaller. Receptacle flat, naked. Florets all ligulate. Style-arms pilose. Anther-cells tailless. Achenes terete, grooved, shortly beaked, slightly rugose. Pappus-hairs rigid, in 1 or 2 series;

outer setaceous, persistent; inner longer, plumose, dilated at the base; or the outer of toothed scales and the inner of 1 row of plumose hairs. Perennial scapigerous herbs, with radical entire or pinnatifid leaves.

Leaves hispid. Buds drooping.

Pappus feathery on the inner achenes, scaly on the outer .. .. .. * *L. hirtus.*
Pappus of all the achenes 2-seriate; outer short, scabrid; inner longer, plumose * *L. hispidus.*

Leaves glabrous. Buds erect.

Pappus 1-seriate, plumose .. .. * *L. autumnalis.*

* **L. hirtus,** *L., Syst. ed. X.* 1194. Leaves 2in.–6in. long, hispid, oblong or lanceolate, almost entire, sinuate or lobed, petiolate. Scapes slender, simple, 3in.–9in. long. Heads ½in.–¾in. in diameter; involucral bracts lanceolate, keeled, glabrous, downy or ciliate on the margins. Achenes striate, muricate. Pappus of the outer achenes of toothed scales; of the inner 2-seriate, plumose.—*Thrincia hirta,* Roth.

NORTH Island: naturalised near Auckland, Napier, and Wellington, &c. SOUTH Island: Canterbury Plains. Dec. to Feb.

* **L. hispidus,** *L., Sp. Pl.* 799. Perennial. Rootstock truncate. Leaves 3in.–8in. long, oblong-lanceolate, almost entire sinuate or runcinate, hispid with forked hairs. Scapes simple, naked or with 1 or 2 scale-like bracts, hispid, 4in.–15in. long. Buds drooping. Heads 1in.–1½in. in diameter; involucre obconic; bracts linear-lanceolate, hispid or slightly woolly. Florets glandular at the apex. Achenes ribbed, muricate; beak slender. Pappus of all the florets 2-seriate, plumose; of the inner slightly dilated at the base.—*Apargia hispida,* Scop.

NORTH and SOUTH Islands: naturalised in pastures and waste places; not uncommon. Sea-level to 3,000ft. Dec. to Feb. Europe.

* **L. autumnalis,** *L., Sp. Pl.* 798. Perennial. Rootstock truncate, 4in.–10in. long. Leaves glabrous or glabrate, sometimes hispid above, lanceolate, toothed or pinnatifid, narrowed at the base. Scapes usually branched and solitary, rarely simple, bracteate. Peduncles slender, swollen upwards. Buds erect. Heads ½in.–1½in. in diameter; outer involucral bracts subulate, glabrous or more or less hairy; inner shorter, linear-obtuse. Achenes very slender, ribbed and slightly muricate. Pappus of all the florets in 1 row, plumose, dilated at the base, brownish-white.—*Oporinia autumnalis,* D. Don.

NORTH Island: naturalised; Auckland, Waikato, Wellington. Probably not uncommon. March. Europe.

26. TARAXACUM, Linn.

Heads ½in.–2in. in diameter, terminating naked fistulose scapes; involucral bracts herbaceous, subimbricate, the inner in 1 row, linear, erect, the outer much shorter. Florets all ligulate, very numerous, yellow. Anther-cells tailless. Achenes oblong-ovate or fusiform, 4–5-ribbed, compressed, muricate above, the apex produced into a long slender beak. Pappus-hairs simple, in many series. Perennial scapigerous herbs, with copious milky juice. Leaves all radical, entire or pinnatifid.

SPECIES, about 25, distributed through all temperate countries. Regarded by many botanists as forms of one variable plant.

NAME, from the Greek, in reference to its properties as an alterative.

1. **T. officinale,** *Weber in Wigg. Prim. Fl. Holsat.* 56. Root stout, fleshy. Leaves extremely variable, 2in.–8in. long, pubescent or glabrous, narrow-oblong oblong-obovate or spathulate, sinuate-toothed or runcinate-pinnatifid. Involucres campanulate; outer bracts usually recurved, inner erect. Achenes striate or ribbed, the ribs rough.

Var. **glabratus.** Smaller. Leaves 2in.–4in. long, slender, linear-oblong. Heads ¼in.–½in. in diameter, subcylindric. Achenes with prominent ribs and raised minute points, muricate.—*Leontodon glabratum*, Banks and Sol. MSS.

NORTH and SOUTH Islands: the large form naturalised in pastures throughout the colony. Var. *glabratus* truly indigenous. Matakana (1865), *T. K.* Mercury Bay; Tolaga Bay; *Banks* and *Sol.!* Cape Terawhiti, *T. K.* SOUTH Island: more frequent in the mountains. Amuri, *T. K.* Sea-level to 4,000ft. Dec. to Feb.; var. *glabratus*, Nov., Dec., speedily disappearing. *Dandelion.*

After fertilisation the inner involucral bracts close over the florets, forming a cone, and the pappus is raised up by the slender beak. After the fruit is formed all the bracts are reflexed, when the exposed fruits, with the radiate pappus, form a globose head.

* LACTUCA, Tourn.

Heads small; involucres narrow, cylindrical, conical in fruit; bracts few, very unequal, in 2–4 series, imbricate. Receptacle flat, naked. Florets usually few, all ligulate. Anther-cells shortly tailed. Achenes plane, compressed, suddenly contracted and produced into a slender beak, dilated at the apex. Pappus short soft caducous, hairs falling separately. Erect annual or perennial herbs, simple or branched, with copious milky juice. Leaves alternate. Heads corymbose or paniculate.

Leaves runcinate with linear segments, entire or hastate. Beak longer than the grey achene .. .. * *L. saligna.*
Leaves lyrate-pinnatifid or entire. Beak shorter than the terete achene .. .. * *L. muralis.*

* **L. saligna,** *L., Sp. Pl.* 796. Erect, 1ft.–2ft. high. Stems strict, slender, slightly branched. Leaves almost glabrous; lower runcinate-pinnatifid or entire, segments narrow-linear, acute; cauline linear, entire, with a sagittate amplexicaul base. Inner involucral bracts much longer than the others, linear-lanceolate. Achene grey, with a narrow border, shorter than the long slender beak.

NORTH Island: Petane, Hawke's Bay, *A. Hamilton!* Jan., Feb.

* **L. muralis,** *E. Mey., Chlor. Hannov.* 431. Annual, erect, 1ft.–3ft. high, very slender. Leaves glabrous, lyrate-runcinate; segments angled and toothed, terminal lobe largest; cauline amplexicaul, auricled. Heads panicled; branches of panicle slender, spreading. Florets about 5. Achene ribbed and muricate, black, longer than the beak.

SOUTH Island: abundantly naturalised in the Kaikoura Mountains; *T. K.* Dec., Jan. Europe.

27. SONCHUS, Tourn.

Involucre ovoid; bracts imbricate, in 2–3 series, unequal, becoming conical after flowering. Receptacle flat, naked. Florets numerous, all ligulate, yellow or purple. Upper part of style and style-arms hairy. Achenes much compressed, grooved, ribbed or transversely rugose. Pappus-hairs in many series, simple. Annual or perennial succulent herbs, with hollow stems and copious milky juice. Heads in loose corymbs or panicles.

Species, about 28, chiefly distributed through the temperate regions of the Northern Hemisphere. *S. asper* has become cosmopolitan, although it must be considered indigenous in Australia and New Zealand. *S. grandifolius* is endemic in the Chatham Islands.

* *Annual.*

Auricles rounded. Achenes longitudinally ribbed .. .. 1. *S. asper.*
Auricles acute. Achenes longitudinally ribbed and transversely rugose .. * *S. oleraceus.*

** *Perennial. Rootstock stoloniferous.*

Leaves 6in.–8in. long, runcinate. Heads glandular, yellow .. * *S. arvensis.*
Leaves 1ft.–2ft. long. Heads purple, cottony .. .. 2. *S. grandifolius.*

1. **S. asper,** *Hill, Herb. Britt.* 147. Leaves entire or pinnatifid or runcinate-pinnatifid, broadly oblong, waved or crisped, spinous-toothed; lower narrowed into winged petioles; upper amplexicaul, auricles rounded, toothed or spinous. Heads subumbellate; involucral bracts acute, glabrous. Achenes ribbed, but not transversely wrinkled.—*S. oleraceus,* L., G. Forst., Prod. n. 282; A. Rich., Fl. N.Z. 230; A. Cunn., Precurs. n. 431. *S. oleraceus β,* Hook. f., Handbk. 166.

Common throughout the colony. STEWART Island; CHATHAM Islands. *Sow-thistle. Rauroroa.* Dec. to March.

The extreme states of *S. asper* and *S. oleraceus* appear widely different, but intermediate forms have been developed by intercrossing, and are often difficult to determine. *S. asper* is certainly the plant collected by Banks and Solander, as is proved by the fruits of their specimens.

Var. **littoralis,** T. Kirk in Trans. N.Z.I. xxvi. (1893) 265. Biennial or perennial. Root stout. Stems 1ft.–1½ft. high, robust, sparingly branched. Radical leaves rosulate, 4in.–7in. long, closely appressed to the ground, ovate-oblong, somewhat fleshy, obtuse, often waved, finely or coarsely toothed; cauline leaves few, acute, amplexicaul, auricles rounded or subacute. Outer involucral bracts acute; inner obtuse. Achenes glabrous, 3–5-ribbed.

On maritime cliffs from Auckland to STEWART Island, but often local; rarely abundant.

This appears very different from the type, but, as the fruits are similar, it does not seem advisable to give it specific rank.

* **S. oleraceus,** *L., Sp. Pl.* 794. Erect, glabrous, branched, 1ft.–3ft. high or more. Leaves undivided or pinnatifid, sharply toothed, the terminal lobe broadest; lower narrowed into winged petioles; upper often entire, half-amplexicaul, auricles spreading, arrow-shaped. Heads ¾in.–1in. in diameter, subumbellate, often cottony at base; involucral bracts imbricating, often with a few glandular hairs, but usually glabrous. Achenes flat, longitudinally ribbed and transversely rugose.—*S. oleraceus a,* Hook. f., Handbk. 166.

Naturalised in cultivated ground throughout the colony. STEWART Island; CHATHAM Islands. *Sow-thistle. Puroha. Pororua. Rauriki. Taweke. Wekeweke.* Nov. to March. Europe.

* **S. arvensis,** *L., Sp. Pl.* 893. Stoloniferous. Stems 2ft.–4ft. high, hollow, stout, angular, simple or sparingly branched, leafy. Leaves variable, spreading, lanceolate, sharply toothed, runcinate; margins waved, almost spinous; upper amplexicaul, auricles obtuse; uppermost entire. Heads subcorymbose, glandular-hispid, 1in.–2in. in diameter. Achenes ribbed and transversely wrinkled.

NORTH Island: Auckland: naturalised, 1865, *T. K. Sow-thistle.* Europe.

2. **S. grandifolius,** *T. Kirk in Trans. N.Z.I.* xxvi. (1893) 266. Rhizomes stout, fleshy, rarely tuberous, sometimes 2½in. in diameter. Stem succulent, 2ft.–4ft. high. Radical leaves erect, 1ft.–2ft. long, 4in.–7in. broad; petiole

6in.–9in. long, stout, dilated at base but not clasping; blade oblong or ovate-oblong, irregularly pinnatifid or pinnate; segments 4–6 on each side, broad, lobulate, often overlapping, coarsely doubly serrate or dentate, almost coriaceous, scabrid above. Lower cauline leaves petiolate; upper broadly sessile, not auriculate. Heads subcorymbose, 1in.–1½in. broad. Peduncles white, with cottony wool. Involucral bracts in 3–4 series, broad, subacute, outer with a median line of spinous or almost foliaceous processes. Florets numerous, purple. Achenes large, broad, with 3–6 longitudinal ribs; margins broad.

CHATHAM Islands, *Enys!* *Cox!* Jan., Feb.

* TRAGOPOGON, Tourn.

Involucral bracts in 1 series, linear-acuminate, connate at the base, equal or subequal. Receptacle convex, naked. Florets numerous, all ligulate, yellow or purple. Achenes elongated, fusiform, with a long slender beak, 5–10-ribbed, muricate. Pappus-bristles in many series, plumose; tips naked. Erect annual or biennial simple or branched glabrous herbs, with alternate wholly amplexicaul entire leaves. Heads solitary, terminal.

* **T. porrifolius,** *L.*, *Sp. Pl.* 789. Stem erect, 3ft.–4ft. high, branched. Leaves elongate, acuminate, slightly dilated at the base and gradually narrowed upwards to an acute point. Peduncles thickened upwards, fistulose below. Involucres usually one-third longer than the purple florets, rarely shorter. Achenes and pappus 1½in.–2in. long. Marginal achenes with rough scale-like tubercles.

NORTH Island: sparingly naturalised in Auckland. SOUTH Island: Akaroa; more plentiful. *Salsify.* Dec., Jan. Europe.

GLOSSARY OF THE PRINCIPAL BOTANICAL TERMS USED IN THIS WORK.

Accumbent. Lying against something, as the cotyledons of an embryo are *accumbent* when they lie against the radicle by their edges

Acerose. Needle-shaped; linear or slender, with a sharp stiff point.

Achene. A single-seeded indehiscent point, having the pericarp thin and coriaceous, not woody, as in the nut.

Acicular. Very slender, but stiff and pointed like needles.

Aculeate. Having prickles.

Acuminate. Applied to a leaf suddenly narrowed at the top and then prolonged into a point.

Acute. Applied to the apex of a leaf when it forms an acute angle or tapers to a point.

Adherent. A term applied to unlike organs when they are united, as calyx with ovary.

Adnate. Applied to an organ that grows in attachment to another for its whole length: stipules, for example, may be adnate to the petiole, anthers to the filament.

Adpressed. See *Appressed.*

Aestivation. The relative position in the unexpanded bud of petals, sepals, and bud-scales.

Afoliate. Without leaves.

Albumen. In botany this term is applied especially to the endosperm, a tissue developed in the embryo-sac, and which may or may not be wholly absorbed by the embryo. In the former case the seed matures as an ex-albuminous seed, in the latter as an albuminous seed. Albumen is generally a mealy, oily, fleshy, or horn-like substance.

Alternate. One after another; applied to leaves when placed singly instead of in pairs, also to stamens when alternating with petals instead of standing before them.

Alveolate. Pitted, so as to resemble a honeycomb.

Amorphous. Without determinate shape.

Amphitropous. Half inverted; applied to ovules attached laterally.

Amplexicaul. Stem-clasping; applied to leaves when the sessile base of the blade clasps the stem.

Anatropous. Inverted; applied to ovules.

Annular. (1) Ring-shaped; (2) banded or marked with rings.

Anther. The pollen-bearing part of a stamen.

Anthesis. The period of expansion of a flower; also the act of expansion of a flower.

Apetalous. Without corolla.

Apiculate. Ending in a small point.

Apiculus. A short pointed tip.

Apocarpous. Applied to the pistil when the carpels or ovaries are all free and distinct.

Appendiculate. (1) Having appendages, as a leaf; (2) of the nature of, or forming, an appendage.

Appressed. Applied to objects closely pressed to each other for their entire length, as hairs against the back of leaves, leaves against the stem, &c.

Arcuate. Arc-like in form; bowed; arched.

Areole. A term applied to one of the meshes of cellular tissue, or to a small distinct angular space such as those made by the "veins" of leaves.

Aril. A membranous pulpy or fleshy appendage growing from the funicle, and enclosing the whole or a portion of a seed.

Arillate. Applied to seeds having an aril.

Arillus. See *Aril*.

Aristate. Bearing awns or bristle-like processes.

Ascending. Slanting or curving upwards; applied to stems, especially to stems of which the lower part rests on the ground.

Asepalous. Having no calyx.

Auricles. The two lateral lobes formed by the indenture or notch at the base of a cordate leaf; more particularly applied to the lobes of sessile and stem-clasping leaves. The term is also applied to any similar lobes, as in the petals of many leguminous plants.

Auriculate. Having auricles; more particularly applied to a leaf having pointed auricles.

Awn. A slender spine or process.

Axil. The angle between the stem and the upper surface of a leaf or other organ.

Axile. Belonging to the axis; applied to placentation when the ovules are attached to the axis of an ovary with two or more cells.

Axillary. Growing in the axil.

Baccate. (1.) Having the form or nature of a berry; pulpy throughout. (2.) Bearing berries.

Barbellate. Studded with short, stiff hairs or bristles.

Base. The end by which a leaf, a part of the flower, a seed, or any other organ is attached to the stem or other organ.

Basilar. Pertaining to the base.

Berry. A fruit in which the whole pericarp is fleshy or pulpy, except the outer skin or rind.

Bifarious. Directed in two ways, or arranged in two opposite rows, as in the case of leaves growing only on opposite sides of a branch.

Bilabiate. (1) Applied to the aestivation of a five-petalled corolla when two adjoining petals are inside or outside the three others; (2) two-lipped, applied to a four- or five-lobed calyx or corolla when the two or three upper lobes stand obviously apart from the others, like an upper lip.

Biternate. Twice ternately divided.

Bract. An appendage of the stem, usually differing in form from an ordinary leaf, and being a modified leaf from the axil of which a flower, or a flower peduncle or pedicel, springs. The involucre in *Compositae* and in *Umbelliferae* is formed of bracts: the glumes of grasses are bracts; the "scales" of cones are lignified bracts. There are, of course, many other modifications of the bract.

Bracteate. Having bracts.

Bracteolate. Having bracteoles.

Bracteoles. The one or two last bracts under each flower, when they differ materially in size, shape, or arrangement from the other bracts.

Caducous. Dropping or falling off, especially at an early stage of development.

Caespitose. Tufted.

Callous. Hardened, and usually thickened.

Calycine. (1) Pertaining to the calyx; (2) of the nature of, situated on, or like a calyx.

Calyptrate. (1) Hood-like; (2) having a hood or lid.

Calyx. The outer series of floral leaves.

Campanulate. Bell-shaped.

Campylotropous. Incurved; applied to ovules.

Canaliculate. Minutely channelled or grooved.

Capillary. Hair-like.

Capitate. Head-shaped, or collected into a head.

Capsule. A dry fruit-vessel of one or more cells, splitting in a regular fashion, but according to various plans.

Carpel. A pistil, or such an organically complete portion of a compound pistil as is comparable to a simple pistil, whether separable or not from the other similar portions.

Carpophore. In flowering plants a portion of the receptacle prolonged between the carpels, as in the geraniums and many *Umbelliferae.*

Catkin. A flower-spike of unisexual flowers. It is usually pendulous, and it separates from the plant when flowering is over.

Cauline. Applied to leaves carried on the stem, as distinct from those that spring from near the root.

Cellular tissue. Consists usually of thin-walled cells, more or less round in form, or with their length not much exceeding their breadth, and not tapering at the ends.

Ciliate. Fringed with hairs, or with fine hair-like teeth.

Ciliolate. Fringed with fine hairs.

Cinereous. Of an ash-grey colour.

Circinate. Rolled inwards from the apex into a coil.

Clavate. Club-shaped.

Cleistogamic. Having flowers which never expand, and are systematically self-fertilised.

Coccus. One of the portions into which a lobed fruit with single-seeded cells separates when ripe.

Coherent. Term applied to organs of the same kind when they are united together, as petals may be coherent with petals, stamens with stamens.

Coma. (1) The empty leaves or bracts that terminate, in a brush or tuft, the flowering stems of some plants; (2) a tuft of hair, cotton, or wool on the outer coat of a seed.

Commissure. The place of meeting or point of union of two bodies, parts, or organs.

Compressed. More or less flattened laterally.

Conduplicate. (1.) Folded together along the midrib so that the halves are face to face, the upper face of the leaf being within; said of a leaf in vernation. (2.) Folded upon themselves and around the caulicle; said of the cotyledons in a seed.

Conglobate. A term applied to a compound flower growing in the form of a sphere or globe.

Connate. Applied to organs of the same kind when united by growth, as opposite leaves sometimes become completely united at the base.

Connective. That portion of the filament of a stamen that is inserted between the anther-lobes, whether these are near to each other or far apart.

Connivent. Nearer together at the summit than the base.

Cordate. Applied to heart-shaped leaves with the petiole at the broader and notched end.

Coriaceous. Leathery in texture.
Corolla. The inner series of the perianth, composed of petals which are usually coloured.
Corymb. An inflorescence in which the branches and pedicels, though starting from different points, attain the same level, forming a flat-topped panicle.
Costate. Ribbed.
Cotyledon. The first leaf or first pair of leaves of the embryo.
Crenate. With regular blunt rounded teeth.
Crenulate. With very small rounded teeth or crenatures.
Crustaceous. Hard and brittle in texture.
Cucullate. Hood-shaped.
Cuneate. Wedge-shaped.
Cupular. Cup-shaped.
Cupule. A small cup.
Cuspidate. Applied to leaves that are tipped with a spine, or that have a short sharp point.
Cyme. A branched flower-cluster, especially if broad and flattish, in which the middle flowers open first.

Deciduous. Falling or subject to fall in season, as petals and bracts after flowering, or leaves in autumn.
Decompound. A term applied to a leaf in which the primary petiole gives off a number of secondary petioles, each supporting a compound leaflet.
Decumbent. Spreading horizontally, or nearly so, at the base, then turning upwards and becoming erect.
Decurrent. Applied to a leaf when its edges are continued down the stem so as to form raised lines or narrow appendages, called wings.
Decussate. In pairs, alternately crossing at right angles.
Deflexed. Sharply turned or bent downwards.
Dehiscence. The opening of an anther or of a capsule to liberate pollen or seed—usually by means of valves, slits, or by splitting along regular lines.
Dehiscent. Opening by valves or along regular lines to liberate pollen or seed.
Dentate. Toothed; having the margin only cut a little way in.
Denticulate. Finely dentate or toothed.
Depauperated. Diminutive or imperfectly developed, as if starved.
Depressed. More or less flattened vertically, or at any rate at the top.
Diadelphous. Applied to stamens united by their filaments into two clusters; or, as in certain *Leguminosae*, consisting of a cluster of nine and one free stamen.
Diandrous. Having two stamens.
Diaphanous. Transparent, or nearly so.
Dichotomous. Dividing repeatedly into two branches of equal thickness.
Dicotyledon. A plant having two cotyledons, or seed-leaves.
Didymous. Twin; found in pairs.
Didynamous. Applied to stamens arranged in two pairs, one pair being shorter than the other.
Diffuse. Applied to stems spreading along the ground for the whole or the greater portion of their length, when they are at the same time rather loosely and very much branched.
Dimidiate. Applied to a normally double organ when one part of it is so much smaller than the other as to appear to be missing; also, split into two on one side, as in the calyptra of some mosses.
Dimorphic. Occurring under two forms.
Dioecious. Having the male and female flowers on separate plants.

Disk. (1) A modification of the floral receptacle within the calyx, or within the corolla or stamens; (2) the expanded base of the style which surmounts the ovary in the *Umbelliferae;* (3) the central part of a radiate compound flower; (4) round openings in the walls of the vessels, such as occur in the *Coniferae*, are sometimes termed disks; (5) in a bulb, the solid base of the stem, around which the scales are arranged.
Dissected. Applied to a leaf that is divided into numerous irregular portions.
Dissepiments. The partitions (*septa*) of a compound ovary.
Distichous. Arranged one above the other in two rows.
Divaricate. Branching off from a stem or axis at an angle of about 90°.
Dorsal. On the back.
Drupe. A fruit consisting of three layers, enclosing a single seed, as in the peach. The first layer, the *epicarp*, is the outer skin; the middle layer is the pulpy *sarcocarp;* and the third layer is the hard, stony *endocarp.*
Drupaceous. (1) Producing or bearing drupes; (2) resembling a drupe.

Ebracteate. Without bracts.
Ebracteolate. Without bracteoles.
Echinate. Covered with long, sharp, almost prickly protuberances.
Eglandular. Destitute of glands.
Emarginate. Notched; decidedly indented at the end of the midrib.
Embryo. The rudimentary plant formed in the ovule.
Endemic. Having its habitat in a certain region or country; peculiar to a region or country.
Endocarp. The inner layer of the ovary; the stony layer surrounding the seed, as in the peach or plum.
Endosperm. The albumen or nutritive matter in which the embryo is often imbedded in the seed. The term is usually restricted to the albumen formed within the embryo-sac, as in the majority of seeds; when formed outside the sac it is termed *perisperm.*
Ensiform. Sword-shaped.
Entire. Consisting of a single piece, with the margin nowhere indented.
Epicalyx. An external involucel or accessory calyx outside of the true calyx.
Epicarp. The outer skin or rind of a fruit.
Epidermis. The outer skin.
Epigynous. Inserted on the ovary.
Epipetalous. Inserted on the petals.
Epiphytic. Growing on other plants by way of support, but not parasitic.
Erecto-patent, or *erect-patent.* Erect, or nearly so, and at the same time spreading.
Erose. Irregularly toothed, as if gnawed.
Erosulate. Slightly erose.
Exarillate. Without an arillus.
Exotic. Brought from abroad; foreign.
Exserted. Protruding beyond, as stamens beyond the perianth.
Exstipulate. Destitute of stipules.
Extrorse. Directed outwards.
Extruded. Thrust out.

Falcate. Curved like the blade of a scythe.
Farinose. Covered with very short, intricate, white hairs, and having the appearance of meal or dust.
Fascicle. A bundle or cluster.
Fastigiate. Having the branchlets of almost equal height; tapering to an apex, like a pyramid.

Ferruginous. Rust-coloured.
Filament. (1) The stalk or support of an anther; (2) any fibre-shaped or thread-like body.
Filiform. Thread-shaped.
Fimbriate. Fringed.
Fistular. See *Fistulose*.
Fistulose. Tubular; hollow, like a reed.
Flabellate. Fan-shaped.
Flaccid. Relaxed, flabby.
Flexuous. Bending gently to and fro in opposite directions.
Floccose. Applied to tomentum when closely intricate and readily detached, like fleece.
Floret. A small flower, one of a cluster forming a compound flower.
Floriferous. Bearing flowers.
Foliaceous. Leaf-like in texture.
Foliolate. Having leaflets.
Follicle. A fruit of a single carpel opening along the ventral suture.
Fugacious. Falling or perishing very early.
Fulvous. Tawny; a mixture of orange-yellow and grey.
Funicle. See *Funiculus*.
Funiculus. The stalk by which the ovule or seed is attached to the placenta.
Furrowed. Applied to the epidermis of any organ, when it is thrown into parallel ridges with intervals between them.
Fuscous. Of a brown colour.
Fusiform. Spindle-shaped; tapering at both ends.

Gamopetalous. Having the petals united, either entirely or at the base only.
Gamosepalous. Having the sepals more or less united.
Geminate. Two side by side; in pairs; twin.
Geniculate. Having knee-like joints or protuberances.
Genus. A special group of closely related species. The name of the genus with that of the particular species constitute the systematic name of the plant.
Glabrate. Nearly glabrous; properly, becoming glabrous from age.
Glabrescent. Slightly glabrous.
Glabrous. Smooth, without hairs of any kind.
Glandular. Bearing glands, or of the nature of glands.
Glandular-hispid. Having hairs that exercise a secretory function.
Glandular-pubescent. Having a coating of soft glandular hairs.
Glandular-setose. Having bristles that terminate in a minute resinous head or drop, or that exercise a secretory function.
Glaucous. Of a pale bluish green.
Glochidiate. Applied to hairs that are hooked at the top.
Glomerule. A cymose flower-cluster which is condensed into a head-like form, and provided with a common involucre.
Granulated. Rough with small grains.

Hastate. Applied to a cordate leaf when the points diverge horizontally; compared to a halbert.
Hermaphrodite. Applied to a flower when both stamens and pistil are present and perfect.
Heterogamous. Applied to flower-heads containing both disk and ray florets.
Hilum. (1) The point of attachment of a seed with the funicle or the placenta; (2) the scar left on the seed where it separates from the funicle or the placenta.

Hirsute. Covered with rather coarse hairs.
Hispid. Thickly covered with rather stiff hairs.
Hispidulous. Minutely hispid.
Homogamous. Applied to flower-heads containing one kind of floret only.
Hyaline. Glassy. The term is also used to denote a pellucid substance in which, as some writers consider, the cell nucleus originates.
Hypogynous. Inserted on the floral receptacle or beneath the pistil.

Imbricate. Overlapping, like the scales of a fish.
Incised. Notched or cut irregularly but sharply.
Included. Stamens or style not projecting beyond the corolla.
Incumbent. Resting or weighing either wholly or partially upon something, as an anther against the inner side of a filament. The term is applied also to an embryo in which the radicle is folded down upon the back of the cotyledons.
Indehiscent. Applied to fruits which do not open along regular lines to liberate the seed.
Indigenous. Native, not exotic; said of a plant originating in the place or country in which it is found.
Inferior. (1.) Applied to an ovary when the outer parts of the flower, especially the corolla and calyx, are attached to it and are *superior;* applied to calyx and corolla that are not attached to the ovary, which is the *superior.* (2.) Applied to the radicle of the embryo when pointing towards the base of the fruit.
Inflorescence. The arrangement of the flowering branches and of the flowers upon them.
Innate. Applied to anthers when firmly attached by their bases to the filaments.
Introrse. Turned inwards.
Involucel. The involucre of a partial umbel.
Involucellate. Having an involucel.
Involucrate. Having or forming an involucre.
Involucre. A series of bracts closely placed round a flower-clustre, and arranged in one or several series.
Involute. Having the margins rolled inwards.

Keel. The two lower more or less combined petals of a papilionaceous flower.

Labiate. Lipped; usually two-lipped or bilabiate.
Lacerate. Having the edges jagged or irregularly cut.
Laciniate. Cut into long, narrow, irregular lobes.
Lacuna. (1) A small pit or depression; (2) a hollow cavity.
Lanceolate. Shaped like a spear-head; applied to a leaf that is three or four times as long as broad, and that tapers to its ends.
Lanuginous. Covered with soft woolly or cottony hairs.
Legume. A one-celled, two-valved seed-vessel or pod, formed of a simple pistil, having the seeds arranged along the inner or ventral suture.
Lenticel. A loose, lens-shaped mass of cells belonging to the corky layer of plants.
Lenticellate. Having lenticels.
Lenticular. Lens-shaped.
Ligulate. Strap-shaped.
Linear. Narrow, with parallel margins, at least four or five times as long as broad.
Lobulate. Separated into lobules.
Loculicidal. Applied to dehiscence when the valves of the capsule split open between the septa, which remain attached to the valves.
Lunate. Crescent-shaped.

Membranous. Like a membrane—thin, rather soft and pliable.
Mericarp. One of the two carpels that compose the fruit of an umbelliferous plant.
Micropyle. The opening or orifice in the primary integuments of the ovule, and, later, the point in a seed which indicates the position of the original orifice.
Monadelphous. Applied to stamens united by their filaments into one set or cluster, thus forming a tube, usually sheathing the pistil.
Moniliform. Beaded; much contracted at regular intervals.
Monoecious. Having the male and female flowers distinct, but on the same tree.
Monogamous. Having flowers distinct from each other, not collected into a head: having flowers with the anthers united.
Mucro. A short, straight, stiff, sharp point.
Mucronate. Applied to leaves having the midrib produced beyond the apex in the form of a small point.
Mucronulate. Slightly mucronate.
Multifid. Cut many times, so as to form several teeth.
Muricate. Covered with protuberances, raised and pointed, but yet short and hard.
Muricatulate. Slightly muricate.
Muriculate. Minutely muricate.

Naked. Applied to flowers having no perianth, and to seeds having no pericarp.
Nectariferous. Honey-bearing.
Node. The part of a stem or its branches from which leaves, branches, or leaf-buds are given off.
Nodose. Having swollen joints.

Obconic. Tapering downwards, when the transverse section shows a circle.
Obcordate. Applied to inverted heart-shaped leaves with a notch at the apex.
Oblanceolate. Lance-shaped, but tapering towards the base more than towards the apex.
Oblique. Unequal-sided. A term applied to leaves in which the development of tissue is greater on one side than on the other.
Obovate. Inversely ovate; applied to leaves that are ovate in shape, but that have the broader part near the apex.
Obovoid. Egg-shaped, with the broad end upwards.
Obsolete. Imperfectly developed.
Obtuse. Blunt or rounded at the end.
Orbicular. Applied to a leaf with a circular outline.
Operculum. A lid, or lid-like part or organ.
Ovary. The enlarged base of the pistil; the portion that contains the ovules.
Ovate. The shape of the longitudinal section of a hen's egg, applied to a leaf; the base of the leaf is the broader part.
Ovule. A small body contained in the cell of the ovary, and ultimately forming the seed.

Paleaceous. Of a chaffy consistence.
Palmate. Applied to leaves when five lobes diverge from the same point; compared to the fingers of the hand.
Panicle. A flower-cluster in which the axis is irregularly divided into branches bearing two or more flowers.
Papilionaceous. Butterfly-shaped; applied to flowers whose petals are the standard, or vexillum, which encloses the others in the bud, two lateral wings and two petals forming the carina or keel, opposite the standard, and more or less united.
Papillose. Pimply or warty.

Pappus. The crown or ring of hairs or of scales produced from the calyx-limb of a composite flower, and surmounting the fruit.

Papyraceous. Resembling paper.

Parietal. Applied to ovules borne on the inner wall of the ovary, or on projections from it.

Partite. Divided nearly to the base.

Patent. Spreading widely.

Pectinate. Applied to leaves when the lateral lobes are numerous, narrow, and regular, like the teeth of a comb.

Pedicel. The stalk supporting a single flower of a raceme, or other compound form of inflorescence.

Pedicellate. On a pedicel, or having pedicels.

Peduncle. The common stalk of an inflorescence.

Pedunculate. On a peduncle.

Pellicle. The outer cuticular covering of plants.

Peltate. Applied to leaves attached to the petiole by the undersurface instead of by the base.

Penicillate. Divided into a tuft of hair-like branches, or into long, slender arms.

Pentandrous. Having five stamens not connected with the pistil.

Perfect. Applied to flowers containing both stamens and pistil.

Perfoliate. Applied to leaves when the base of the blade not only clasps the stem, but closes round it on the opposite side, so that the stem appears to pierce through the blade.

Perianth. The outer envelope of a flower, whether consisting of calyx or corolla or both.

Pericarp. The portion of the fruit formed of the ovary and whatever adheres to it, exclusive of and outside of the seed or seeds, exclusive also of the persistent receptacle, or of whatever portion of the calyx persists round the ovary without adhering to it.

Perigynous. Inserted round the ovary.

Persistent. Applied to any part of the flower that remains until the fruit is more or less mature.

Petal. A corolla leaf.

Petaloid. Resembling petals, or having a petal-like perianth.

Petiole. The footstalk of a leaf.

Petiolulate: Having separate stalks to each leaflet.

Phyllotaxis. The arrangement of leaves in definite numbers in cycles on a stem.

Piliferous. Bearing or tipped with hairs.

Pilose. Hairy, with soft distinct hairs.

Pinna. The primary or secondary division of a bi- or tri-pinnate leaf, with the leaflets it comprises.

Pinnate. Applied to compound leaves having the several leaflets arranged one after the other on either side of the midrib or petiole.

Pinnatifid. Cut into lobes by incisions extending about half-way from the margin to the midrib.

Pinnatipartite. Pinnately parted, with the lobes extending rather more than half-way to the midrib.

Pinnatisect. Applied to lobes cut into segments by incisions extending very nearly to the midrib.

Pinnule. One of the smaller or ultimate divisions of a pinnate leaf or frond.

Pistil. The female or seed-bearing organ of flowering plants; usually composed of the ovary, with its contained ovules, and the stigma, with generally an intervening style.

Pistillate. Applied to flowers having a pistil, but either no stamens at all or imperfect ones.

Placenta. That part of the ovary which carries the ovules.
Placentation. The arrangement of the ovules within the ovary.
Plano-convex. Flat or plane on one side and convex on the other.
Pollen. The fertilising grains or cells contained in the anthers.
Polygamous. Having male, female, and perfect flowers on the same or different plants.
Polygonous. Having many angles.
Polygynous. Having many carpels.
Polymorphic. Having many forms or stages.
Procumbent. Applied to stems spreading along the ground for the whole or the greater portion of their length.
Proliferous. (1) Producing organs in an abnormal fashion, as leaf-buds in the place of flower-buds or seeds; (2) having an excessive development of parts.
Proterogynous. When the stigma matures earlier than the anthers of the same flower.
Puberulous. Minutely pubescent.
Pubescence. A coating of short, soft, downy hairs.
Pubescent. Covered with pubescence.
Pulvinate. Cushion-like.
Punctate. Covered with dots, points, or translucent internal glands.
Punctiform. Like a point or dot.
Punctulate. Dotted with minute pits.
Pungent. Terminating in a stiff, sharp point, like a prickle.
Putamen. The dry endocarp of a drupaceous fruit.
Pyrenes. The name given to the distinct stones or nuts, each enclosing a seed, of which the putamen sometimes consists.
Pyriform. Pear-shaped.

Quadrifarious. Arranged in four vertical ranks.
Quaternary. Consisting of four parts.
Quinate. Arranged in fives.

Raceme. An inflorescence in which the flowers are borne on pedicels along a single undivided axis.
Radical. Applied to leaves and flowers inserted on a rhizome or stock, or so close to the base of the stem as to appear to proceed from the root, rhizome, or stock.
Radicle. The future root of the embryo.
Raphe. See *Rhaphe.*
Receptacle. (1) The extremity of the peduncle (above the calyx) upon which the corolla, stamens, and ovary are inserted; (2) also applied to the short, flat, convex or conical axis on which the flowers of a capitate inflorescence are seated.
Reduplicate. Having the margins projecting outwards into salient angles.
Reflexed. Curved backwards.
Reniform. Kidney-shaped.
Repand. Having a wavy or uneven outline.
Replum. The false septum, framework, or process left after the valves of a capsule or other dehiscent fruit have fallen.
Reticulate. Resembling a network.
Retrorse. Turned or bent backward.
Retuse. Having a shallow notch at a rounded apex.
Revolute. Having the margins rolled outwards.
Rhachis or *Rachis.* The axis of an inflorescence or of a compound leaf.
Rhaphe or *Raphe.* The fibrovascular cord that connects the hilum of an anatropous or amphitropous ovule with the chalaza; the point at which the nucleus and integuments of an ovule coalesce.

Rhizome. A prostrate or subterranean stem that produces roots below and leaves above, sending up annual aerial shoots from its extremity ; a rootstock.
Rostrate. Beaked.
Rosulate. Applied to radical leaves when they spread on the ground in a rose-like cluster.
Rotate. Applied to a calyx or corolla when the sepals or petals are spread out horizontally, or nearly so, from the base, like a wheel or star.
Rufous. Yellowish or brownish red.
Rugose. Wrinkled, or marked with irregular raised or depressed lines.
Rugulose. Marked with fine wrinkles.
Runcinate. A term applied to a pinnatifid leaf when the points of the large central lobes are reflexed.

Saccate. Applied to a spurred corolla or calyx when the spur is short and round like a little bag.
Sagittate. Applied to an auriculate leaf when the points are directed downwards and the leaf is like an arrow-head.
Sarcocarp. The succulent or fleshy part of a stone-fruit.
Sarmentose. Applied to woody stems when the branches are long and weak, though scarcely climbing.
Scaberulous. Somewhat scabrous.
Scabrid. Slightly rough to the touch.
Scabrous. Rough to the touch ; having short points or little asperities.
Scandent. Climbing.
Scape. A leafless flower-peduncle, springing from the stock or from near the base of the stem, or apparently from the root itself.
Scapigerous. Scape-bearing.
Scarious. Membranous, thin, dry, or shrivelled ; not green.
Scions. A name given to young plants formed at the end or at the nodes of branches or stocks creeping wholly or partially above ground, or sometimes to the creeping stocks themselves.
Sepals. A leaf or segment of a calyx.
Septate. Divided by partitions or septa.
Septicidal. Applied to the dehiscence of a capsule whose valves separate at the line of junction of the carpels—that is, along the line of the placentas or dissepiments.
Septum. A partition.
Serrate. With regular pointed teeth, like a saw.
Serrulate. With very small, fine teeth.
Sessile. Inserted without a stalk of any kind.
Setaceous. Bristle-shaped, having the character of setae or bristles.
Setiferous. Bearing setae (fine bristles).
Setose. Bearing very stiff, erect, straight hairs, or having the surface set with fine bristles.
Simple. Consisting of a single piece, as applied to an umbel without secondary umbels.
Sinuate. Having broad, not deep, rounded incisions ; wavy.
Sinus. The curved or rounded hollow between two projecting lobes.
Soboliferous. Producing soboles, or rooting underground stems.
Spathulate. Oblong, with the lower part narrow and tapering, resembling in shape a chemist's spatula.
Species. An assemblage of individual plants bearing a sufficient resemblance to each other to warrant the conclusion of their descent from a common ancestor.
Spinescent. Slightly spiny.

Spinulose. Furnished with diminutive spines.
Squarrose. Rough, with spreading or projecting points.
Stamen. The anther with its filament, or the anther alone when sessile.
Staminodium. An abortive stamen.
Standard. The upper petal of a papilionaceous corolla.
Stellate. Arranged like the rays or points of a star.
Stigma. That part of the carpel which receives the pollen for the fertilisation of the ovules; it is usually situate at the top of the style.
Stigmatiferous. Bearing stigmatic cells.
Stigmatose. Of or relating to a stigma.
Stipella. A secondary stipule, sometimes found on compound leaves at the points where the leaflets are inserted.
Stipellate. Having stipellæ.
Stipe or *Stipes*. A stalk-like support.
Stipitate. Borne on a special stipe or stalk.
Stipulate. Having stipules.
Stipule. A leaf or scale-like appendage at the base of the leaf-stalk.
Stolon. A trailing basal branch that is disposed to take root at the tip, or at intervals.
Striæ. Fine superficial furrows or thread-like lines.
Striate. Marked with parallel longitudinal lines, either slightly raised or merely discoloured.
Strict. Very straight, narrow, and upright, or close.
Strigillose. Slightly strigose.
Strigose. Covered with rather short, stiff hairs, which lie close along the surface, all in the same direction.
Strophiolate. Having a strophiole.
Strophiole. An appendage similar to an aril, originating at or near the micropyle.
Style. That portion of a pistil or carpel between the ovary and the stigma.
Stylopodium. A style expanded at its base into a disk which crowns the ovary.
Sub (in composition) has the force of "almost," "not quite," "imperfectly."
Subulate. Awl-shaped.
Succulent. Thick, soft, and juicy.
Suffrutescent. Having the lower part slightly woody.
Suffruticose. Applied to undershrubs or plants in which the flowering branches, forming a considerable portion of the plant, die down after flowering, but leave a more or less prominent perennial and woody base.
Superior. Growing or placed above, as the upper petal in the corolla of a lateral flower. The usual application of the term is to the ovary when free from the calyx; to the calyx when its limb is united with the ovary, so that the calyx appears to be above the ovary; to the radicle of the embryo when pointing towards the apex of the fruit.
Suture. Applied to the line or seam formed by the union of two margins in any part of a plant; more especially applied to the seam of a dehiscent carpel where the valves unite.
Syngenesious. Applied to anthers cohering longitudinally by their margins.

Terete. Cylindrical, or nearly so.
Ternate. (1) Applied to branches when three proceed from the same node on the same side of the stem; (2) applied to leaflets, segments, lobes, or veins of leaves when three in number and starting from the same point.
Ternatisect. Cut into three lobes or divisions.
Testa. The outer coat of the seed.

Tetradynamous. Applied to stamens when (as in *Cruciferæ*) there are six, four of them longer than the two others.
Tetragonous. Four-angled.
Tetramerous. Consisting of four parts.
Tetrandrous. Having four stamens, free from the pistil.
Thyrsoid. Applied to an inflorescence that takes the form of a narrow panicle, widest about the middle.
Thyrsus. A narrow panicle, with its widest part about midway between its base and its apex.
Tomentose. Covered with tomentum.
Tomentum. Short, soft, dense, cottony hairs.
Torose. Swollen at intervals.
Tortuous. Irregularly bent and twisted.
Torulose. Slightly swollen at intervals.
Torus. The extremity of the peduncle (above the calyx) upon which the corolla, stamens, and ovary are inserted. It is sometimes little more than a mere point or minute hemisphere, but is often more or less elongated, thickened, or otherwise enlarged. It must not be confounded with the receptacle of inflorescence in a composite flower.
Translucent. Semi-transparent.
Triandrous. Having three stamens that are free from the pistil.
Trichotomous. Divided into three nearly equal branches springing from a common point.
Trifid. Applied to a leaf that is divided into three lobes by linear sinuses extending about half-way to the base.
Trifoliolate. Having three leaflets.
Trigonous. Obtusely three-angled.
Trimerous. Composed of three similar parts.
Trimorphic. Having, or passing through, three different forms or stages.
Triquetrous. Acutely three-angled, the sides of the triangle having concave faces.
Truncate. Abruptly terminated, as if the extremity were cut off.
Tuber. A short, thickened internode or portion of an underground stem, beset with modified axillary buds or eyes.
Tubercle. A minute swelling.
Tuberculate. Having small knobby projections or excrescences.
Tumid. Having a blistered appearance; swollen.
Turbinate. Top-shaped.
Turgid. Swollen or distended.

Uliginal. Growing in swamps or muddy places.
Umbel. A flower-cluster in which several pedicels spring from the same point and are of about the same length. An umbel is said to be *simple* when each of its branches or rays bears a single flower; *compound* when each ray bears a partial umbel or umbellule.
Umbellate. Applied to an inflorescence when it takes the form of an umbel.
Umbellule. A partial or secondary umbel.
Uncinate. Hooked.
Undulate. Wavy; waved obtusely up and down.
Unguiculate. Applied to petals when the lower part forms a stalk or claw.
Unifoliolate. Having one leaf.
Unisexual. Applied to flowers in which either stamens or pistil are wanting.
Urceolate. Hollow and contracted at or below the mouth, like an urn.
Utricle. A thin, rather loose pericarp, containing a single seed; also any thin bottle-like or bladder-like body.

Valvate. (1) Opening as if by valves, as do some anthers and most dehiscent fruits; (2) also applied to sepals, petals, &c., when they are in contact without overlapping.
Variety. An individual or group of individuals differing from the rest of the species to which it belongs in some constant peculiarities that are not essential in the differentiation of the species.
Ventral. Pertaining to the anterior face of an organ; opposed to *dorsal.*
Ventricose. Swelling out on one side or in the middle.
Vernation. The mode in which unexpanded leaves are disposed in the leaf-bud.
Versatile. Turning freely on a support.
Verticillate. Arranged in whorls.
Villous. Having long, soft hairs, not matted.
Virgate. Twiggy.
Viscid. Sticky, adhesive.
Viscous. Covered with a sticky or clammy exudation.
Vittæ. Small globular oblong or even linear vesicles filled with oil, imbedded in the substance of the pericarp of *Umbelliferæ.*

Whorl. A ring of organs all originating in the same plane.

SYNOPSIS OF THE NATURAL ORDERS.

(Taken from Hooker's "Handbook of the New Zealand Flora" and from Le Maout and Decaisne's "System of Botany.")

A. *Phaenogamic or Flowering Plants.*

Class I.—DICOTYLEDONS.

Stem, when perennial, furnished with pith, concentric layers of wood and bark. Leaves usually with netted venation. Organs of the flower generally 4 or 5 each, or multiples of those numbers. Seeds having an embryo with 2 cotyledons. In germination the radicle lengthens, and forks or branches.

The exceptions to one or other of these characters are too numerous to be mentioned.

Subdivision I.—ANGIOSPERMAE.

Ovules enclosed in an ovary, and the seeds in a seed-vessel.

Subclass I. **Thalamiflorae.** Flowers with both calyx and corolla. Petals free, and stamens usually inserted immediately beneath the pistil or ovaries. Ovary always superior.

Exceptions: Petals 0 in *Clematis, Myosurus, Caltha,* some species of *Cruciferae, Colobanthus,* and *Stellaria;* united at the base in some *Portulaceae* and *Malvaceae.*
Sepals petaloid in *Clematis* and *Caltha.*
Stamens perigynous in some *Stellariae* and *Colobanthus.*

§ 1. *Anthers adnate, opening by lateral slits. Pistil apocarpous.*

1. **Ranunculaceae.** Sepals 3–10, often petaloid. Petals 0 or 5–20. Stamens indefinite. Fruit of many or few achenes or follicles.—Herbs or opposite-leaved climbers. (p. 1.)

Of the four genera, three have petaloid sepals and no petals.

2. **Magnoliaceae.** Sepals and petals forming together three or more series, imbricate in aestivation. Carpel 1 or more.—Trees, with alternate exstipulate leaves. (p. 21.)

§ 2. *Anthers opening towards the stigma (inwards). Pistil syncarpous. Placentas parietal (rarely axile in* Pittosporeae*).*

* **Papaveraceae.** Flowers regular. Stamens ∞, free. Ovary 1-celled.—Herbs, rarely shrubs, with alternate leaves and milky juice. (p. 22.)

* **Fumariaceae.** Flowers irregular. Stamens 6, in two bundles; rarely 4, free. Ovary 1-celled.—Herbs, erect or climbing. Leaves alternate, usually multifid. (p. 23.)

3. **Cruciferae.** Sepals and petals 4. Stamens 6, 4 longer than the others.—Herbs, with alternate or rosulate exstipulate leaves. (p. 24.)

One *Lepidium* is rather shrubby. *Nasturtium* sometimes wants petals, and two or more of the stamens.

* **Resedaceae.** Calyx 4–8-partite. Petals 4, 8, 2, or 0, often cut. Stamens 3–40. Fruit various.—Herbs or shrubs. Leaves alternate. (p. 39.)

4. **Violarieae.** Sepals and petals 5. Anthers 5, their connectives enlarged or produced upwards, often connate. Placentas usually 3.—Herbs or shrubs, with alternate stipulate leaves. (p. 40.)

5. **Pittosporeae.** Sepals, petals, and stamens 4 or 5. Placentas usually 2. Capsule coriaceous or woody, 2-valved.—Shrubs or trees, with usually opposite evergreen exstipulate leaves. (p. 46.)

Ovary sometimes 2–5-celled.

* **Polygaleae.** Flowers irregular. Sepals 5. Petals 3–5. Stamens usually 8, monadelphous.—Herbs, shrubs, or trees. Leaves usually alternate, simple, exstipulate. (p. 52.)

§ 3. *Pistil syncarpous, 1-celled. Placenta basal.*

6. **Caryophylleae.** Sepals 4 or 5. Petals 4 or 5 or 0, free. Stamens 4 or 5, 8 or 10, hypogynous or perigynous.—Herbs, with opposite entire leaves. Flowers white or green. (p. 53.)

Petals absent in some *Stellariae* and *Colobanthus.* Stamens perigynous in *Colobanthus.*

7. **Portulaceae.** Sepals 2. Petals 5, usually united at the base. Stamens 5, usually opposite and adherent to the bases of the petals.—Herbs, with opposite, alternate, or imbricate leaves. Flowers white. (p. 64.)

§ 4. *Pistil more or less syncarpous, 2 or more celled. Placentas axile. Disk 0, or a raised torus.*

8. **Elatineae.** Sepals and petals 2–5, all free, imbricate. Stamens definite, hypogynous, free. Ovary 2–5-celled; ovules many.—Small creeping water-herb. Leaves opposite, stipulate, pellucid-dotted. Flowers minute, solitary, axillary. (p. 66.)

9. **Hypericineae.** Sepals and petals 5, hypogynous, free, imbricate. Stamens indefinite, hypogynous, free or polyadelphous. Ovary 3–5-celled; styles 3–5, free or connate; ovules numerous.—Herbs, shrubs, or trees, with opposite exstipulate pellucid-dotted leaves. Flowers yellow, in 3-chotomous cymes. (p. 67.)

10. **Malvaceae.** Calyx-lobes 5, valvate. Petals 5, usually connate at the base and adnate to the staminal tube, contorted. Stamens indefinite; filaments monadelphous; anthers 1-celled. Ovary of 1 or more free or connate carpels.—Herbs, shrubs, and trees, with often stellate down. Leaves alternate, stipulate. (p. 68.)

11. **Tiliaceae.** Sepals 4 or 5, valvate. Petals 4 or 5, often lobed or cut, imbricate. Stamens numerous, on a raised torus; filaments free; anthers 2-celled, often with terminal pores. Ovary 2–10-celled.—Trees or shrubs. Leaves alternate or opposite, stipulate. (p. 74.)

12. **Lineae.** Sepals 5, free, imbricate. Petals 5, free, contorted. Stamens 5, hypogynous; filaments united at the base into a cup. Ovary 3–5-celled; styles 3–5, free or connate.—Herbs. Leaves alternate, small, exstipulate. Flowers large, usually corymbose. (p. 77.)

13. **Geraniaceae.** Sepals 5, free, imbricate. Petals 5, equal or unequal, free, imbricate. Stamens 10, hypogynous, all fertile or some without anthers; filaments free or united at the base. Ovary 3–5-lobed; cells usually 1-seeded.—Herbs. Leaves alternate or opposite, stipulate or exstipulate. Flowers usually axillary, solitary, geminate or umbelled. (p. 78.)

SUBCLASS II. **Disciflorae.** Flowers with both calyx and corolla. Petals free, and stamens usually inserted upon the surface or at the base of a thickened hypogynous disk. Ovary rarely inferior.

Exceptions: Petals absent in *Dodonaea*, in one *Pomaderris*, *Discaria*, and *Alectryon*.
Stamens hypogynous in *Pennantia*, *Coriaria*, and *Dodonaea*.
Ovary inferior in *Pomaderris*.

14. **Rutaceae.** Sepals and petals 4 or 5, free, imbricate, or valvate. Stamens 8 or 10, rising from the outer base of an hypogynous disk. Ovary of 4 or 5 free or united 2-ovuled carpels, separating when ripe into as many 1-seeded 2-valved cocci. (p. 84.)

15. **Meliaceae.** Calyx small, 4- or 5-lobed, imbricate. Petals 4 or 5, linear, usually valvate and adnate at the base with the staminal tube. Stamens united into a thick tube, usually inserted below an annular or tubular disk. Ovary 3–5-celled.—Trees. Leaves exstipulate, compound. (p. 86.)

16. **Olacineae.** Calyx small, 4- or 5-lobed. Petals 4 or 5, free or connate, valvate. Stamens 4 or 5, hypogynous or surrounding an annular disk. Ovary 1-celled or imperfectly 2- or 3-celled; style 1; ovules 1–3, pendulous.—Shrubs or trees. Leaves alternate, exstipulate. (p. 87.)

17. **Celastrineae.** Calyx small, imbricate. Petals 4–5, sub-perigynous, spreading, imbricate. Stamens 4–5, inserted at the outer base of the disk; filaments subulate.—Shrubs or trees. Leaves usually alternate, stipulate. (p. 88.)

18. **Stackhousieae.** Calyx 5-lobed. Petals 5, linear, erect, free or connate above the base. Stamens 5, 2 shorter. Ovary 2–5-celled. Fruit of 2–5 globose cocci.—Herbs. Leaves alternate, small, quite entire. Flowers small, greenish, racemose. (p. 89.)

19. **Rhamneae.** Calyx superior or inferior, 4- or 5-lobed. Petals 4 or 5, minute or 0. Stamens inserted on the edges of a disk, as many as the petals and opposite them, small, incurved. Ovary 3-celled; style 1; ovule in each cell 1, erect. Fruit of 3 cocci.—Shrubs or trees, with often stellate down. Leaves alternate, rarely opposite, stipulate, or 0. Flowers small. (p. 90.)

* **Ampelideae.** Calyx small; lobes imbricate. Petals valvate, caducous. Stamens inserted outside the disk and opposite to the petals. Ovary free.—Shrubs, usually climbers. Leaves simple or compound, stipulate or not. (p. 93.)

20. **Sapindaceae.** Calyx 2–5-sepalled. Petals 0 in the New Zealand species. Stamens 5–8, hypogynous or inserted within a disk. Ovary 2- or 3-celled; style 1; ovules 1 or 2 in each cell, pendulous.—Trees. Leaves exstipulate, simple or compound. Flowers racemose. (p. 94.)

21. **Anacardiaceae.** Calyx 3–7-lobed. Petals 5, imbricate. Stamens 5, inserted at the base of a lobed disk. Ovary 1-celled; style 1; ovule 1, erect or pendulous.—Shrubs or trees. Leaves usually alternate, exstipulate. Flowers usually small and panicled. (p. 96.)

22. **Coriarieae.** Sepals 5. Petals 5, free, becoming fleshy after flowering. Stamens 10, hypogynous, all free, or 5 with the filaments adnate to the petals. Carpels 5–10, 1-celled, 1-ovuled, whorled round a fleshy disk; styles 5–10. Fruit of dry achenes enclosed in the fleshy petals.—Shrubs or herbs. Leaves opposite, exstipulate. Flowers racemose. (p. 97.)

This order is a very anomalous one, whose affinities have never yet been discovered.

Subclass III. **Calyciflorae.** Flowers with both calyx and corolla. Petals usually free, and stamens inserted on the tube of the calyx or top of the ovary, which is often inferior.

Exceptions: Perianth apparently absent in some *Halorageae*.

Petals absent in *Meryta*, *Fuchsia*, *Tetragonia*, and some *Halorageae*. Petals united at the base in *Acaena*, obscurely in *Tillaea*, united into a tubular corolla in some *Loranthi*.

Stamens hypogynous in some *Droserae* and *Tillaeae*.

§ 1. *Corolla very irregular or regular. Pistil apocarpous. Albumen 0.*

23. **Leguminosae.** Calyx tubular or campanulate. Petals papilionaceous in the New Zealand species. Stamens 10, sheathing the ovary. Ovary of one 1-celled carpel. Fruit a legume. (p. 98.)

Carmichaelia has a very exceptional pod.

24. **Rosaceae.** Calyx tubular or expanded. Petals regular. Stamens numerous (free in *Acaena*). Carpels 2 or more. Fruit various. (p. 124.)

§ 2. *Corolla regular. Stamens definite. Pistil syncarpous or apocarpous, inferior or superior. Albumen fleshy.*

25. **Saxifrageae.** Calyx inferior or superior, 5-cleft. Stamens 5 or 10. Pistil syncarpous; placentas axile; ovules numerous.—Herbs, shrubs, or trees. Leaves opposite or alternate, simple or compound. (p. 136.)

Stamens 2 or 3 in *Donatia*.

26. **Crassulaceae.** Calyx 3- or 5-sepalled, inferior, free. Petals and stamens sub-hypogynous. Pistil apocarpous. Follicles 1–∞-seeded.—Small herbs. Leaves opposite, quite entire. (p. 141.)

27. **Droseraceae.** Calyx 5-cleft, inferior, free. Petals and stamens 4 or 5, usually hypogynous. Ovary 1-celled, with parietal placentas; ovules numerous.—Small herbs. Leaves radical or alternate, covered with long glandular hairs. (p. 144.)

28. **Halorageae.** Calyx-tube adnate to the ovary; limb 4-toothed or 0. Petals 2, 4 or 0. Stamens 2 or 4, epigynous. Ovary 1- 2- or 4-celled, with 1 pendulous ovule in each cell. Fruit small, indehiscent.—Herbs. Leaves radical or opposite or whorled. Flowers minute, often unisexual. (p. 147.)

Callitriche has no perianth and 1 stamen in the male flower. *Myriophyllum* has an incomplete perianth in the male flower, as have some *Gunnerae*.

§ 3. *Corolla regular. Pistil syncarpous, wholly inferior; ovules numerous, on axile placentas. Albumen 0. Petals 0 in one* Fuchsia.

29. **Myrtaceae.** Calyx-lobes 4 or 5, valvate or imbricate. Stamens indefinite. Ovules few or many.—Trees or shrubs. Leaves evergreen, opposite, with pellucid dots. (p. 156.)

* **Lythrarieae.** Calyx-lobes valvate. Petals perigynous, rarely epigynous, usually crumpled in bud. Stamens usually definite. Ovary free, rarely adnate to the calyx-tube, 2–∞-celled; cells ∞-ovuled. Seeds small, exalbuminous.—Herbs, trees, or shrubs. Leaves usually opposite, simple, exstipulate. (p. 166.)

30. **Onagrariae.** Calyx-lobes 4, valvate. Stamens 8; ovules indefinite.—Herbs, shrubs, or trees. Leaves opposite or alternate. (p. 167.)

§ 4. *Corolla regular. Pistil syncarpous, wholly inferior; ovules few or numerous, on parietal placentas. Embryo straight.*

Ovary 1-celled, 1-ovuled, in the only New Zealand genus of *Cucurbitaceae*.

31. **Passifloreae.** Petals persistent with the sepals. Stamens definite, adnate with the stalk of the ovary; albumen fleshy.—Climber, with tendrils. Leaves alternate. (p. 181.)

32. **Cucurbitaceae.** Flowers unisexual. Stamens 3 or 5, usually variously combined by their anthers. Ovary 1-celled and 1-ovuled in the only New Zealand genus; albumen 0.—Climbers or trailers with tendrils. Leaves alternate. (p. 182.)

§ 5. *Corolla regular or 0. Pistil syncarpous, inferior; ovules few or numerous, on axile placentas. Albumen fleshy. Embryo curved.*

33. **Ficoideae.** Calyx 3–∞-lobed. Petals in the New Zealand genera numerous or 0. Stamens definite or indefinite. Fruit an indehiscent drupe, or fleshy below and bursting at the top by many valves within the calyx.—Fleshy herbs. Leaves opposite or alternate. (p. 183.)

§ 6. *Petals regular, often small, deciduous. Stamens 4 or 5, epigynous. Pistil syncarpous. Ovary wholly inferior, with an epigynous disk; ovules solitary in each cell.*
Petals 0 in *Meryta.*

34. **Umbelliferae.** Petals 5, usually imbricate. Stamens 5. Styles always 2. Fruit separating into two 1-seeded carpels.—Herbs, rarely undershrubs. Leaves alternate, simple or compound. Flowers small, umbelled or capitate. (p. 185.)

35. **Araliaceae.** Petals 5, usually valvate. Stamens 5. Styles 2–5. Fruit not separating, drupaceous or dry, 2–∞-celled.—Shrubs or trees, rarely herbs. Leaves simple, or 1–7-foliolate. (p. 214.)

Meryta has no petals and has anomalous inflorescence.

36. **Corneae.** Petals 5, valvate. Stamens 5. Style 1. Ovary 1–2-celled. Fruit ovoid, 1–2-celled.—Shrubs or trees. Leaves simple. (p. 223.)

Subclass IV. **Corolliflorae** or **Monopetalae.** Flowers with both calyx and corolla. Petals combined into a lobed corolla. Stamens inserted on the tube of the corolla.

Exceptions: Corolla absent in *Jasmineae.* Petals free or almost free in some *Campanulaceae* and *Myrsineae.*

Stamens epigynous in *Stylidieae,* some *Campanulaceae* and *Ericeae*; hypogynous in some *Jasmineae.*

§ 1. *Corolla epigynous, bearing the stamens.*

37. **Caprifoliaceae.** Flowers panicled or solitary. Anthers free. Ovary 2-celled. Leaves opposite or alternate, exstipulate. (p. 226.)

38. **Rubiaceae.** Flowers panicled, capitate, or solitary. Anthers free. Ovary 2-celled. Leaves opposite and stipulate, or whorled and exstipulate. (p. 228.)

* **Valerianeae.** Flowers irregular. Corolla-lobes imbricate. Stamens as many as the corolla-lobes or fewer. Ovary 3-celled, two cells 1-ovuled, one empty; style 1; ovule pendulous. Seed exalbuminous.—Herbs. Leaves opposite, simple or divided, exstipulate. (p. 250.)

* **Dipsaceae.** Flowers hermaphrodite, irregular. Corolla-lobes imbricate. Stamens usually 4. Ovary 1-celled; ovule 1, pendulous. Seed albuminous.—Herbs. Leaves opposite, exstipulate. Flowers usually in involucrate heads, rarely in axillary whorls, involucrate. (p. 251.)

39. **Compositae.** Flowers collected in involucrate heads. Anthers combined. Ovary 1-celled; ovule erect.—Herbs, shrubs, or trees. Leaves usually alternate or radical. (p. 252.)

CORRIGENDA.

p. 6. No. 12, for "*R. pauciflorus*" read "*R. paucifolius.*"

p. 7. Under 2. *Echinella*, after *R. philonotis*, read "(*R. sardous*, Crantz)."

p. 15. Under *R. lappaceus*, var. *villosus*, line 3, for "*subscapasus*" read "*subscaposus*," and for "333" read "335."

p. 19. *R. Limosella*, for "T. Kirk ex F. Muell." read "F. Muell. ex T. Kirk."

p. 33. Under *Capsella*, line 5, for "capsula" read "capsa."

p. 44. Under *H. dentata*, line 7, for "germinate" read "geminate"; and in first line of third paragraph, for "lenticillate" read "lenticellate."

p. 48. Under *P. obcordatum*, line 5, for "germinate" read "geminate."

p. 49. Under *P. pimeleoides*, var. *reflexum*, for "*pimelioides*" (*bis*) read "*pimeleoides.*"

p. 53. In key, after * *Saponaria*, enter "*Silene*. Capsule 6-valved, or rarely 3-valved at the apex."

p. 55. Under *Lychnis*, for "*vespartina*" read "*vespertina.*"

p. 91. Under *Pomaderris*, line 4, for "ovulary" read "ovary."

p. 106. Under *Corallospartium*, add "Corolla persistent."

p. 109. Under *C. Monroi*, line 10, for "teeth" read "beak."

p. 110. Under *C. Williamsii*, line 7, for "cuneate" read "arcuate." For "Te Kaha Bay and Raukokore Bay, Bay of Plenty" read "Te Kaha, Omaio, and Raukokore, in the Bay of Plenty." In last line, for "two teeth of the calyx" read "the calyx teeth."

p. 111. Under *C. grandiflora*, var. *alba*, insert "as" after "disgustingly."

p. 132. At head of page, for "*Alchmilla*," read "*Alchemilla.*"

p. 225. Under *G. lucida*, last line, for "Poukater" read "Poukatea."

p. 302. Under *R. australis*, line 10, delete "*R. McKayi*, Buch. in Trans. N.Z.I. xiv. 354, t. 34."

INDEX OF NATURAL ORDERS.

[NOTE.—An asterisk indicates that an order is not represented by species native to New Zealand.]

INDEX OF GENERA AND SPECIES.

[NOTE.—Synonyms are printed in italics. An asterisk against the name of a species indicates that the plant is naturalised or introduced. Names placed between inverted commas are names suggested for adoption in cases in which specific value is not quite ascertained.]

INDEX OF NATIVE AND OF POPULAR NAMES.

John Mackay, Government Printer, Wellington.—1890.